AF846768

Global Climate Change and Environmental Policy

V. Venkatramanan • Shachi Shah
Ram Prasad
Editors

Global Climate Change and Environmental Policy

Agriculture Perspectives

Springer

Editors
V. Venkatramanan
School of Interdisciplinary
and Transdisciplinary Studies
Indira Gandhi National Open University
New Delhi, Delhi, India

Shachi Shah
School of Interdisciplinary
and Transdisciplinary Studies
Indira Gandhi National Open University
New Delhi, Delhi, India

Ram Prasad
School of Environmental Science &
Engineering
Sun Yat-sen University
Guangzhou, Guangdong, China

ISBN 978-981-13-9569-7 ISBN 978-981-13-9570-3 (eBook)
https://doi.org/10.1007/978-981-13-9570-3

© Springer Nature Singapore Pte Ltd. 2020
This work is subject to copyright. All rights are reserved by the Publisher, whether the whole or part of the material is concerned, specifically the rights of translation, reprinting, reuse of illustrations, recitation, broadcasting, reproduction on microfilms or in any other physical way, and transmission or information storage and retrieval, electronic adaptation, computer software, or by similar or dissimilar methodology now known or hereafter developed.
The use of general descriptive names, registered names, trademarks, service marks, etc. in this publication does not imply, even in the absence of a specific statement, that such names are exempt from the relevant protective laws and regulations and therefore free for general use.
The publisher, the authors, and the editors are safe to assume that the advice and information in this book are believed to be true and accurate at the date of publication. Neither the publisher nor the authors or the editors give a warranty, express or implied, with respect to the material contained herein or for any errors or omissions that may have been made. The publisher remains neutral with regard to jurisdictional claims in published maps and institutional affiliations.

This Springer imprint is published by the registered company Springer Nature Singapore Pte Ltd.
The registered company address is: 152 Beach Road, #21-01/04 Gateway East, Singapore 189721, Singapore

Preface

Agriculture has come a long way since the agricultural revolution that took place 10000 years ago. The transformation from food gathering stage to food production provided human societies sufficient impetus for the meteoric growth of agriculture so much so that the agricultural activity is considered as the most important human enterprise that transformed global landscape. Technological advancement, industrial mode of production, and ever-increasing urge of the industrious cultivators engendered green revolution which aided in augmenting world agriculture production. Nevertheless, the last two centuries of the second millennium witnessed the human population growth from one billion to six billion, and currently, human population hovers around 7.7 billion. Human population growth indeed poses a couple of intriguing questions: Firstly, how to transform the present-day agriculture sector to achieve food and nutritional security through innovative solutions and technology adoption? Secondly, how to adapt the food system against human-induced climate change and checkmate the negative impacts of climate change on the agriculture production system? Since agriculture has immense mitigation potential and can substantially help in achieving our commitment to keep the increase in surface air temperature well below two-degree centigrade, it is pertinent to dovetail and harmonize the adaptation strategies and mitigation approaches of climate change in the agricultural policy-making. Agriculture policy-making through the lens of climate change must factor in ecological thinking and agricultural sustainability; vulnerability assessment of agro-based households; gender perspectives; policy measures to harmonize the demands of food production, feed and fodder production, and biofuel generation; strategies to improve the farmers or cultivators and other stakeholders through climate-smart adaptation practices like organic farming and agroforestry; capacitating the farmers in climate risk management, food value chain transformation through stakeholder-driven policy planning; and provision of agricultural inputs and services that include weather-based automated agro-advisories, crop insurance, and social security. The book endeavours to present the broad contours of global climate change, climate policy, and agriculture. The book targets the scientists, researchers, academicians, graduates, and doctoral students working on environmental science, environmental biology, and agricultural sciences. It also caters to the needs of policy-makers to frame policies on climate change, food security, agricultural resources, integrity of the food supply chain, and gender equity. We are deeply honoured to receive chapters from leading scientists and professors with rich experience and expertise in the field of global climate change, sustainable agriculture, and climate policy. The chapters provide an in-depth analysis of climate policy, sustainability of agroecosystem, vulnerability assessment of stakeholders, climate-smart farming, water footprint, policies to mitigate GHGs from agriculture and animal husbandry, agro-advisories, gender policy dimensions in agriculture, and biofuel policy.

Our sincere gratitude goes to the contributors for their insights at the intersection of global climate change and agriculture. We sincerely thank Dr. Mamta Kapila, Senior Editor, Springer, and Ms. Raman Shukla, Mr. Ashok Kumar, and Ms. Raagai Priya Chandra Sekaran for their generous assistance, constant support, and patience in finalizing this book.

New Delhi, India — V. Venkatramanan
New Delhi, India — Shachi Shah
Guangzhou, China — Ram Prasad

Contents

About the Editors and Contributors

Editors

V. Venkatramanan is an Assistant Professor at the School of Interdisciplinary and Transdisciplinary Studies, Indira Gandhi National Open University, New Delhi. His interests include climate change studies, biodegradation and green technologies for environmental management. He has published more than 20 research papers in peer-reviewed journals and book chapters.

Shachi Shah is an environmentalist with nearly two decades of teaching and research experience at various respected universities and institutes. She is an Associate Professor (Environmental Studies) at the School of Interdisciplinary and Transdisciplinary Studies, IGNOU, New Delhi. Her research interests include green technologies for waste management and energy generation, bioremediation, waste valorization, plant growth-promoting organisms, and biodiversity conservation. Moreover, she has authored more than 50 publications.

Ram Prasad has been associated with Amity Institute of Microbial Technology, Amity University, India, since 2005. His research interests include plant-microbe interactions, sustainable agriculture and microbial nanobiotechnology. He has published more than 150 research papers, review articles and book chapters and 5 patents and edited or authored several books. He has 12 years of teaching experience and has been awarded the Young Scientist Award (2007) and Prof. J.S. Datta Munshi Gold Medal (2009) by the International Society for Ecological Communications; FSAB Fellowship (2010) by the Society for Applied Biotechnology; American Cancer Society UICC International Fellowship for Beginning Investigators, USA (2014); Outstanding Scientist Award (2015) in the field of Microbiology by Venus International Foundation; BRICPL Science Investigator Award (ICAABT-2017); and Research Excellence Award (2018). He has also served as a Visiting Assistant Professor, Whiting School of Engineering, Department of Mechanical Engineering, Johns Hopkins University, USA, and is currently a Research Associate Professor at the School of Environmental Sciences and Engineering, Sun Yat-Sen University, Guangzhou, China.

Contributors

Muhammad Adnan Department of Agricultural Sciences, The University of Swabi, Swabi, Pakistan

Tripti Agarwal National Institute of Food Technology Entrepreneurship and Management, Kundli, Sonipat, Haryana, India

Amanullah Department of Agronomy, The University of Agriculture, Peshawar, Pakistan

A. Nambi Appadurai World Resources Institute, New Delhi, India

Muhammad Ardiansyah Center for Climate Risk and Opportunity Management in Southeast Asia and Pacific, Bogor Agricultural University, Bogor, Indonesia

Department of Soil Science and Land Resources, Faculty of Agriculture, Bogor Agricultural University, Bogor, Indonesia

Muhammad Arif Department of Agronomy, The University of Agriculture, Peshawar, Pakistan

Gerhard R. Backeberg Water Research Commission, Pretoria, South Africa

Balasubramanian T. N Agricultural Meteorology Department, Tamil Nadu Agricultural University, Coimbatore, India

Louis Bockel FAO Economist, Policy Officer in Regional Office for Africa, Accra, Ghana

Rizaldi Boer Center for Climate Risk and Opportunity Management in Southeast Asia and Pacific, Bogor Agricultural University, Bogor, Indonesia

Department of Geophysics and Meteorology, Faculty of Mathematic and Natural Sciences, Bogor Agricultural University, Bogor, Indonesia

Richard J. Culas School of Agricultural and Wine Sciences, Charles Sturt University, Orange, NSW, Australia

Rashila Deshar Central Department of Environmental Science, Tribhuvan University, Kathmandu, Nepal

Ga. Dheebakaran Agro Climate Research Centre, Tamil Nadu Agricultural University, Coimbatore, India

Christiaan Philippus Du Plooy Agricultural Research Council – Vegetable and Ornamental Plants, Pretoria, South Africa

Shah Fahad Department of Agricultural Sciences, The University of Swabi, Swabi, Pakistan

Roselyn Furoc-Paelmo Institute of Crop Science, College of Agriculture and Food Science, University of the Philippines Los Baños (UPLB), College, Laguna, Philippines

V. Geethalakshmi Agro Climate Research Centre, Tamil Nadu Agricultural University, Coimbatore, India

V. C. Goyal National Institute of Hydrology, Roorkee, India

Meeta Gupta National Institute of Hydrology, Roorkee, India

Purnama Hidayat Department of Crop Protection, Faculty of Agriculture, Bogor Agricultural University, Bogor, Indonesia

Syamsu Dwi Jadmiko Center for Climate Risk and Opportunity Management in Southeast Asia and Pacific, Bogor Agricultural University, Bogor, Indonesia

Talha Jan Department of Agronomy, The University of Agriculture, Peshawar, Pakistan

Urban B. Kalimba Department of Agriculture and Cooperatives, Iringa District Council, Iringa, Tanzania

Durba Kashyap National Institute of Food Technology Entrepreneurship and Management, Kundli, Sonipat, Haryana, India

Madan Koirala Central Department of Environmental Science, Tribhuvan University, Kathmandu, Nepal

S. Kokilavani Agro Climate Research Centre, Tamil Nadu Agricultural University, Coimbatore, India

Sandeep Kumar Centre for Environmental Sciences and Climate Resilient Agriculture (CESCRA), ICAR-Indian Agricultural Research Institute, New Delhi, India

Hazel Anne Makahiya Institute of Crop Science, College of Agriculture and Food Science, University of the Philippines Los Baños (UPLB), College, Laguna, Philippines

Phokele Maponya Agricultural Research Council – Vegetable and Ornamental Plants, Pretoria, South Africa

Bernadette C. Mendoza Institute of Biological Sciences, College of Arts and Sciences, University of the Philippines Los Baños (UPLB), College, Laguna, Philippines

Teodoro C. Mendoza Institute of Crop Science, College of Agriculture and Food Science, University of the Philippines Los Baños (UPLB), College, Laguna, Philippines

Ishaq Ahmad Mian Department of Soil and Environmental Sciences, The University of Agriculture, Peshawar, Pakistan

Sylvester Mpandeli Water Research Commission, Pretoria, South Africa

S. Mulla India Meteorological Department, Pune, Maharashtra, India

Hassan Munir Department of Agronomy, University of Agriculture, Faisalabad, Pakistan

Edward Nesamvuni University of Venda, Thohoyandou, South Africa

S. Panneerselvam Agro Climate Research Centre, Tamil Nadu Agricultural University, Coimbatore, India

Jyoti P. Patil National Institute of Hydrology, Roorkee, India

Shiv Prasad Centre for Environmental Sciences and Climate Resilient Agriculture (CESCRA), ICAR-Indian Agricultural Research Institute, New Delhi, India

Bushra Praveen IIT Indore, Indore, Madhya Pradesh, India

Anantanarayanan Raman Charles Sturt University & Graham Centre for Agricultural Innovation, Orange, NSW, Australia

Fahd Rasul Department of Agronomy, University of Agriculture, Faisalabad, Pakistan

Muhammad Riaz Department of Environmental Sciences and Engineering, Government College University, Faisalabad, Pakistan

Laure Sophie Schiettecatte FAO EX-ACT expert, ESA Division, FAO HQ, Rome, Italy

K. R. Sheetal Central Arid Zone Research Institute (CAZRI), Regional Research Station, Bikaner, Rajasthan, India

K. K. Singh India Meteorological Department, New Delhi, India

Sudhir Kumar Singh K. Banerjee Centre of Atmospheric and Ocean Studies, University of Allahabad, Prayagraj, Uttar Pradesh, India

Anter Parulian Situmorang Center for Climate Risk and Opportunity Management in Southeast Asia and Pacific, Bogor Agricultural University, Bogor, Indonesia

Dewi Sulistyowati Center for Climate Risk and Opportunity Management in Southeast Asia and Pacific, Bogor Agricultural University, Bogor, Indonesia

C. Valli Institute of Animal Nutrition, Tamil Nadu Veterinary and Animal Sciences University, Chennai, Tamil Nadu, India

V. Venkatramanan School of Interdisciplinary and Transdisciplinary Studies, Indira Gandhi National Open University, New Delhi, Delhi, India

Sonja L. Venter Agricultural Research Council – Vegetable and Ornamental Plants, Pretoria, South Africa

Ade Wachjar Department of Agronomy and Horticulture, Faculty of Agriculture, Bogor Agricultural University, Bogor, Indonesia

Ecological Thinking and Agricultural Sustainability

1

Anantanarayanan Raman

Abstract

Ecological or ecocentric thinking emerges from our appreciation of oneness with nature. Technocentric perception driven by scientific and empirical thinking builds on Charles Darwin's Theory of Evolution and Adam Smith's Wealth of the Nations. Those who can empathize with the ecocentric thinking can see the 'big' picture and understand the illusion of human mastery over nature. Nature has its precise mechanism of constant renewal and replenishment of materials, operating in a cyclical manner. When we humans thought that we have gained mastery over technology, we started interfering with the cycles of nature. Eventually, we damaged them to that extent that we have made them go berserk and turn linear. Consequently, we are currently facing stunning problems, such as pollution and other similar displeasing developments on Earth. In today's highly technocentric environment, where economic paradigms rule the roost, ecological paradigms are seen as 'primitive' and 'conservative'. To a few others, ecological paradigms appear daunting, challenging, and difficult to practice. The term 'sustainable development' refers to something more than, simply, growth. A change in the kind of growth is needed, a kind of development that is less material- and energy-intensive and more equitable in the distribution of its benefits. This emphasizes that changes are necessary and that the security, well-being, and the survival of the planet should be mutualistic with those changes. Sustainable development is not about giving priority to environmental concerns, but it is about incorporating environmental strengths into the economic system. Sustainability represents ideas of stability, equilibrium, and harmony with nature. Sustainable development is an attempt to reduce the politics in decision-making by artificially replacing conflict with consensus. Ecological thinking and its derivative ecological

A. Raman (✉)
Charles Sturt University & Graham Centre for Agricultural Innovation,
Orange, NSW, Australia
e-mail: araman@csu.edu.au; anant@raman.id.au

© Springer Nature Singapore Pte Ltd. 2020

V. Venkatramanan et al. (eds.), *Global Climate Change and Environmental Policy*,
https://doi.org/10.1007/978-981-13-9570-3_1

agriculture are practices that spin around simplicity and modesty. Aggressive dollar-driven thinking has no place in ecological thinking. Climate change, for example, is a problem created by us humans because of our badly thought-out and hasty practices of land use. If we realize this weakness and remedy it, then we still have hope to leave a cleaner and better world for the future generations of humans as well as other organisms that are as important as *H. sapiens*! We think that speed and rapid turnarounds of events are the norms of today. Is speed the root cause of present-day ecological–environmental malady, which has pushed us to think of sustainability?

Keywords
Agricultural sustainability · Ecological agriculture · Ecological thinking · Organic farming · Biodynamic farming · Natural farming · Permaculture · System of rice intensification

1.1 The Present Agricultural Scene

Over millennia, or perhaps for even more, our human ancestors lived hunting wild animals and gathering wild plants. Somewhere between 4500 and 10,000 years ago, the hunter-gatherer societies, in at least seven regions of the world, independently domesticated specific animal and plant species, which subsequently developed into agricultural economies.

One major human intervention of nature was the establishment of settlements, which involved the disturbance of soil and associated vegetation. Humans cleared vegetation to build residences. As long as humans remained hunter-gatherers, the disturbance to the natural environment was minimal, given the vastness of time. Once they moved to other localities establishing new settlements, the previously occupied sites regenerated back to near-natural near-original state. Such a recovery never eventuated — and could never happen — with humans settling permanently in specific places (Raman 2019). Clearing vegetation for building residences had its own other forms of consequences: The cleared sites encouraged aggressive, invasive plants to colonize and occupy vacant spaces due to either deliberate introductions or natural migrations. When humans moved from one place to another, they carried seeds of certain plants either deliberately or inadvertently and 'introduced' them into newer environments. One recent-time example would be the deliberate introduction of mango trees (*Mangifera indica*, Anacardiaceae) by humans into a new biogeographical locality—West Africa in 1824, from where this plant was spread to other warm regions of the world (Rey et al. 2004). Rivers are one other critical source that distributes seeds and vegetative material propagating them in new environments. Thus various reasons explain colonization of cleared areas by plants that do not usually occur in (or belong to) a particular region. The best examples for the natural colonization of plant material into the Indian landmass are the plant species that were domesticated by early Holocene 'farmers' of the Fertile

Crescent nearly 12,000 years ago. Those introduced plants, later, in 9000–10,000 years ago, stimulated the beginnings of systematic agriculture in southern Asia (Singh et al. 2016).

The transition from a foraging to a farming way of life was a major event in the evolutionary history of humankind. During this period, humans tried various techniques and primitive technology, thus making efforts to achieve better outcomes. Technology played a key role in enhancing agricultural capabilities of humans. The industrial revolution in Europe in the late seventeenth century ushered in new techniques and technologies that changed the global profile of agriculture. Since World War II, some nations have produced grains and other agricultural crops at around two-and-a-half times more than what they really required. Advancing technology and the urge to produce more in that period placed immense pressure on national economies to push agricultural production to greater levels. By responding to this economic pressure — by manipulating land and water to our advantage — we humans have inflicted substantial disruption to functional ecosystems on which the whole fabric of civilization depends. Through such behaviour we have pushed the world to a new, hitherto unperceived crisis. We have placed the Earth and its cycles of natural materials under stress, similar to the way we would strain a truck by simply loading it with 2–3 times more than its recommended load-carrying capacity. Such an action has resulted in what we today simplistically describe as 'environmental problems'. Some examples would be human population increase and consequently changed demographics, air and water pollution, overexploitation and depletion of natural materials such as plants and animals, and accumulation of non-degradable wastes, which turn into toxic over time.

Land came under severe stress in the last few decades (Fig. 1.1) (United Nations Environment Programme 1999). One highly serious issue that arose from unplanned utilization and overstressing of the environment is the widespread and unprecedented rate of recurrence of famines and droughts and eventual impoverishment in many parts of the world (The Brundtland Commission 1987). The Green Revolution was a concerted human effort in the 1960s to enhance agricultural productivity by altering several traditional practices, such as the use of high-yielding varieties, injudicious use of chemicals as fertilizers and pesticides, and heavy mechanization of farmland. The concept, developed by the American wheat geneticist Norman Borlaug, was trialled first in Mexico and subsequently followed religiously in many developing countries. The reality is that only 25% of the total land area of the Earth is suitable for farming activities. The remainder, which is either too dry or extremely harsh, experiences an adverse climate unsuitable for farming, or a permutation of these. Of this 25%, barely 3% is highly fertile, therefore productive land, 6% yields modestly, and the remaining 13% the output is poor. These are natural limiting factors to agriculture, but processes such as deforestation, desertification, and erosion — the results of mismanaged human activities — are further shrinking the area appropriate for agriculture. For example, of the *c.* 300 M ha of total land area in India, more than 50% is highly degraded and is beyond any redemption. One of the consequences of such mismanagement is the dramatic slowing down of per capita food production. Especially in these parts of the world, concern is mounting on the sustainability of

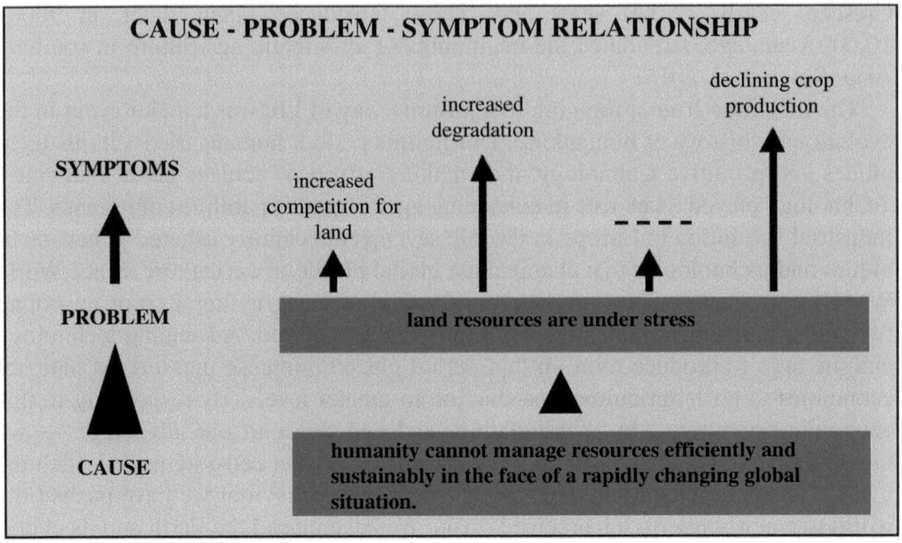

Fig. 1.1 Cause–problem–symptom relationship in stressed land-use pattern. (Adapted from: http://www.fao.org/docrep/004/).

'green revolution', because of continuing and accelerating degradation and destruction of the agricultural material base. In Africa, on an average, 10 times the value of plant nutrients are being removed annually from soil than that is being returned to soil. Overall, more than a third of the total land available on the Earth is being exploited for agriculture. Even in industrialized nations, for example the USA, soil is being eroded close to 20 times faster than it is being replaced (Hall and Hall 1993).

One possible solution for such crises lies in our ability to recognize farmland as an ecosystem: 'the agricultural ecosystem'. Such a perception alone can help us salvage whatever little materials (which will be, according to agricultural economists, 'resources') we still have with us, so that the world can sustain itself productively and usefully to humans and other organisms in the long run. Before the advent of inorganic fertilizers in the nineteenth century, farming depended solely on natural materials for nutrients. Will it be possible to combine the well-established indigenous practices with innovative methods such as new-crop breeds that will respond to low-chemical inputs? To achieve this, we will need models that will suit the circumstances of a particular region's economic and geographic profiles. Those models must also be sensitive to the social and environmental conditions at micro level (Woodwell 1990).

The Agriculture & Environment Conference of 1911 clarified that conventional agricultural practice has been the sole reason for the present environmental degradation (Edwards et al. 1993). The same conference also cautioned that traditional agricultural practices — some of which impress as sustainable — are rapidly disappearing and are replaced by farming practices that depend heavily on finite fossil fuels and associated technologies. To meet the needs for an effective and efficient management of soil, water, and other natural materials, can we aim for and work towards sustaining our production of food, fodder, fibre, and fuel on a per capita basis? Such a refined approach would minimize our dependency on petrochemical and other finite materials, currently used, overused, and abused in conventional agriculture, contributing to the improvement of the quality of soil, water, and other natural materials. Such an approach will improve per capita income and achieve greater equity in distribution. To achieve true sustainability, the human family needs to embrace an understanding in profound clarity that we, humans, are not beyond, but an integral part of nature. Embracing this understanding will involve changing the way we live and how we could organize ourselves sociologically and politically (Edwards et al. 1993; Schaller 1993).

1.2 Ecological Thinking

Ecological thinking arises out of our appreciation of our oneness with nature. Technocentric perception driven by scientific and empirical thinking emphasizes Darwin's Theory of Evolution, laced by the 'survival of the fittest' concept, which has led to human dominance perception. Those who can empathize with the ecological thinking can relate to the 'big' picture and also understand the illusion of human mastery over nature. In reality, we lose sight of the 'self' when we fail to perceive the wholesomeness of nature; also when we fail to perceive that, we are a mere component in the great scheme of things. We need to recognize that nature has its precise mechanism of constant renewal and replenishment of materials, which operates in a cyclical manner. When we humans thought that we have gained mastery — through science and its offspring, the technology — and have thus turned intensely technocentric, we started interfering with the cycles of nature. Eventually, we damaged them to go berserk in many instances and in some to turn linear. Consequent to this transformation from cyclicity to linearity, we are currently facing stunning problems, such as what we identify as 'pollution' and similar, not-so-desirable developments on the Earth. In today's technocentric environment, where economic paradigms rule the roost, ecological paradigms are seen as either 'difficult to practice' or 'primitive' or 'conservative'. To a few others, ecological paradigms appear daunting and challenging.

Movements endorsing ecological paradigms have been occurring throughout the world in different points of time. For example, in Australia, in the second half of the twentieth century, several thinkers have been contributing towards this end. For example, William Mollison and David Holmgren have created the unique 'nature–design system', which has come to stay as permaculture. Customarily, we see ecology as a hardcore science relating to the understanding of interactions of organisms with nature's factors. The offshoot of ecology — environmental science — speaks of strategies that would mitigate issues created by us humans (e.g. climate change). The value of perceiving ecology as a science-based empirical discourse, however, gradually came under close scrutiny in the middle decades of the twentieth century. The borders between ecology as a science and ecology as an art eventually turned obscure in the minds of several eminent ecologist-thinkers, who had previously practiced ecology as a pure, empirical science. This obscurity eventuated in the melding of philosophy on the one hand and ecology on the other (Naess 2008). However, the seeds for eco-philosophical thinking were indeed sown earlier by Aldo Leopold, an American ecologist-forester, who spoke of 'land ethic' in his *Sand County Almanac* (1949). Eco-philosophical thinking would be hard to perceive and compartmentalize, but when seen as a major advantage and emotional strength, it enables those that have succeeded to become more intensely creative and innovative. The obscure edges of eco-philosophical thinking — hereafter, ecological thinking — link the measurable scientific dimension of ecology and the immeasurable abstract (plus the partly measurable social) dimension of ecology. Ecological thinking as a distinct paradigm empowers humans with an ability to look within and outside. It is a powerful instrument that bridges empiricism and the abstract, thus providing intelligent and thinking humans a capacity to acquire a powerful vision.

1.3 Development of Agriculture Through Millennia

The oldest evidence of organized farming practice comes from Jericho, presently in the Jordan valley. Circa 10,000 years ago, at the spring-fed oases of Jericho, strains of the eventual direction of civilization's advances manifested. A few other smaller farming communities also flourished near the present city of Damascus and along the Euphrates. Over the next 1000 years in the Near East, domesticated plants and animals provided new and dependable food. These materials were considered dependable, because they could be stored for future needs and had the potential of ever-expanding yields. With the emergence of such agricultural societies, complex human social systems, namely villages, towns, cities, and city states, began to emerge. These systems exercised control over natural landscapes and gradually converted them into agricultural landscapes to feed human populations. It is noted that these modified landscapes produced grains for their populations only: an early presentation of what we today euphemistically call 'self-sufficiency'.

The Near East and China provide early evidences for 'organized' farming. Agriculture, as a practice, seems to have evolved not once or twice, but several times in human history, since different animals and plants have been domesticated separately and independently in different segments of the world. These early agricultural societies expanded to adjacent regions and emerged as independent cultures, because of their confluence with the natural world around them. Against this natural and cultural growth of human societies, we need to contrast the agricultural landscapes of today, when we may realize how close we are, potentially, to the end of nature and its materials that are finite. Recent satellite pictures suggest that close to 20 M ha of rainforests are being degraded and lost annually in several of the tropical countries such as India, Cameroon, Myanmar, and Costa Rica. We also need to realize that the process of trying to transform natural landscapes into economically productive agricultural landscapes — accelerated after the industrial revolution in Europe — received further impetus with developments in agricultural machinery in the 1950s. However, we need to keep in focus that all of this was the continuation of a process that began 8000–10,000 years ago in Asia, and probably in the Western Hemisphere as well, when humans first domesticated animals and plants. Long before the European industrial revolution, humans have been trying to simplify their way of life. Some elementary technologies had been developed, perfected, and used thousands of years before (Valiulis 2014).

How did agriculture begin? What was the sequence in which plants and animals were first domesticated? Why only particular plants and animals were domesticated and not others? What were the wild ancestors of these domesticated plants and animals? Why did agriculture emerge in some regions of the world and not in others? Answers to such questions came from the investigations of the Russian biologist and geneticist Nikolai Vavilov (Portrait 1.1) and the American archaeologist Robert Braidwood (Portrait 1.2). Vavilov (1992), after extensive travels and collecting seed samples from different countries, drew the following conclusions: Because hundreds of varieties of ancient wheat existed in a small, isolated pocket on the Ethiopian Plateau, diversity in cultivated forms resulted from experimentation and deliberate human selection over time. The longer a crop is grown, the more extensively it gets used, and the greater the genetic variety that eventuates within a species. The greater its use by humans, the greater its resistance to pests and diseases. In essence, Vavilov indicated that the geographical area where a crop plant had the greatest diversity of forms would also be the place where it was first domesticated. By locating the centre of a crop's genetic diversity, we can know its epicentre. Vavilov argued that determination of a species' epicentre is critical for biological and genetic research on domesticated plants. However, we know today that the Vavilov theory has at least one flaw: domesticated organisms can, and did, originate in one geographical region and develop their diversity in another. The best examples that illustrate the

Portrait 1.1 Nikolay Vavilov (1887–1946). (Source: https://russiapedia.rt.com/prominent-russians/science-and-technology/nikolay-vavilov/)

Portrait 1.2 Robert Braidwood (1907–2003). (Source: https://msu-anthropology.github.io/deoa-ss16/braidwood/braidwood.html)

weakness in the Vavilov theory are cattle and pigs, which have a much broader distribution than their ranges where they were first domesticated. Another problem would be to locate the wild relative of the domesticated organism. Robert Braidwood studied the Fertile Crescent in the Near East (Braidwood et al. 1983) and indicated the following: (i) Archaeological evidence of the transition to an agricultural way of life in the Near East corresponded with the natural habitat zone for all potential domesticates. This argument was based on his findings that all the wild ancestors of seven major Near Eastern domesticates — barley, emmer, einkorn wheat, goats, sheep, pigs, and cattle — were sourced to the Zagros Mountains in Iraq. (ii) Discovery of the archaeological remains of a farming village at Jarmo dates back by 8000–9000 years ago. The Braidwood team reconstructed the climate when Jarmo flourished, based on sound scientific reasoning. That led to the assembly of evidence for the evolution of a very different way of life, from hunter-gatherer to settlements that later evolved into societies. (iii) Establishment of a human–cultural context is absolutely critical for understanding the evolution of agriculture.

The Vavilov theory based on the present-day plant-distribution patterns and the Braidwood theory based on the past and its reconstruction partly clarify issues of a complex jigsaw puzzle. However, their contributions have provoked several biologists, archaeologists, and historians to investigate the unresolved tiles of the of the gigantic jigsaw puzzle.

The Fertile Crescent in the Near East flourished and developed into a strong agricultural economy around 10,000 years ago. When fully formed *c.* 8000 years ago, it was already the home to plants and animals (e.g. barley, wheat, lentils, sheep, goat, cattle, and pigs) that would form the basis of many agricultural economies flourishing down the ages. In China, the earliest known farming settlements existed along the Yellow River in the north and along the Yangtze River in the south, well before agricultural development in the rest of the world. It is biological evidence rather than archaeological evidence that has contributed to the full picture of the agricultural evolution of Sub-Saharan Africa and Central America. Nonetheless, little is known about the evolution of farming practice in Southeast Asia and South America, although speculation continues on the root-crop agriculture in these regions. A satisfying explanation for the transformation of nomadic hunter-gatherers into organized farming communities is yet to be found. Scholars realize that they need to explain what was different about those particular hunter-gatherer societies where domestication of wild species occurred (Harlan 1992; Smith 1998).

Population growth has been one critical external factor that forced the hunter-gatherer groups to establish into settlements, drawn as they were towards an agricultural lifestyle. Modern interpretations partly reject the population theory and see overpopulation as one of the several unexplainable but interrelated factors. Modern interpretations value regional explanation more than a general, global explanation. Regional explanation more often tends to recognize the transition from nomadic hunting-gathering groups to established societies through a sequence of unresolved developmental puzzles. In the Fertile Crescent, for example, domestication of cereals and goats and the subsequent development of agricultural economies were part of a complex and long-term transformation. This can be better appreciated when we compare and contrast the Levantine Corridor (the narrow strip of land between the Mediterranean Sea and the North African desert), Southern Sahara, and the eastern segment of North America. In all these three regions, seed plants, and not root crops, were domesticated (e.g. barley, einkorn wheat, and emmer wheat in Levantine Corridor; millet, sorghum, and African rice in the southern Sahara; marsh elder, sunflower, chenopods, and squash in eastern North America); wild ancestors of these domesticates were key food items before their domestication; the regional human societies had developed efficient technologies for harvesting and processing seeds; the people who domesticated these plants lived in relatively large, permanent communities, leading a sedentary way of life; and the seed plants in question were cultivated near lakes and rivers ensuring predictable water supply.

1.4 Modern Agriculture: Evolved on the Principles of Technology, Economics, and Management

Agriculture is presently driven by the urge to produce more in small land spaces. Technology's ever-widening capabilities have enabled us to go crazy with this initiative. In the last few decades, we have witnessed tremendous success. Countries that have not been self-insufficient in food production in the 1950s have achieved self-sufficiency in the 1970s and have even started exporting grains. Norman Borlaug sketched the grand design for this landmark achievement. Many developing nations adopted that design and realized self-sufficiency in agriculture. Many developed nations captured the Borlaug design and improved their agriculture significantly and substantially. In numerous instances, nations achieved remarkable monetary gains, as they combined technical and chemical innovations with entrepreneurial opportunism. However, two factors still remained outside the realm of human manipulation: the climate and market. The gains derived from improved technologies were strongly constrained by these two. Simply said, vagaries of climate and market influence and swing agricultural production immensely.

However, contemporary agricultural practice has somewhat understood the roles of climate and market. Developed nations use natural sciences to predict the short- and long-term climate behaviour. They apply management science to predict and understand market in both shorter and longer terms. The guarantee of these predications is, of course, debatable. Nevertheless, achieving greater clarity in these enterprises has empowered developed nations to perform better, given that the other variables in the agricultural enterprise had already been brought under human control. Thus we humans have learnt to fit agriculture into human context. We are fully convinced that the science of agriculture and the business of agriculture need to go hand in hand to achieve better results in production and profitability. Developed nations focused on extensive cropping practice, whereas a majority of developing nations resorted to intensive cropping practice. Developed nations, because of their innate economic capability, attempted producing more and more by employing new science and novel technology (e.g. use of combines, mechanical sowers, harvesters). The developing nations, on the other hand, invariably, use the massive human-power base available to them at low cost and therefore use less-efficient, or sometimes even obsolete, technology. It was, in each of such starkly different contexts, a case of recognizing and then capitalizing on one's competitive advantage that has grown.

To recap what we have seen before, contemporary agricultural practice involves efficient incorporation of animal and plant sciences, agricultural economics, business management, and marketing. The notion of agriculture in developed nations is 'whole-farm business', subscribing to the dictum 'better to solve the whole problem in an approximate way rather than to solve part of a problem in a precise way'.

Management is an integral part of the agricultural enterprise today. It is a powerful tool to remain productive and profitable. Sound agricultural management depends on sound knowledge about farming processes. But fundamentally it requires a skill in juggling diverse components — the intricately intertwined biological, economic, and human components — of a whole farm. However, we need to remember that each component is unique, with its own special characteristics. The success of a farm business relies on the ability of the farmer in achieving his/her goals through efficiency in technical production and sound financial management, targeting profit. Problem-solving skill is another critical dimension of effective management, since different kinds of problems can easily arise in farming and surprise (occasionally 'shock' as well) the farmer. Such surprises and shocks are inevitable in farming, simply because so much of the farm system consists of living material: crop plants, cattle, sheep, and even pests, pathogens, and weeds. Their behaviour as living systems is unpredictable. At least until this point of time, we have no wherewithal to predict them. Each life form thrives in its own set of specific conditions aiming the best performance (e.g. growth, reproduction). But we need to recognize that the farm ecosystem is a fragile system and conditions will usually be suboptimal. Climate is yet another element of that which can spring surprises, since with all our modern technology (e.g. satellite imagery), we still cannot forecast the weather with 100% guarantee.

The gist of agricultural management will be to make the system work at its greatest efficiency with minimum inputs achieving maximum outputs. In contemporary agricultural contexts, outputs will be production and profit, whereas inputs will be a range of biological, economic, and human investments. An efficient farm manager will aim to put together all the inputs so that he/she will achieve as many of the desired, which will become the prescribed objectives within a determined time frame. A clever manager will also keep in view the fact that not all of one's objectives can be realized fully, and there will be inevitable trade-offs. But that clever manager will also remember that he/she will make every possible effort to minimize trade-offs, by judiciously assessing the risks involved and implementing appropriate remedial measures at appropriate times. A thorough manager will also make right judgements by analysing and assimilating the past information and experience, along with incorporating current research information. Right judgements have always enabled good decisions. A sound understanding of scientific principles always predisposes a manager to making more well-founded management decisions. Science is an intellectual procedure that seeks to explain the cause and effect relationships between two aggressive variables in the contextual ambience of several related, less-aggressive variables. Scientific thinking and ability enable the farmer to perceive the role of either an individual or multiple factors that influence a process. Scientific reasoning operates either by simplification and excluding the factors except those being investigated (reductionistic practice) or by looking at large parts of production systems and by measuring the performance of various parts collectively and cohesively (holistic practice) (Raman 2013a).

Recognition of the finiteness of materials (e.g. natural, human, and financial) is the force that drives Environmental Economics, which seeks to explain their distribution on the basis of how governments provide options of their management via support and subsidies. Is this a reasonable approach? Today, Environmental Economics is considered a scientific discipline, yet one innate strength (or weakness?) is that it traditionally considers social outcomes more aggressively than environmental outcomes. Humans are fundamentally driven by their emotions, and science tends not to get involved in asking questions about this. Environmental economists, in their effort to offer solutions to complex social problems, generally tend to simplify the complexities of the human world into variables that can be decoupled from the rest for measurement, losing the organic interconnectivity of a society's living processes.

The critical thing to recognize here is that the dynamics of farm practice involves the appropriate blend of science and economics, so that the most desired outcomes — productivity and profitability — are realizable. A purely scientific approach to deal with farm practice and agricultural problems, divorced from economic necessities and realities, will only provide a partial solution, and a purely economic approach that ignores scientific trials and practice will be just as flawed.

1.5 Precipitation of Crises

Thus far, I have drawn your attention to three closely knit, complex matrix of industrialized (conventional) agriculture: (i) adoption of efficient technology, (ii) clear economic goals, and (iii) clever management strategies. This matrix contributed in a major way to a rapid degradation of the Earth's limited land, for instance. The magnitude of the problem is clearly demonstrated in Australia. From the time of the arrival of the Europeans to 1975, 45% of agricultural land badly needed remediation, because of various forms of degradation inflicted to that land because of farming practices. For example, soil erosion increased dramatically with the introduction of European-farming practices in Australia during that time (Woods 1984). Extensive removal of native vegetation, to make room for farmlands, exposed the land, destabilized soil structure, and contributed to soil erosion by water and wind. Intensive cultivation practices have resulted in loss of organic matter on topsoil and damaged the soil structure resulting in reduced capacity for infiltration and surface waterlogging. Degradation of vegetation was another obvious result of European-farming practices. Natural plant-population clusters became scanty, losing their density and vigour. The proportion of native perennials, which mostly constituted the natural vegetation, declined in a substantial manner, resulting in vast tracts of vacant land, thus making it vulnerable to invasion by undesirable plant elements. A fabulous example comes from the distribution of the most-dreaded *Parthenium hysterophorus* (Asteraceae), which has spread across almost all of the vast tracts of erstwhile rich pastureland of Queensland (e.g. Mitchell Grass Downs and Brigalow Belt) (Fig. 1.2) (Dhileepan et al. 2018). One dramatic and far-reaching outcome of clearing of native trees for agriculture in Australia is dryland salinity (Fig. 1.3). Native vegetation included long-living (perennial) woody-tree species that transpired large volumes of water and maintained groundwater far below the soil surface. In the wake of modern agriculture, vast tracts of scrub and forest land have been cleared and replaced with short-lived herbaceous plants, which utilize less volumes of water. Eventually, groundwater moved upwards, bringing the deep-seated Na and K salts up to the surface (Cocks 1992). Soil acidity, due to land mismanagement — injudicious use of fertilizers such as super-phosphate and other synthetic fertilizers and absolute removal of crop residues from the land which has the capacity to neutralize the soil's acidic content — has risen (Cumming and Elliot 1991). Available soil nutrients have tapered to micro quantities. Uninterrupted cultivation drained them and that in turn induced decline in the quality of soil structure. To replenish nutrients, farmers started injecting synthetic fertilizers (Derrick and Dann 1997). Similar to the recent surge in the use of synthetic fertilizers to strengthen the weakening soil, we have been using violent and aggressive chemicals, such as dieldrin, heptachlor, and DDT, to keep pestiferous arthropods and pathogens under control. Although these applications did offer immediate benefits, we now realize that they have caused more harm to the soil. Residues persist in the soil and build up exponentially, which in turn have been damaging the soil biota and their fascinating diversity. Such issues arising out of badly thought of land-use patterns occur plentifully throughout the world.

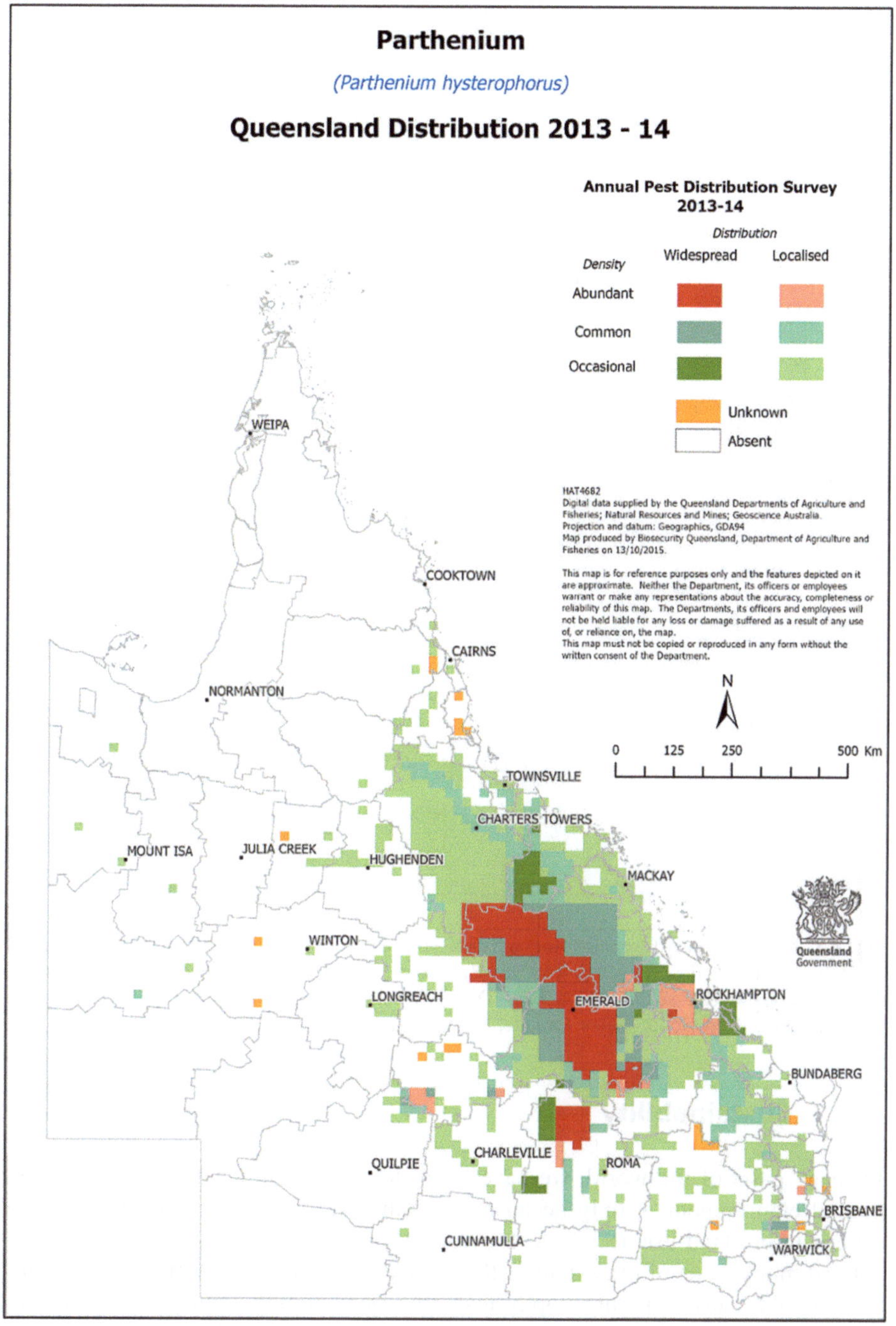

Fig. 1.2 Distribution of *Parthenium hysterophorus* in Queensland (Australia). (Source: https://www.daf.qld.gov.au/__data/assets/pdf_file/0003/790491/Parthenium_2013.pdf [Courtesy: K. Dhileepan, Queensland Department of Agriculture & Fisheries, Brisbane and the Queensland Department of Agriculture & Fisheries, Brisbane, Australia])

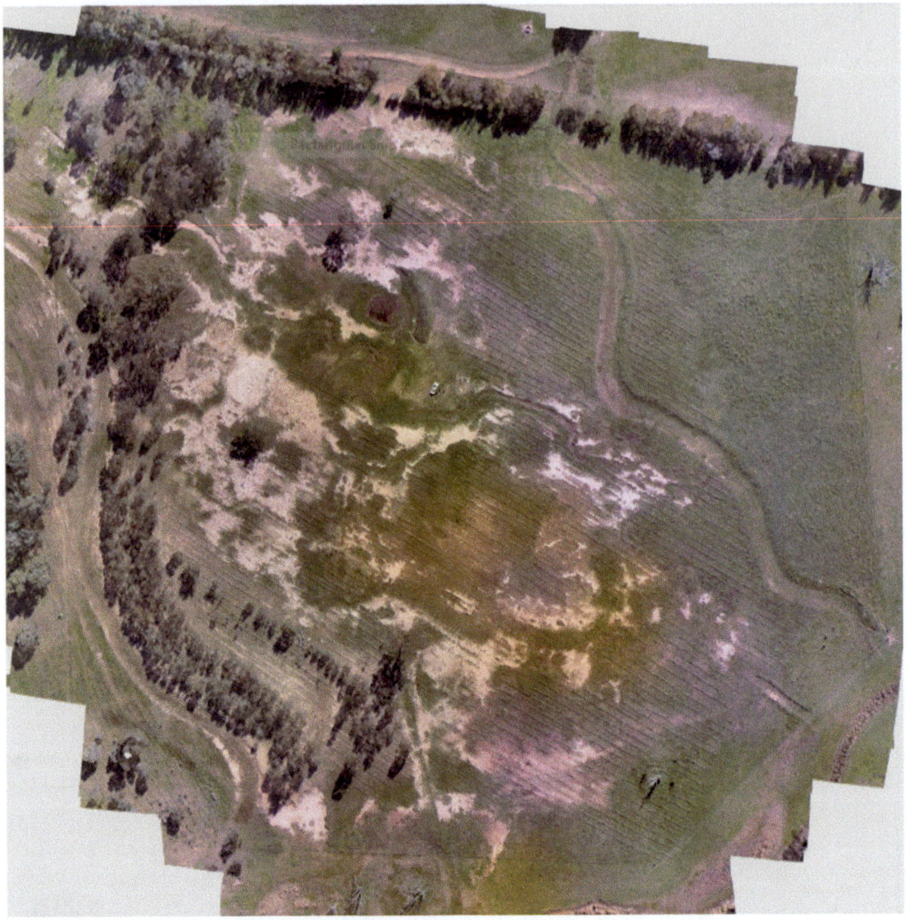

Fig. 1.3 Aerial imagery (drone photography) of a representative landsite (Sloanes Creek, Central-West New South Wales, 32°85′ S, 148°93° E, head-water catchment: *c.* 580 ha) showing prominent salinity-induced scalds. (Courtesy: David Mitchell, Department of Primary Industries, Government of New South Wales, Orange, NSW, Australia)

1.6 Sustainability

In many parts of the world, we have been inducing significant alterations to our natural environment, simply because economic criteria dominate over ecological criteria in our land-use decisions. The clarion call of Rachel Carson (1963) conscientized many of us of the extent of the critical and long-term damage such alterations could cause to our biophysical environment and, consequently, to our agricultural efforts. We began searching for a model that would enable us to achieve both economic and ecological goals — a sort of 'win–win' situation — so that the Earth can provision the materials for a longer period than our present economics-driven agricultural practices will allow.

At this point, we need to reflect on the question 'how did our ancestors manage their environment?' We need to know how pre-industrial civilizations coped with climate change, which certainly existed in those times as well (Pain 1994): (i) It was not so much climate change that caused problems, but the entrenched modes of adaptations to change. (ii) Such responsiveness depended on individual and collective choices, which were, of necessity, shaped by the past. (iii) The development and use of knowledge was the main mechanism for survival in conditions of rapid change. This means that adaptation to changing conditions depended on the perception and interpretation of the signs of impending change, on the timely development of knowledge, technology, and organization in reaction to those signs. (iv) By virtue of their privileged position, the elite who had a formal and social mandate to lead were often shut off from direct or even indirect experience of the signs of change. They had the power to maintain their lifestyles and the way things were when it was no longer prudent to do so. These perceptions enabled growing numbers of people to accept the concept of sustainability, a concept that could help us develop a set of guiding principles and goals to promote equity between and within generations of humans. Working for these outcomes will involve us in (i) maintaining the Earth's life-support systems and (ii) improving individual and community well-being.

In contrast to such a broad-based ecological perception of sustainability, economists would generally think of sustainability in narrower terms of maintaining consumption at a constant level forever. Unfortunately, economic thinkers seldom recognize that degradation of the biosphere will eventually dry up society's spending power. If we were really living in ways that would secure food consumption in the future, we would also be monitoring and regulating the use as well as abuse of our biophysical resources (Diesendorf 1997). Carson's *Silent Spring* received several follow-up commentaries in the 1960s (e.g. Boulding 1966; Mishan 1967), which stirred public, scientific, and political debates on achieving sustainability. The United Nations convened a conference on the Human Environment in Stockholm in 1972. Perhaps the most significant of the then prevailing thoughts was that of Meadows et al. consolidated in the publications *The Limits to Growth* (1972) and *Beyond the Limits* (1992). Meadows et al. illustrated the material and environmental limits to future growth in the way we use materials and energy. Their global model suggested that industrial capital would depreciate faster than any new investment could rebuild it. In brief, the global outlook was painted bleak and catastrophic. However, the Meadows et al. viewpoints have been received with considerable resentment from practitioners of economics.

In the 1980s, public consciousness of environmental issues, such as changes in global climate patterns, deforestation, and pollution, increased substantially. The newly formed green political parties gained representations in local, regional, and national governments. The United Nations set up the World Commission on Environment and Development in 1982 under the chairmanship of Gro Harlem Brundtland. The Brundtland Report, titled *Our Common Future* (1987), heralded in the concept of sustainable development, defining it as 'the development that meets the needs of the present without compromising the ability of future generations to meet their own needs'. The weakness of this definition is that it does not explain either development or needs. The word 'needs' is confusing, since it relates quite generically to both ecological sustainability and economic wants. However, the body of the report does refer to issues such as equity between and within generations, conservation of biodiversity and ecosystems, dealing cautiously with risk and uncertainty, economic development and well-being, and community participation. This report emphasizes economic development, but suggesting a different meaning from economic growth. As a follow-up of the Stockholm Conference, the Earth Summit held in Rio de Janeiro in 1992 facilitated several international negotiations and decisions related to the environmental security of the Earth, its biophysical materials, and the people (Grubb et al. 1993; Turbayne 1993).

1.7 Sustainability and the Natural-step Framework

The stark reality is that agriculture is a commercial enterprise. How can a commercial enterprise design its activities for a sustainable future? The natural-step framework (Nattrass and Altomare 2001) offers a cohesive linkage between commercial enterprise management and environmental management by exploring a concept called corporate sustainability, building on the following: (i) The whole structure of industrial society is based on a faulty design. Ours is a take–make–waste society that violates the conditions for sustainable human life on the Earth. To understand the problem, we need to take a natural systems view of our society and its relationship to the environment. (ii) Although the elements of the problem are complex in their many dimensions, the core issues are easy to understand through the conceptual framework. (iii) It may not be too late for industrial society to take action, if we act now. There is no more time for business as usual. It is not necessary or important to assign blame. It is necessary to take action, to change our present unsustainable course. (iv) Humanity is now able to take its evolution into its own hands by conscious choice and design. Some innovative companies are already taking conscious evolutionary action, and some of those are using the natural-step framework in that process. The natural-step framework provides an elegant and simple design to integrate environmental issues into the frame of business reality and to move the enterprise towards sustainable development. It includes four core processes: (i) perceiving the nature of unsustainable direction of business and society and the self-interest implicit in shifting to a sustainable direction; (ii) understanding the first-order principles of sustainability, i.e. the four system conditions; (iii) strategic visioning through backcasting from a desired sustainable future; and (iv) identifying strategic steps to move the company from its current reality towards its desired vision.

1.8 The Challenge of Ecological Sustainability

We earlier saw that the term 'sustainable development' gained global acceptance after the recommendations made by the UN-sponsored World Commission on Environment and Development (UNCED) (The Brundtland Commission 1987).

From that time on 'environmental sustainability' and 'sustainable development' are being discussed in many quarters around the world.

The environmental sustainability concept originated essentially out of social concerns: seeking improvement in human welfare by protecting the raw materials used by humans and by ensuring that the sinks for human wastes do not overflow (Goodland 1995). Environmental sustainability is a social goal that will only be realized when humanity learns to live within the limitations of the biophysical environment, by purposefully maintaining the Earth's natural-material capital both as biomass and as a sink for wastes. Achieving this will also preserve the economic subsystem of the Earth's ecosystem. The critical factor will be to strike a judicious balance between production and consumption. Of course, any depletion of non-renewable resources is unsustainable in the strict sense of the term; however some conservationists argue that modest use of materials will be acceptable, provided their depletion rates are somewhat equal to the rate at which renewable substances can be created.

Sustainable agriculture incorporates three arms of sustainability: social, environmental, and economic. Any development activity should not only be socially acceptable and economically viable but also environmentally sensitive. The broad focus in the context of sustainable agricultural development will be the overall improvement of the well-being of humans by reducing poverty, hunger, and eventually disease, simultaneously maintaining the human-support system, the natural capital, which includes the environment's sink (for the waste materials) and source (natural materials). Global human consciousness has now evolved into a widespread appreciation that our assets as a species include natural-material capital. The notion of environmental sustainability builds on this awareness and focuses our concern onto the present state of our soil, atmosphere, water, forests, and wetlands. Our ecosystems need, at the very least, to be conserved, or better still, conserved and given the security of a global commitment to their not being put at risk again, plus strategies and works on the ground to make this more than rhetoric.

1.9 Has Economics, as a Discipline, Responded Positively?

The answer perhaps is 'not', as wholesomely as the occasion would demand. Prevailing models of economic analysis treat consumption of natural capital as income, and such an approach promotes unsustainable patterns of economic activity. Consumption at the cost of natural capital is not income. Common sense prompts us to recognize that our means of producing income needs to be sustainable, but at the present rate of consumption, natural capital is becoming slimmer, scarcer, and scantier. Consumption of natural capital will lead to liquidation. Environmental sustainability needs thought, effort, and action that has a conservation focus. It is time that we accepted that natural capital is no longer a commodity to be used indiscriminately and injudiciously, but to be used with extreme prudence and care.

The view that environmental sustainability is critical only for (and in) developing countries is a myth. It is the developed countries — not only in per capita terms but also in absolute terms — that have precipitated so much of the Earth's environmental changes. Developed countries need to adopt environmental sustainability measures more quickly and vigorously than the developing countries (Goodland et al. 1992). However, Goodland's argument does not extend to the issue of the conservation of biodiversity, because a majority of the Earth's biological materials remain in tropical ecosystems that are generally distributed in the poorer belt of the Earth. In the longer run, developing countries will gain for themselves and contribute to the Earth as a whole only as they apply the precautionary principle (Raman 1998).

Environmentally sustainable development does not require us to make a simplistic choice between socially desirable activities and extractive policies. It is a combination of both; more than that, it seeks to strike a desirable balance between the two, which is hard in practical terms. The level of hardness will be intensely experienced in developing countries, because the conservation of natural materials for future needs will need to be viewed against the people's immediate necessities to lead a decent life of minimum comfort. Nevertheless, what would be 'minimum comfort' is a debatable point. One major strength of the human species is the ability to act with hindsight and forethought. Yet, the major weakness is that many of our current development activities are self-centred, despite a greater understanding of ecology and entropy. We seem to be oblivious to the antithetical relationship between the fast-shrinking natural-material capital and the ever-growing human population. We need to realize that ignorance of such an inverse relationship between two such critical variables is threatening and thwarting our survival. On the one hand, there is an urgent need for action among the world's governments and corporations and, on the other, just as urgent a need for an individual and community response!

1.10 Conservation of Biological Diversity and Ecological Integrity

Biological diversity, also referred to as biodiversity (Wilson 2010), refers to the variety of life on the Earth and as such is an essential aspect of the basic life-support systems for humans and other living beings. Normally, we recognize (i) genetic biodiversity, (ii) species biodiversity, and (iii) ecosystem biodiversity. Genetic biodiversity is the information contained in the genes of different organisms (e.g. microbes, plants, animals) on the Earth. Species diversity refers to different 'kinds' (variety) of organisms on the Earth. Estimates indicate that there are anywhere between 5 and 50 M species on the Earth, although only close to 1.4 M have been formally known. Ecosystem diversity refers to the variety of habitats, surviving communities, and the interconnecting ecological processes in the biosphere. One major influencing reason to recognize biological diversity is the rate at which organisms are becoming extinct. For example, since the colonization of Australia by Europeans nearly two centuries ago, c. 10% of mammalian species have become extinct (Woinarski et al. 2015). Of the known 20,000 plant species, close to 100 have become extinct and about 3000 are currently under the 'severely threatened' status (Fitzsimons et al. 2010). The principal threat to biological diversity is either the destruction or the fragmentation of habitats due to development activity.

Introduction of feral animals and plants and injudicious application of synthetic fertilizers and chemicals into the land have indeed accelerated the process. We need to keep in view that diversity of organisms provides us with materials that supply food, clothing, pharmaceuticals, fuel, building materials, and further to many other products and services, which can withstand measurement and valuing in economic terms.

Ecological integrity orchestrates biological diversity, and biological diversity contributes to ecological integrity. Balance in the ecological functions of ecosystems (e.g. carbon fixation, nutrient cycling, regulation of microclimate) is possible because of biological diversity. Ecological integrity also enables the maintenance of the evolutionary potential necessary for adaptations in organisms to changing environmental conditions. Economists interpret this as the conservation of a representative sample of each existing species, gene types and gene flows, and ecosystem somewhere on the Earth. Economists' perspective entails a high probability of an ecological collapse, because it accepts the dispensability of all but that one, single surviving remaining 'representative' ecosystem. The net loss of biodiversity under this regime cannot be measured, but estimated to be vast, extensive. We know minimum populations are necessary to maintain survival, but we cannot quantify what those minima are.

1.11 Other Related Influencing Dimensions of Ecological Integrity

1.11.1 Cultural Diversity and Its Conservation

Colonialism of various countries by different European kingdoms in the seventeenth and eighteenth centuries contributed tremendously to loss of cultural diversity across most parts of the world. This loss was accelerated because of growth and spread of mass media in the twentieth century. Dominant societies were destroyed (Wolfe 2006). Some attempted to suppress cultural diversity of invaded landscapes and some even tried to homogenize them with theirs (Fernandez 2014). This resulted in the failure to protect languages, social structures, and economic and political beliefs of native groups of people. We need to remember that these less-known cultures and social systems in many instances have evolved more ecologically sustainable and socially equitable ways of living, and we may need to draw upon the knowledge and experience of such groups and model them (Haverkort and Hiemstra 1999). In other words, sustaining cultural diversity is a key tactic in sustaining biodiversity and thus ecological sustainability of agriculture.

1.11.2 Individual and Community Well-being

This transcends conservation towards enhancement, because its goal is the improvement of individual and community welfare by following a path of economic progress that does not impair the welfare of future generations. Despite the ambiguity with the terms 'economic progress' and 'growth', this will aim at valuing a form of development in which society does not reach an irrevocable stage of either universal hardship or universal poverty. The term well-being has a wider connotation than being purely economic, because a set of ecological, social, political, and economic meanings can easily be drawn as inferences from it.

1.11.3 Intergenerational Equity

Equitable treatment of future generations is fundamental. However, what remains under debate is what should be handed down to future generations: the health or the diversity or the productivity of the environment? Or, all the three together? Earlier cited principles offer some leads to answering these questions. We should be able to pass on not only the primary, minimalist aspects of biological and cultural diversities but also the enhanced qualities of those principles as well. In effect, the principle of intergenerational equity plays a key role in tying together the minimalist and enhanced outcomes. Nevertheless, two ethical questions arise at this context, on which we need to reflect with extreme caution:

- What are the present generation's responsibilities to future generations?
- Since we cannot survey the views of our children's children and beyond, how best can we help them?

1.11.4 Community Participation in Decision-Making

The value for this emerges from the recognition that decisions about technological and industrial development not only involve the proponents and the government but also, and more importantly, other community stakeholders, such as the direct and indirect consumers, environmental activists, and health workers. These stakeholders hold a body of knowledge and experience that often collectively exceeds that of the dominant stakeholders.

1.12 The Precautionary Principle

The precautionary principle proposed by O'Riordan and Cameron (1994) recognizes and validates 'uncertainty' and recommends that this element of uncertainty should not be used as an excuse for doing nothing, shifting the onus of proof away from opponents to proponents, and taking anticipatory and preventive action. In the context of ecological–agricultural sustainability, the precautionary principle can be extrapolated with varying degrees of strength, depending on how either serious or significant the problems triggered by the application of this principle would be. Varied interpretations and extrapolations exist in the context of ecological sustainability and sustainable agriculture. For example, the Australian Intergovernmental Agreement on the Environment (Government of Australia 1992) indicates:

> Where there are threats of serious or irreversible environmental damage, lack of full scientific certainty should not be used as a reason for postponing measures to prevent environmental degradation. In the application of the precautionary principle, public and private decisions should be guided by careful evaluation to avoid, wherever practicable, serious or irreversible damage to the environment and an assessment of the risk-weighted consequences.

On the other hand, the British Government's environmental white paper (Chancellor of the Duchy of Lancaster 1993) indicates:

> Where the state of our planet is at stake, the risks can be so high and the costs of corrective action so great, that prevention is better and cheaper than cure… Where there are significant risks of damage to the environment, the government will be prepared to take precautionary action to limit the use of potentially dangerous materials or the spread of potentially dangerous pollutants, even where scientific knowledge is not conclusive, if the balance of likely costs and benefits justifies it.

Many see the precautionary principle as an environmental insurance, taken out by those who are disinclined to gamble on ecosystem malaise or even shutdown. The benefits coming from a precautionary approach will not necessarily be recognized however, simply because nature-driven ecological processes have –follow-on effects beyond the limits of our powers of observation. This explains why so many land developers and progress-oriented politicians never see any value in O'Riordan–Cameron's precautionary principle. We recognize, as a fundamental, that no matter who takes a position on the principle of precaution, no matter what the position is, views are informed by one's value judgements and thus are the subject to challenge.

1.13 The Scale of Ecological Sustainability

As we discuss ecological sustainability, it will be important that we place the discussion in the context of a scale — scale of time and space — so that we can consider the 'magnitude of change' and the 'velocity at which change would occur'. Change has always been occurring in the context of every living being. It is implicit in the notion of evolution. And the same fluidity is built into the notion of ecosystem: interactions and life forms are equally susceptible to change. As species reach the limit of their capacity to change, they become extinct. But we need to recognize and realize that what we described earlier as behavioural change in organisms and the consequent evolution within a species would have occurred in fractions of time for microbes, such as bacteria, and in massive periods of time for larger animals and humans. Because a given ecosystem includes species of greatly varying complexity — the more diverse an ecosystem, the more resilient it is — we need to be cautious not to oversimplify our picture of the kind of change that a given ecosystem may be undergoing, because many different timescales are involved, and many of them overlaid in highly complex manner. The human-induced greenhouse effect violates ecological sustainability because it is causing changes within several decades in the climate system whose natural changes of similar magnitude would require several millennia or more. Similarly, soils which took thousands of years to build up are being lost in a few decades. The concept of ecological sustainability needs to be applied on the spatial scales of the whole planet, continents, individual countries, and even at the scales of ecosystems such as biological regions tapering down to far smaller natural reserves. Our modern understanding of the economic concept of sustainability relies heavily on the Hartwick model (Common 1995). The Hartwick model proposes that if a given, constant population consumes a non-renewable energy for which a human-made substitute exists, then economic sustainability can be achieved if the profits of this consumption are invested. However, this model only works under several restrictive assumptions, and many of them have not been tested yet. More research is necessary to produce a practical and implementable version of economic sustainability. Within the existing economic models, sustainability is concerned with maintaining consumption at some constant level for a long, but unspecified period of time. Economists generally seek the maximum constant rate per capita consumption that can be maintained indefinitely. The interesting paradox of this approach is that once the sustainable level of consumption is achieved, economic growth will be superfluous and unnecessary. It remains to be seen whether economic sustainability defined in this way entails ecological sustainability and vice versa.

1.14 Alternative Practices of Agriculture

In earlier sections, we saw how agriculture evolved over thousands of years and how the industrial revolution in Europe changed human perceptions of agriculture rather dramatically, and, with the onset of growing concerns on the exploitation of land and other natural materials indiscriminately, how we began to look at options of sustainability. We also considered briefly the ongoing debates on the terminology and meanings of the twentieth-century terms namely sustainability, sustainable development, environmentally sustainable development, and ecologically sustainable development. Here we will look at the forces that encouraged some people to think differently from the conventional practice of agriculture. We will see the origins and evolution of alternative farming practices developed by some of the thinkers who thought outside the box. We need to remember here that the effort at this stage is only to illustrate how the thought processes began and why such leaders thought differently.

1.14.1 Organic Farming

Organic farming evolved with time by European farmers who were not convinced of the conventional farming practices that deeply engaged in manipulative techniques to reduce human effort and increase yield. The stimulus for this paradigm arose from Humphrey Davy's (1813) and Justus von Liebig's (1840) explanation that plants absorbed those minerals in humus and manure, and not from soil organic matter. Both Davy and von Liebig argued that inorganic mineral fertilizers could replace natural manures and bring agriculture under science-based management (i.e. technocentric approach), enhancing production and efficiency. This breakthrough led to agricultural revolution facilitating industrialized agriculture. In brief, the following can be seen as the goals of organic farming — sustainable methods of farming, avoidance of pollution, welfare of animals, and use of renewable resources. However, on a commercial scale at least, organic farming is an approach that seeks to secure financial viability, as well as the output of quality products within predictable time frames (Kristiansen and Merfield 2006). This can be achieved by skilful farmers drawing on scientific wisdom (Murphy 1992). Farmers by and large like to produce better food or fibre than their neighbours, or at least as well. But of course, the quality and quantity of production are constrained by various factors, including environmental conditions, and so comparisons with one's neighbour's performance have to be drawn with care. In an age where 'Landcare' as in Australia (https://landcareaustralia.org.au/) has become so effectively implanted on the psyche of farmers, healthy competition between producers is also extending beyond the economic bottom-line to environmental and social outcomes. Increasingly, farmers on both sides of the 'organic divide' are concerned about long-term impacts of agriculture on the farm and the catchment (landscape, watershed) and about other value-added issues, such as farm-animal welfare (Newton 1995).

1.14.2 Rudolf Steiner's Biodynamic Farming

One other alternative practice developed in the 1920s is biodynamic farming. Thoughts of biodynamic farming coincided with the onset of agricultural intensification. However, the alternative modes of farming — better designated as eco-farming — became excellent models for sustainable farming (Rat der Sachverständigen für Umweltfragen 1985). Biodynamic farming, with the development of its knowledge system and its diffusion, provides an example of the driving and inhibiting forces that govern any sustainable innovation that affects and involves society (Gerber and Hoffman 1998). The knowledge base of biodynamic farming includes:

- All the know-how and facilities necessary for producing, processing, marketing, and consuming products within the farming system
- The epistemology (theory of the methods or grounds of knowledge) and its influence on sociocultural context
- The institutions supporting the promotion of these processes. Can the spece among the three dot points be reduced?

Biodynamic farming voluntarily restricts itself to the use of certain management options. This is a special feature of its epistemology and a key reason for its environmental compatibility. Mainstream natural science proposes that nature is a more complex version of inanimate matter; this view flows from the premise that the total can be described as the sum of its parts. Biodynamic farming, on the contrary, represents the view that only the relationships between elements in a biological system can be described by chemical and physical laws, not the organism because it functions as a whole. An organism has a history which influences its behaviour, has activities and variability, and has inbuilt means of organizing its living processes, all of which indicate a higher order. It was the German scientist, poet, and philosopher Johann Wolfgang von Göthe who offered this explanation. His thought was picked up later by Rudolf Steiner of Austria and developed into biodynamic farming, the oldest variant of ecological farming (Gerber and Hoffman 1998).

Steiner (1924) departed from the organic understanding of nature. He proposed that farming needs to be seen as the task of the shaping and managing the farm as an organism. The farm should develop an agricultural individuality in its physical location, taking into consideration the economic and social conditions of that location. In brief, biodynamic farming follows the principle that the farm is a goal-oriented organization of agricultural production which is to a large extent self-sufficient and internally balanced.

Conventional modern farming seeks to suppress the so-called undesirable elements in the agro-ecosystem and introduce the so-called desirable elements, for example, by manipulating the species mix with synthetic chemicals. In contrast, biodynamic farmers seek to make use of even a partial knowledge of — but with a fulsome respect for — ecological processes and relationships and intervene by stimulating natural processes (Schaumann 1977; Köpcke 1994). Ecologically sound interventions, even only when implemented in a modest area, are nevertheless seen as a significant step in the realization of their cosmovision (Beekman and de Jonge 1999). Eco-farming, proclaimed as biodynamic farming by Steiner, has evolved since then, through a range of processes involving farmers, advisers, processors, and consumers, into a significant alternative food supply system in many parts of the world. It has become a movement, in the sense that supporters of Steiner's biodynamic principles see them as not only channelling their lifestyle but also informing their worldviews. Thus, for instance, in some countries, schools are set up to provide primary and secondary schooling, following a Steiner-inspired curriculum.

Some guiding principles of biodynamic thought are shown below (Dewes and Schmitt 1995):

- Biodynamic farming offers a perspective for future sustainable agriculture.
- The knowledge system of biodynamic farming is threatened by external pressures.
- Biodynamic farming will only remain as an alternative for action, if it defends its knowledge system.
- Biodynamic farming will only be sustainable if its knowledge system develops further.
- The knowledge system of biodynamic farming offers an ethical model for action.

1.14.3 Masanobu Fukuoka's Natural Farming

Fukuoka's natural farming aims at reversing the degenerative momentum of modern agriculture. Natural farming needs no machines, no chemicals, little weeding, no ploughing, nor application of prepared compost. Fukuoka distinguishes his concept of natural farming as one that cooperates with nature rather than trying to improve it by what he terms as 'assault' and 'conquest'. Fukuoka's inspiration to develop natural farming arose from his perception of the degeneration suffered by the land and Japanese society from the 1940s. He was determined to remain committed to Japanese traditional farming practice; however, he refined the process in such a way that his natural farming would demand less labour and less disruption of nature than any other, while maintaining same yields as his peers.

Fukuoka's natural farming is constructed upon the interconnections of farming practice with aspects of Japanese culture and the spiritual health of the individual. His perceptions arise from the belief that healing of the land and purification of the human spirit are the same process; his proposal reinforces the relationship between the two. He proclaimed that natural farming is, in principle, a way of life that is marked by an ongoing process of attitudinal change. As a young man, Fukuoka left his rural home to pursue a career as a microbiologist in the city of Yokohoma. Working as a plant pathologist for some years in a laboratory, Fukuoka was successful in applying principles of Western agricultural science; however, that success frustrated him since his career remained steady and uneventful. It was in 1938, when he was 25 years old, that he went through a rigorous period of introspection, during which he questioned the long-term viability and validity of western, technocentric agricultural science. In a dawn of vision, he realized that all accomplishments of human civilization are meaningless in front of the wholesomeness of nature (Fukuoka 1987). This realization formed the message of his life's work. He returned to his native village to test the soundness of his ideas by applying them in his own fields. The basic idea came to him as he happened to pass an old field, which had remained unploughed and unused for many years. In that old field, he saw healthy rice seedlings sprouting through a tangle of grasses and weeds. From this clue, he stopped flooding his field to grow rice. He stopped sowing rice seed in spring and, instead, put the seed out in autumn, sowing it directly onto the surface of the field. He learnt to manage weeds with a permanent ground cover of white clover and a mulch of rice and barley straw. Once the crop was established, he learnt not to interfere with the plant and animal communities in his field (Korn 1978). In the 1970s, Fukuoka harvested 1100–1300 lbs. (c. 400–500 kg) of rice per quarter acre which was approximately equal to the rice production systems following chemical or conventional farming methods. Although the results were comparable, the impact each of these methods left on the soil was different. Fukuoka's soils always improved in quality (e.g. fertility, structure, and capillarity and retentivity of soil) with time. Disease agents and damaging insects are always to be expected in farming, but Fukuoka thought the crops can never be devastated, because only the weak plants are affected; the best management strategy for disease and insect control is to grow crops in a healthy environment.

1.14.4 Bill Mollison's Permaculture

The Australian concept of permaculture evolved on the organic farming principles of Fukuoka. Bill Mollison founded the permaculture movement with part support from David Holmgren. Mollison regards permaculture as a philosophy for working with, rather than against nature, an approach more characterized by extended and thoughtful observation, rather than protracted and thoughtless labour, an observation of plants and animals in all their functions, rather than as separate elements of single-product systems. Permaculture also emphasized the idea that the human species is in no way superior to other life forms. A culture that cannot relate to this dictum is capable of destroying any living thing. In passionate terms, permaculture encourages farmers to see opportunities, not threats, to see solutions, not problems (Mollison and Slay 1991).

Proponents argue that permaculture is a system that allows us to exist on the Earth by using energy that is naturally in flux and relatively harmless and by using food and natural resources that are abundant, in such a way that we do not continually destroy life on Earth. Permaculture, a holistic concept, aims at decreasing energy consumption and at encouraging humans to take part in food production, at least indirectly by supporting a responsible food crop grower. It also aims at meeting all energy needs from within the system. When this is not achieved, we pay the price in energy for consumption and consequent pollution. True costs of agriculture become no longer viable and affordable, and therefore, in effect, we kill the world and the world in turn will kill us. We need to strike a chord of cooperation with the living and non-living objects that surround us. Cooperation entails harmony; opposition brings chaos and disaster.

Permaculture advocates the care of Earth (Table 1.1), which means care of all living beings and non-living objects. It implies harmless and rehabilitative activities, active conservation, ethical and frugal use of resources, and the right livelihood. This philosophy also recognizes and values the intrinsic worth of every living thing. A tree is something of value in itself, even if it has no commercial value for us. It is alive and functioning is important. It is performing its part in nature. For example, it is recycling nutrients and biomass, providing oxygen and utilizing carbon dioxide for the region, sheltering animals, and building water and soil. Thus permaculture pervades all aspects of environmental, social, and economic systems. Cooperation and not competition is the key.

1.14.5 Henri de Laulanié's System of Rice Intensification

Henri de Laulanié SJ, a Jesuit priest in Madagascar, developed and perfected a technique that is popularly known today as the *System of Rice Intensification* (SRI), which reinforces the prudent and economical use of water in rice paddies, which was hitherto perceived as an 'aquatic' plant. Laulanié's report (1993) published in *Tropicultura* (Brussels, Belgium) supplies comprehensive details of his multiple trials made in the rice paddies of Madagascar. Overall, SRI minimizes

Table 1.1 Earthcare Ethics

Think about the long-term consequences of your actions. Plan for sustainability
Where possible use species native to the area, or those naturalized species known to be beneficial The thoughtless introduction of potentially invasive species may upset natural balances in your home area
Cultivate the smallest possible land area. Plan for small-scale, energy-efficient intensive systems rather than large-scale, energy-consuming extensive systems
Be diverse, polycultural (as opposed to monocultural). This provides stability and helps us to be ready for change, whether environmental or social
Increase the sum of yields: Look at the total yield of the system provided by annuals, perennials, crops, trees, and animals. Also regard energy saved as a yield
Use low-energy environmental (solar, wind, and water) and biological (plant and animal) systems to conserve and generate energy
Bring food-growing back into the cities and towns, where it has always traditionally been in sustainable societies
Assist people to become self-reliant, and promote community responsibility
Reafforest the Earth and restore fertility to the soil
Use everything at its optimal level and recycle all wastes
See solutions, not problems
Work where it counts (plant a tree where it will survive; assist people who want to learn)

Source: Modified from Patrick Whitefield (2004)

farmer dependence on external inputs. Norman Uphoff (2003), a world leader in propagating the philosophy of sustainable agriculture, comments that SRI is a profound lesson for scientists, extension personnel, and farmers to remain open to new ideas. He further comments that not every proposed change in agricultural practice warrants undiluted attention. But if a possible innovation would have many benefits, it should be subjected to empirical rather than just logical tests, because our scientific knowledge is not (and never will be) perfect or complete. In the SRI instance, a paradigm shift was involved, one that is not yet fully understood and certainly not universally accepted.

1.15 A Dawning Realization of Ecological Agriculture

So far, we saw briefly how we humans made efforts to bring plants and animals under our control, how that need gradually developed into greed, how science and technology satiated that greed, and the major and serious consequences of that greedy action. We also saw, taking a set of specific examples, how some people began to think differently and how they interpreted the shades of difference between need and greed. Will it be ever possible for us to live in harmony with the remainder of the world, through cooperation, and not through conflicts?

To quote Rachel Carson (1963):

> Have we fallen into a mesmerized state that makes us accept as inevitable that which is inferior or detrimental, as though having lost the will or the vision to demand that which is good? Such thinking in the words of the ecologist Paul Shepard, 'idealises life with only its

> head out of water, inches above the limits of toleration of the corruption of its own environment ... Why should we tolerate a diet of weak poisons, a home in insipid surroundings, a circle of acquaintances who are not quite our enemies, the noise of motors with just enough relief to prevent insanity? Who would want to live in a world which is just not quite fatal?' Yet such a world is pressed on us.

Modern agriculture has been a critical instrument in the dramatic transformation of the world and its environment. An industry that evolved primarily as a food production program in the last few decades has grown into a nature destruction program, simply because we are guided more by greed than by realistic need. Have we lost our perception of the land? And, have we forgotten that we humans were born out of the great land which patiently supports us and tolerates all our misdeeds? Have we unconsciously forgotten (or do we deliberately ignore) the call of Archie Roach to listen to the people and listen to the land?

Ecological thinking drives ecological agriculture. It seeks and calls for a change—a dramatic and significant one—in humans, which will value every organism with respect and regard. As long as we see every organism other than *Homo sapiens* as something meant for the service of human species, then we are inviting trouble. This may not occur in the immediate future, but in long run, it will. We, then, would need to think of mitigations and resolutions. Simply said, human arrogance as the most superior organism is the key cause for our problems today. Contentment and respect for the Earth as a whole is the sum and substance of ecological thinking, the axle of the wheel of ecological agriculture.

1.16 What Needs to Change?

Against this challenge, will it be possible for us to perceive, articulate, develop, evaluate, and apply an ecologically sound agricultural practice? Will it be ever possible for us to examine the implications of ecologically sound agriculture for landholders, farmers, and other stakeholders? The pure and applied scientists who practise as agronomists and soil scientists want to make agriculture a sustainable process by dealing with the scientific principles and the social dynamics that will underpin such a practice. The second group of people involved with agriculture are interested in managing change in the participants, market forces and mechanisms, incentive structures, and regulatory policies. We need to be clear that ecological agriculture is a complex system. It not only involves the biophysical elements such as soil, crops, animals, farming practices, and the intricate interactions among them, but also it needs to validate human knowledge, learning, institutions, and policies. The human factor cannot be modelled in a predictive sense, but because of what it is, it needs to be factored into the larger questions of societal renewal for a sustainable future. Consider these arguments by Röling and Jiggins (1998):

- The change to ecological agriculture is not only a question of sound scientific claims with respect to its appropriateness and feasibility, but especially also one of widely shared learning and social reconstruction of the environment.

- The change to ecologically sound farming is not only the outcome of technical intervention but especially also a negotiated outcome based on accommodation among paradigms, coalitions, institutional interests, and politics.
- Ecological agriculture will not be achieved by merely introducing different methods and technologies to individual farmers. A transformation of the entire soft system which can be called conventional agriculture is imperative, developing into an equally complex, but different soft system which can be called ecological agriculture.
- This transition requires a management of change that goes beyond providing policymakers with scenario studies based on computer simulation, in that it must forge ways forward which emerge from interaction among stakeholders, based on shared perspectives, shared ways of making visible the state of the environment, shared strategies, and collective decision-making.

Human effort, at least in the last two centuries, has been concentrating narrowly on maximizing the value of one target variable, such as crop yield, by manipulating a certain combination of variables present in the ecosystem. This effort has led to the partial destruction or total collapse of numerous ecosystems. It is impossible to control all the feedback loops. The environment is inherently unknowable and continuously changing. It cannot be suppressed to serve a limited goal. Hence an adaptive management strategy would be in order. The adaptive management strategy depends on flexible, diverse, and redundant regulation, monitoring for responsiveness, and experimental probing to appreciate and practise 'new' ways of perceiving the environment (Holling 1995).

What will be the characteristics of a general adoption of ecological agriculture?

- Farmers will recognize—that is understand and embrace—the idea of the farm as an ecosystem in itself and an integral component of the natural ecosystem.
- This simple, powerful proposition creates the thirst for a fuller understanding of the mystery of the living system of the farm. Farmers will seek to better understand and respect the role and functions of watersheds, biomes and biotopes, and landscapes, concepts that help explain the workings of the natural ecosystem, and factors that are part of the challenge of managing the farm ecosystem.
- As farmers reach a more complete understanding of the farm ecosystem, they will become progressively more critical of their current practices. They will seek ways to minimize external inputs to improve productivity. Instead, they will seek to maximize the natural capacity of the farm ecosystem to work in harmony with the vitality and fecundity of nature.

The most critical element in this process will be the need for each of us to let go the attitudes and convictions that are deeply embedded. From a value framework in which humans are considered the world's superior organism, mainly due to scientific skills and technological advancements, we need to see that we are an infinitesimally small part of the natural world. We need to accept that we have to shed our sense of superiority and learn with utmost modesty to live in harmony with the natural world and not in conflict. Such a realization is indeed hard to achieve, because we are guided by our selfishness and ego. In practical terms, we need to develop sensitivity and work towards achieving that realization. The shift to ecological agriculture coincides with or is predicated upon a shift from seeing everything around us as being individual or broken-down elements, to accepting everything around us as part of a larger, more complex whole. This is part of the Copernican Revolution which commenced with the realization that humans are not the centre of the universe, but rather a temporary and marginal event in an immense space. (Note: the thinking that fuelled the Copernican Revolution is now finding a fuller expression in the New Science—growing acceptance of multiple equivalent realities, as opposed to one solid objective truth, and of the inherent complexity of nature, as opposed to the certainty that knowledge can be accumulated about it.) Such a shift fits well with the shift from the arrogant focus on our efforts to monitor and control of nature to a realization that humankind is inherently part of nature (Tarnas 1991).

1.17 An Ecological 'Knowledge System'

One step towards achieving ecologically sound agriculture will be to accept the vitality of ecological knowledge. A potentially fruitful approach here is the knowledge system perspective, which looks at the 'institutional actors' within the arbitrary boundary of what can be considered a 'theatre of innovation', a drama capable of interpretation as a soft system. But how useful is this line of thought? One cannot say that such actors as research, extension, and farmers are a system. In all likelihood, they are not, in that there is no synergy among their potentially complementary contributions to innovative performance. Notwithstanding this problem, however, if we look at them as potentially forming a soft system, one starts to admit the possibility of facilitating their collaboration and hence the possibility of enhancing their synergy and innovative performance. Innovation is an emergent property of a soft system (Wilson and Morren 1990). Perhaps, it is premature to dismiss the knowledge system approach.

The following words of Funtowicz and Ravetz (1994) are worthy of reflection:

> The future looks less like the past than ever before and has in some ways become very threatening. As a species we are no longer guaranteed survival, even in the short term—and this is a consequence of our own doings, as collective humanity. We are living in a risk society.

1.18 Conclusion

The term 'sustainable development' refers to something more than simply growth; we talk of development here which is strikingly, subtly, and intricately different from growth. A change in the kind of growth is needed, a kind of development that is less material- and energy-intensive and more equitable in the distribution of its benefits. This emphasizes that changes are needed and that the security, well-being, and the survival of the planet depend on such changes (The Brundtland Commission 1987). However, some environmental activists reject the concept of sustainable development and even ESD as meaningless (ESD–Working Groups 1990–1994); they argue that these terms represent a way of thinking that seeks to find ways of allowing vested business interests to achieve their financial objectives. Sustainable development is not about giving priority to environmental concerns; it is about incorporating environmental assets into the economic system. Sustainable development encompasses the idea that the loss of environmental amenity can be substituted by wealth creation that placing a price tag on the environment will help us protect it unless degrading it is more profitable, that economic growth is necessary for environmental protection and therefore should take a priority over it. Sustainable development represents an adoption of the term sustainability, which once represented ideas of stability and equilibrium and harmony with nature. Sustainable development is an attempt to reduce the politics in decision-making by artificially replacing conflict with consensus, by emphasizing technocratic decision-making process such as cost–benefit analysis and economic instruments, and by ensuring environmental conflicts decided by the market.

At this juncture, it would not only be relevant but also be appropriate to refresh our minds with the visionary thoughts and words of Joseph Chelladurai Cornelius Kumarappa (JCK) (1892-1960), a committed and ardent disciple of Mahãtmã Gandhi. A couple of statements from JCK (1962), cited below, are adequate to kindle some refreshed thinking in us. On purpose, I will refrain from either elaborating or commenting on JCK's words, which squarely challenge Adam Smithian thinking and its consequence of technocentrism be it in agriculture or elsewhere:

> There is no such thing as the principles of Economics of Gandhiji. With Gandhiji, economics is a part of a way of life. There are no governing principles as are applied in the case of ordinary laws that have been enunciated in text books on Economics. Only two life principles govern all Gandhiji's economic, social, political and other considerations, viz. Truth and Non-violence. Anything that cannot be satisfactorily tested on these touch-stones, as it were, cannot be regarded as Gandhian. If a scheme of things leads to violence or necessitates untruth, then we may regard that as non-Gandhian. (p. 5)

> The agricultural economy which had been practised in our country in olden days is an instance of the enterprising economy. It is a self-sufficient economy. (p. 7)

> The American method of producing grains makes agriculture an industry as distinct from an occupation. We need to make that distinction clear because a great many of the differences that arise and the methods that are used and the principles followed come out of that distinction. An industry is not concerned with the question whet her people are starving in India or

> in any part of the world, its sole concern is to maintain high prices. Human considerations do not prevail at all. All that is needed is keep down the supply so that the demand will put-up the prices. … Ours is an ancient land with a teeming population. Land requires time to recuperate between crops. The more we extract from the land the longer it takes to regain its fertility. Our methods of agriculture have, through the experience 0f centuries, reached a stage where production is balanced with its recuperative power. If we use the Mithaiwala methods[1] and attempt to exploit the land to the fullest extent by stimulating it, we shall be exhausting its fertility faster than its ability to regain itself and finally we shall be turning good cultivable soil into barren deserts. (p. 18)

Has the Mahãtmã said anything pertinent to sustainability, when that word never had the connotation and usage, as it presently has? Mahãtmã Gandhi's primary concern was to alleviate poverty in India. He was clearly a human-centred person. He was all the time concerned about the poorest Indian living in the remotest India (Lal 2000). Yet, while thinking of wholesome and worthwhile paradigms to improve the lot of that Indian, he drew generously from nature, which, to him, was plentiful and bountiful. In 1909 he envisioned that an insatiable and unending pursuit of wealth accruing—and its derivative material benefits—that predominated the Western culture is a key threat to the Earth and natural materials. He forewarned how much such an imprudent lifestyle can bring about destruction to nature. By his remarks on freedom, self-reliance, and personal physical effort, he reiterated his recognition of nature and its vastness.

Ecological thinking and its derivative ecological agriculture are practices that spin around simplicity and modesty. Aggressive dollar-driven thinking has no place. Climate change for example is an issue created by us humans because of our badly thought-out and hasty, changed practices of land use (Stabinsky and Ching 2012). If we realize this weakness and remedy it, then we still have hope to leave a cleaner and better world for later generations of humans as well as other organisms, which are as important as *H. sapiens*! We are in a crazily fast era. We think that speed and rapid turnarounds of events are the norms of today. Is speed the root cause of present-day ecological–environmental malady, which has pushed us to think of sustainability (Raman 2013b)? I have no answers.

Acknowledgements I am highly grateful to David Mitchell (Department of Primary Industries, Government of New South Wales, Orange) and Kunjithapatham Dhileepan (Department of Agriculture and Fisheries, Government of Queensland, Brisbane) for sparing me images for use in this article.

[1] In page 13 of this monograph, JCK clarifies what he means by Mithaiwala economics and the alternative he suggests as the mother's method. According to him, the Mithaiwala economics creates false standards and violence, while the mother's method of production develops her love and truthfulness, but entails hard work.

References

Beekman W, de Jonge J (1999) Biodynamic agriculture and spirituality. In: Haverkort B, Hiemstra W (eds) Food for thought: ancient visions and new experiences of rural people. ETC–COMPAS, Leusden, pp 199–211

Boulding K (1966) The economics of the coming spaceship earth. In: Jarrett H (ed) Environmental quality in a growing economy. Johns Hopkins University Press & Resources for the Future, Baltimore, pp 3–14

Braidwood L, Braidwood R, Howe B, Reed C, Watson PJ (1983) Prehistoric archaeology along the Zagros flanks. The University of Chicago Oriental Institute Publications, Chicago

Carson R (1963) Silent spring. Hamish Hamilton, London

Chancellor of the Duchy of Lancaster (1993) Realising our potential: a strategy for science, engineering and technology. Her Majesty's Stationery Office, London

Cocks KD (1992) Use with care. University of New South Wales Press, Sydney

Common M (1995) Sustainability and policy. Cambridge University Press, Cambridge

Cumming RW, Elliot GL (1991) Soil chemical properties. In: Charman PEV, Murphy BW (eds) Soils: their properties and management. Sydney University Press, Sydney, pp 193–205

Davy H (1813) Elements of agricultural chemistry. In a course of lectures for the Board of Agriculture. W. Bulwer, London

de Laulanié H (1993) Le système de riziculture intensive malgache. Tropicult (Brussels) 11:110–114

Derrick J, Dann P (1997) Soils and agriculture. In: Diesendorf M, Hamilton C (eds) Human ecology, human economy. Allen & Unwin, St Leonards

Dewes T, Schmitt L (eds) (1995) Beiträge zur Wissenschaftsagung zum Ökologischen Landbau. Wissenschaftlicher Fachverlag, Gießen

Dhileepan K, Callander J, Shi B, Osunkoya OO (2018) Biological control of parthenium (*Parthenium hysterophorus*): the Australian experience. Biocontrol Sci Tech 28:970–988

Diesendorf MO (1997) Ecologically sustainable development principles. In: Diesendorf MO, Hamilton C (eds) Human ecology, human economy. Allen & Unwin, St Leonards, pp 64–97

Edwards CA, Wali MK, Horn DJ, Miller E (eds) (1993) Agriculture and the environment. Elsevier, Amsterdam

ESD—Working Groups (1990–1994) Several reports. The Australian Government Publication Service, Canberra

Fernandez E (2014) Child protection and vulnerable families: trends and issues in the Australian context. Soc Sci 3:785–808

Fitzsimons J, Legge S, Traill B, Woinarski J (2010) Into oblivion? The disappearing native mammals of northern Australia. The Nature Conservancy, Melbourne

Fukuoka M (1987) The road back to nature: regaining the paradise lost. Japan Publications, Osaka

Funtowicz SO, Ravetz JR (1994) Science for the post-normal age. Futures 25:739–755

Gerber A, Hoffman V (1998) The diffusion of eco-farming in Germany. In: Rölling NG, Wagemakers MAE (eds) Facilitating sustainable agriculture. Cambridge University Press, Cambridge, pp 134–152

Goodland R (1995) The concept of environmental sustainability. Annu Rev Ecol Syst 26:1–24

Goodland R, Daly HE, El-Serafy S (1992) Population, technology, and life style: the transition to sustainability. The Islands Press, Washington, DC

Government of Australia (1992) Intergovernmental agreement on the environment. Commonwealth of Australia, Canberra http://www.environment.gov.au/about-us/esd/publications/intergovernmental-agreement. Accessed 7 Nov 2018

Grubb M, Koch M, Thomson K, Munson A, Sullivan F (1993) The Earth Summit agreements: a guide and assessment. Earthscan & the Royal Institute of International Affairs, London

Hall CAS, Hall MHP (1993) The efficiency of land and energy use in tropical economies and agriculture. Agric Ecosyst Environ 46:1–30

Harlan JR (1992) Crops and man. The American Society of Agronomy, Madison

Haverkort B, Hiemstra W (eds) (1999) Food for thought: ancient visions and new experiences of rural people. ETC–COMPAS, Leusden
Holling CS (1995) What barriers, what bridges? In: Gunderson LH, Holling CS, Light SS (eds) Barriers and bridges to the renewal of ecosystems and institutions. Columbia University Press, New York, pp 6–37
Köpcke U (1994) Nährstoffkreislauf und Nährstoffmanagement unter dem Aspekt des Betriebsorganismus. In: Mayer J et al (eds) Ökologischer Landbau—Perspektif für die Zukunft! Stiftung Ökologie und Landbau, Bad Dürkheim
Korn L (1978) Introduction. In: Fukuoka M (ed) The one-straw revolution: an introduction to natural farming. Rodale Press, Emmaus
Kristiansen P, Merfield C (2006) Overview of organic agriculture. In: Kristiansen P, Taji A, Reganold J (eds) Organic agriculture: a global perspective. CSIRO Publishing, Collingwood, pp 1–24
Kumarappa JC (1962) Gandhian economic thought. Sarva Seva Sangh Prakashan, Varanasi
Lal V (2000) Too deep for deep ecology: Gandhi and the ecological vision of life. In: Chapple CK, Tucker ME (eds) Hinduism and ecology: the intersection of earth, sky, and water. Centre for the Study of World Religions, Harvard Divinity School, Cambridge, MA, pp 183–212
Leopold A (1949) A Sand County almanac and sketches here and there. Oxford University Press, Oxford
Mishan EJ (1967) The costs of economic growth. Staples Press, London
Mollison B, Slay RM (1991) Introduction to permaculture. Tagari Publications, Tyalgum
Murphy MC (1992) Organic farming as a business in Great Britain. University of Cambridge, Cambridge
Naess A (2008) The shallow and deep ecology movement. Trumpeter 24:59–67
Nattrass B, Altomare M (2001) The natural step for business. New Society Publishers, Gabriola Island
Newton J (1995) Profitable organic farming. Blackwell Science, Oxford
O'Riordan T, Cameron J (eds) (1994) Interpreting the precautionary principle. Earthscan, London
Pain S (1994) Rigid cultures caught out by climate change. New Sci 141(1915):13
Raman A (1998) The WISDOM initiative for environmentally-sound development activity: a worthwhile model for developing countries. Int J Ecol Environ Sci 24:421–430
Raman A (2013a) Linking holistic and reductionistic approaches: teaching of the undergraduate subject 'introduction to ecological agriculture'. Agric Educ Mag 85:22–24
Raman A (2013b) Ecological management of rice agriculture in Southern India. Int J Ecol Environ Sci 39:37–49
Raman A (2019) Plant domestication and evolution of agriculture in India with special reference to peninsular India. In: Chakrabarti R (ed) Environmental history of India. Indian Council for Historical Research, New Delhi. (in press)
Rat der Sachverständigen für Umweltfragen (1985) Umweltprobleme der Landwirtschaft. Kohlhammer Verlag, Stuttgart
Rey J-Y, Diallo TM, Vannière H, Didier C, Keita S, Sangaré M (2004) The mango in French-speaking West Africa. Fruits 61:121–129
Röling NG, Jiggins J (1998) The ecological knowledge system. In: Röling NG, Wagemakers MAE (eds) Facilitating sustainable agriculture. Cambridge University Press, Cambridge, pp 284–311
Schaller N (1993) The concept of agricultural sustainability. Agric Ecosyst Environ 46:89–97
Schaumann W (1977) Der Biologisch-Dynamische Landbau, Ökologischer Landbau—eine Eurpäische Aufgabe. Stiftung Ökologie und Landbau, Verlag CF Müller, Karlsruhe
Singh S, Singh A, Rajkumar R, Kumar KS, Samy SK, Nizamuddin S, Singh A, Sheikh SA, Peddada V, Khanna V, Veeraiah P, Pandit A, Chaubey G, Singh L, Thangaraj K (2016) Dissecting the influence of Neolithic demic diffusion on Indian γ-chromosome pool through J2–M172 haplogroup. Sci Rep 6:19157. https://doi.org/10.1038/srep19157
Smith BD (1998) The emergence of agriculture. Scientific American Library, WH Freeman, New York

Stabinsky R, Ching LL (2012) Ecological agriculture, climate resilience and a roadmap to get there. Third World Network, Penang
Steiner R (1924) Geisteswissenschaftliche Grundlagen zum Gedeihen der Landwirtschaft (Landwirtschaftlicher Kurs). Rudolf Steiner Online Archive. http://anthroposophie.byu.edu. Accessed 7 Nov 2018
Tarnas R (1991) The passion of the western mind: understanding the ideas that have shaped our world view. Ballentine Books, New York
The Brundtland Commission (1987) Our common future. Oxford University Press, Oxford
Turbayne D (1993) To the summit and beyond. The Australian Council for Overseas Aid, Canberra
United Nations Environment Programme (1999) The future of our land: facing the challenge. Food and Agriculture Organisation of the United Nations, Rome
Uphoff N (2003) Higher yields with fewer external inputs? The system of rice intensification and potential contributions to agricultural sustainability. Int J Agric Sustain 1:38–50
Valiulis AV (2014) A history of materials and technologies development. VGTU Press Technika, Vilnius
Vavilov NI (1992) The origin and geography of cultivated plants. Cambridge University Press, Cambridge
von Liebig J (1840) Die Organische Chemie in ihre Anwendung auf Agricultur und Physiologie. Verlag von Friedrich Vieweg & Sohn, Braunschweig
Whitefield P (2004) Earth care manual: a permaculture handbook for Britain and other temperate climates. Permanent Publications, East Meon
Wilson EO (2010) The diversity of life. Harvard University Press, Harvard
Wilson K, Morren G (1990) System approaches for improvement in agriculture and resource management. Macmillan, New York
Woinarski JCZ, Burbidge AA, Harrison PL (2015) Ongoing unraveling of a continental fauna: decline and extinction of Australian mammals since European settlement. PNAS 112:4531–4540
Wolfe P (2006) Settler colonialism and the elimination of the native. J Genocide Res 8:387–409
Woods LE (1984) Land degradation in Australia. The Australian Government Publication Service, Canberra
Woodwell GM (ed) (1990) The earth in transition: patterns and processes of biotic impoverishment. Cambridge University Press, Cambridge

[illegible] (1999) [illegible]

[illegible]

[illegible]

Climate Policy

2

Balasubramanian T. N and A. Nambi Appadurai

Abstract

The World Meteorological Organization (WMO) as early as in 1976 talked about the potential ill effects of the increased accumulation of greenhouse gases (GHGs) in the atmosphere on the future climate and weather. The WMO along with the United Nations Environment Programme (UNEP) did establish the Intergovernmental Panel on Climate Change (IPCC) in 1988 with a mandate to provide scientific information to governments on the risks associated with climate change and its impacts on natural and human systems. No doubt climate change is one of the defining challenges of the twenty-first century having a profound impact on the needs of the global population, poverty alleviation, sustenance of natural ecosystems and food security. Climate change is no longer considered to be just an environmental concern but also a development problem affecting both the developing and developed countries. One of the central policy issues in the context of climate change is how countries of the world should allocate resources. After a few years of intense engagement and international negotiation processes, the world community has signed the historic Paris climate agreement in 2015, which calls for substantive domestic and international climate actions to tackle climate issues. Individual countries are making efforts to strengthen their "Intended Nationally Determined Contributions" (INDCs) by streamlining their mitigation and adaptation initiatives. Some of the areas where investments have been made to reshape policies and actions include (i) renewable energy, (ii) building of rural and urban resilience capacity, (iii) poverty alleviation and (iv) sectoral priorities. The environmental governance focus is also shifting from National to the sub-National and local level. Internationally, there

Balasubramanian T. N (✉)
Agricultural Meteorology Department, Tamil Nadu Agricultural University, Coimbatore, India

A. N. Appadurai
World Resources Institute, New Delhi, India

© Springer Nature Singapore Pte Ltd. 2020

V. Venkatramanan et al. (eds.), *Global Climate Change and Environmental Policy*,
https://doi.org/10.1007/978-981-13-9570-3_2

have been considerable efforts in synchronising the INDC commitments with the Sustainability Development Goals (SDGs). One of the key policy agreement that Parties entered under the Paris agreement is to stabilise rising temperature below 2 °C preferably to limit to 1.5 °C. At the local level, policy actions should focus on assessing the vulnerability of both physical and social systems to climate change, development of best bet technology to reduce the impacts of climate change and measures to enhance the adaptive capacities of local communities and enhance overall climate risk management capacity of the region. Transformative changes could be achieved by developing and executing proactive environmental and climate risk management policies, promoting effective implementation strategies and linking responsive institutions across scales for achieving activities effectively. Though developed countries are fundamentally responsible for global warming, the major impacts are borne by poor and marginalised countries. Hence, the international climate regime needs innovative climate policies and institutional structures that would help promote international cooperation. It is also imperative to design and implement appropriate national climate policies that would contribute to individual countries' green growth and promote climate-sensitive development. This chapter focuses on the elements that go into the making of climate policy and their relevance to the emerging global and national scenarios.

Keywords
Climate policy · UNFCCC · Kyoto protocol · Copenhagen agreement · Paris agreement · Climate policy-making · COP negotiations · Per capita CO_2 emissions

2.1 Introduction

Climate change is a threat to both human and natural systems. The world has come to realise this a few decades ago, with the scientific evidence shared by WMO. Though the global climate policy is predominantly shaped by the scientific pieces of evidence, there have been implicit gaps in the policies. The UNFCC plays a significant role in facilitating the negotiated agreements that set the tenor of international climate policies. Scientific bodies like IPCC and WMO provide the scientific base to make informed decisions. Over the years, the mainstay of climate negotiations is focused on emission reduction, technology transfer, equity, capacity building and climate finance. Bitter battles between developed and developing countries finally reached a negotiated settlement in the year 2015 with the advent of the Paris agreement. The Paris Agreement's intent is 'to central aim is to strengthen the global response to the threat of climate change by keeping a global temperature rise in this century well below 2 °C above pre-industrial levels and to pursue efforts to limit the temperature increase even further to 1.5 °C'. Further, the international climate agreement (Paris agreement) endeavours to augment the capabilities of countries to manage the impacts of climate change and also at synchronising the climate finance with climate-resilient development pathway.

The "United Nations" and IPCC have taken many decisions on reducing GHGs emissions, finance support to developing countries by developed countries and enhancing adaptive capacity to meet the challenges from climate change from Earth summit to Paris agreement. But concrete action is still hanging. This indicates that the policy discussed in the negotiations have not been developed with the fullest involvement of the countries participated. The policy also lacks flexibility in general to meet the development goals of the countries participated.

2.2 Overview of Policies, Accords and Agreements

2.2.1 Key Conventions and Summits on Environment and Climate Issues

The conferences, viz. "the United Nations Conference on the Human Environment" (Stockholm, 1972); "the United Nations Conference on Environment and Development" (Rio de Janeiro, 1992); "The World Summit on Sustainable Development in Johannesburg" (South Africa, 2002) and "the United Nations Conference on Sustainable Development" or "Rio + 20" (Rio de Janeiro, 2012) are milestones in the environmental history. The birth of the "United Nations Framework Convention on Climate Change" (UNFCCC) is a significant moment indeed. UNFCCC is an international treaty, an outcome of "the United Nations Conference on Environment and Development" (UNCED), also known as the "Earth Summit" held at Rio de Janeiro from 3 to 14 June 1992. The treaty was aimed to stabilise greenhouse gas concentrations in the Earth's atmosphere at a level that would prevent malevolent impact on the existing climate system. The treaty had no legal binding on any of the countries participated in the emission of greenhouse gases. Rather, the treaty did include provisions for updates (protocols) that would set mandatory emission limits in future.

Many issues had been discussed in the 1992 "UNCED", but the important ones are switching over to alternate energy resources from fossil fuel-based energy and focus on public transport so as to minimise the vehicular emissions. Further, the highlight of the summit is the UNFCCC which paved way for Kyoto protocol and Paris agreement alternate source of energy to replace fossil fuels and a new reliance on the public transportation system in order to reduce vehicle emissions. An important achievement of the summit was an agreement on the climate change convention which in turn led to the Kyoto Protocol and Paris agreement.

2.2.2 Kyoto Protocol

The Kyoto Protocol is an international agreement linked to the UNFCCC (https://en.wikipedia.org/wiki/Kyoto_protocol). The extracts are presented here under:

- The Kyoto Protocol put in place "legally binding commitments for the reduction of six greenhouse gases (CO_2, CH_4, N_2O, SF_6, HFC, perfluorocarbons) produced by the listed Nations in Annex I" (industrialised) as well as the general commitment for all member countries.
- Though adopted on 11 December 1997 in Kyoto, entered into force on 16 February 2005. As of now, 183 Parties ratified the protocol.
- Under the Kyoto Protocol, industrialised countries agreed to reduce their collective GHGs emission by 5.2% compared to the year 1990 by 2008 to 2012.
- Kyoto protocol includes defined "flexible mechanisms" such as International Emissions Trading, Carbon Development Mechanism and Joint Implementation to allow Annex I countries to meet their GHG emission limitations by purchasing GHG emission reductions credits from elsewhere, through financial exchanges.

The detailed rules for the implementation of the Kyoto Protocol were adopted at the COP7 in Marrakesh, Morocco in 2001 and are referred to as Marrakesh Accords. Its first commitment period started in 2008 and ended in 2012.

The United Nations encouraged governments to ratify (Doha Amendment, UNFCCC) amendments relating to the second commitment period of the Kyoto Protocol, the international emissions reduction treaty. Ratification of the Doha Amendment to the Kyoto Protocol was a valuable part of the momentum for global climate action for the years leading up to 2020 (https://unfccc.int/process/the_kyoto_protocol/the_doha_amendmend). Parties to the Kyoto Protocol adopted an amendment to the Kyoto Protocol at COP8 held in Doha on 8 December 2012. At Doha, the adopted amendments were:

- New commitments for Annex I Parties to the Kyoto Protocol who agreed to take on commitments in a second commitment period from 1 January 2013 to 31 December 2020
- A revised list of greenhouse gases (GHGs) to be reported on by Parties in the second commitment period
- Amendments to several articles of the Kyoto Protocol which specifically referenced issues pertaining to the first commitment period and which needed to be updated for the second commitment period

During the first commitment period, 37 industrialised countries and the European Community committed to reducing GHG emissions to an average of 5% against 1990 levels. During the second commitment period, Parties committed to reducing GHG emissions by at least 18% below 1990 levels in the 8-year period from 2013 to 2020; however, the composition of Parties in the second commitment period is different from the first.

2.2.3 Copenhagen Agreement

The Copenhagen Agreement is an outcome document of COP15 to UNFCCC. Here are some excerpts of the Copenhagen accord (https://en.wikiapedia.org/wiki/copenhagen_Accord).

- Endorses the continuation of the Kyoto Protocol.
- Emphasises a "strong political will to combat climate change in accordance with the principle of common but differentiated responsibilities and respective capabilities".
- To prevent anthropogenic interference with the climate system, an increase in temperature should be well below 2 °C in a context of sustainable development to combat climate change.
- Vulnerable countries should have a comprehensive adaptation programme including from international community.
- Deep cuts in global emission are required according to science of climate change (AR4 report of IPCC) to reduce emission as early as possible and a low emission development strategy is essential to sustainable development.
- States that act for adaptation is urgently required to reduce vulnerability and build resilience in developing countries.
- Developed Nations shall provide adequate predictable and sustainable financial resources, technology and capacity building.
- Developed countries (Annexure I) would commit to economy-wide emissions target to 2020 to be submitted by 31 January 2010.
- Delivery of reduction and finance by developed countries will be measured and verified.
- Developing Nations would implement mitigation actions and report once in 2 years.
- Need to enhance removal of GHG emission by forest.
- Developed Nations should enable the developing Nations to reduce emissions.
- Low emission developing countries will be given incentives to continue the low emission pathway.
- A goal for the world to raise $100 billion every year by 2020 from a wide variety of resources to help developing countries cut carbon emissions.
- Establishes the Copenhagen Green Climate Fund.
- Establish technology mechanisms to accelerate technology development and transfer.
- Calls for an assessment of the implementation of the Accord to be completed by 2015.

The primary concerns associated with the accord: The accord itself is not legally binding

- No decision was taken on whether to agree a legally binding successor or complement to the Kyoto Protocol.

- The accord sets no real targets to achieve emissions reductions.
- The accord was drafted by only five countries·
- The deadline for an assessment of the accord was drafted for 6 years, by 2015.
- The mobilisation of 100 billion per year to developing countries will not be fully in place until 2020.
- There is no agreement on how much individual countries would contribute to or benefit from any funds.
- COP delegates only "took note" of the accord rather than for adopting it.
- There is not an international approach to technology.
- Forgets fundamental sectoral mitigation, such as transportation.
- It shows biases in silent ways such as the promotion of incentives on low gas-emitting countries.

2.2.4 Paris Agreement (https://en.wikipedia.org/wiki/Paris_Agreement)

At COP21 in Paris, on 12 December 2015, Parties to the UNFCCC reached a landmark agreement "to combat climate change and to accelerate and intensify the actions and investments needed for a sustainable low carbon future", and this agreement mobilised almost all countries to take effective steps to check climate change and adapt to the climate variability and climate change. As such, it charts a new course in the global climate effort. The Paris Agreement's central aim is "to strengthen the global response to the threat of climate change by keeping the global temperature rise in this century well below 2 °C above pre-industrial levels and to pursue efforts to limit the temperature increase even further to 1.5 °C". Also, the international climate agreement endeavours to upscale the capabilities of countries to manage the impacts of climate change and also at synchronising the climate finance with climate-resilient development pathway.

To reach these ambitious goals, "appropriate mobilisation and provision of financial resources, an innovative technology framework and enhanced capacity-building are essential. Further, the agreement demands the developed countries to provide enabling condition and pro-active support to developing and vulnerable countries to meet their climate specific national commitments and objectives".

The Paris Agreement demands all Parties to propose the "intended nationally determined contributions" and take active measures to achieve nation-specific commitments. This includes requirements that all Parties report regularly on their emissions and on their implementation efforts. The global stock-taking exercise conducted every 5 years is aimed at assessing the progress with respect to climate action and to provide productive insights so as to aid the Parties to achieve the GHG reduction commitments.

2.3 Limitations of the UNFCCC COP Meetings

Out of 195 countries in the world, 193 countries are member states to the UN and 2 countries are non-member observer states. World population had crossed 7.7 billion (2018) and the population density was 50.79 people/km^2 (2015). Per capita per annum emission of CO_2 during 2015 was 4.9 tonnes for the world, while it was highest for Qatar (39.7 tonnes) and the least was with the country Pakistan (0.9 tonnes), the Philippines (1.1 tonnes) and Indonesia (2.0 tonnes). As per WMO annual greenhouse bulletin of 2012, since the industrial revolution, about 375 billion tonnes of carbon was emitted by the human in the atmosphere as CO_2. Some estimate indicated that world nations annually released 38.2 tonnes of CO_2 into the air by burning of fossil fuels such as coals excluding natural sources like oceans, soils, plants animals, volcanoes, etc.

From Earth Summit (1992) to Paris Agreement (2015), over 23 years, negotiations were going on at the Conference of Parties meetings on the reduction of GHGs emission firstly by developed countries so as to reduce/stabilise rise in air temperature with reference to pre-industrial period and also on the adaptation measures to be undertaken to reduce the impact from climate change across countries of the world. But concrete results could not be obtained from the parties, may be due to weak inter- and intra-cooperation between COP's parties. All promoted actions from the COP's whether legal or non-legal are only optional and not obligatory. Hence, the participation of countries is very weak. All countries talk about climate change, but not acted in a real sense to take strong policy decisions against the impact of climate change. This is the biggest weakness among the countries of the world. Hereunder, selected rules/regulations of COP's meeting are discussed for their validity.

2.3.1 Per capita CO_2 Emission of Different Countries

During the pre-industrial period, there was a natural balance between the source (fossil fuels) and sink (forest and ocean) for the released CO_2. But from the beginning of the industrial period, CO_2 emissions were on the rise in the absence of regulations accelerating the development of industrial countries. On the other hand, the size of the sink, especially the net forest cover, started declining as a result of deforestation, commercial agriculture and expansion of human settlement and industrial development in the industrialised countries. It is stated that the ideal size of the forest cover has to be maintained at 33% of the geographical area to balance the land degradation and to enlarge sink capacity to CO_2. Further, forest cover and population are negatively related. Similarly, forest cover and accumulation of CO_2 in the atmosphere are negatively related. Since the atmosphere is an open common resource, countries will not have exclusive right to emit GHGs. Country's population are taken into consideration for computing per capita emission. As stated earlier in this chapter, the global per capita per annum emission of CO_2 during 2015 was 4.9 tonnes, while it was highest for Qatar (39.7 tonnes), and the least was with the country like Pakistan (0.9 tonnes), the Philippines (1.1 tonnes) and Indonesia (2.0 tonnes). In the case of Qatar, the CO_2 emission was more despite lesser population while there are few countries like Indonesia and Pakistan that are brought under the lower emission category as they have low emissions and low population. Hence, there is an argument that emission must be reported on a country basis rather than on population basis. It is logical to understand that when the population is more, the per capita emission would be lesser. Individual countries are making efforts to strengthen their Intended Nationally Determined Contributions (INDCs) by streamlining their mitigation and adaptation initiatives and actions. Internationally, there has been considerable efforts in synchronising the NDC commitments with the Sustainability Development Goals (SDGs). Emission is an integral part of economic development but in the long run, it might prove to be a deterrent to sustainable development. Hence, it is imperative to deploy innovative green technologies that will help reduce emissions without having a larger impact on the development agenda of nations. The INDC goals set by countries will help this process.

2.3.2 International Emission Trading

According to Article 17 (Kyoto Protocol) on emissions trading, countries having emission units that can be spared, can sell the emission units to other countries that have incidentally exceeded their emission limits. This is governed by an international market mechanism called "carbon market". With the financial meltdown, the carbon trading was very low and losing its significance. There should be concerted efforts to revive this market mechanism and promote carbon trading among countries across continents of nearest.

2.3.3 Strong Political Will

Political will is key to the success of international climate accord. Despite sound warnings and scientific evidence generated by international scientific communities, politicians seem to be unprepared or underprepared to deal with the consequences of climate change. To have any chance of meeting the Paris commitments every country has to introduce tough policy measures and put the INDCs to effective use to show tangible results. This needs a strong commitment by the policymakers to use innovative policy instrument like carbon taxing, incentivising green technologies and enhancing general awareness about climate change and its consequences among concerned stakeholders besides channelising much needed financial assistance.

2.3.4 Finance Mobilisation

As per the Copenhagen Agreement, it was agreed that developed countries would raise funds of about $30 billion from 2010 to 2012 of new and additional resources and agreed a "goal for the world to raise $100 billion per year by 2020", from "a wide variety of sources", to help developing countries cut carbon emissions (mitigation measures). It was also agreed to strengthen multilateral funding mechanisms to support adaptation activities more intensively. But unfortunately, there has been not much forward momentum in this and there is no clear agreement even in the Paris deal about how and when this target will be achieved. There is no clear commitment from developed countries on the quantum and timelines of contributions. The progress of global emission reduction is very much dependent on financial mobilisation and the help extended to developing countries by the developed nations. After several rounds of negotiations, it is more compelling and desirable for developing countries to adopt policy frameworks that will respond to their domestic exigencies first before those of the international system. It is therefore strongly suggested that environmental policies must be people-friendly, people-centred and people-driven (Maikasuwa 2013).

2.4 Review of Climate Policy Actions of Select Countries Across the Continents

Based on the outcomes of successive COP meetings, individual countries have framed their national policies to meet the challenge of climate change and those are discussed hereunder for selected countries to understand the climate policy perspective.

2.4.1 India (Asia)

India's National Action Plan on Climate Change (NAPCC 2008) which was unveiled on 30 June 2008 recognises and lays emphasis on the importance of social equity, poverty reduction and gender inclusivity as the best form of adaptation to climate change. NAPCC underscores the importance of adaptation strategies at the sub-national and local level, while acting on the reduction of GHG emissions through mitigation actions. There are eight core national missions under NAPCC, namely, "the solar mission, enhanced energy efficiency, sustainable habitat, water, sustaining the Himalayan ecosystem, green India, sustainable agriculture and strategic knowledge for climate change". Accordingly, each State of India has prepared its own State Action Plan for Climate Change (SAPCC). The progress of SAPCC is very limited and sporadic because of various factors including lack of funding. Arivudai Nambi et al. (2010) from their final report on the project "Vulnerability

Assessment and Enhancing Adaptive Capacity to Climate Change in Semi-Arid Areas in India", based on community-level adaptation to climate change and its relevance to national action plan reported the following policy needs at the community level for their adaptation against climate change impact especially for the States of Andhra Pradesh and Rajasthan of India. The experimentation at the local level had significant lessons and insights for the national action plan and policies which are stated below:

1. Promotion of climate information services and dissemination of right information at the right time for farmers at the local level. This could be achieved through the establishment of a network of agro-met weather stations at the local level and connect it to the national weather data network and provide location-specific advisories to reduce the climate-related risks in crop production.
2. Promotion of rural energy efficiency at the local level using improvised cook-stoves and better management of fuelwood without impacting the forests.
3. Provision of state-level input subsidies for enhancing water use efficiency through the promotion of technologies like the System of Rice Intensification (SRI).
4. Support appropriate technologies through the use of traditional knowledge, and blend it with scientific knowledge to tailor-made adaptation measures suiting to the needs (e.g. traditional gravity-based water irrigation system, sloppy land treatment through the promotion of local trees which have better soil retaining and water holding capacity).
5. Appropriate livelihood diversification measures, especially for rural women as to manage climate risks at local level.

2.4.2 Bhutan (Asia)

Bhutan has high potential to increase carbon sequestration. Bhutan is carbon negative country, mainly because of the vast forest cover (about 70%). Sonam Lhaden Khandu (2015) notes that the impact of climate change in Bhutan will affect major sectors like water resources, agriculture, forestry and biodiversity and human health. Bhutan has invested in vulnerability assessment of its region and devising relevant adaptation strategies to suit its needs. The adaptation strategies include but not limited to "weather forecast", and "management of flood, land slide, forest fire, and disasters" (https://adaptation-undp.org/explore/bhutan). Bhutan developed its National Adaptation Programme of Action (NAPA) as early as 2006. The NAPA formed a set of objectives, which included identifying immediate projects and activities that can help communities adapt and to integrate climate change risks into the national planning process. The NAPA projects and profiles were updated in 2012. Environmental protection is a priority for the Kingdom of Bhutan, and this is mandated by the Constitution (Sonam Lhaden Khandu 2015). Under Intended Nationally

Determined Contribution (INDC), Bhutan has proposed ten adaptation measures (https://reliefweb.in/sites/reliefweb.int/files/.../Bhutan_NAP_country_briefing.pdf) and those are:

- "Water security through Integrated Water Resource Management.
- Climate resilient agriculture to achieve food and nutritional security.
- Sustainable forest management and conservation of bio-diversity.
- Strengthened resilience to climate change hazards through early warning system.
- Minimising climate related health risks.
- Climate proof transport infrastructure against landslides and flash floods.
- Promoting climate resilient livestock farming towards poverty alleviation and self sufficiency.
- Promote clean renewable and climate resilient energy generation.
- Enhancing climate information services for vulnerability and adaptation assessment.
- Integrate climate resilient and low emission strategies in urban and rural settlements".

2.4.3 United States of America (North America)

Daniel Bodansky (2009) proposed that US climate policy should be guided by ten fundamental precepts.

- "U.S. foreign policy on climate change must grow out of domestic policy. Thus, developing a strong domestic climate policy is a top foreign policy priority.
- The United States must ensure that the next phase of the international climate change regime covers a significantly greater share of global emissions than Kyoto, including emissions from major developing countries such as China and India.
- Although adopting a long-term greenhouse gas (GHG) concentration target would be useful for internal policy development purposes, international agreement on a long-term target is not essential, and seeking such an agreement may be counter-productive.
- Despite the UNFCCC's status as the official forum for the negotiations, the United States will need to use unofficial channels outside of this process to reach a deal on a new climate agreement.
- The United States should pursue a more flexible international agreement than Kyoto, one which allows states to take action along multiple tracks.
- Although binding international commitments to reduce emissions would be the best negotiating outcome, the United States should explore more informal arrangements as well, given the difficulty of obtaining consent to treaty commitments both from developing countries and from the U.S. Senate.

- The United States should actively pursue opportunities to address climate change in other forums and institutions, rather than relying solely on the UNFCCC process.
- The United States should give greater priority to the problem of adaptation.
- The United States should use financial assistance as an incentive for greater action by developing countries.
- The United States should consider, as a last resort, imposing trade measures against countries that fail to participate in or comply with the international climate change regime".

In order to position itself in global arena in terms of credibility, and climate action, there is a need for the United States to enact legislations on climate change at the domestic sphere. However, whatever guidance is given for policy decisions above, it did not happen in reality in the United States, due to the change in leadership in the United States. The United States has contrary view always on the results of COPs meeting.

2.4.4 Australia (Oceania)

The policy of the Australian government on climate change covers six sectors with specified purposes (www.environment.gov.au/climate-change) and those are:

- "Government and international initiatives: The government policy to reduce emissions and adapt to climate change
- Local government and communities: Assistance and enabling environment to adopt more efficient and cost effective practices and also adapt to the climate change.
- Business and Industry: Incentives provided to enable innovation and adapt to smarter technologies.
- Individuals and households: Adopting change at the households level through energy efficiency and consumer choice.
- Farmers and Managers: Helping the land and agriculture sectors to reduce GHG emission and adapting to climate change.
- Climate science and data: The science and impact of climate change as well as Australia's current and future estimates of emissions".

2.4.5 Germany (Europe)

Germany has also set itself ambitious climate targets. With the 2010 energy concept, which builds on the Integrated Energy and Climate Programme of 2007, targets were set out for reducing greenhouse gas emissions, expanding renewable energy sources and increasing energy efficiency. The main objective of the energy concept is to ensure a climate-compatible, reliable and affordable energy supply for Germany. This goal was also agreed on by the German government in the 2013 coalition agreement. The German government is also advocating the continued integration of this triad of targets at EU level: greenhouse gas reduction, expanding renewable energies and increasing energy efficiency. Germany has also provided substantial financial assistance to developing countries to promote both mitigation and adaptation actions at different levels, besides investing many resources on domestic climate actions (https://www.cop23de/en/bmu/german-climate-policy).

2.4.6 Switzerland (Europe)

Switzerland follows meticulously the three pillars of the Paris Agreement namely, limiting temperature increase, strengthening adaptive capacity and climate compatible investments (Swiss Confederation 2018). The government employs active policy to reduce GHGs limiting global warming to significantly under 2 °C. The applicable CO_2 Act focuses on reducing Switzerland's domestic emissions. This country intends to reduce domestic GHG by at least 20% from their 1990 levels by 2020. The CO_2 Act concerns fossil heating and motor fuels but also includes other important GHGs in addition to CO_2. The country assigns the Federal government the role in adaptation to climate change (https://www.bafu.admin.ch/bafu/en/home/topics/climate/info.../climate-policy.html). The key policy activities include CO_2 levy, emission trading, buildings, CO_2 emission regulation for vehicles and compensation for CO_2 emissions, climate training and communication programme, technology fund, sector agreement under Environment Protection Act and CO_2 Act and strong adaptation strategies.

2.4.7 Russia (77% Under Asia and 23% Under Europe)

Global warming in Russia includes climate politics, contribution to global warming and the influence of global warming in Russia. In 2009, Russia was ready to reduce emissions 20–25% from its 1990 emission levels by the year 2020 (https://en.wikipedia.org/wiki/climate_chnage_in_Russia). Perelet et al. (2008) indicated that a response to climate change challenges requires designing and adopting special mitigation and adaptation policies, an early warning system for climate change relating natural disasters and abrupt ecosystem changes. In their paper, they discussed climate change impacts on Russia's economy, health and wellbeing of its people as well as policies with a view to implementing its international commitments to the Kyoto Protocol and adapting to new environment. Sustainable development has potential to lessen the climate change vulnerability through augmenting resilience and improving the adaptive capacity.

2.4.8 Canada (North America)

Climate change is being addressed very seriously by the provinces than the federal government. It has pledged actions based on scientific evidence and advice (https://en.wikipedia.org/wiki/climate_change_in_canada). In Canada, the rate of temperature rise is twice the rate of global warming. Canada would not meet the Kyoto Protocol commitment, i.e. 6% emission reduction below 1990 level for the year 2008–2012 as a signatory to the Kyoto Protocol. At province level, levying carbon tax per tonne of carbon, phasing out coal-fired power plant by 2030 (Alberta); adaptation strategy and its integration to government business, building adaptation strategies to key sectors like agriculture and industry (British Columbia); reducing transport emission through public transport, incentives to purchase electric vehicles, climate change adaptation as climate ready (Ontario) and public transport, energy recovery from biomass, land use pattern reform (Quebec) were the public policy considered. In general, the Canadian government has the policy of reducing GHG's emission (https://www.canada.ca/en/services/environment/../climatechange/climate.action.html).

2.4.9 New Zealand (Oceania) (https://en.wikipedia.org/wiki/climate_change_in_New_Zealand)

New Zealand ratified the UNFCCC in September 1993 and signed the Kyoto Protocol to the UNFCC on 22 May 1998 and ratified it on 19 December 2002. New Zealand deals with climate change impacts through varied approaches. Since July 2010, the sectors constituting energy sector, fossil fuel and few industries are warranted to report the emissions. New Zealand has an emission trading scheme and from July 2010, the energy and liquid fossil fuel and some industry sectors had obligations to report emissions. In July 1994, 4 months after UNFCCC came in to force, the fourth National Government announced the following policies (https://en.wikipedia.org/wiki/climate_change_in_New_Zealand):

- "The target of reducing net emissions to 1990 volumes by the year 2020
- A target of slowing growth of gross emissions by 20%.
- Increased storage of carbon in plantation forests.
- Energy sector reforms.
- An energy efficiency strategy.
- Renewable energy sources.
- Use of Resource Management Act.
- The voluntary agreement with Industry".

2.4.10 Kenya (Africa)

Kenya is an active member of UNFCCC and also the Kyoto Protocol. Interestingly, Kenya structures its national policy in tune with strategic interests of the country and also sync with the regional and global approaches in climate management (Session paper 2016). Kenya did launch its National Climate Action Plan on Wednesday, 27 March 2013. The constitution of Kenya (2010) provides ground for the formulation of adaptation and mitigation legislation, policies and strategies. The

2010 National Climate Change Response Strategies (NCCRS) recognised the importance of climate change impacts for Kenya's development (Government of Kenya 2013). The identified policy in general includes "Resilient low carbon development; Mainstreaming climate change; Education, public awareness, information access, research and technology; Knowledge management; Climate change governance including climate finance, mainstreaming the issues pertaining to climate change like gender, vulnerable population; climate change measurement, reporting and benefit measurement; and framework implementation" (Session paper 2016).

2.5 Building Blocks for Climate Policy-Making

Three principal approaches to understanding climate change regime include exposure of the region to climate change and its intensity, the sensitivity of the region to climate change and coping capacity of the region. These are to be considered for assessing the vulnerability of any region to climate change. Sustainable development measures would reduce vulnerability and thereby enhance resilience capacity. Some of the key elements that would go into the making of climate policy may include:

- Benchmark assessment of the present development to meet the needs of people
- Assessment of future development activities to sustain the country's economy
- The present GHG emission profile
- Future GHG emission considering development needs
- Regional vulnerability assessment to climate change
- Regional assessment of present-day adaptive capacity at different levels
- Present resilience capacity at the regional level
- Technology availability
- Assessment of access to domestic and international resources
- Country's gold reserve
- Population growth and demographic changes
- State of urbanisation
- Political will and political stability
- Sector-wise emission details
- Potential for carbon sequestration and avenues for it
- General awareness and perceptions on climate change issues

2.6 Anticipated Impact Based Models on Physical, Biophysical and Social Resources

Katharine Ricke (2018) has indicated that economic growth in India will be slow because of the impact of climate change on its resource base. Several studies have indicated that India will be impacted profoundly by climate change. The impacts will be far more pronounced because of the large coastal area that India has, the population density and the large marginalised population.

As per the IPCC report, several studies point out that the Paris agreement of 2015 may not be sufficient enough to reduce the vulnerability of the countries to climate change. Instant reduction of emission of CO_2, enhancing the availability of technology and regulating both production and use of fossil-based energy are critically important. In the past 150 years, the world has witnessed an increase in mean temperature to a level of 0.8 to 1.0 °C. The sustained increase in temperature may lead to severe disasters and natural calamities that would result in the extinction of vital organisms from earth. Two-pronged strategy is imperative (i) to reduce GHG emissions substantially and (ii) to manage the already accumulated emissions in the atmosphere effectively. If GHGs accumulation is not controlled or managed effectively, there will be unprecedented increase in temperature up to 2 to 3 °C compared to the present level which will result in unprecedented sea level rise, destroying coastal areas to the point of no return. This would impact the small island countries in a big way. With an increase in sea level, about 50 million people will get affected/displaced in India, China, Egypt, Bangladesh, Indonesia, Japan, the Philippines, America and Vietnam. Climate change will also result in an increase in malaria and other health hazards over Asia, Africa and South American continents. Policies that promote the withdrawal of coal-based power plants and investments in alternate energy regimes are important. If the much needed financial resources are not mobilised to support developing countries, it would be difficult to honour the commitments under the Paris climate deal. It is estimated that annually the developing countries must allot substantially about 2.4 trillion in the energy sector to reduce GHG emissions and contain temperature rise.

As per the report of Economic Losses, Poverty and Disasters 1998–2017 of the UN Office for Disaster Risk Reduction (Deccan Chronicle 2018a), India suffered a whopping $79.5 billion economic loss due to climate-related disasters in the last 20 years (according the UN report that highlights the impact of extreme weather events on the global economy). The greatest economic losses were experienced by US ($944.8 billion) followed by China ($492.2 billion), Japan ($376.3 billion), India ($79.5 billion) and Puerto Rico ($71.7 billion).

As per the recent AR5 IPCC report (Deccan Chronicle 2018b), at the current level of GHGs emissions, we could pass the 1.5 °C marker as early as 2030 and no later than the mid-century. The United States of late, is disinclined to the formation of coalition to act against climate change. In such circumstances, India should take a leadership role and aid in the formation of an active coalition, which would work for achieving Paris agreement on climate change, and rescue the vulnerable population from the negative impacts of climate change. The recent IPCC report "Global warming of 1.5 °C" is indeed a wake-up call for every one of us.

2.7 Conclusion

The atmosphere is common to the planet Earth. The atmosphere does not differentiate between countries. The accumulation of GHG emissions in the atmosphere affects all the community of countries to a varying degree depending on their level of vulnerabilities, affecting human well-being. Effective, innovative and grounded policies to manage climate risks prudently is the need of the hour. Strong political will and international co-operation are two important pillars of climate policy. Every country has its responsibility to cut out. Here are some key action strategies that would help better manage climate risks: Moving from thermal energy and switching to other forms of renewable energy would make a huge difference to contain temperature rise. Both legal and other regulatory measures will help achieve that goal. Vulnerability assessment is key to understand the scale of impacts. Appropriate assessment tools have to be used to understand climate risks at different levels. This would be a good starting point to device appropriate actions. A standard protocol could help to compare and contrast the progress made in reducing emissions and the level of capacities attained to manage climate risks. It would also help countries to understand their resource needs. Based on the intensity of vulnerability, suitable capacity building measures can be put in place to improve resilience capacity. Further, a heavy carbon tax must be levied to all activities which involve fossil fuel usage. Additionally, policies that facilitate green technology transfer would help the developing countries to maintain economic growth and not pollute the atmosphere. Emission profiles of the various sector have to be studied to prioritise actions. Monitoring of the progress in reduction of GHG is important. Appropriate tools and methodologies have to be used to measure and monitor GHG emissions from each sector. Strict regulatory measures will help achieve the targets set by individual countries.

Both autonomous adaptation and planned adaptation are needed to manage climate risks at the local level. Countries should work to implement various adaptation measures that fit the situation. The irony is that although more than 80% of the human causes of adverse climate change is attributed to development activities undertaken by the advanced industrial nations, almost 85% of the negative consequences of climate change are pitiably borne by the developing countries of the world. Irrespective of the international support extended under the international climate agreements, it is important that developing countries should start looking internally to streamline their policies and processes and divert their precious domestic financial endowments to climate risk management activities, which will lead to a win-win situation.

References

Arivudai Nambi A, Balasubramanian TN, Manuel RI (2010) Vulnerability assessment and enhancing adaptive capacity to climate change in semi-arid areas of Andhra Pradesh and Rajasthan in India. Submitted to Swiss Agency for Development and Cooperation, New Delhi, by M.S. Swaminathan Research Foundation, Chennai, August 2010, p 187

Daniel Bodansky (2009) Climate change: top 10 percepts for US foreign policy. US global leadership: an initiative of the climate change policy program of RFF. Resource for the future. Issue Brief#09-01, p 21

Deccan Chronicle (2018a) India lost $79.5 billion in natural disasters: UN. Coimbatore edition, Tamil Nadu, India, p 10

Deccan Chronicle (2018b) Impact of 2°C warming is catastrophic to India. Deccan Chronicle, Coimbatore edition, Tamil Nadu, India, p 7

Government of Kenya (2013) National climate change action plan 2013–2017. Vision, 2030. Ministry of Environment and Mineral Resources, Nairobi

Katharine Ricke (2018) CO_2 emissions cost India $210 billion every year. Extract of publication from J Nat Clim Chang as published in Deccan Chronicle. Coimbatore edition, India, p 10. (27th September 2018)

Maikasuwa SA (2013) Climate change and developing countries-issues and policy implications. J Res Dev 1(2):15–24

NAPCC (2008) National action plan for climate change. Prime Ministers Council on Climate Change, Government of India, 30th June 2008

Perelet R, Pegov S, Yulkin M (2008) Climate change. Russia country paper. Human development report 2007/2008, fighting climate change: human solidarity in a divided world. Human Development Report Office occasional paper, p 30. https://core.ac.uk/download/pdf/6248664.pdf

Session Paper (2016) Executive summary of session paper No3 of 2016 on National climate change framework policy. Ministry of Environment and Natural Resources. State Department of Environment, Kenya

Sonam Lhaden Khandu (2015) National adaptation plan (NAP's), Bhutan. Climate Change Division. National Environmental Commission. Royal Government of Bhutan

Swiss Confederation (2018) Switzerland's climate policy. Ul-1803-E. Federal Office for Environment (FOEN). Series: State of the Environment, p 28. https://www.bafu.admin.ch/bafu/en/home/.../climate/.../kimapolitik-der-schweiz.html

Web Links

https://en.wikipedia.org/wiki/Earth_Summit
https://en.wikipedia.org/wiki/Kyoto_protocol
https://unfccc.int/process/the_kyoto_protocol/the_doha_amendmend
https://en.wikiapedia.org/wiki/copenhagen_Accord
https://en.wikipedia.org/wiki/Paris_Agreement
https://unfccc.int/process/the-kyoto-protocol/mechanisms/emissions-trading
https://www.cop23de/en/bmu/german-climate-policy
https://en.wikipedia.org/wiki/climate_change_in_canada
https://en.wikipedia.org/wiki/climate_change_in_New_Zealand
https://en.wikipedia.org/wiki/climate_chnage_in_Russia
https://adaptation-undp.org/explore/bhutan
https://reliefweb.in/sites/reliefweb.int/files/.../Bhutan_NAP_country_briefing.pdf
https://www.bafu.admin.ch/bafu/en/home/topics/climate/info.../climate-policy.html
www.environment.gov.au/climate-change
www.canada.ca/en/services/environment/../climatechange/climate.action.html

Vulnerability Assessment of the Agro-Based Households to Climate Change in the Bundelkhand Region and Suggesting Adaptation Strategies

3

Meeta Gupta, Jyoti P. Patil, and V. C. Goyal

Abstract

Natural disasters like droughts have worsened the conditions of the villages of Bundelkhand region, India. Droughts have caused a diverse impact on the economic, environmental and social conditions of the districts. Therefore, in order to identify the variability of vulnerability, Livelihood Vulnerability Index (LVI) was calculated to assess the vulnerability, with the purpose of identifying relevant adaptation response mechanisms. The index is applied in a comparative study of four selected watersheds of Bundelkhand region, that is Ur watershed (Tikamgarh district, Madhya Pradesh), Kathan watershed (Chhatarpur–Sagar district, Madhya Pradesh), Patrahi–Lakheri watershed (Jhansi district, Uttar Pradesh) and Sajnam watershed (Lalitpur district, Uttar Pradesh). The sub-watershed-based classification was used to assess the vulnerability of people, livelihood and ecosystem to climate change, using primary and secondary data, to identify highly vulnerable sub-watersheds within a watershed, and a comparative analysis was done amongst the four districts. The LVI–IPCC approach was used to reflect the vulnerability based on 39 environmental and socio-economic sub-indicators, through IPCC-identified components: exposure, sensitivity and adaptive capacity. The overall vulnerability results reveal that the Ur watershed in Tikamgarh district was the most vulnerable to climate change than the rest due to high sensitivity and less adaptive capacity. The findings helped in suggesting sector specific as well as overall drought management and adaptation strategies to cope up with the climate change. These can be implemented by the state government and the local bodies to reduce the vulnerability and enhance adaptive capacity of all the four drought-prone districts.

Keywords

Climate change · Adaptation · Bundelkhand · Drought · LVI–IPCC · Watershed

M. Gupta (✉) · J. P. Patil · V. C. Goyal
National Institute of Hydrology, Roorkee, India

© Springer Nature Singapore Pte Ltd. 2020

V. Venkatramanan et al. (eds.), *Global Climate Change and Environmental Policy*,
https://doi.org/10.1007/978-981-13-9570-3_3

Abbreviations

BPL	Below poverty line
FGD	Focused group discussion
IMR	Infant mortality rate
IPCC	Intergovernmental Panel on Climate Change
LVI	Livelihood Vulnerability Index
MoWR	Ministry of Water Resources
MP	Madhya Pradesh
MPCST	Madhya Pradesh Council of Science and Technology
NIH	National Institute of Hydrology
NP	Nagar Panchayat
PDS	Public distribution shop
SC	Scheduled Castes
SHG	Self-help group
SRR	Seed replacement rate
ST	Scheduled Tribes
SW	Sub-watershed
TIFAC	Technology Information, Forecasting and Assessment Council
UP	Uttar Pradesh
UPRSAC	Remote Sensing Application Centre, Uttar Pradesh

3.1 Introduction

The Bundelkhand region of Central India is comprised of 13 districts: six districts in Madhya Pradesh (MP) and seven in Uttar Pradesh (UP). The region comes under the semi-arid zone and is significantly sensitive to climate change and prone to droughts. The region is known for its socio-economic backwardness (Development Alternatives 2007). The Planning Commission (Government of India) has identified most of the districts of the region as "the poorest districts of the country" (Development Alternatives 2007).

The region faces a variable climate condition that has been worsened due to irregular rainfall, high evapotranspiration losses, increased run-off and poor soil water retention and large areas of barren and uncultivable land (Chandramauli 2016). The population of Bundelkhand is primarily dependent on agriculture and farm activities for livelihood. Agriculture, livestock rearing and seasonal out-migration provide more than 90% of the rural income in the Bundelkhand region (Samra 2008). Irrigation activities are heavily depended on the water availability. The issues like gradual disappearance of "traditional water management practices" and inadequate "water-harvesting infrastructures" have further added to the stress in the region (Bhisht et al. 2014). The ever-growing population and a parallel increase in the demand for natural resources have left agricultural and water resources in the region susceptible to increasing climate change risks, affecting livelihoods of local communities. Crop productivity is influenced by the changing weather conditions all throughout the year. South-west monsoon plays an important role during the sowing time in the fields; however, it has been fluctuating drastically in the past few years, causing huge losses to the farmers. These regions experience both development challenges as well as climate-change uncertainties. Thus, focusing on climate adaptation interventions becomes a high priority. In order to assess the livelihood-related vulnerabilities and derive the strategies and solutions for mitigating the impacts of climate change, a vulnerability assessment was conducted in the proposed watershed area.

The concept of vulnerability has been evolving over the years. Vulnerability is "the degree to which a system is susceptible to, and unable to cope with, adverse effects of climate change, including climate variability and extremes" (Parry et al. 2007). Different methodologies have been proposed by scientists to dissect the vulnerabilities to climate change. However, the definition proposed by the Intergovernmental Panel on Climate Change (IPCC) that views vulnerability as a function of exposure, sensitivity and adaptive capacity is widely adopted (McCarthy et al. 2001). The vulnerability framework as proposed by the IPCC is viewed as one of the most powerful analytical tools for assessments (Turner et al. 2003).

The LVI–IPCC vulnerability approach includes the major indicators into three contributing factors to vulnerability and these are exposure, sensitivity and adaptive capacity. According to the IPCC report, exposure is "the magnitude and duration of climate related exposure, such as drought temperature variability or change in precipitation" (Parry et al. 2007). Sensitivity is defined as "the degree to which a system can be affected, negatively or positively, by a change in climate". This includes the change in mean climate and the frequency and magnitude of extremes. The effect may be direct or indirect (Parry et al. 2007). Adaptive capacity is a "system's ability to adjust to climate change (including climate variability and extremes), to moderate potential damage, to take advantage of the opportunities or cope with the consequences" (Parry et al. 2007). The more adaptive a system is, the less vulnerable it is.

The purpose of this study is to assess the vulnerability of people, livelihood and ecosystem with the purpose of identifying relevant adaptation response mechanisms, in the four selected watersheds of Bundelkhand region, that is Ur watershed (Tikamgarh district, MP), Kathan watershed (Chhatarpur–Sagar district, MP), Patrahi–Lakheri watershed (Jhansi district, UP) and Sajnam watershed (Lalitpur district, UP) shown in Fig. 3.1. These selected watersheds give a representation of the overall water scenario in these districts. The area of the watersheds selected is approximately 1000 km^2. For our study, the selected watersheds have been further divided into sub-watersheds on the basis of drainage network, topography and soil types for carrying out detailed study. The sub-watershed level vulnerability assessment was done, mostly on secondary data, to identify highly vulnerable sub-watersheds for further assessment and implementation of pilot adaptation measures.

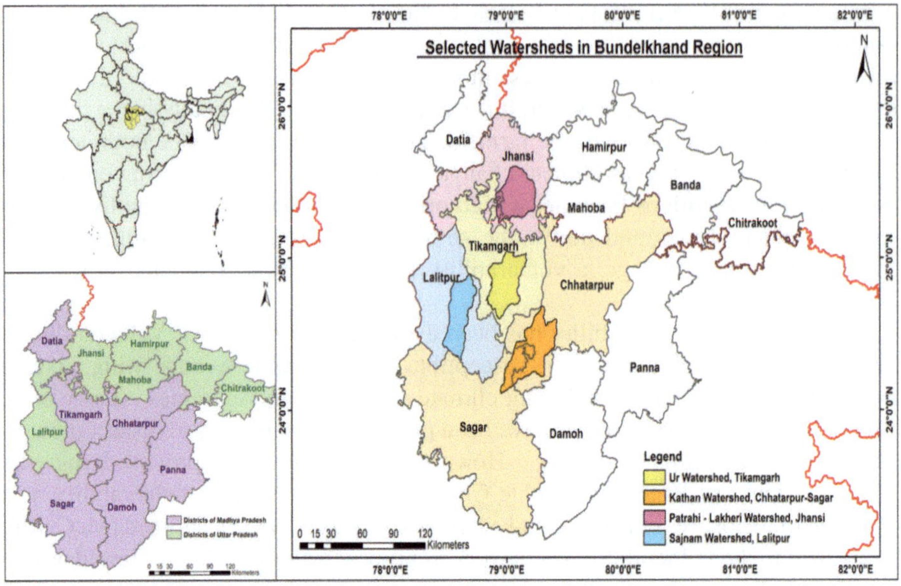

Fig. 3.1 Index map and location map of the selected watersheds

3.2 Physical and Socio-economic Description of Study Area

3.2.1 Ur Watershed, Tikamgarh District, Madhya Pradesh

Tikamgarh District is located at the centre of the historical and geographical region of Bundelkhand of which physical boundaries are formed by Betwa River and two of its tributaries, Jamni and Dhasan. The study area selected for our research is Ur watershed which is located in Tikamgarh district. It is bounded by Chhatarpur in the east and south, while the western and northern boundaries run along Lalitpur and Jhansi districts of Uttar Pradesh, respectively. The mainland watershed area extends between latitudes 24°35′0″ N and 25°05′0″ N and between 78°50′0″ E and 79°10′0″ E longitudes. The total geographical area of the Ur watershed is 991 km^2 and has an elevation of 400 m above the main sea level. The Ur watershed has a maximum length of 119 km from north to south with an average width of about 80 km. The Ur River flows in a north to north-east direction. The watershed has been divided into eight sub-watersheds which is shown in Fig. 3.2.

The Ur watershed area falls under four development blocks of Tikamgarh district, that is Jatara (32.81%), Palera (7.76%), Baldeogarh (27.53%) and Tikamgarh (31.90%). The study area comprises 190 villages with a population of 296,204, which is 20% of the total population of the district (Directorate of Census Operations, Madhya Pradesh 2011a). The population density of the watershed is 299 persons per sq. km. The Scheduled Castes and Scheduled Tribes population is reported as 24% and 6% of the total population of the area. The literacy rate in the watershed area is 51%, out of which 38% is female which is very low than the average literacy

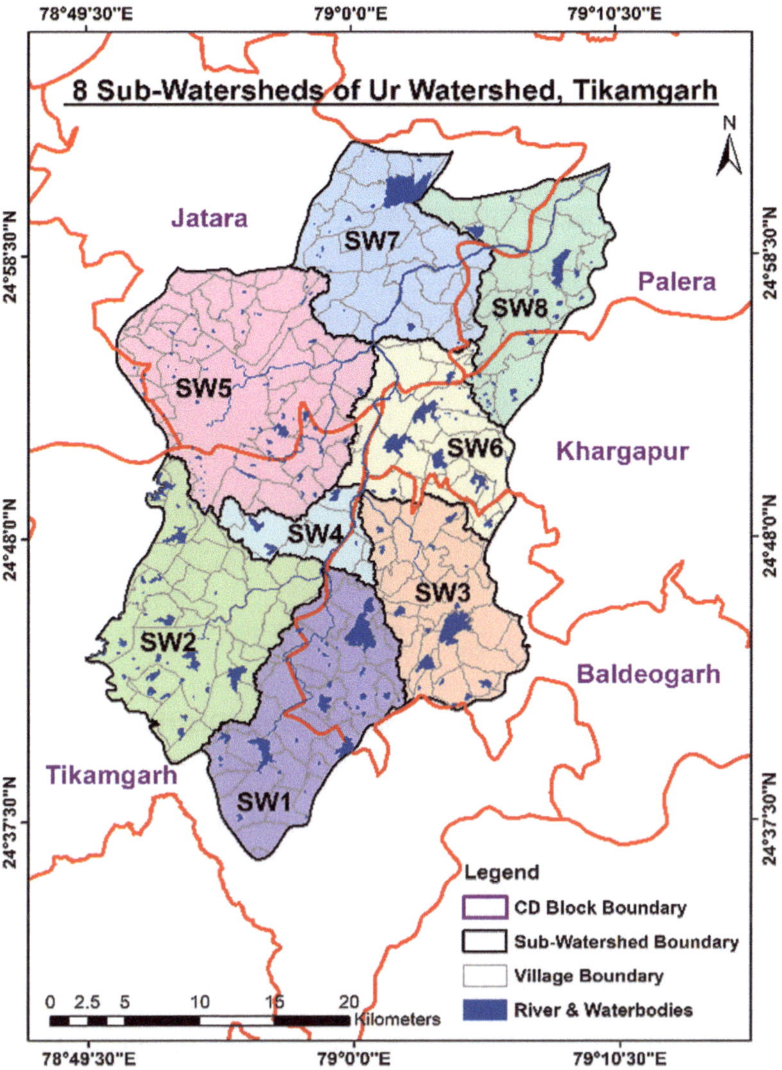

Fig. 3.2 8Sub-watersheds of Ur watershed, Tikamgarh. (Source: MPCST)

rate of Tikamgarh district (61.43%). The watershed suffers from high level of poverty, that is 35.35% of total rural families are below poverty line and also report a low percentage of working population.

Economy of this watershed is primarily based on agriculture. The per capita income generated by farmers per annum ranges from Rupees 5000 to 20,000. Apart from agriculture, people practice livestock rearing and fishery and generate livelihood out of it. The watershed reports a very high migration rate because of almost dry and less productive land. Sub-watershed-wise composition is given in Table 3.1, and the list of villages covered in each sub-watershed is given in Appendix A.

Table 3.1 Sub-watershed-wise administrative description of Ur watershed

S. no.	Sub-watershed	Total geographical area (km^2)	Number of villages	Number of households	Total population
1.	Sub-watershed 1 (SW1)	140	33	9600	45,287
2.	Sub-watershed 2 (SW2)	150	30	6457	33,530
3.	Sub-watershed 3 (SW3)	106	27	8731	38,936
4.	Sub-watershed 4 (SW4)	38	9	3118	14,984
5.	Sub-watershed 5 (SW5)	213	43	9132	54,879
6.	Sub-watershed 6 (SW6)	103	20	7495	33,590
7.	Sub-watershed 7 (SW7)	129	21	8525	43,170
8	Sub-watershed 8 (SW8)	111	18	6279	31,828
	Total watershed	**991**	**201**	**59,337**	**296,204**

Note: 11 villages are common between one or more sub-watersheds; thus, the total number of villages falling under this watershed is 190

3.2.2 Kathan Watershed, Chhatarpur–Sagar District, Madhya Pradesh

Chhatarpur district is located in the central portion of plateau of Bundelkhand region in Madhya Pradesh. It is surrounded by the districts of Mahoba (Uttar Pradesh), Panna, Tikamgarh, Sagar and Damoh. Rivers Ken and Dhasan form the physical boundaries on east and the west, respectively. The study area chosen for the research is Kathan watershed, which is located in Chhatarpur district and partially covers the Sagar district. River Kathan flows in a north to north-east direction. The mainland watershed area extends between latitudes 24°05′0″ N and 24°45′0″ N and between 78°55′0″ E and 79°30′0″ E longitudes. The total geographical area of the Kathan watershed is 1345.08 km^2. Bila dam, a major irrigation project, also exists on River Kathan and is located in Shahgarh block in Sagar district. The watershed is divided into eight sub-watersheds, which is shown in the Fig. 3.3.

The Kathan watershed area falls under four development blocks of Chhatarpur district, that is Bada Malhera (26.84%), Buxwaha (35.43%), Ghuwara (4.86%) and Bijawar (1.97%), and one development block of Sagar district, that is Shahgarh (30.90%). The distribution of the various blocks is represented graphically in Fig. 3.3. The study area comprises 231 villages and 3 towns, with a population of 246,098. The population density of the watershed is 182 persons per sq. km. The Scheduled Castes and Scheduled Tribes population are reported as 23% and 9% of the total population of the area. The literacy rate in the watershed area is 59%, out

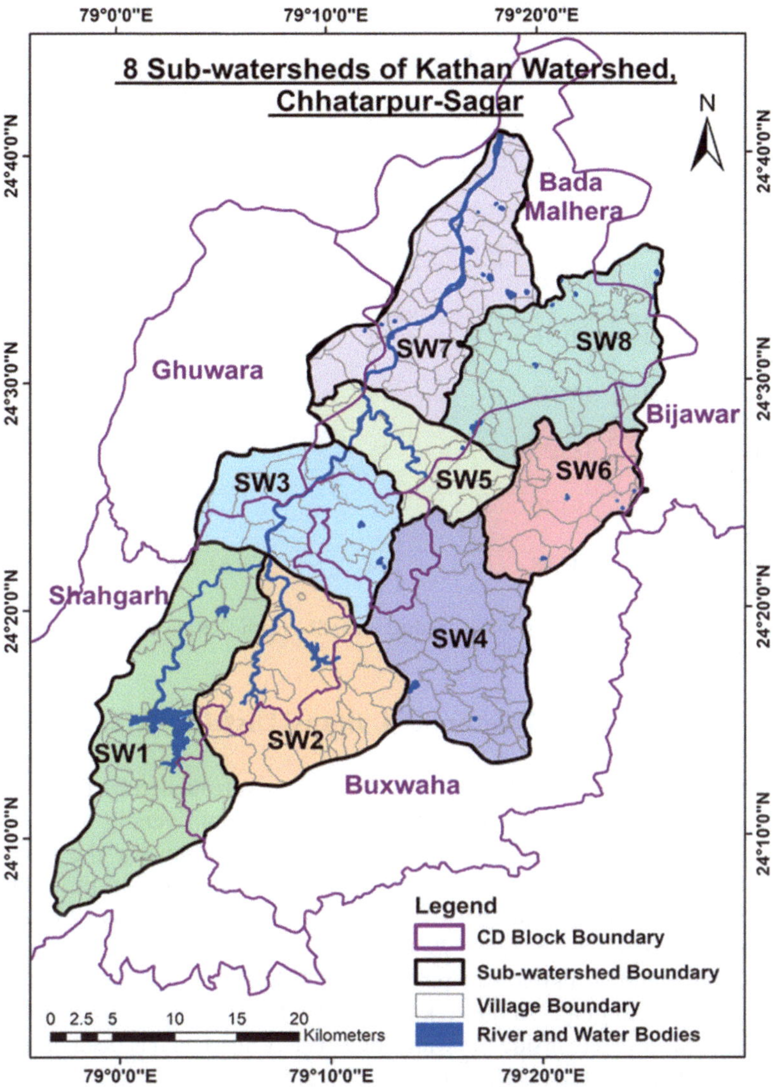

Fig. 3.3 8 Sub-watersheds of Kathan watershed, Chhatarpur–Sagar. (Source: MPCST)

of which 39% is female which is lower than the average literacy rate of Chhatarpur district (63.74%) (Directorate of Census Operations, Madhya Pradesh 2011a).

Agriculture is the prime earning activity for the people in the district, which is not very remunerative. Besides agriculture, farmers practice animal husbandry and fishery and are able to generate annual income in the range of Rupees 49,500–75,000. The landless families migrate to towns and work in bidi making, furniture making, bamboo craft, leather products and incense stick rolling. Sub-watershed-wise composition is given in Table 3.2, and the list of villages covered in each sub-watershed is given in Appendix B.

Table 3.2 Sub-watershed-wise administrative description of Kathan watershed

S. no.	Sub-watershed	Total geographical area (km^2)	Number of villages/ towns	Number of households	Total population
1.	Sub-watershed 1 (SW1)	244.09	55	14,225	63,944
2.	Sub-watershed 2 (SW2)	185.78	29	4669	19,581
3.	Sub-watershed 3 (SW3)	150.44	18	6864	30,860
4.	Sub-watershed 4 (SW4)	186.93	27	4770	22,887
5.	Sub-watershed 5 (SW5)	91.16	10	2156	10,083
6.	Sub-watershed 6 (SW6)	106.83	18	2471	11,428
7.	Sub-watershed 7 (SW7)	185.37	36	11,820	54,984
8.	Sub-watershed 8 (SW8)	194.50	41	7296	32,331
	Total watershed	**1345.08**	**234**	**54,271**	**246,098**

There are 3 towns covered within the watershed area, Shahgarh (NP) in SW1, Buxwaha (NP) in SW4 and Bada Malhera (NP) in SW7

3.2.3 Patrahi–Lakheri Watershed, Jhansi District, Uttar Pradesh

Jhansi district, located in the Bundelkhand region, is surrounded by the districts of Jalaun, Tikamgarh, Hamirpur and Mahoba. The important rivers of the district are Betwa, Pahuj, Dhasan and Lakheri. The study area chosen for the research is Patrahi–Lakheri watershed. Both Rivers Patrahi and Lakheri flow in a north-east direction. The mainland extends between latitudes 25°36′05.04″ N and 25°16′00.62″ N and between 79°03′26.86″ E and 79°07′28.72″ E longitudes. The total geographical area of the Patrahi–Lakheri watershed is 965.38 km^2, out of which 886.40 sq. km is in UP and the remaining 78.60 sq. km is in MP. There is an irrigation project and two medium-sized tanks inside the watershed. The watershed has been divided into four sub-watersheds which is shown in the Fig. 3.4.

The Patrahi–Lakheri watershed area falls under four development blocks of Jhansi district: Mauranipur (22.52%), Bangara (28.40%), Gursarai (38.85%) and Bamour (1.17%), and covers a very small portion of two development blocks of Tikamgarh district: Niwari (8.60%) and Palera (0.47%). The study area comprises 192 villages and 3 towns with a population of 357,666, which is 18% of the total population of the Jhansi district (Directorate of Census Operations, Uttar Pradesh 2011a). The population density of the watershed is 370 persons per sq. km. Scheduled Castes population is reported as 35% of the total population of the area, whereas the watershed has almost nil percentage of Scheduled Tribes population. The literacy rate in the watershed area is 70%, out of which 37% is female, which is lower than the average literacy rate of Jhansi district (75.05%).

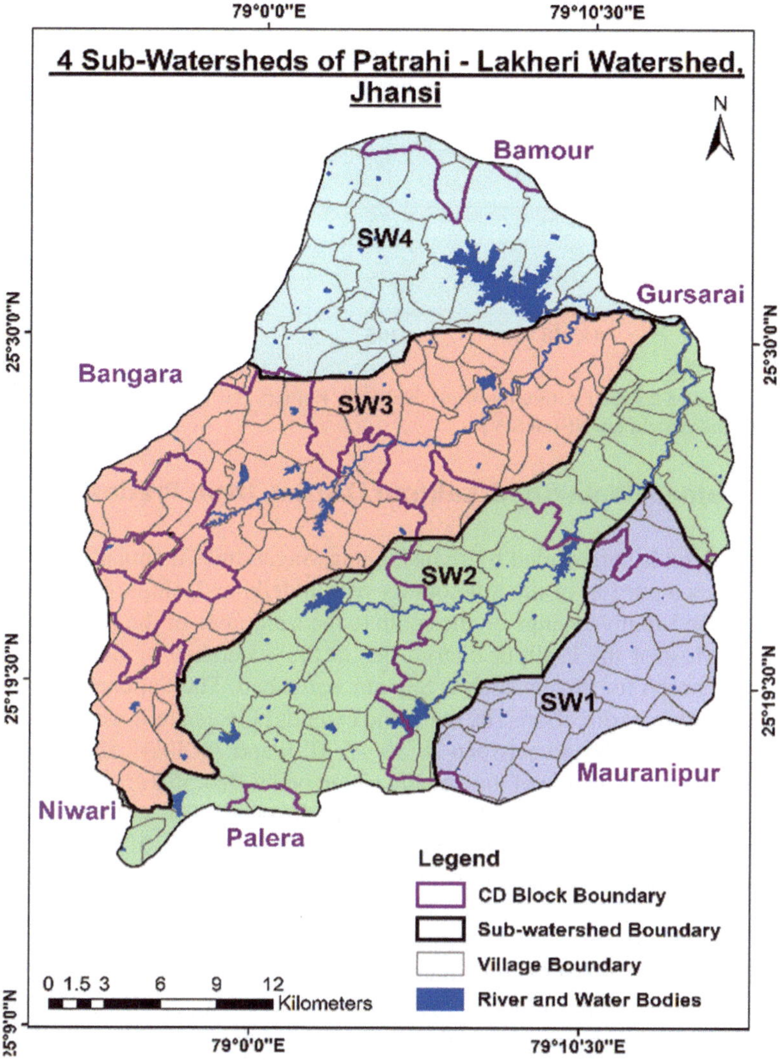

Fig. 3.4 4Sub-watersheds of Patrahi–Lakheri watershed, Jhansi. (Source: UPRSAC)

There is a high-level dependency on agriculture in the rural sector despite having many industries in the district. The farmers generally have an average landholding size of 2–5 hectares. Besides agriculture, farmers practice animal husbandry; fishery and landless families generate income for their livelihood mainly from labour and batai cultivation. The per capita income generated by farmers per annum ranges from Rupees 14,000 to 50,000. There is an increased trend of migration observed. The people usually migrate as they get a higher wage in cities, that is Rupees 250–300/day. The sub-watershed-wise composition is given in Table 3.3 and the list of villages covered in each sub-watershed is given in Appendix C.

Table 3.3 Sub-watershed-wise administrative description of Patrahi–Lakheri watershed

S. no.	Sub-watershed	Total geographical area (km²)	Number of villages/towns	Number of households	Total population
1.	Sub-watershed 1 (SW1)	124.69	29	11,006	59,482
2.	Sub-watershed 2 (SW2)	320.97	56	23,833	1,27,205
3.	Sub-watershed 3 (SW3)	346.20	75	22,017	1,13,061
4.	Sub-watershed 4 (SW4)	173.52	35	10,891	57,918
	Total watershed	**965.38**	**195**	**67,747**	**357,666**

There are 3 towns covered within the watershed area, Kathera (NP) in SW 2 and Tarichar Kalan (NP) and Tondi Fatehpur (NP) in SW3

3.2.4 Sajnam Watershed, Lalitpur District, Uttar Pradesh

Lalitpur district is surrounded by Jhansi, Guna, Sagar and Tikamgarh districts. The district forms a portion of the hill country of Bundelkhand, sloping down from the outliers of the Vindhya Range on the south to the tributaries of the River Yamuna on the north. Most of the Lalitpur district lies in the watershed of Betwa River. The study area chosen for the research is Sajnam watershed. The Sajnam River flows in a north to north-east direction. The mainland extends between latitudes 24°23′09.87″ N and 24°23′15.66″ N and between 78°30′47.00″ E and 78°30′45.55″ E longitudes. The total geographical area of the Sajnam watershed is 964.55 km². Only two irrigation projects exist in the watershed with very few tanks. The watershed is further divided in into four sub-watersheds which is shown in the Fig. 3.5.

The Sajnam watershed area falls under five development blocks of Lalitpur district: Mandwara (18.31%), Birdha (38.27%), Mahroni (7.51%), Bar (35.56%) and Jakhaura (0.35%). The study area comprises 136 villages with a population of 213,161, which is 17% of the total population of the Lalitpur district (Directorate of Census Operations, Uttar Pradesh 2011a). The population density of the watershed is 220 persons per sq. km. Scheduled Castes and Scheduled Tribes population is reported as 21% and 4% of the total population of the area. The literacy rate in the watershed area is 56%, out of which 37% is female, which is relatively lower than the average literacy rate of Lalitpur district (63%).

A vast majority of the population in the watershed relies primarily on agriculture for its livelihood. Landholdings are dominated by small and marginal farmers. More than 73% of the holdings are less than one or two hectares. Besides agriculture, animal husbandry is an integral part of the rural economy. It provides milk and milk products and helps in supplementary income of the people living in villages. Despite the presence of a major power plant falling in the watershed area, people migrate to towns and bigger cities as they are not provided employment. The per capita income generated by farmers per annum ranges from rupees 12,000 to 72,000. The sub-watershed-wise composition is given in Table 3.4 and the list of villages covered in each sub-watershed is given in Appendix D.

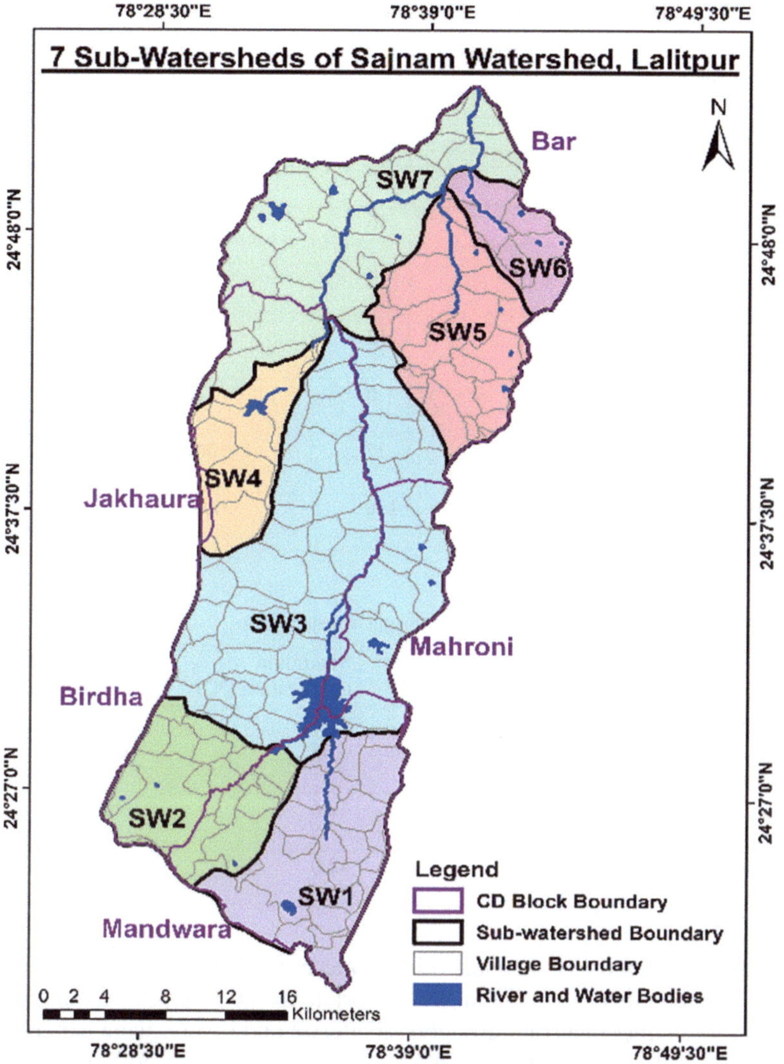

Fig. 3.5 7 Sub-watersheds of Sajnam watershed, Lalitpur. (Source: UPRSAC)

3.3 Climate Description

The climate of the district is characterized by scorching dry summers, erratic rainfall and cold winter. The district receives rainfall from the south-west monsoon during months of June to September. The rainfall pattern is erratic, irregular, scanty and uncertain which causes drought and is a common feature of the district. The details about temperature and rainfall for all the four watersheds have been provided in Table 3.5. The climate data used in this study was obtained from India Meteorological Department.

Table 3.4 Sub-watershed-wise administrative description of Sajnam watershed

S. no.	Sub-watershed	Total geographical area (km²)	Number of villages/towns	Number of households	Total population
1.	Sub-watershed 1 (SW1)	121.64	19	4732	27,013
2.	Sub-watershed 2 (SW2)	103.64	21	3284	19,085
3.	Sub-watershed 3 (SW3)	337.55	46	13,481	77,830
4.	Sub-watershed 4 (SW4)	66.76	5	2027	12,261
5.	Sub-watershed 5 (SW5)	113.13	12	3453	19,627
6.	Sub-watershed 6 (SW5)	41.39	6	1700	9448
7.	Sub-watershed 7 (SW7)	180.44	27	8626	47,897
	Total watershed	**964.55**	**136**	**37,303**	**213,161**

Table 3.5 Temperature and rainfall details for the 4 watersheds

Watersheds	Minimum temperature (°C)	Maximum temperature (°C)	Average annual rainfall (mm)
Ur watershed, Tikamgarh	7.0	41.8	854.00
Kathan watershed, Chhatarpur-Sagar	7.1	42.3	1068.33
Patrahi–Lakheri watershed, Jhansi	24.1	42.6	837.00
Sajnam watershed, Lalitpur	21.0	42.0	1204.50

The departure analysis of annual rainfall was carried out for all the districts using climate data obtained from India Water Portal, which is given in Fig. 3.6b. The analysis reveals that recurrent droughts, moderate and severe, were experienced in 2006–2007, 2007–2008 and 2010–2011. Moderate drought conditions prevailed in most of the drought years in all the four districts. The average drought frequency varies between 1 in 4 years.

3.4 Methodology

3.4.1 Data and Analytical Tool

For the analysis, village level, block level and district level secondary information have been obtained from the district statistical handbooks and, additionally, validated through in-depth interviews, focused group discussion (FGDs) and observing

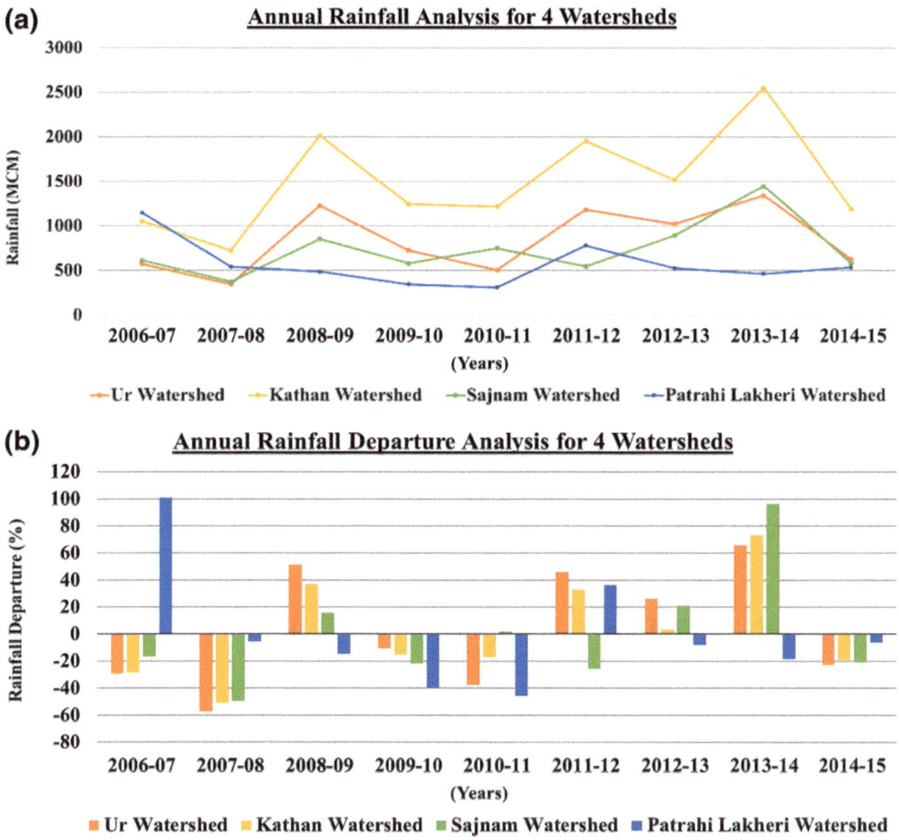

Fig. 3.6 (**a**) Annual rainfall analysis, (**b**) annual rainfall departure analysis

onsite conditions using questionnaires prepared for the purpose. Table 3.6 gives the list of villages that were selected for conducting FGDs, interviews and surveys for the study during the year 2016 and 2017. Each of the FGD composed of 25–30 participants from all age groups and gender.

3.4.2 Selection of Indicators

For the purpose of conducting vulnerability assessment, Livelihood Vulnerability Index was computed for all the four watersheds, and the sub-watersheds of each watershed and were used to derive vulnerability contributing factors, namely "exposure (E), sensitivity (S) and adaptive capacity (A)". These are represented through a number of indicators that would reflect these major components. The description along with the source of data for different indicators chosen is presented in Table 3.7. These indicators were chosen from a broader list of sub-indicators. These are based on literature review, discussions with the experts and nature of relationship with the

Table 3.6 List of villages selected for conducting FGDs/interviews/surveys

S. no.	District	Villages visited for FGDs / interviews/ surveys
1.	Tikamgarh	Bad Madai, Baisa Ugad, Baiwarkhas, Banpura Khurd, Bilgaon, Ghooratal, Gudanwara, Karmasan Hata, Karmaura, Lar Khurd, Lidhouratal, Mahendra Maheva, Mamaun, Manchi, Moramanna, Nainwari, Pali, Papawani, Patar Khera, Patha Khas, Pipra, Samarra, Shyampura, Sujanpur and Vikrampura
2.	Chhatarpur	Amarmau, Bada Malhera, Bamni, Bamnora Kalan, Bhojpura, Bhanguwan, Bilagram, Dalipur, Hirapur, Majhora and Phutwari
3.	Jhansi	Atrauli, Akseo, Barwar, Bacchera, Bamhauri, Bhadokhar, Dhamna Payak, Durgapur, Imlia, Kachaneo, Lidhora, Magawara, Pacchwara, Rajwara, Tondi Fatehpur and Uldan
4.	Lalitpur	Anaura, Bacchrawani, Bagoni, Dulawan, Gangchari, Gona, Jariya, Jharkon, Mirchwara, Narahat, Samogar, Talgaon and Tila

three major components of vulnerability. Through LVI–IPCC, sub-watershed level vulnerabilities were assessed, after which strategies were suggested to cope up with climate change problems.

3.4.3 Calculation of LVI-IPCC Framework Approach

The LVI–IPCC major components, that is exposure, sensitivity and adaptive capacity, were further categorized into 9 indicators with 39 sub-indicators under them. Table 3.8 shows the organization of the 9 indicators in the LVI–IPCC framework.

Firstly, the raw data is collected for the selected 39 sub-indicators which is then transformed into appropriate data measurement units such as percentages, ratios and indices. Secondly, the sub-indicators are standardized as an index using Eq. (3.1) since, each of the sub-indicators is measured on a different scale.

$$\text{Indicator Index}\left(Ix\right) = \left(I_b - I_{\min}\right) / \left(I_{\max} - I_{\min}\right) \tag{3.1}$$

where I_x is the standardized value for the sub-indicator, I_b is the value for the sub-indicator I for a particular sub-watershed and $I_{\min}$ and $I_{\max}$ are the minimum and maximum values, respectively, for each sub-indicator across the sub-watershed.

After standardization, the sub-indicators were averaged using Eq. (3.2) to calculate the profile value of each indicator.

$$\text{Profile Value}\left(P\right) = \sum_{n_{i=1}} \text{Indicator Index}_{I_x i} / n \tag{3.2}$$

where P is one of the 9 indicators for the sub-watershed and index $I_x i$ represents the sub-components, indexed by i, that make up each indicator, and n is the number of sub-indicators in each indicator. The profile values calculated highlight the vulnerability arising through that particular indicator. This is depicted with the help of spider diagrams where the scale ranges from 0 (least vulnerable) at the centre of the web to 1.00 (more vulnerable) at the edge, increasing in increments of 0.20.

Table 3.7 List of indicators and sub-indicators included in computation LVI

Indicators	Sub-indicators	Source of data	Rationale
Climate change and natural disasters (CCND)	Average annual rainfall (mm)	India Meteorological Department, 1991–2013	Higher rainfall is better for the crop growth and also in case of water availability
	Average number of rainy days		
	Inter-annual variation in rainfall		Higher inter-annual variation indicates high probability for occurrence of drought events
	Average number of droughts within 24 years		Increase in such events can further worsen the problems of water availability and water quality, thereby affecting the life of people
	Average number of floods within 24 years		
Demographics (D)	Density of population	Directorate of Census Operations (2011a, b, c, d)	High density is an indication of population pressure on natural resources
	Percentage of child population (0–6 age)		Children are the most dependent section of the society because of limited adaptive capacities
	Percentage of Schedule Caste population		This population is relatively poorer and less educated and depends heavily on natural resources for their livelihood
	Percentage of Schedule Tribe population		
	Sex ratio		The social and cultural structure and sector-specific employment make it harder for women to recover from any tragic event due to limited access to the facilities
	Percentage of marginal workers		These work for less than 183 days, so in order to meet the expenditure for the rest of the year, the villagers migrate to towns in order generate a livelihood
	Percentage of below poverty line (BPL) families	National Resource Cell for Decentralised District Planning (2007–2012), BPL Survey 2002	This population is already deprived of the basic facilities and has limited access to resources to sustain because of which they are less adapted to the climate change stressors

(continued)

Table 3.7 (continued)

Indicators	Sub-indicators	Source of data	Rationale
Health (H)	Infant mortality rate (IMR)	Annual Health Survey (2011–12)	Temperature and rainfall changes may change the geographic range of the vector borne diseases exposing new populations to it
	Per capita public health expenditure (Rs)	Choudary and Amarnath (2012)	Lower expenditure can limit the people to avail the health facilities, thus delayed treatment for the ailments
Agriculture (A)	Percentage of agricultural main workers	Agriculture Census (2011)	A high level of agricultural dependency will increase district's vulnerability to climate change and fluctuation in terms of agriculture trade
	Percentage of cultivators main workers		
	Percentage of marginal agricultural workers		
	Percentage of non-agricultural labour		This population doesn't have stable occupation, and they mostly migrate to towns and cities in search for better opportunities and livelihood
	Cropping intensity		This indicates pressure on the same land for farming. Extensive cropping will require extensive irrigation facilities, making it sensitive to climate change
	Density of poultry population	Ministry of Agriculture, Department of Animal Husbandry, Dairying and Fisheries (2014)	Adverse impacts of climate change can severely affect the livestock and poultry population. Their sensitivities can be increased due to the occurrences of newly unidentified diseases, heat strokes, scarcity of water, non-availability of fodder and low productivity
	Density of livestock population		

Ecosystem (E)	Per cent forest cover	MPCST/UPRSAC (2011)	Forests are highly sensitive to the impacts of climate change as it affects the composition and distribution of forest resources. It can disturb the balance of bio-geochemical cycle, making the forests prone to degradation. It can also affect the forest productivity. Lastly, this may result in habitat shifting of fauna in the region
	Per cent wasteland area	Directorate of Census Operations (2011a)	Wasteland in the region increases the sensitivity due to the loss in land fertility, thus decreasing the land area fit for farming or grazing in the region
	Surface water availability (m^3)	MPCST/UPRSAC (2011)	There is a high reliance on surface water for irrigation, drinking water and other purposes. Variation in rainfall may affect the recharging of ponds and talabs
	Groundwater availability (m^3)	Ministry of Water Resources (2013a, b)	There is a high reliance on groundwater for irrigation, drinking water and other purposes. Variation in rainfall may affect the recharging of groundwater resulting in over-extraction of groundwater resources
	Soil organic carbon (% weight)	Soil and land use survey of India (2007–08)	Low levels can affect the fertility of the soil further, impacting the crop yield
Human capital (HC)	Percentage of women panchayat representatives	MP/UP State Election Commission (2014–2019)	More number of women at administrative positions will understand the problems of other women and, thus, help in the uplifting their status
	Literacy rate	Directorate of Census Operations (2011b)	Increased literacy levels increasing people's capabilities and access to information and thus their ability to cope with adversities
	Number of educational centres		Education level of a population is seen as an important determinant of its quality of life Higher education trains them to cope up with adversities
	Number of health-care centres		Health-care facilities increase the adaptive capacity by providing infrastructures to respond to the health impacts of climate variability
	Number of veterinary care centres		Health-care facilities are important for the livestock as it is an alternative occupation practiced, so it is necessary to maintain the livestock healthy

(continued)

Table 3.7 (continued)

Indicators	Sub-indicators	Source of data	Rationale
Physical capital (PC)	Percentage of households with access to drinking water, within the premises	Directorate of Census Operations (2011b)	Accessibility to adequate water facilities within the village provides easy access to drinking water and carrying out household chores and, thus, reduces time lost for women travelling distances to fetch water
	Percentage of households with access to sanitation facilities		Accessibility to adequate excreta disposal facilities is fundamental to decrease the frequency of associated disease to climate change
	Percentage of households with access to electricity		Access to electricity is fundamental to daily life and for quality life because it is central to all aspects of lives—lighting, heating, pumping, agricultural productivity, refrigeration of food and medicines, etc.
	Percentage of functioning hand pumps	National Rural Drinking Water Programme (2017)	Provision of hand pumps is there in every village, but very few are in working state. For better access and availability, more number of functioning hand pumps should be installed in accordance to the number of households of a village
Social capital (SC)	Number of self-help groups (SHGs)	Directorate of Census Operations (2011b)	Provision of SHGs increased social connectivity and contributes in poverty alleviation and women empowerment
	Number of public distribution shops (PDS)		Availability and accessibility to PDS is important as it enables the supply of food to the poor at subsidized prices. It also ensures that there is equal distribution amongst the villagers and makes them equipped in situations of food shortages
Financial capital (FC)	Access to credit (number of banks/MFI/VC)	Directorate of Census Operations (2011b)	Greater number of these banks in a district implies easy credit to small and marginal framers; these play a pivotal role in the development and transformation of the rural and agrarian economy
	Number of agricultural credit societies		

Table 3.8 IPCC contributing factors to vulnerability and major indictors

S. no.	IPCC major components	Indicators
1.	Exposure	Climate change and natural disasters
2.	Sensitivity	Demographics, health, agriculture and ecosystem
3.	Adaptive capacity	Human capital, physical capital
		Social capital and financial capital

Once the values for each of the 9 indicators for each sub-watershed were calculated, the weights of each major indicator W_P are determined by the number of sub-indicators that make up each indicator and are included to ensure that all sub-indicators contribute equally to vulnerability measurement (Sullivan et al. 2002).

Following the calculation of the profile values of the indicators, they are further combined according to the categorization scheme stated in Table 3.8 into IPCC contributing factors using Eq. (3.3) for calculation of the LVI-IPCC.

$$\text{Contributing Factor}\left(\text{CF}\right) = \sum_{i=1}^{n} W_{P_i}\, x\, P_i \,/\, \sum_{i=1}^{n} W_{P_i} \tag{3.3}$$

where W_{Pi} is the weightage of the Profile I and P_i are the major indicators for the sub-watershed.

Once exposure, sensitivity and adaptive capacity were calculated, the three contributing components were combined and LVI-IPCC of a sub-watershed using the Eq. (3.4):

$$\text{LVI} - \text{IPCC} = \left(\text{Exposure} - \text{Adaptive Capacity}\right) \times \text{Sensitivity} \tag{3.4}$$

Scaling is done from −1 to +1 indicating low to high vulnerability. This is done for each area and then they are ranked in a decreasing order of vulnerability.

3.5 Results and Discussion

3.5.1 Ur Watershed, Tikamgarh District, Madhya Pradesh

The vulnerability indices of the indicators ranged from 0.07 to 0.70 and the differences in vulnerability of the 8 sub-watersheds are presented individually in the spider diagrams (Fig. 3.7).

The vulnerability spider diagram shows that all the sub-watersheds are more vulnerable due to climate change and natural disasters, agriculture and health problems. The climate data indicates that villages have witnessed increasing temperature and decreasing rainfall over the last 24 years that has directly affected the crops, fodder, land, water and forest resources. With such variations in the climatic conditions, the indicator that makes all the 8 sub-watersheds vulnerable is the occurrence of natural disasters. Tikamgarh district exists in the semi-arid regions and is one of

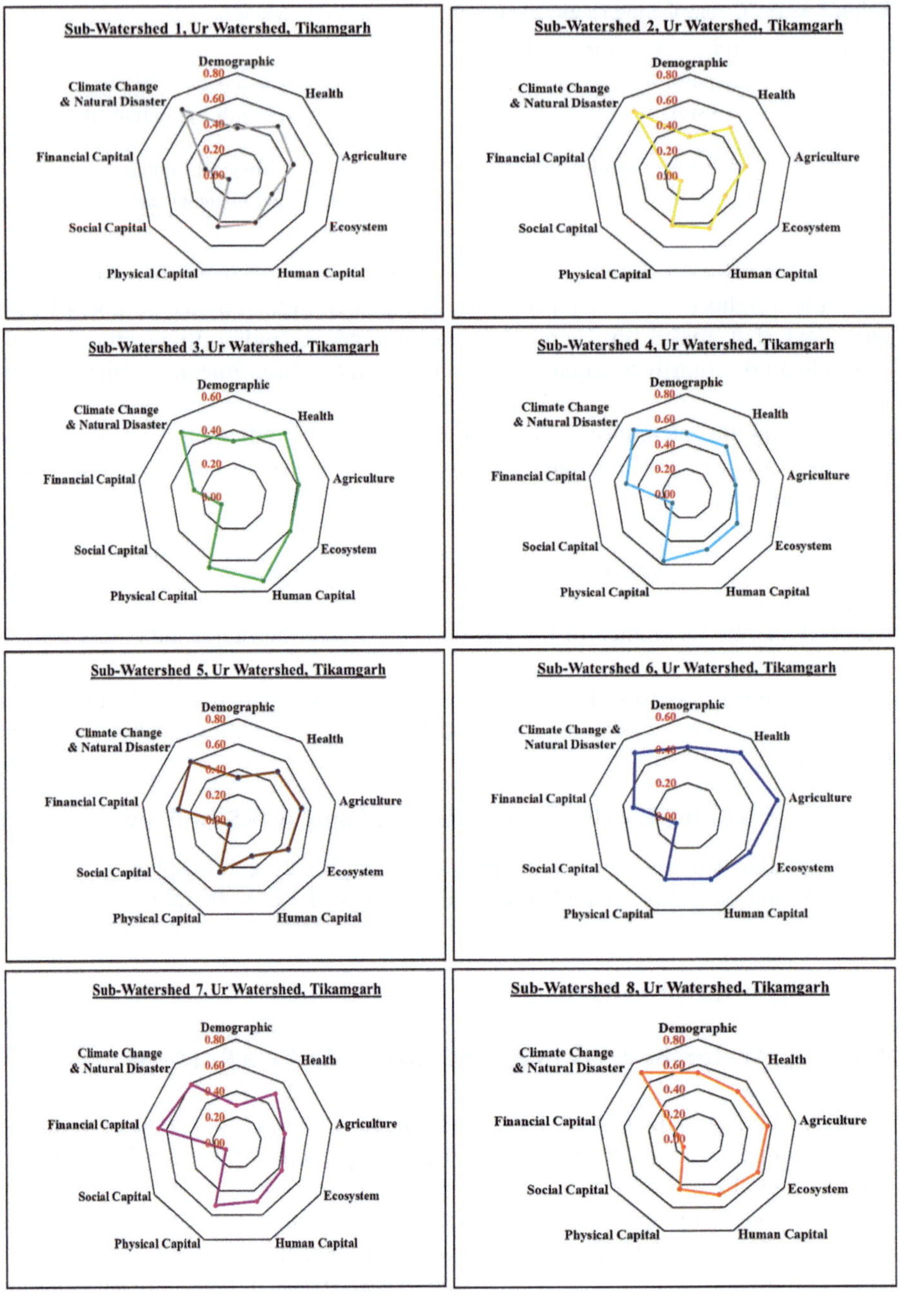

Fig. 3.7 Spider diagrams for the major indicators for 8 sub-watersheds of Ur watershed

the drought-affected districts in Madhya Pradesh. The frequency of occurrence of droughts is once in 4 years. The most drought-prone block is Jatara where the critical dry spells last for more than 17 days and least is Tikamgarh (Songhua 2014). The drought is manifested in the form of late onset and early withdrawal, drying of reservoirs, crop failure, etc. The cumulative build-up of meteorological droughts ripples into hydrological drought with a complex set of highly differentiated adverse impacts and trade-offs. Tikamgarh district is not known as a flood-prone region; however floods are most likely to occur due to excessive rains and release of water from the Matatila dam situated in Uttar Pradesh. The most flood-prone block is Palera and the least is Tikamgarh. Thus, flood becomes one of the major disasters during heavy downpour.

In regards to health services, Tikamgarh is construed as one of the backward districts of Madhya Pradesh. The value of IMR is high against the state value. The district lags behind in providing basic facilities. There are very few health centres present because of which the municipal health workers have to stay in the villages and have to perform deliveries at the patients' houses. They also attend to minor health problems as the district lacks professional doctors and paramedic staff.

With already low levels of health status in Madhya Pradesh, further economic vulnerabilities have worsened the situation, pushing the poor into the trap of poverty. The overall health expenditure is comparatively very low for Madhya Pradesh in comparison with other states, which is a matter of concern. Also it is observed that there is a very slow rise in the health spending both by the central and state government. This can be supported by the fact that there is unequal distribution of resources for health across different states. The result of this is that Madhya Pradesh is unable to generate enough resources through revenue generation and thus is also unable to allocate a greater proportion of budgetary resources for health care.

The district lacks in the access to human, social, financial and physical resources, thereby making them more vulnerable to the adversities of climate change events. Lack of infrastructure to support high school and higher secondary level education has resulted in deterioration in quality of education and hence human resource development. This has further resulted in lower literacy rates and situation far more poor in case of female literacy. The area suffers from low accessibility to drinking water and sanitation facilities. Scarcity of drinking water increases drudgery among women as they have to travel long distance to fetch water in order to carry out the daily household chores. Infrastructural constraints, particularly in the electricity sector, disallow farmers to tap groundwater resources for agricultural irrigation.

Table 3.9 represents the results for the LVI–IPCC, focusing on the three factors contributing to vulnerability: exposure, sensitivity and adaptation. Since the value is larger than 0, all the 8 sub-watersheds are found to be exposed to climate extremes than its adaptive capacity.

The vulnerability triangle diagram (Fig. 3.8) shows the contributing factors for vulnerability index based on the LVI-IPCC approach. The diagram clearly shifts towards right, indicating that the exposure factor contributes majorly to the vulnerability for the entire watershed. As we can see from Table 3.9, the exposure values are higher than adaptive capacity and sensitivity values in all the sub-watersheds,

Table 3.9 LVI-IPCC results for 8 sub-watersheds of Ur watershed

LVI-IPCC components	SW1	SW2	SW3	SW4	SW5	SW6	SW7	SW8
Exposure	0.67	0.67	0.51	0.67	0.60	0.50	0.59	0.70
Sensitivity	0.39	0.38	0.39	0.46	0.45	0.47	0.38	0.55
Adaptive capacity	0.34	0.33	0.39	0.45	0.34	0.34	0.46	0.37
LVI–IPCC	0.13	0.13	0.05	0.10	0.12	0.07	0.05	0.18

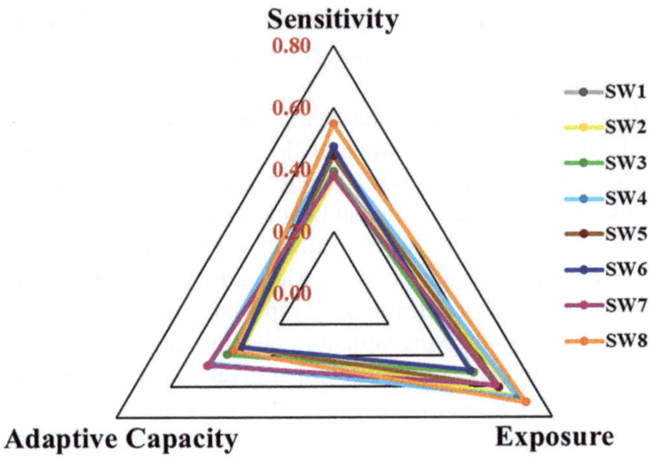

Fig. 3.8 Vulnerability triangle diagram for 8 sub-watersheds of Ur watershed

thereby, indicating that all the sub-watersheds are vulnerable to climate change events in present as well as future.

The following is the order of vulnerability of the eight sub-watersheds of Ur watershed, as shown in Fig. 3.9:

$$SW8 > SW1 > SW2 > SW5 > SW4 > SW6 > SW7 > SW3$$

As per the LVI–IPCC results, the most vulnerable sub-watershed in the Ur watershed is SW8, which falls majorly in the Palera block. The high vulnerability is because the exposure (0.70) and sensitivity (0.55) values are the highest among all the sub-watershed and also are more than the adaptive capacity values (0.37).

At the sub-indicator level, high inter-annual variation in rainfall makes SW8 most exposed to drought and extreme climate change conditions. This indicates high probability of drought years. SW8 falls in the Palera block which also experiences floods occasionally. Villages falling in this sub-watershed, such as Mahendra Maheva and Budaur, are affected by floods (Songhua 2014).

The sub-watershed reports high percentage of marginal agricultural labour, that is workers who work less than 183 days. So in order to meet the expenditure for the rest of the year, the villagers migrate to towns in order generate a livelihood.

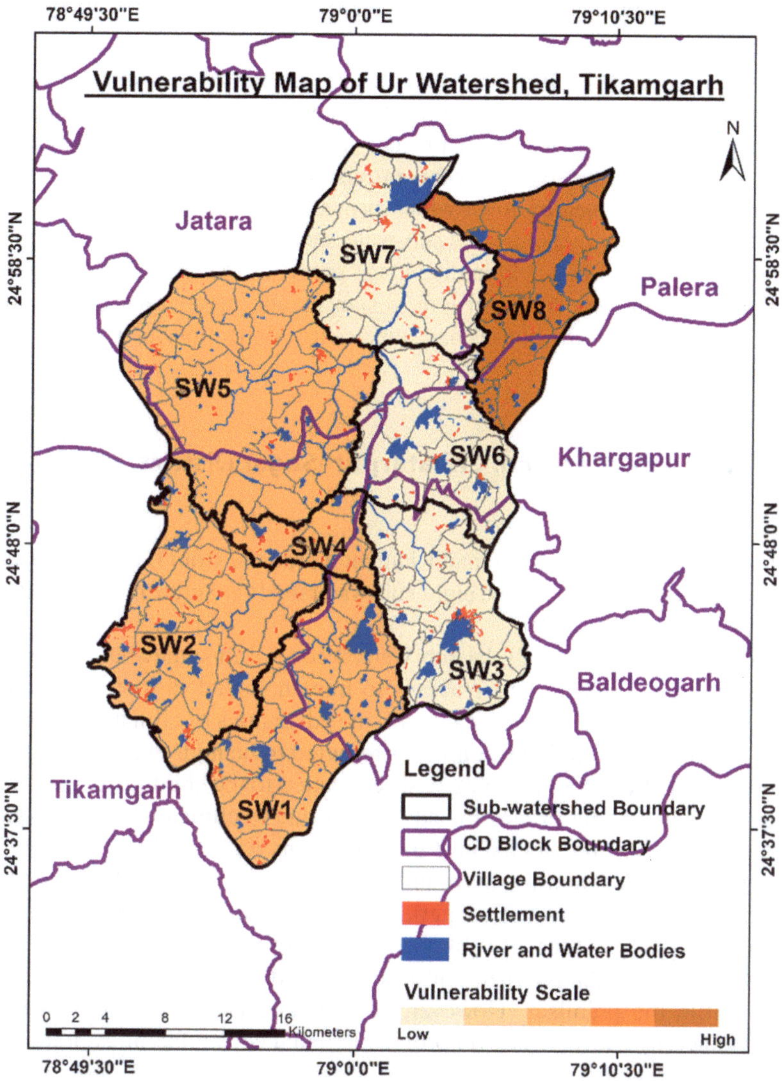

Fig. 3.9 Vulnerability map of Ur watershed

The natural resources of this sub-watershed need to be protected as it can hamper the forest productivity and soil quality to greater extent. The sub-watershed has the highest percentage of wastelands which are left unutilized.

More concentrated efforts are required to tackle the increasing health problems. Financially, more facilities and support are required to be provided to the villagers as they majorly depend on agriculture for their livelihood.

The overall LVI-IPCC scores indicate that SW7 falling in Jatara block is comparatively less vulnerable than other sub-watersheds because of better adaptive capacity (0.44) and less sensitivity (0.42). The sub-watershed is relatively equipped with infrastructure facilities in comparison to the rest of the sub-watersheds which makes it less vulnerable. It has a higher access to human resources and physical resources that allows them to cope up with the adversities of climate change events. The villagers also have access to primary health centres and medicine centres located within the premises. However, provision of government/private hospitals was there only in the nearby town which provided ambulance van facilities in the villages in case of emergencies. The sub-watershed has a relatively better provision for drinking water facilities within the village premises as result of which women of the villages do not have to travel long distances and lesser time is consumed in fetching water. Most of the houses have electricity connections and receive light for like 12 h a day.

3.5.2 Kathan Watershed, Chhatarpur–Sagar District, Madhya Pradesh

The vulnerability indices of the major indicators ranged from 0.05 to 1.00, and the differences in vulnerability of the eight sub-watersheds are presented individually in the spider diagram (Fig. 3.10).

The diagrams show that all the sub-watersheds fall in somewhat mildly vulnerable zone which is due to better access to human, social, financial and physical resources that allows them to cope up with the adversities of climate change events. Availability of human capital resources is reported to be high in the study area. The villages have facilities for providing basic education to the children and also making them aware and capable to face the adversities due to climate change. Along with the schools, most of the blocks have the provisions for skill development centres. The centres provide trainings to youth in different domains such as information technology, management services, agriculture, telecom, etc., and help them in getting good jobs and generate a livelihood for themselves.

The villages have the provision of primary health centres and medicine centres, located within the premises. However, government/ private hospitals were there only in the nearby towns which do provide ambulance van facilities in the villages in case of emergency.

Considering the financial infrastructures, the villages have provision for cooperative banks, rural development banks and primary agricultural credit societies which deals with the agricultural farmers and help in providing loans to the farmers and rural artisans.

The physical capital resources comprising the basic infrastructure and services are more equally distributed in this region. The villagers have access to water resources for drinking and domestic purposes within the premises of their villages which reduces their time and distances travelled for fetching water. Some households also have electricity connections and receive light for like 12 h a day. Under the Swachh Bharat Mission (implemented by the Ministry of Drinking Water and Sanitation in 2014), toilets have also been constructed within the villages with the

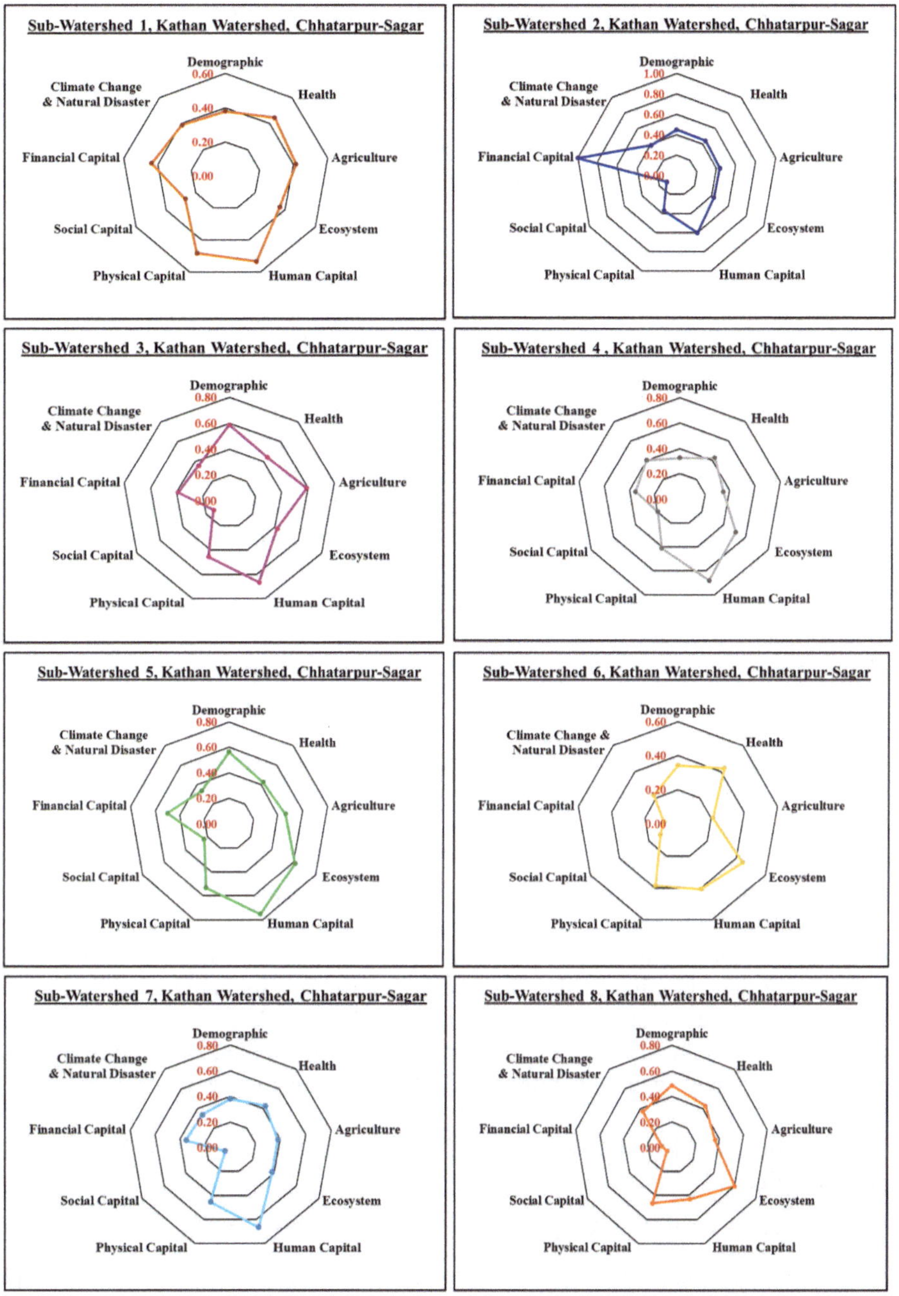

Fig. 3.10 Spider diagram of the major indicators for 8 sub-watersheds of Kathan watershed

Table 3.10 LVI–IPCC results for 8 sub-watersheds of Kathan watershed

LVI-IPCC components	SW1	SW2	SW3	SW4	SW5	SW6	SW7	SW8
Exposure	0.39	0.39	0.36	0.40	0.34	0.22	0.34	0.37
Sensitivity	0.40	0.44	0.53	0.39	0.53	0.33	0.39	0.47
Adaptive capacity	0.46	0.51	0.48	0.47	0.56	0.31	0.45	0.33
LVI–IPCC	−0.03	−0.05	−0.06	−0.03	−0.12	−0.03	−0.04	0.02

aim of reducing the problems of open defecation and other unhygienic practices. The availability of basic facilities and the provision of important infrastructure facilities have thereby increased the adaptive capacity to any climatic stressor.

The vulnerability is mainly arising due to higher agriculture dependence. Agriculture is the prime earning activity for the people in the district, which is not very remunerative. Majority of the population in the district is of unskilled agriculture labour. There is a lack of proper irrigation facilities and advanced agricultural practices due to which most of the land is used for single cropping. Due to poverty and traditional agricultural inputs, output is not able to meet the needs of growing population. Forward linkages are not up to the need of farmers. Due to lack of market and credit facilities, farmers are facing various problems.

The situation of health has also been evaluated to be quite bad for the district. Due to unhygienic surroundings and open sewage, various diseases like filarial, elephantiasis and malaria are prevalent. Considering a limited availability of medical facilities, many people are not able to get the proper medical attention at the right time.

Table 3.10 presents the results for the LVI–IPCC, focusing on the three factors contributing to vulnerability: exposure, sensitivity and adaptation. The values of adaptive capacity are higher than the exposure and sensitivity values, except the SW8, which indicate that the rest of the sub-watersheds have a higher capacity to adapt or overcome the adverse situations of climate change.

The vulnerability triangle diagram (Fig. 3.11) shows the contributing factors for vulnerability index based on the LVI–IPCC approach. The diagram clearly shifts towards left, which indicates that the adaptive capacity values are higher than exposure and sensitivity to climate change in all the sub-watersheds. This further implies that all the sub-watersheds have a better access to the human, social, physical and financial resources which mobilize them to build resilience to climate change impacts.

The order of vulnerability of the eight sub-watersheds of Kathan watershed is shown in Fig. 3.12:

$$SW8 > SW4 > SW6 > SW1 > SW7 > SW2 > SW3 > SW5$$

As per the results, the most vulnerable watershed is SW8, falling under the Bada Malhera block of the Chhatarpur district. The sub-watershed reports higher exposure (0.37) and sensitivity (0.47) values and low adaptive capacity (0.33) value indicating that the sub-watershed area is more exposed to and prone to climate changes and its impacts. The condition of drought develops mainly during March–May and

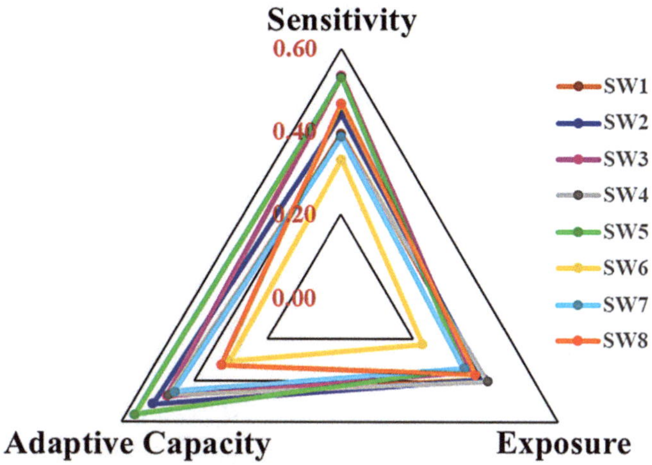

Fig. 3.11 Vulnerability triangle diagram for 8 sub-watersheds of Kathan watershed

is found to be a temporary reduction in water and moisture availability. It has also been reported that bore wells and tube wells dry up in the scorching heat during summer posing severe threats to life.

At the sub-indicator level, the sub-watershed has recorded the highest percentage of Scheduled Castes population. This section of the society, in particular, is characterized by a lack of education and its facilities, an absence of ownership of productive resources, extreme dependence for livelihood on agriculture, economic indebtedness and poor participation in the secondary and tertiary sector. Without education and access to resources, they are not able to realize their strength and opportunities to develop themselves.

The vulnerability is also arising due to high agriculture dependence. The sub-watershed reports the highest percentage of cultivators and lowest percentage of non-agricultural labour. The area suffers from successive dry spells causing heavy amount of crop failure leading to low and unstable agricultural production; as a result, the people are under financial stress.

There is an increasing pressure on the natural resources of this sub-watershed. The sub-watershed reports the lowest forest cover and surface water availability. The area has very few ponds and talabs which have also shrunk in the size over the years due to erratic rainfall and excessive siltation. The natural resources, thus, need to be rejuvenated and protected as it can hamper the forest productivity, soil quality and water availability to greater extent.

On the vulnerability scale, SW5 (which falls in Bada Malhera block and partially in Buxwaha and Ghuwara block) is the least vulnerable amongst the eight sub-watersheds because of highest adaptive capacity value (0.56). The area receives the highest average annual rainfall that helps in the recharging of surface and groundwater resources. In terms of natural resources, the sub-watershed reports the highest percentage of dense forest cover which helps in meeting the requirements of timber,

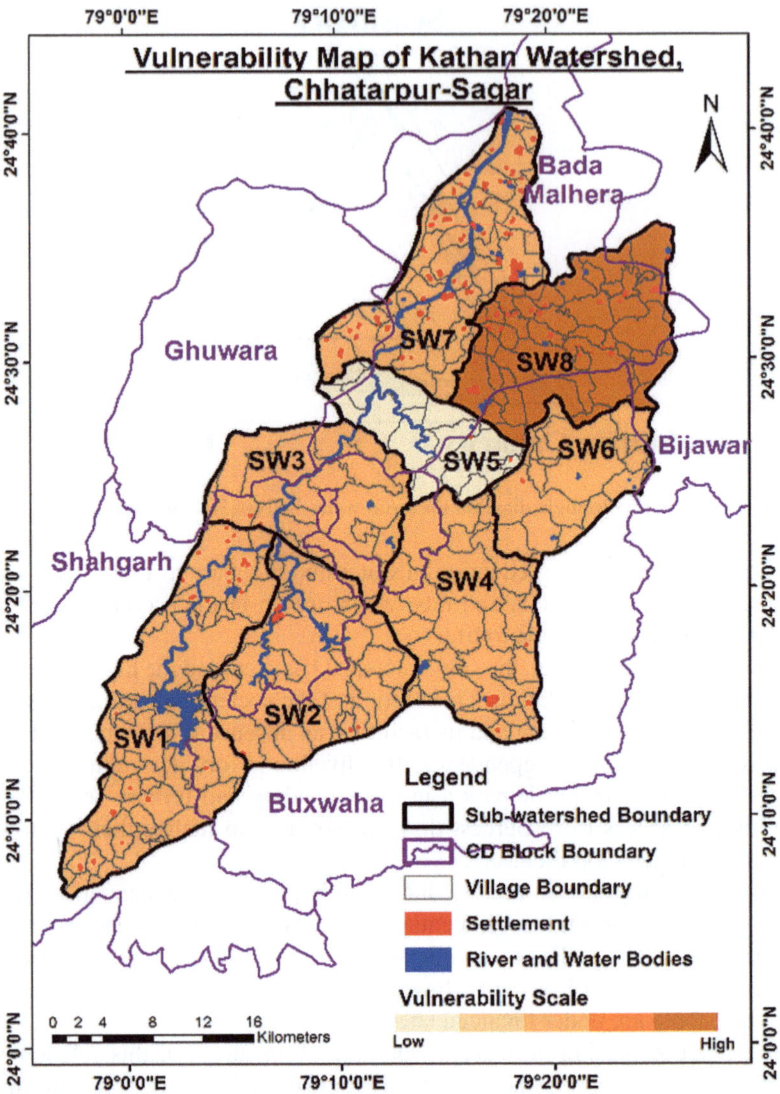

Fig. 3.12 Vulnerability map for Kathan watershed

poles, bamboo and fuel wood for the villagers. The sub-watershed is well equipped with infrastructure facilities in comparison to the rest of the sub-watersheds which makes it less vulnerable. The area is equipped with various facilities within the village premises, such as educational centres up to secondary level, primary health centres, community health centres and medicine shops, animal care centres, etc. Around 66% households have electricity connections in their homes and receive electricity for like 12 h a day. Under the Swachh Bharat Mission, private and

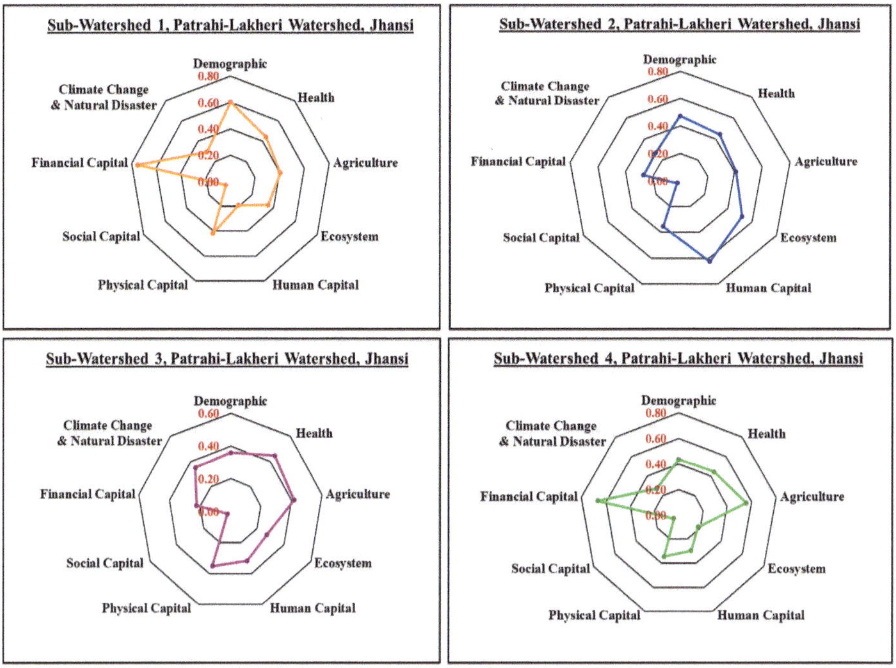

Fig. 3.13 Spider diagrams of the major indicators for 4 sub-watersheds of Patrahi–Lakheri watershed

community toilets have also been constructed in order to reduce the problems of open defecation and other unhygienic practices. The availability of basic facilities and the provision of important infrastructure facilities allow them to cope up with the adversities of climate change events.

3.5.3 Patrahi–Lakheri Watershed, Jhansi District, Uttar Pradesh

The vulnerability indices of the major indicators ranged from 0.02 to 0.75, and the differences in vulnerability of the four sub-watersheds are presented individually in the spider diagram (Fig. 3.13).

The diagram demonstrates that all the sub-watersheds fall in mildly vulnerable zone, which is due to higher access to human resources and financial resources that enable them to deal with climate change impacts. The villages have government schools up to secondary level that provide the students with the fundamental education and make them conscious about the events related to climate change and their impacts. In addition to the schools, many skill-training centres have opened up that deliver vocational training and skilling to help individuals with better employment opportunities and make them financially independent.

Considering the health sector, there are PHCs and medical shops in the villages that provide an integrated curative and preventive health care to the villagers. In case of the severe or chronic medical conditions, there are government/ private hospitals in the nearby towns that can provide proper treatment. To assist the villagers in financial matters, there is provision of agricultural credit societies that provide short term and medium term loans to the farmers and local artisans to meet their financial requirements.

The vulnerability of the sub-watersheds is mainly arising due to the negligence in the health sector. The value of IMR of Jhansi district is less than the state average, that is 70. Some of the reasons responsible for increasing number of infant deaths are malnutrition and lack of facilities on addressing ailments of infants and proving immunization injections. Since the region is mostly rural, there is high level of gender discrimination where girl child is not preferred or very neglected. Further, the existing health infrastructure facilities are the least in the Bundelkhand region. With already low levels of health status in Jhansi, further economic vulnerabilities have worsened the situation, by increasing the miseries of the poor.

The people of the watershed are highly dependent on agriculture as a main source for livelihood. The farmers are small scale with an average landholding size of 2–5 hectares which means that it is mainly them who will be impacted in case of any climate change implications. The cropping pattern is dominated by Rabi crops which occupy major gross cropped area. Cereals like paddy, wheat and pulses like black gram, masoor, gram, peas and oil seeds are the main crops, whereas sesame seeds and vegetables are grown mainly for subsistence. The area suffers from drought leading to recurrent crop failure, due to which there is low and unstable agricultural production. As a result, the people are under financial stress. Moreover, there is a decrease in crop production due to insufficient irrigation, undulated rocky terrain, erratic rainfall, excessive use of fertilizers, inadequate nursery and shortage of cold storages, soil erosion, low carbon and humus content and low seed replacement rate (SRR) for pulses. It has been experienced by the farmers that the onset of monsoons has shifted from June end to July end which has led to delayed sowing and subsequent reduction in crop yield. Apart from these issues, another issue that the farmers face is regarding fixing the price of their produce. Their decision on agriculture inputs is influenced either by large farmers or middlemen in deciding prices as there are no government shops to sell their produce.

Table 3.11 presents the results for the LVI-IPCC, focusing on the three factors contributing to vulnerability, exposure, sensitivity and adaptation, with the help of which a vulnerability map for the watershed was prepared (Fig. 3.14).

Table 3.11 LVI-IPCC results for 4 sub-watersheds of Patrahi–Lakheri watershed

LVI-IPCC components	SW1	SW2	SW3	SW4
Exposure	0.30	0.27	0.35	0.27
Sensitivity	0.46	0.46	0.37	0.42
Adaptive capacity	0.32	0.39	0.27	0.33
LVI–IPCC	−0.01	−0.06	0.03	−0.02

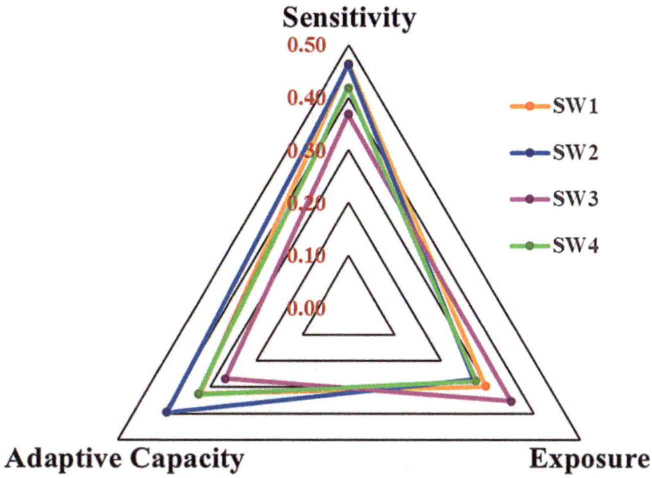

Fig. 3.14 Vulnerability triangle diagram for 4 sub-watersheds of Patrahi–Lakheri watershed

The vulnerability triangle diagram (Fig. 3.14) shows the contributing factors for vulnerability index based on the LVI–IPCC approach. The diagram clearly shifts towards top, which indicates that the all the sub-watersheds are highly sensitive to climate change. The changing environmental conditions have a significant impact on the people, natural resources and the biodiversity. Climate stresses and shocks in drought-prone areas lead to high-scale migration of people as their livelihoods heavily depend on natural resources. The diagram also shows that all the sub-watersheds show high adaptive capacity values indicating that the area is well equipped with facilities and infrastructure which help them to fight back and cope up with the adverse conditions arising due to the climate change impacts.

The following is the order of vulnerability of the four sub-watersheds of Patrahi–Lakheri watershed (Fig. 3.15):

$$SW3 > SW1 > SW4 > SW2$$

According to the results, the most vulnerable sub-watershed is SW3, which falls in two developmental blocks of Jhansi district, that is Bangara and Gursarai, and very little portion of Tikamgarh district (i.e. Niwari). The sub-watershed experiences high exposure (0.35) values and lowest adaptive capacity values (0.27) in comparison to the rest of the sub-watersheds.

At the sub-indicator level, the sub-watershed receives the lowest average annual rainfall amongst the four sub-watersheds, thus causing impacts on the agriculture as well as on the availability of surface and groundwater. Low groundwater recharge levels have been reported over the years. The situation has further aggravated due to low availability of the functioning hand pumps in the sub-watershed area.

In demographic terms, the sub-watershed reports a high percentage of Scheduled Castes and Scheduled Tribes population. This population is relatively poorer and less educated and deprived of the basic necessities of life, for example they are not allowed

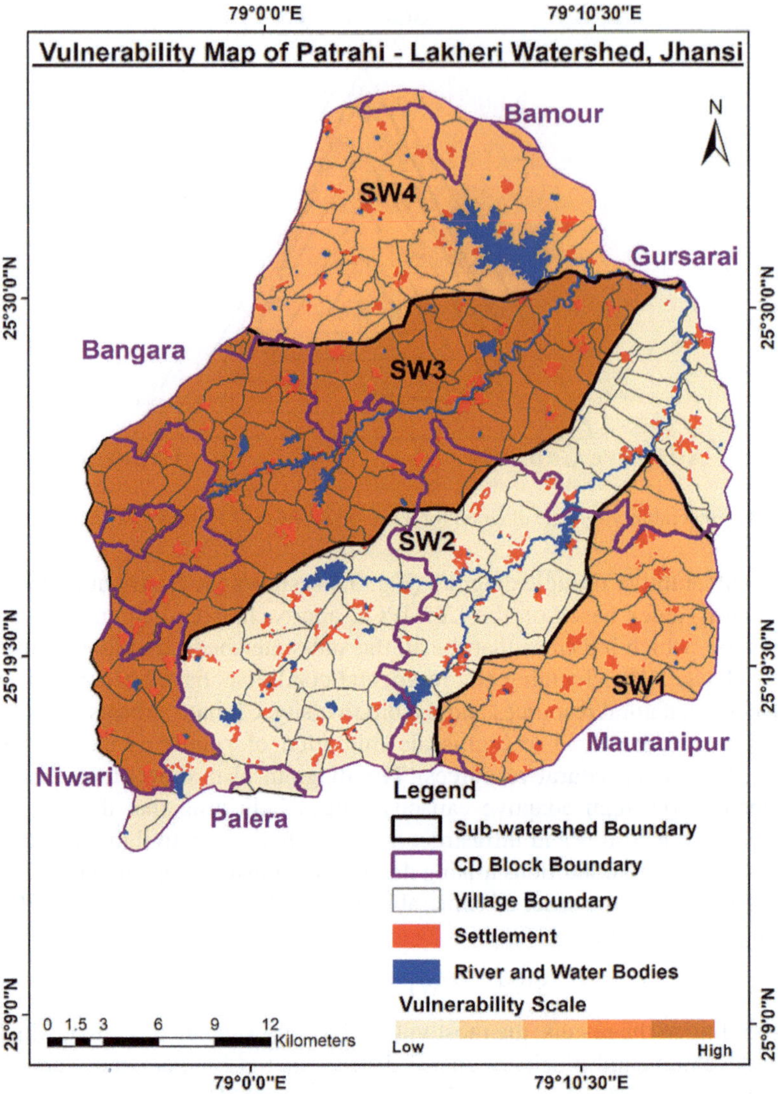

Fig. 3.15 Vulnerability map for Patrahi–Lakheri watershed

to use common village wells and tanks for fetching water and they are not much involved in village meetings and discussions as a result of which the population is left uninformed and unaware. This situation worsens their condition and makes them more sensitive to climate change. The sub-watershed observes the lowest literacy rate, which makes the people less informed and capable in comparison to others and, thus, decreases their ability to cope with adversities. There is a higher level of agricultural dependency in this sub-watershed, which increases sub-watershed's vulnerability to climate change and fluctuation in terms of agriculture trade.

On the contrary, SW2 is the least vulnerable amongst the four sub-watersheds because of better adaptive capacity (0.39) which gives them a better access to resources and facilities that allows them to cope up with the adversities of climate change events. The sub-watershed covers three developmental blocks of Jhansi district (i.e. Bangara, Mauranipur and Gursarai). It reports the lowest percentage of Scheduled Castes and Scheduled Tribes population. There is a lower percentage of cultivators and marginal agricultural labours which highlight that the population under this sub-watershed isn't solely dependent on agriculture for their livelihood. Also the percentage of marginal agricultural workers, that is workers who work less than 183 days, is also low indicating that they practice other occupations by staying in villages and do not migrate too frequently. Considering the natural resources, this sub-watershed reports the highest percentage of dense forest cover which helps in meeting requirements of timber, poles, bamboo and fuel wood for the villagers. The sub-watershed also shows high availability of human capital and social capital resources such as schools, skill development centres, health centres, public distribution shops, etc.

3.5.4 Sajnam Watershed, Lalitpur District, Uttar Pradesh

The vulnerability indices of the major indicators ranged from 0.00 to 0.82, and the differences in vulnerability of the seven sub-watersheds are presented individually in the spider diagram (Fig. 3.16). The diagram shows that all the sub-watersheds fall in mildly vulnerable zone which is due to better access to human, social, physical and financial resources that allows them to cope up with the adversities of climate change events.

The villages have primary and secondary level education centres. Along with these, most of the blocks have the provision for skill development centres that provide a chance to the youth and the rural women to hone their skills and earn a steady income. Skills imparted help in improving employability and livelihood opportunities, reduces poverty, enhances productivity and promotes environmental sustainable development. There is provision for medical facilities available within the district. A good number of sub centres and primary health centres are present in each block as well as one community health centre in each block. The area is well equipped with a number of commercial and cooperative banks, rural banks and agricultural credit societies that have helped the farmers in accessing financial assistance largely. They have also encouraged the rural artisans by providing them loans that have helped them in expanding their work.

The vulnerability of the sub-watersheds is mainly arising due to poor state of the rural health. Although the medical facilities are available in each block, most of the villages still lack the access to the health-care facilities. There is also lack of female doctors and paramedic staff due to which lot of females find difficulty in discussing their medical issues and at times prefer not to consult. Since the region is mainly rural, there have been evidences of high female child mortality, which highlights gender discrimination. This has further added to a higher value of IMR for the

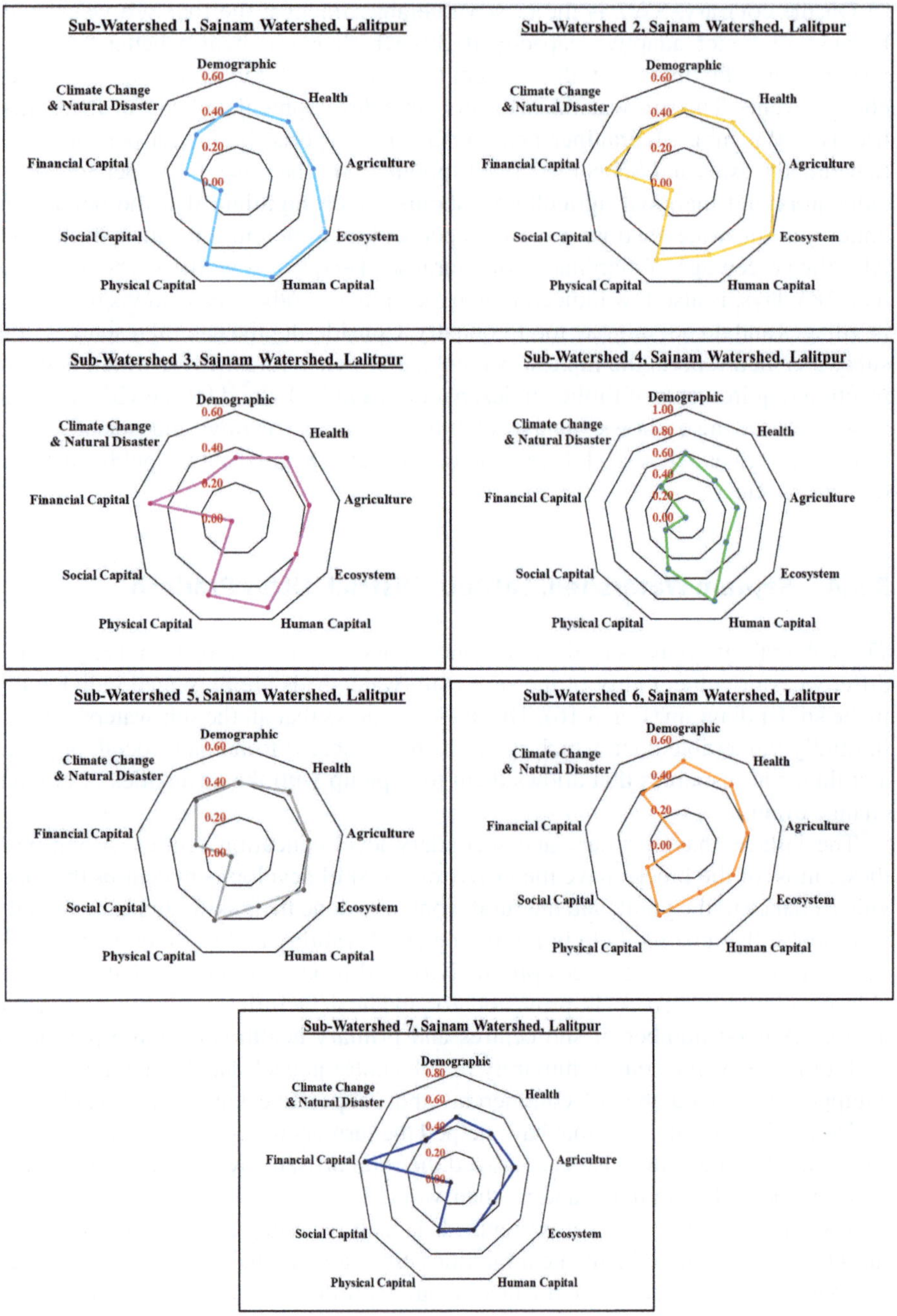

Fig. 3.16 Spider diagrams of the major indicators for 7 sub-watersheds of Sajnam watershed

Table 3.12 LVI-IPCC results for 7 sub-watersheds of Sajnam watershed

LVI-IPCC components	SW1	SW2	SW3	SW4	SW5	SW6	SW7
Exposure	0.35	0.37	0.27	0.38	0.38	0.38	0.38
Sensitivity	0.47	0.50	0.39	0.52	0.41	0.41	0.44
Adaptive capacity	0.43	0.40	0.42	0.51	0.30	0.28	0.41
LVI–IPCC	−0.04	−0.01	−0.06	−0.07	0.04	0.04	−0.01

district. In analysing infant mortality, some of the other dominant factors are maternal factors in the reproduction process, environmental contaminations, nutritional deficiency, injuries to child and practices in the health care of the child.

The economy of the study area is mainly based on agriculture. Cropping pattern is dominated by Rabi crops. The region suffers from drought or flood consistently due to which there is low and unstable agricultural production, and consequently the farmers are under financial stress. Moreover, crop production experiences severe problems in district due to poor irrigation, undulated rocky terrain and erratic rainfall, thereby making the watershed vulnerable.

Table 3.12 presents the results for the LVI–IPCC, focusing on the three factors contributing to vulnerability: exposure, sensitivity and adaptation.

The vulnerability triangle diagram (Fig. 3.17) shows the contributing factors for vulnerability index based on the LVI–IPCC approach. The diagram clearly shifts towards top, which indicates that all the sub-watersheds are highly sensitive to climate change. The changing environmental conditions have a significant impact on the people, natural resources and the biodiversity. Climate stresses and shocks in drought-prone areas lead to high-scale migration of people as their livelihoods heavily depend on natural resources.

The diagram also shows that all the sub-watersheds show high adaptive capacity values indicating greater access to human, social, physical and financial facilities and infrastructures to the villagers which together enable people achieve their livelihood objectives and help in uplifting their poverty status. These services also make them aware and prepare them to face and cope up with climate change impacts.

The following is the order of vulnerability of the seven sub-watersheds of Sajnam watershed, as shown in Fig. 3.18:

$$SW6 > SW5 > SW7 > SW2 > SW1 > SW4 > SW3$$

The most vulnerable sub-watershed is SW6, which falls in Bar block of Lalitpur district because the sub-watershed reports the highest exposure value (0.38) which is more than the adaptive capacity (0.28). In fact, the adaptive capacity value for the sub-watershed is the lowest amongst all the sub-watersheds.

At the sub-indicator level, the sub-watershed reports a higher percentage of Scheduled Caste population. This population continues to be oppressed,

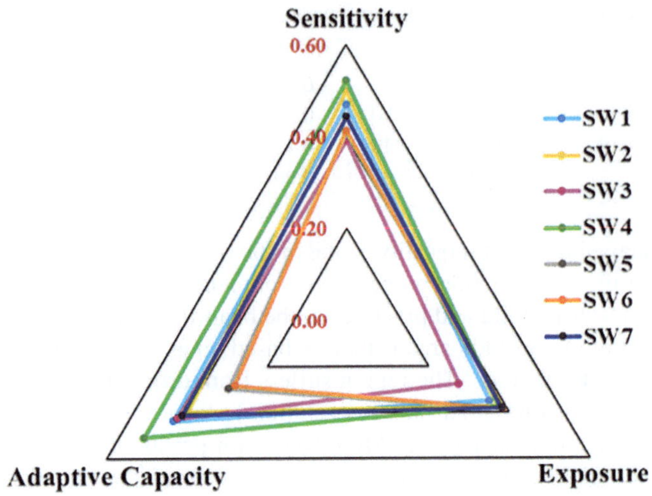

Fig. 3.17 Vulnerability triangle diagram for 7 sub-watersheds of Sajnam watershed

discriminated against in villages, in educational institutions, in job market etc. leaving them uninformed and unaware thus, making them more susceptible to climate change.

In terms of the natural resources, the sub-watershed does not support any forest cover and reports a lower soil quality. The natural resources, thus, need to be restored and sustainably managed as it can hamper the productivity of the forests and soil status to a greater extent.

Another important reason that makes the region vulnerable is due to the presence of coal-based thermal power plant, Lalitpur Power Generation Company Limited, that produces 1980 megawatts of power in 2015. For the setup of the plant, agricultural lands of close by villages were taken away in return of very less compensation and promises of employment in the plant and better infrastructural facilities in the village. But as promised, none of the things were provided to the villagers, making them more vulnerable to shocks and reducing their adaptive capacity to recover from those shocks. Also, the plant has been affecting their existing and main source of livelihood, that is, agriculture. The fly ash from the plant settles down on their crop and destroys the crop which affects crop productivity and puts a financial pressure on the people. Another problem is the alarming level of groundwater. After the setup of the plant, the people drill the ground till 700 ft. for irrigation purposes which was earlier just 200 ft.

Considering the human capital component, the sub-watershed completely lacks infrastructure supporting education which has resulted in the lowest literacy rate of the sub-watershed amongst the seven sub-watersheds. The people are less informed and less capable in comparison to others which have ultimately decreased their ability to cope with adversities. The sub-watershed does not even have facility of healthcare centres within the premises, and the villagers have to go to the neighbouring

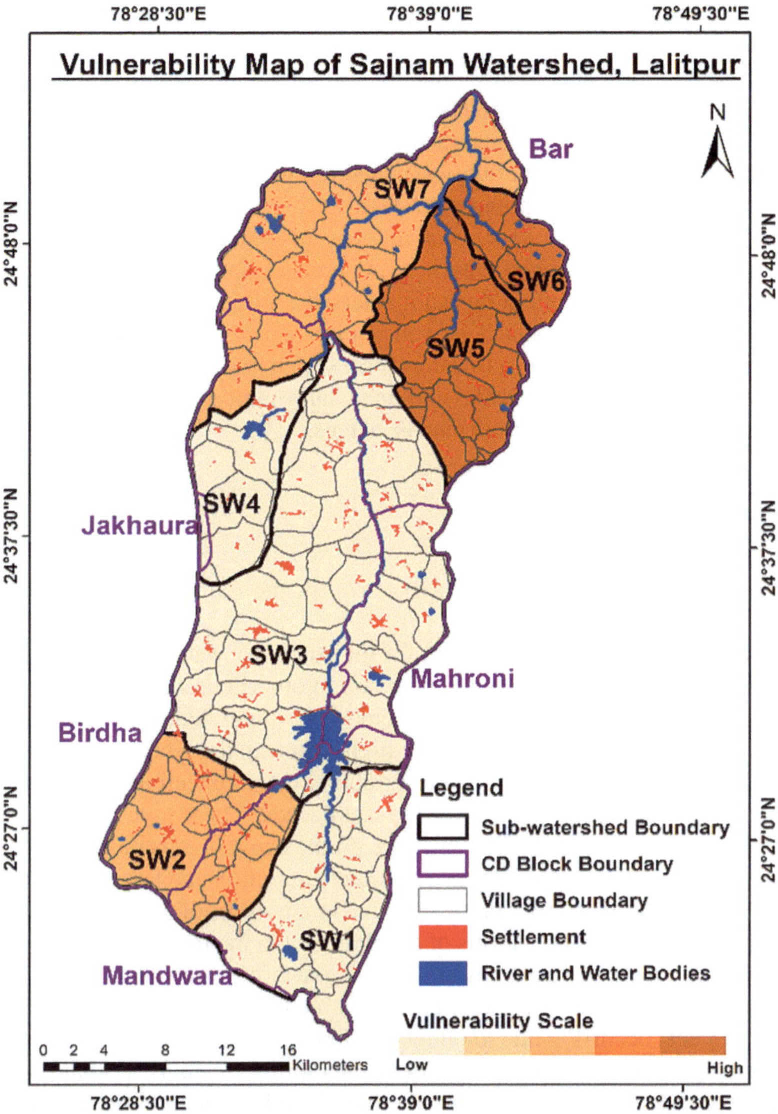

Fig. 3.18 Vulnerability map for Sajnam watershed

villages/towns for treatment. In case of emergency, there is no immediate facility available for the villagers. The provision of self-help groups in this sub-watershed is at a very nascent stage in comparison to the others. Few SHGs that have come into existence are helping in empowering the women of the villages. The financial situation seems to be a little grim in this sub-watershed. There are no banking facilities available within the watershed. So the villagers have to go to the neighbouring villages or nearby towns in case if they have to take loans or any financial assistance.

On the contrary, SW3 is the least vulnerable amongst the seven sub-watersheds because of lowest exposure (0.27) and sensitivity (0.39) values. The sub-watershed covers four developmental blocks, majorly Birdha and Mahrauni and partially Bar and Mandwara. The sub-watershed exhibits low inter-annual rainfall variation which indicates the low occurrence of droughts in the region. Considering the demographics component, the sub-watershed reports low percentage of child population and Scheduled Caste population. This section of society is highly sensitive to climate change as they are dependent and less informed and have limited adaptive capacities. The sub-watershed reports a low percentage of wasteland which shows that the land is well utilized in this sub-watershed for farming or grazing purposes. Additionally, the sub-watershed is well equipped with infrastructure facilities in comparison to the rest of the sub-watersheds, which makes it less vulnerable. It has a higher access to human resources and financial resources that allows them to cope up with the adversities of climate change events.

3.6 Results

The overall results for LVI–IPCC assessment for the four watersheds are given in Table 3.13.

Out of the four watersheds chosen for the study, Ur watershed (Tikamgarh district) shows high vulnerability in comparison to other four watersheds. Mostly, the sub-watersheds in the Ur watershed are somewhat vulnerable which is because their exposure and sensitivity levels are very high, whereas the adaptive capacity levels are low. There has been more climatic variability due to uncertain precipitation pattern and increasing temperature over the last decade. These together have resulted in very high exposure values. The pressure on the ecosystem is more in this district with more land utilization, higher groundwater extraction and larger area under irrigation, which has made them more sensitive to any form of impacts in the context of climate variability. Lower levels of development in the form of infrastructure and low levels of access to resources as well as assets have resulted in lower coping capacity of the people in these districts which makes them more vulnerable to any form of impacts occurring due to climate change.

Followed by Ur watershed on the vulnerability scale is Kathan watershed (Chhatarpur–Sagar district) due to relatively high exposure values and lower

Table 3.13 Overall LVI–IPCC results for the 4 watersheds in Bundelkhand region

LVI–IPCC components	Ur Watershed, Tikamgarh	Kathan Watershed, Chhatarpur-Sagar	Patrahi–Lakheri Watershed, Jhansi	Sajnam Watershed, Lalitpur
Exposure	0.52	0.37	0.29	0.35
Sensitivity	0.31	0.35	0.35	0.35
Adaptive capacity	0.20	0.23	0.24	0.29
LVI–IPCC	0.10	0.05	0.02	0.02

adaptive capacity values making it prone to more climate change impacts such as droughts and floods. The high vulnerability is reported from the sub-watershed located in the Bada Malhera block of the Chhatarpur district. The block reports higher exposure and sensitivity values and low adaptive capacity value indicating that the sub-watershed area is more exposed to and prone to climate changes and its impacts. The rest of the sub-watersheds are somewhat mildly vulnerable due to reported higher adaptive capacity which gives them a better access to human, social, financial and physical resources, thus making them more adaptive to cope up with the adversities of climate change events.

Next to Kathan watershed is the Sajnam watershed in Lalitpur District. The high vulnerability is reported from the sub-watershed that falls in the Bar block. This sub-watershed reports the lowest adaptive capacity value amongst all the sub-watersheds. The problems are further aggravated due to the presence of coal-based thermal power plant that has acquired agricultural lands for the setup and has reduced the groundwater levels due to excessive extraction. The rest of the sub-watersheds are relatively less vulnerable due to low exposure to climate change stressors and shocks and also have better access to the human, physical, social and financial resources.

Patrahi–Lakheri watershed in Jhansi district is comparatively less vulnerable because of lowest exposure values and relatively high adaptive capacity values. The sub-watersheds have an access to and control over the human, social, physical and financial resources which mobilize them to build resilience to climate change impacts.

3.7 Community Adaptations to Coping with Climate Change Impacts

Bundelkhand's agriculture scenario is at crossroads. It has to find a way to feed the growing population while being environmentally, socially and economically sustainable. To achieve this, the Bundelkhand region needs to focus on developing and deploying technologies that improve water, fertilizer, labour and energy use efficiency while simultaneously improving soil, ecosystem and social resilience, restoring degraded agro-ecosystems and creating alternate sources of income for farmers. In addition to these formidable goals, the issue of climate change and its potential impacts on agriculture and food security must be addressed appropriately through smart agricultural interventions.

The long-term nature of climate change and its imminent impact on agriculture warrant an agricultural development policy and practices that incorporates both short- and long-term planning perspectives. As regards climate change, there are two different types of adaptation—reactive adaptation and anticipatory adaptation. As the names suggest, reactive adaptation responds to changes or impacts after they have occurred, while anticipatory adaptation responds to these changes before they occur.

Based on the assessment and consultation, the National Institute of Hydrology, under its two projects based in Bundelkhand region, funded by Technology Information, Forecasting and Assessment Council (TIFAC) and Ministry of Water Resources (MoWR), designed an adaptation plan, based on both reactive and anticipatory approaches. The plan provides adaptation strategies in the selected study areas catering to sustainable agricultural practices, soil and water conservation, water purification technology, alternative livelihood occupations and market linkages and land resource planning for improving the productivity and income of the farmers and providing safe drinking water to the villagers. The proposed strategies have been piloted in some villages of the Ur watershed and Kathan watershed, and these are some of the best strategies and easily accepted by the farmers. Similar activities have been proposed and planned for the other two watersheds also.

3.7.1 Sustainable Agricultural Practices

- Under the project, demonstration on line sowing technique with seed drill and specific seed varieties was tested in three villages, involving seven progressive farmers in Manchi, Baisa and Samara villages, Ur watershed. Sites were selected through community participation. The seeds for major crops such as soybean, black gram (AZAD 1 and PU35) and groundnut (IG20) were provided in order to ensure livelihood, nutritional security and economic empowerment of farmer at faster pace.
- Understanding the importance of agro-horti model to sustainable agriculture practices, demonstration on the implementation of the Wadi model was provided in two progressive farmer's land, one in Baisa village, Baldeogarh, and another one in Chaturkari village, Jatara, Ur watershed (Fig. 3.19). The Wadi model presents a sustainable solution that makes farming profitable even on small plots and ensures nutritional security of the households by the provisioning of cereals, pulses, vegetables and fruits. This agri-horti-based model reduces climate risks, regenerates production potential of the land and ensures that farmers enjoy a regular flow of income due to diversification of production.

3.7.2 Soil and Water Conservation

Promotion of check dams, Gabion structure and farm pond was started as an appropriate intervention to enhance water storage capacity and recharge groundwater in the drought-prone Bundelkhand region in central India. This is proving to be a most cost-effective, smart and sustainable solutions as the farmers in this rain-fed agricultural area are able to take three crops annually. These structures help reduce erosion and increase soil water infiltration; the retained water can be used for irrigation. They also reduce the required channel maintenance and thus increase groundwater levels and recharge rate. To identify the areas where recharge structures are highly essential and to know their suitable locations, the thematic layers such as

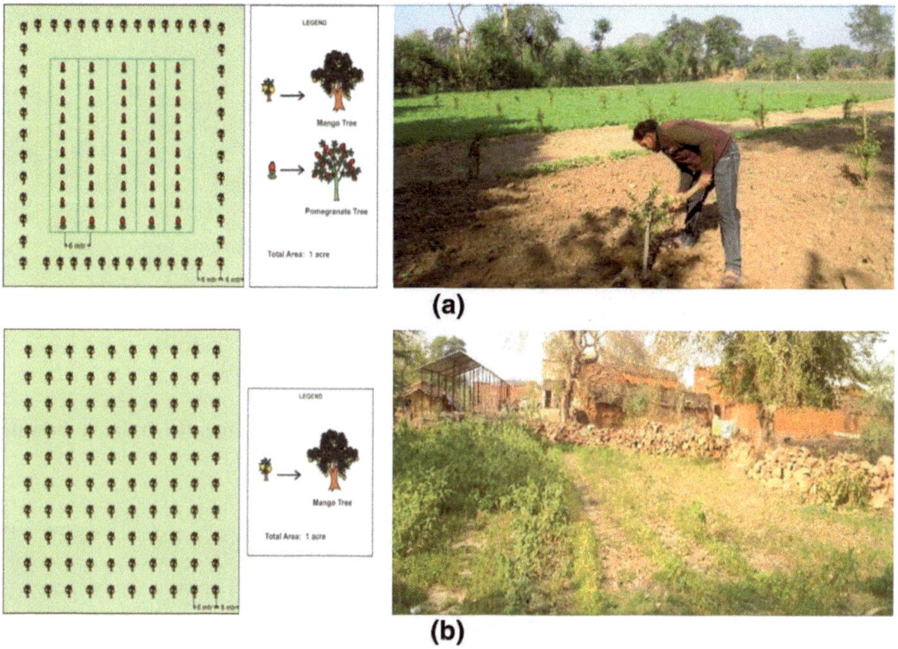

Fig. 3.19 (**a**) Demo plot (1 acre) of a farmer from Baldeogarh for mango and pomegranate tree. (**b**) Demo plot (1 acre) of a farmer from Jatara for mango tree

hydro-geomorphology, land use, soils, etc., were generated by MPCOST and UPRSAC and were integrated to locate suitable areas and structures for artificial recharge.

3.7.3 Water Purification Technology

TIFAC supported the technology demonstration programme for the access to safe water where one school was selected in Baisa village, Baldeogarh block, Ur watershed, based on the student strength and status of water contamination (faecal coliform—highly positive; TDS—148; absence of chemical contamination) and availability of electricity. Jal TARA filter has been installed within the primary school premises. The filter is designed to treat drinking water contaminated with pathogenic bacteria and turbidity using slow sand filtration technique. Training was given to the administration on how to use the filter and conduct the water quality testing (Fig. 3.20).

(a) (b)

Fig. 3.20 (**a**) Jal TARA filter installed in middle school, Samarra village. (**b**) Training on water quality testing

Fig. 3.21 Trainings provided to villagers on market-friendly products

3.7.4 Alternative Livelihood Occupations and Market Linkages

Small-scale and landless farmers are quite prevalent in villages of the selected study areas. These farmers do not have any fixed source of income and often migrate when weather conditions disrupt economic activities in the region. So in order to generate additional income for the villagers and stop their migration to nearby cities in search of livelihood, trainings were provided for market-friendly products. Some of the activities for which trainings were provided are listed below and shown in Fig. 3.21.

- Vermicomposting and biofertilizer/biopesticide production training
- Low-cost edible mushroom production training
- Jute items making with natural colours training
- Bamboo craft training

- Bee keeping training
- Pisciculture

3.7.4.1 Success Story: Jute and Bamboo Craft Training in Tikamgarh and Market Linkage

There is always a huge demand for products made from jute, and it also provides an alternative source of income and also helps in restoring the traditional art. Samarra village, Ur watershed, was chosen where hands-on training on making jute and bamboo products was given to a group of 30 villagers, which included men, women and young girls. After the completion of jute and bamboo craft trainings, a marketing exposure visit of artisans at Vigyan Mela Bhopal and Antyoday Mela Tikamgarh was organized (Fig. 3.22). The artisans displayed their finished craft items at stalls and sold the item. During this marketing exposure period, both male and female artisans got an opportunity to learn and understand about the customers' choice, negotiations and the demand for items along with the exposure to other craftsmen and their products/designs.

3.7.5 Land Resource Planning

Land resources plan is similar to land use plan. The land use plan is a scheme of rational land use worked out on the basis of economic and social development objectives. The purpose of this land use planning is to use land resources rationally to coordinate and rationally allocate land use amongst different sections of the national economy. In the light of the extent of the planning and the aim of the plan, the land use plan can be divided into three kinds, that is overall land use planning, detailed land use planning and special land use planning. The suggested land resources management actions for the study areas are:

1. Action for double cropping
2. Action suggested for ley farming and fuel wood plantation
3. Action suggested for gap plantation
4. Action suggested for agroforestry

Adoption of appropriate adaptation strategies in all the four watersheds would result in the empowerment of rural communities and sustainable livelihoods. Some of the measures based on vulnerability assessment are detailed in Table 3.14 sector-wise.

Fig. 3.22 Market exposure to jute and bamboo craft artisans

Table 3.14 Sector-wise short-term and long-term measures based on the assessment

Sectors	Key risks	Short term	Long term
Demographics	Imbalanced male-female ratio	Awareness on SHGs	Encourage girl child birth
	Increased migration	Engage in non-farm occupations (pisciculture, fishing, handicrafts)	Strengthening of technology dissemination centres
	Increased poverty	Livestock and poultry rearing	E-governance in agriculture, finance, health
		Increase in microfinance facilities	
Health	Food and water shortage causing malnourishment	Strengthen primary health care	Public health surveillance
	Heat strokes and loss of life	Increased public health expenditure	Implement heat warning systems
Agriculture	Reduced crop productivity	Use of improved seeds	Agro-forestry and agro-horticulture models, Wadi
		Mulching	
		Organic farming	Multi-cropping
		In situ moisture conservation	Crop intensification
	Livestock migration in search of water	Dry sowing and line sowing	Drip irrigation
		Ridge and furrow irrigation methods	Drought-tolerant and short-duration crops
		Strategic weaning of calves	Extension of crop insurance
		Supplementary fodder during dry season	Relocating the herds
Ecosystem	Stress on water resources	Rejuvenation of village tanks and water bodies	Construction of farm ponds and storage tanks
	Reduction in soil moisture	Soil carbon sequestration	Construction of Nala bunds, check dams, percolation ponds and stop dams
	Increase of forest fires	Increase in forest cover	
	Desertification of land	Conversion of wasteland to pastureland, gap plantation, etc.	Early warning capability of forest fire danger
Human capital	Withdrawal of children from schools to have more earning hands	Raising community awareness on climate change issues	Encouraging vocational training courses
	Not sufficient amount of health-care facilities	Training on health and knowledge skills	Building health and education infrastructure
	Reduced quality of life		

(continued)

Table 3.14 (continued)

Sectors	Key risks	Short term	Long term
Physical capital	Scarcity of water and conflicts over sharing of water	Rooftop rainwater harvesting	Efficient water distribution system to avoid water loss and wastage
		Water supply tanker services	
	Sanitation and hygiene problems	Repairing of existing and introduction of new hand pumps to utilize groundwater	Infrastructure to sustain greater number of electricity connections
	Daily household chores come to standstill due to no electricity		
Social capital	Increased prices of food and other products	Subsidized prices of goods	Increased storage facilities of agriculture produce
		Availability of substitutes	Transparency in distribution system from farm to consumer
		Organizing recreational activities	
	Community networks are broken and social interaction decreases	Capacity building programmes for government schemes and facilities	
		Establishment of farmers adaptation clubs/clusters	Weekly mandatory meetings for updated interventions
Financial capital	Reduced income	Access of loans at lower interest rates	Subsidies on agricultural inputs
	Economic losses	Awareness on available government schemes like Sampoorna Gramin Rojgar Yojna	Awareness on SHGs
	Increased credit risk for financial institutions		Provision of finance for setting up agro-processing industries
	Loss of tax revenue generation	Facilitate microfinancing support	
	Rural indebtedness		

Along with this, there is a need to follow a systematic approach to completely eliminate the problem, which may consist of:

- Conducting research in the field of agriculture so as to identify better strategies to adapt to climate change, for example determining crop mix that can resist the impacts of climate change in different parts/block of the state.
- Establishing better meteorological stations that will help in dissemination of local weather information and provide forecast and disaster warnings to the farmers.
- Long-term capacity building on climate change adaptation will play an important role in coping up and adapting to climate variability.

- Review of state procurement policies to incorporate measures for purchase of alternate crops that are grown in the drought-prone regions.

3.8 Conclusion

The main objective of this study was to identify climate change vulnerable watersheds and sub-watersheds along with the villages falling under the selected study area. This was done in order to attain knowledge and understanding on the contributing factors of vulnerability and to prioritize the area which needs immediate action for the agro-based households of the region.

The findings of the study are based on the LVI–IPCC approach, which reveals that Ur watershed is the most vulnerable out of the four watersheds. The difference in the vulnerability arises because of differences in the watershed's sensitivity, adaptive capacity and exposure to disaster and climate change. The agro-based households of the region are majorly exposed to extreme climate change, rising temperature and decreasing rainfall leading to occurrence of successive droughts. Some of the other reasons leading to high vulnerability of the regions are dependence on rain-fed agriculture which is highly sensitive to climate change, increasing pressure on natural resources, acute water shortage, over-extraction of groundwater, lack of infrastructure facilities, extreme poverty and high-scale migration from villages to towns. Based on the vulnerability assessment, the findings and the suggested adaptive strategies would enable the villagers of Bundelkhand to better adapt to the impacts of the climate change and enable government and local bodies to develop programmes and take initiatives to strengthen the most vulnerable villages.

Acknowledgement This work was carried out at the National Institute of Hydrology (NIH), under the projects supported by Technology Information, Forecasting and Assessment Council (TIFAC), Department of Science and Technology and Ministry of Water Resources, RD & GR, Government of India. The authors would like to acknowledge the team of Madhya Pradesh Council of Science and Technology (MPCST) and Remote Sensing Applications Centre, Uttar Pradesh (UPRSAC), for generating all GIS database required for the study.

Appendices

Appendix A: Villages Falling Under each Sub-watershed in Ur Watershed, Tikamgarh District, Madhya Pradesh

Sub-watershed 1		
S. no.	Village	Block
1.	Ahar	Baldeogarh
2.	Duda Khera	Baldeogarh
3.	Ganeshpura	Baldeogarh
4.	Kachhya Khera Khas	Baldeogarh
5.	Karmasan Hata	Baldeogarh
6.	Khushipura	Baldeogarh
7.	Ladwani Khas	Baldeogarh
8.	Luharra	Baldeogarh
9.	Madan Sagar	Baldeogarh
10.	Narayanpur	Baldeogarh
11.	Pratappura	Baldeogarh
12.	Sukora	Baldeogarh
13.	Ajnaur Khas	Tikamgarh
14.	Bad Madai Khas	Tikamgarh
15.	Bakpura	Tikamgarh
16.	Banjarya	Tikamgarh
17.	Basgoi	Tikamgarh
18.	Budki Khera	Tikamgarh
19.	Gudanwara	Tikamgarh
20.	Jamuniya Khera	Tikamgarh
21.	Kachhya Khera	Tikamgarh
22.	Lar Khas	Tikamgarh
23.	Lar Ugad	Tikamgarh
24.	Nainwari	Tikamgarh
25.	Parsuwa	Tikamgarh
26.	Patar Khera	Tikamgarh
27.	Prempura	Tikamgarh
28.	Radhapur	Tikamgarh
29.	Samarra	Tikamgarh
30.	Sapaun Khas	Tikamgarh
31.	Shyag	Tikamgarh
32.	Shyampura	Tikamgarh
33.	Sundarpur	Tikamgarh

Sub-watershed 2		
S. no.	Village	Block
1.	Anantpura	Tikamgarh
2.	Bad Madai	Tikamgarh
3.	Bahadurpur	Tikamgarh
4.	Dhajrai	Tikamgarh
5.	Durgapur	Tikamgarh
6.	Gopalpura Bhata	Tikamgarh
7.	Gopalpura Khas	Tikamgarh
8.	Hajuri Nagar	Tikamgarh
9.	Jatauwa Khas	Tikamgarh
10.	Lakhora	Tikamgarh
11.	Laxmanpura	Tikamgarh
12.	Mabai Bhata	Tikamgarh
13.	Madhumar Gross Form	Tikamgarh
14.	Madhumar Khas	Tikamgarh
15.	Madhuwan	Tikamgarh
16.	Madumar Bhata	Tikamgarh
17.	Mamaun	Tikamgarh
18.	Manik Chowk	Tikamgarh
19.	Matauli	Tikamgarh
20.	Mohanpura	Tikamgarh
21.	Narguda Bhata	Tikamgarh
22.	Narguda Khas	Tikamgarh
23.	Papawani	Tikamgarh
24.	Patha Bhata	Tikamgarh
25.	Patha Khas	Tikamgarh
26.	Ranipura	Tikamgarh
27.	Rigaura	Tikamgarh
28.	Shri Nagar Khas	Tikamgarh
29.	Sundarpur	Tikamgarh
30.	Uttampura	Tikamgarh

Sub-watershed 3		
S. no.	Village	Block
1.	Badaghat	Baldeogarh
2.	Baisa Khas	Baldeogarh
3.	Baisa Ugad	Baldeogarh
4.	Banera	Baldeogarh
5.	Banpura Khurd	Baldeogarh
6.	Bhaira Novar	Baldeogarh
7.	Bhelsi	Baldeogarh
8.	Brishbanpura	Baldeogarh
9.	Chandooli	Baldeogarh
10.	Devi Nagar	Baldeogarh
11.	Dumbar	Baldeogarh

(continued)

Sub-watershed 3		
S. no.	Village	Block
12.	Durga Nagar	Baldeogarh
13.	Gukrai Khas	Baldeogarh
14.	Jhinguwan	Baldeogarh
15.	Jinagarh Jangal	Baldeogarh
16.	Jinagarh Khas	Baldeogarh
17.	Kailpura	Baldeogarh
18.	Karmasan Hata	Baldeogarh
19.	Lakheri	Baldeogarh
20.	Pratappura	Baldeogarh
21.	Raj Nagar	Baldeogarh
22.	Rajpura	Baldeogarh
23.	Sebar Khas	Baldeogarh
24.	Sebar Ugad	Baldeogarh
25.	Sujanpura	Baldeogarh
26.	Talmau	Baldeogarh
27.	Tamora	Baldeogarh

Sub-watershed 4		
S. no.	Village	Block
1.	Karmasan Ghat	Baldeogarh
2.	Karmasan Hata	Baldeogarh
3.	Daryaw Nagar	Tikamgarh
4.	Jaswant Nagar	Tikamgarh
5.	Kater Khera	Tikamgarh
6.	Mabai Bhata	Tikamgarh
7.	Majana Khas	Tikamgarh
8.	Mavai	Tikamgarh
9.	Papawani	Tikamgarh

Sub-watershed 5		
S. no.	Village	Block
1.	Bachhoda Bhata	Jatara
2.	Bachhoda Khas	Jatara
3.	Barmadang Bhata	Jatara
4.	Barmadang Khas	Jatara
5.	Barmamanjh	Jatara
6.	Bilgay Bhata	Jatara
7.	Bilgay Khas	Jatara
8.	Chandaua Khera	Jatara
9.	Chandrapura	Jatara
10.	Chatukari	Jatara
11.	Deokha	Jatara

(continued)

Sub-watershed 5		
S. no.	Village	Block
12.	Deopur	Jatara
13.	Dhamna Bhata	Jatara
14.	Dhamna Khas	Jatara
15.	Harpura	Jatara
16.	Hirdenager	Jatara
17.	Janakpur Bhata	Jatara
18.	Janakpur Khas	Jatara
19.	Kamal Nagar	Jatara
20.	Kurrai	Jatara
21.	Kuwarpura	Jatara
22.	Larkhurd	Jatara
23.	Luharguwa Khas	Jatara
24.	Luharguwa Bhata	Jatara
25.	Mahuabag	Jatara
26.	Mathupura	Jatara
27.	Man Bujurg Bhata	Jatara
28.	Mau Bujurg Khas	Jatara
29.	Panyara Khera	Jatara
30.	Pathara	Jatara
31.	Ramgarh	Jatara
32.	Serai	Jatara
33.	Vaidpur	Jatara
34.	Vikrampura	Jatara
35.	Bamhori Nakiban	Tikamgarh
36.	Bhadra	Tikamgarh
37.	Majana Bhata	Tikamgarh
38.	Mavai	Tikamgarh
39.	Pagara Jangal	Tikamgarh
40.	Raipur	Tikamgarh
41.	Rorai	Tikamgarh
42.	Sunoni	Tikamgarh
43.	Kari (NP)	Tikamgarh

Sub-watershed 6		
S. no.	Village	Block
1.	Balwantpura Khas	Baldeogarh
2.	Balwantpura Ugad	Baldeogarh
3.	Banpura Sapaun	Baldeogarh
4.	Banyani	Baldeogarh
5.	Bhiloni	Baldeogarh
6.	Bhitarwar	Baldeogarh
7.	Chaubara	Baldeogarh
8.	Hirapur Kha	Baldeogarh
9.	Jatera	Baldeogarh

(continued)

Sub-watershed 6		
S. no.	Village	Block
10.	Madori	Baldeogarh
11.	Majguwan	Baldeogarh
12.	Pipra	Baldeogarh
13.	Sarkanpur Khas	Baldeogarh
14.	Sarkanpur Ugad	Baldeogarh
15.	Shyampura	Baldeogarh
16.	Sijaura	Baldeogarh
17.	Jauwa	Jatara
18.	Karmaura	Jatara
19.	Raghunathpura	Jatara
20.	Majana Khas	Tikamgarh

Sub-watershed 7		
S. no.	Village	Block
1.	Garroli	Baldeogarh
2.	Bagoura	Jatara
3.	Baiwarjangal	Jatara
4.	Baiwarkhas	Jatara
5.	Bajetpura	Jatara
6.	Baldeopura	Jatara
7.	Bandarguda	Jatara
8.	Deoraha	Jatara
9.	Gor	Jatara
10.	Hirdenager	Jatara
11.	Kandwa	Jatara
12.	Kitakhera	Jatara
13.	Lidhouratal	Jatara
14.	Machora	Jatara
15.	Manch	Jatara
16.	Muhara Khas	Jatara
17.	Pathara	Jatara
18.	Piprat	Jatara
19.	Shah	Jatara
20.	Simariya	Jatara
21.	Jatara (NP)	Jatara

Sub-watershed 8		
S. no.	Village	Block
1.	Gora	Baldeogarh
2.	Guna	Baldeogarh
3.	Baharo Tal	Jatara
4.	Bamhori Abda	Jatara

(continued)

Sub-watershed 8		
S. no.	Village	Block
5.	Deoraha	Jatara
6.	Hunarganj	Jatara
7.	Tanga	Jatara
8.	Jatara (NP)	Jatara
9.	Bhatgora	Palera
10.	Budaur	Palera
11.	Guda Najdik Pali	Palera
12.	Magrai	Palera
13.	Mahendra Maheva	Palera
14.	Morramanna	Palera
15.	Pali	Palera
16.	Phoolpur	Palera
17.	Ratanguwan	Palera
18.	Toury	Palera

Appendix B: Villages Falling Under Each Sub-watershed in Kathan Watershed, Chhatarpur-Sagar District, Madhya Pradesh

Sub-watershed 1		
S. no.	Village	Block
1.	Bicchaun	Buxwaha
2.	Lahar	Buxwaha
3.	Semra Sekh	Buxwaha
4.	Bagaich	Shahgarh
5.	Baghrohi	Shahgarh
6.	Barkhera Shahgarh	Shahgarh
7.	Basona	Shahgarh
8.	Batwaha	Shahgarh
9.	Bei	Shahgarh
10.	Bilagram	Shahgarh
11.	Bilguwan	Shahgarh
12.	Birthipur	Shahgarh
13.	Chakk Dalpatpur	Shahgarh
14.	Chakk Kaneri	Shahgarh
15.	Chakkmohari	Shahgarh
16.	Chhayan	Shahgarh
17.	Dalpatpur	Shahgarh
18.	Dhurmar	Shahgarh
19.	Dulchipur	Shahgarh
20.	Gadgara	Shahgarh
21.	Ghutrai	Shahgarh

(continued)

Sub-watershed 1		
S. no.	Village	Block
22.	Gomatupur	Shahgarh
23.	Harrai	Shahgarh
24.	Kajrawan	Shahgarh
25.	Kaneri	Shahgarh
26.	Kanikhedi	Shahgarh
27.	Kanikhedi Khurd	Shahgarh
28.	Khatora Kalan	Shahgarh
29.	Khatora Khurd	Shahgarh
30.	Lamnau	Shahgarh
31.	Luharra	Shahgarh
32.	Malkhuwan	Shahgarh
33.	Mohari	Shahgarh
34.	Mudari Bujurg	Shahgarh
35.	Muhli	Shahgarh
36.	Nanakpur	Shahgarh
37.	Narwan	Shahgarh
38.	Niwahi	Shahgarh
39.	Nouraj	Shahgarh
40.	Padrai	Shahgarh
41.	Papet	Shahgarh
42.	Papeti	Shahgarh
43.	Pratppura	Shahgarh
44.	Pura Shahgarh	Shahgarh
45.	Rabara	Shahgarh
46.	Rajoula	Shahgarh
47.	Ranipur	Shahgarh
48.	Rurawan	Shahgarh
49.	Shahgarh (NP)	Shahgarh
50.	Simariya Kalan	Shahgarh
51.	Simariya Khurd	Shahgarh
52.	Simriya Uwari	Shahgarh
53.	Singhpur	Shahgarh
54.	Tatarwara	Shahgarh
55.	Tinsuwa	Shahgarh

Sub-watershed 2		
S. no.	Village	Block
1.	Bebasa	Buxwaha
2.	Beergarh	Buxwaha
3.	Bhadator	Buxwaha
4.	Bijawali	Buxwaha
5.	Chaurai	Buxwaha
6.	Gewlai	Buxwaha
7.	Ghaughara	Buxwaha

(continued)

Sub-watershed 2		
S. no.	Village	Block
8.	Ghaughari	Buxwaha
9.	Govindpura	Buxwaha
10.	Gugwara	Buxwaha
11.	Iashwarmau	Buxwaha
12.	Luharpura	Buxwaha
13.	Madiya Buzurg	Buxwaha
14.	Majhaguwan Batton	Buxwaha
15.	Mandpur	Buxwaha
16.	Nibar	Buxwaha
17.	Sanauda	Buxwaha
18.	Amarmaoh	Shahgarh
19.	Chandola	Shahgarh
20.	Fulwari	Shahgarh
21.	Garera	Shahgarh
22.	Jamuniya	Shahgarh
23.	Karrai	Shahgarh
24.	Khargorani	Shahgarh
25.	Madantala	Shahgarh
26.	Mohanpura	Shahgarh
27.	Neguwan	Shahgarh
28.	Ratanpur Shahgarh	Shahgarh
29.	Sasan	Shahgarh

Sub-watershed 3		
S. no.	Village	Block
1.	Lidhoura	Bada Malhera
2.	Luhani	Buxwaha
3.	Maddeora	Buxwaha
4.	Bamnora Kalan	Ghuwara
5.	Dalipur	Ghuwara
6.	Halawani	Ghuwara
7.	Mawai	Ghuwara
8.	Agara	Shahgarh
9.	Bodanganj	Shahgarh
10.	Garroli	Shahgarh
11.	Hansrai	Shahgarh
12.	Indora	Shahgarh
13.	Kalra	Shahgarh
14.	Khairwaha	Shahgarh
15.	Ludayara	Shahgarh
16.	Rampur	Shahgarh
17.	Sadpur	Shahgarh
18.	Tigoda	Shahgarh

Sub-watershed 4		
S. no.	Village	Block
1.	Amakhuwa	Buxwaha
2.	Birampura	Buxwaha
3.	Buxwaha (NP)	Buxwaha
4.	Dangrai	Buxwaha
5.	Dardoniya	Buxwaha
6.	Dugasara	Buxwaha
7.	Gadhoi	Buxwaha
8.	Harduwa	Buxwaha
9.	Hinota	Buxwaha
10.	Hirdepur	Buxwaha
11.	Jakha	Buxwaha
12.	Jara	Buxwaha
13.	Jujharpura	Buxwaha
14.	Kasera	Buxwaha
15.	Kero	Buxwaha
16.	Kuhi	Buxwaha
17.	Madia Khurd	Buxwaha
18.	Mangarai	Buxwaha
19.	Mara	Buxwaha
20.	Palda	Buxwaha
21.	Pondi	Buxwaha
22.	Sagouriya	Buxwaha
23.	Sahpura	Buxwaha
24.	Teiyamar	Buxwaha
25.	Tilai	Buxwaha
26.	Chouki	Shahgarh
27.	Hirapur	Shahgarh

Sub-watershed 5		
S. no.	Village	Block
1.	Bineda	Bada Malhera
2.	Dhanguan	Bada Malhera
3.	Manakpura	Bada Malhera
4.	Pandajhir	Bada Malhera
5.	Pipariya Kalan	Bada Malhera
6.	Sirbon	Bada Malhera
7.	Surajpura Road	Bada Malhera
8.	Darguwa	Buxwaha
9.	Hatna	Buxwaha
10.	Raipura Kalan	Ghuwara

Sub-watershed 6		
S. no.	Village	Block
1.	Bajna	Buxwaha
2.	Banpura	Buxwaha
3.	Baranand No. 1	Buxwaha
4.	Baranand No. 2	Buxwaha
5.	Basantpura	Buxwaha
6.	Chhayan	Buxwaha
7.	Dhimarwa	Buxwaha
8.	Jagara	Buxwaha
9.	Kachari	Buxwaha
10.	Kanjra	Buxwaha
11.	Kherakhurd	Buxwaha
12.	Kuwapalo	Buxwaha
13.	Majhora	Buxwaha
14.	Malar	Buxwaha
15.	Mudar	Buxwaha
16.	Nimani	Buxwaha
17.	Shobha	Buxwaha
18.	Siddhai	Buxwaha

Sub-watershed 7		
S. no.	Village	Block
1.	Gorakhpura	Ghuwara
2.	Jhigari	Ghuwara
3.	Kheri	Ghuwara
4.	Panwari	Ghuwara
5.	Richhara	Ghuwara
6.	Bada Malhera (NP)	Bada Malhera
7.	Bandha	Bada Malhera
8.	Bankpura	Bada Malhera
9.	Baraj	Bada Malhera
10.	Barma	Bada Malhera
11.	Bhojpura	Bada Malhera
12.	Dharampura	Bada Malhera
13.	Erora	Bada Malhera
14.	Futwari	Bada Malhera
15.	Garkhuwan	Bada Malhera
16.	Ghinochi	Bada Malhera
17.	Hardotha	Bada Malhera
18.	Kanera	Bada Malhera
19.	Kayan	Bada Malhera
20.	Khirkuwa	Bada Malhera
21.	Kuwarpura Kalan	Bada Malhera
22.	Kuwarpura Khurd	Bada Malhera
23.	Lakhanwa	Bada Malhera

(continued)

Sub-watershed 7		
S. no.	Village	Block
24.	Malpura	Bada Malhera
25.	Monpura	Bada Malhera
26.	Morra	Bada Malhera
27.	Nadiya	Bada Malhera
28.	Pathiya	Bada Malhera
29.	Pipra Kalan	Bada Malhera
30.	Rajpura	Bada Malhera
31.	Salaiya	Bada Malhera
32.	Satpara	Bada Malhera
33.	Sedhpa	Bada Malhera
34.	Sijwaha	Bada Malhera
35.	Silaro	Bada Malhera
36.	Sujanpura	Bada Malhera
37.	Tigari	Bada Malhera

Sub-watershed 8		
S. no.	Village	Block
1.	Chopra	Buxwaha
2.	Andhiyara	Bada Malhera
3.	Arol	Bada Malhera
4.	Baman Kola	Bada Malhera
5.	Bamni	Bada Malhera
6.	Barsat	Bada Malhera
7.	Bhanguwan	Bada Malhera
8.	Bharwani	Bada Malhera
9.	Bilwar	Bada Malhera
10.	Chhaikuwan	Bada Malhera
11.	Gopalpura	Bada Malhera
12.	Jasguwan Kalan	Bada Malhera
13.	Kalothar	Bada Malhera
14.	Karri	Bada Malhera
15.	Kewlai	Bada Malhera
16.	Khatola	Bada Malhera
17.	Kheron	Bada Malhera
18.	Maharajganj	Bada Malhera
19.	Maharajpura	Bada Malhera
20.	Mailwar	Bada Malhera
21.	Mawai	Bada Malhera
22.	Moli	Bada Malhera
23.	Murli Kheda	Bada Malhera
24.	Para	Bada Malhera
25.	Pargaspura	Bada Malhera
26.	Partappura	Bada Malhera
27.	Patan	Bada Malhera

(continued)

Sub-watershed 8		
S. no.	Village	Block
28.	Rajapur	Bada Malhera
29.	Ranikheda	Bada Malhera
30.	Ranipura	Bada Malhera
31.	Ranital	Bada Malhera
32.	Sadwa	Bada Malhera
33.	Sigrampura	Bada Malhera
34.	Sironj	Bada Malhera
35.	Suka	Bada Malhera
36.	Surajpura Kalan	Bada Malhera
37.	Surajpura Khurd	Bada Malhera
38.	Tahanga	Bada Malhera
39.	Toriya	Bada Malhera
40.	Udaipura Khurd	Bada Malhera

Appendix C: Villages Falling Under Each Sub-watershed in Patrahi- Lakheri Watershed, Jhansi District, Uttar Pradesh

Sub-watershed 1		
S. no.	Village	Block
1.	Bhakauro	Bangara
2.	Lidhora	Gursarai
3.	Akseo	Mau Ranipur
4.	Bamhauri	Mau Ranipur
5.	Barauri	Mau Ranipur
6.	Chakara	Mau Ranipur
7.	Churari	Mau Ranipur
8.	Dhamna Payak	Mau Ranipur
9.	Dhawakar	Mau Ranipur
10.	Ghatlahchura	Mau Ranipur
11.	Itaial	Mau Ranipur
12.	Jabalpura	Mau Ranipur
13.	Jhankari	Mau Ranipur
14.	Khandarka	Mau Ranipur
15.	Khanuwan	Mau Ranipur
16.	Kharka Sani	Mau Ranipur
17.	Kotra	Mau Ranipur
18.	Lakhesur	Mau Ranipur
19.	Larauni	Mau Ranipur
20.	Madwan	Mau Ranipur
21.	Mailoni	Mau Ranipur
22.	Mau Rural	Mau Ranipur

(continued)

Sub-watershed 1		
S. no.	Village	Block
23.	Merki	Mau Ranipur
24.	Piprokhar	Mau Ranipur
25.	Rora	Mau Ranipur
26.	Rupa Dhaman	Mau Ranipur
27.	Sijari Khurd	Mau Ranipur
28.	Suhagpur	Mau Ranipur
29.	Tilera	Mau Ranipur

Sub-watershed 2		
S. no.	Village	Block
1.	Bagarauni	Bangara
2.	Bangara Dhawa	Bangara
3.	Banhauri Suhaga	Bangara
4.	Bar Urf Rampur	Bangara
5.	Chaukri	Bangara
6.	Chirkana	Bangara
7.	Gairaha	Bangara
8.	Ghurat	Bangara
9.	Gudha	Bangara
10.	Kachaneo	Bangara
11.	Kathera (NP)	Bangara
12.	Kathera Rural	Bangara
13.	Luhar Gaon Ranipu	Bangara
14.	Magarwara	Bangara
15.	Nimoni	Bangara
16.	Pachauro	Bangara
17.	Pachwara	Bangara
18.	Palara	Bangara
19.	Patha Kharka	Bangara
20.	Rajpura	Bangara
21.	Ratosa	Bangara
22.	Sanaura	Bangara
23.	Sewara	Bangara
24.	Sewkara Dhawa	Bangara
25.	Tikari	Bangara
26.	Bagrauni	Gursarai
27.	Bikram Pura Sani	Gursarai
28.	Deora Khurd	Gursarai
29.	Eoni	Gursarai
30.	Ghat Kuwan	Gursarai
31.	Gorpura	Gursarai
32.	Kedar Tai	Gursarai
33.	Khali Pura	Gursarai
34.	Kotra	Gursarai

(continued)

Sub-watershed 2		
S. no.	Village	Block
35.	Kutaura	Gursarai
36.	Maheba	Gursarai
37.	Maru Kachhiyau	Gursarai
38.	Pasaura	Gursarai
39.	Pucchi	Gursarai
40.	Turka Lahchura	Gursarai
41.	Bachera	Mau Ranipur
42.	Berwai	Mau Ranipur
43.	Bihata	Mau Ranipur
44.	Budhai	Mau Ranipur
45.	Chimadwara	Mau Ranipur
46.	Durgapur	Mau Ranipur
47.	Garhwan	Mau Ranipur
48.	Kakwara	Mau Ranipur
49.	Khan Pura	Mau Ranipur
50.	Kishor Pura	Mau Ranipur
51.	Rewan	Mau Ranipur
52.	Saptwara	Mau Ranipur
53.	Siaori	Mau Ranipur
54.	Siyawnikhurd	Mau Ranipur
55.	Jewar	Palera
56.	Mardanpura	Palera

Sub-watershed 3		
S. no.	Village	Block
1.	Amanpura	Bangara
2.	Bagroni Jagir	Bangara
3.	Baman Naiguwan	Bangara
4.	Bansar	Bangara
5.	Bhagaro	Bangara
6.	Bhitaura	Bangara
7.	Bijaigarh	Bangara
8.	Bijna	Bangara
9.	Budhawali	Bangara
10.	Charhrau Dhawari	Bangara
11.	Dadpura	Bangara
12.	Darbatyau	Bangara
13.	Dhonda	Bangara
14.	Ghatrajwara	Bangara
15.	Hanauta	Bangara
16.	Hati	Bangara
17.	Imilia	Bangara
18.	Khajraha	Bangara
19.	Lathesra	Bangara

(continued)

Sub-watershed 3		
S. no.	Village	Block
20.	Luhar Gaon Bhat	Bangara
21.	Luhari	Bangara
22.	Nimghana	Bangara
23.	Nota	Bangara
24.	Pahai Khurd	Bangara
25.	Pura Bujurg	Bangara
26.	Rajgir	Bangara
27.	Satpura	Bangara
28.	Sijara	Bangara
29.	Sijaura	Bangara
30.	Siguwan	Bangara
31.	Uldan	Bangara
32.	Ataniya Dehat	Gursarai
33.	Baswaha	Gursarai
34.	Bela	Gursarai
35.	Birsingh Pura	Gursarai
36.	Chhiraura buzurg	Gursarai
37.	Dabar	Gursarai
38.	Dhan Bilgaon	Gursarai
39.	Dhurwai	Gursarai
40.	Dugara	Gursarai
41.	Garaha	Gursarai
42.	Itawa	Gursarai
43.	Joniya	Gursarai
44.	Karri	Gursarai
45.	Lathawara	Gursarai
46.	Majhara Tai	Gursarai
47.	Mawai	Gursarai
48.	Pandwaha	Gursarai
49.	Rajapur	Gursarai
50.	Rajwara	Gursarai
51.	Semari Kachhiyan	Gursarai
52.	Silori	Gursarai
53.	Sujawan	Gursarai
54.	Tondi Fatehpur (NP)	Gursarai
55.	Tori Fatehpur Deh	Gursarai
56.	Bandh	Mau Ranipur
57.	Madarwas	Mau Ranipur
58.	Bamhori	Niwari
59.	Baraipura	Niwari
60.	Bhamora Khas	Niwari
61.	Chachawali	Niwari
62.	Dhamna	Niwari
63.	Dhimarpur (najumri)	Niwari
64.	Ghugsi Khas	Niwari

(continued)

Sub-watershed 3		
S. no.	Village	Block
65.	Gwavali	Niwari
66.	Jikhangaon	Niwari
67.	Kalothara	Niwari
68.	Majra Chachawali	Niwari
69.	Majra Patharam	Niwari
70.	Patharam	Niwari
71.	Tehar ka Khas	Niwari
72.	Taricharkalan (NP)	Niwari
73.	Uboura	Niwari
74.	Umri	Niwari
75.	Urdora	Niwari

Sub-watershed 4		
S. no.	Village	Block
1.	Atarsuwan	Bamour
2.	Bachheh	Bamour
3.	Baraura	Bamour
4.	Pureniya	Bamour
5.	Sarsenda	Bamour
6.	Atrauli	Gursarai
7.	Baghaira	Gursarai
8.	Bamanwan	Gursarai
9.	Banka Pahari	Gursarai
10.	Barwar	Gursarai
11.	Basari	Gursarai
12.	Bhadokhar	Gursarai
13.	Bhasneh	Gursarai
14.	Bitthaura	Gursarai
15.	Dhawari	Gursarai
16.	Garhi Kargaon	Gursarai
17.	Gata	Gursarai
18.	Ghuraiya	Gursarai
19.	Gundaha	Gursarai
20.	Itaura	Gursarai
21.	Jhala	Gursarai
22.	Khiriya	Gursarai
23.	Lohar Gaon	Gursarai
24.	Londi	Gursarai
25.	Madha Dilwali	Gursarai
26.	Madhura Pura	Gursarai
27.	Maigaon	Gursarai
28.	Pasrai	Gursarai
29.	Puratani	Gursarai
30.	Rana pura	Gursarai

(continued)

Sub-watershed 4		
S. no.	Village	Block
31.	Raniyara	Gursarai
32.	Tahrauli Kalan	Gursarai
33.	Tahrauli Khas	Gursarai
34.	Tai	Gursarai
35.	Tenduwa	Gursarai

Appendix D: Villages Falling Under Each Sub-watershed in Sajnam Watershed, Lalitpur District, Uttar Pradesh

Sub-watershed 1		
S. no.	Village	Block
1.	Arjun Khiriya	Mandwara
2.	Bacchra	Mandwara
3.	Badgana	Mandwara
4.	Banyana	Mandwara
5.	Bareja	Mandwara
6.	Daulatpur	Mandwara
7.	Digwar	Mandwara
8.	Dongra Kalan	Mandwara
9.	Jharaota	Mandwara
10.	Kakaruwa	Mandwara
11.	Khairai	Mandwara
12.	Lalitapur	Mandwara
13.	Machharka	Mandwara
14.	Madawara Range	Mandwara
15.	Maharaj Pura	Mandwara
16.	Mahrauni Range	Mandwara
17.	Makripur	Mandwara
18.	Narahat	Mandwara
19.	Padariya	Mandwara
20.	Parsai	Mandwara
21.	Patna Mahdawara	Mandwara
22.	Piyara	Mandwara
23.	Taraoli	Mandwara

Sub-watershed 2		
S. no.	Village	Block
1.	Bangariya	Birdha
2.	Barkhera	Birdha
3.	Barodiya Raen	Birdha

(continued)

Sub-watershed 2		
S. no.	Village	Block
4.	Betna	Birdha
5.	Dorna	Birdha
6.	Kewlari	Birdha
7.	Kirroda	Birdha
8.	Laltipur Range	Birdha
9.	Maholi	Birdha
10.	Mamda	Birdha
11.	Patauwa	Birdha
12.	Pipariya Donga	Birdha
13.	Riccha	Birdha
14.	Saja	Birdha
15.	Salaiya	Birdha
16.	Sukhpura	Birdha
17.	Umariya Dongra	Birdha
18.	Bamhori Sindwaha	Mandwara
19.	Gona	Mandwara
20.	Imilia Kalan	Mandwara
21.	Jamora	Mandwara
22.	Jilauni	Mandwara
23.	Naya Gaon	Mandwara
24.	Patna Sindwaha	Mandwara
25.	Sarkhadi	Mandwara

Sub-watershed 3		
S. no.	Village	Block
1.	Banoni	Birdha
2.	Bharoni	Birdha
3.	Dailwara	Birdha
4.	Daroni	Birdha
5.	Gangchari	Birdha
6.	Semaria	Birdha
7.	Andela	Birdha
8.	Bajarra	Birdha
9.	Bhadrau	Birdha
10.	Bhagwaha	Birdha
11.	Bhonrda	Birdha
12.	Birdha	Birdha
13.	Charhrau	Birdha
14.	Chhilla	Birdha
15.	Kalothara	Birdha
16.	Kalro	Birdha
17.	Karisa	Birdha
18.	Khajuriya	Birdha
19.	Khiriya Chhatara	Birdha

(continued)

Sub-watershed 3		
S. no.	Village	Block
20.	Khitwans	Birdha
21.	Kokata	Birdha
22.	Kuluwa	Birdha
23.	Menwar	Birdha
24.	Neem Khera	Birdha
25.	Pipriya Pali	Birdha
26.	Rameshra	Birdha
27.	Richh Pura	Birdha
28.	Satarwans	Birdha
29.	Sataura	Birdha
30.	Singepur	Birdha
31.	Tenga	Birdha
32.	Tikra Tiwari	Birdha
33.	Tila	Birdha
34.	Tor	Birdha
35.	Amora	Mahrauni
36.	Baryo	Mahrauni
37.	Chirola	Mahrauni
38.	Jariya	Mahrauni
39.	Kuraura	Mahrauni
40.	Naiguwan	Mahrauni
41.	Samogar	Mahrauni
42.	Silawan	Mahrauni
43.	Silawangrant	Mahrauni
44.	Sindwaha	Mahrauni
45.	Sukadi	Mahrauni
46.	Bagoni	Mandwara
47.	Bairwara	Mandwara
48.	Pathabiaipura	Mandwara

Sub-watershed 4		
S. no.	Village	Block
1.	Anor	Birdha
2.	Hansara Kalan	Birdha
3.	Jharkon	Birdha
5.	Kalyanpura	Birdha
7.	Patsemra	Birdha
9.	Talgaon	Birdha
4.	Jijyawan	Jakhaura
6.	Mirchwara	Jakhaura
8.	Raghunath Pura	Jakhaura

Sub-watershed 5		
S. no.	Village	Block
1.	Ajnora	Bar
2.	Badokhara	Bar
3.	Banpur	Bar
4.	Bar Range	Bar
5.	Billa	Bar
6.	Booti	Bar
7.	Chhilla	Bar
8.	Kuwagaon	Bar
9.	Mirchwara	Bar
10.	Pah	Bar
11.	Sunwahagran	Bar
12.	Suri Khurd	Bar
13.	Surikalan	Bar
14.	Talbehat Range	Bar
15.	Udaipura	Bar
16.	Udya	Bar
17.	Umari	Bar

Sub-watershed 6		
S. no.	Village	Block
1.	Bangra	Bar
2.	Bilata	Bar
3.	Dangrana	Bar
4.	Jaraoli	Bar
5.	Kailoni	Bar
6.	Kakdari	Bar
7.	Mogan	Bar
8.	Pura Dhadkuwa	Bar

Sub-watershed 7		
S. no.	Village	Block
1.	Bachhrawni	Bar
2.	Bahrawani	Bar
3.	Bar	Bar
4.	Basatrawan	Bar
5.	Bhawani	Bar
6.	Bhelonilodh	Bar
7.	Burogaon	Bar
8.	Chandawali	Bar
9.	Dashrara	Bar
10.	Dulawan	Bar
11.	Fadari	Bar

(continued)

Sub-watershed 7		
S. no.	Village	Block
12.	Gadyana	Bar
13.	Gahrao	Bar
14.	Gugarwara	Bar
15.	Karmai	Bar
16.	Kathwar	Bar
17.	Khaira	Bar
18.	Marroli	Bar
19.	Nagara	Bar
20.	Pata Pachaura	Bar
21.	Purapachuani	Bar
22.	Semara Bujurg	Bar
23.	Semrabhag Nagar	Bar
24.	Talbehat Range	Bar
25.	Teela	Bar
26.	Thatkhera	Bar
27.	Todi	Bar
28.	Toria	Bar
29.	Turka	Bar
30.	Birari	Birdha
31.	Kachnoda Kalan	Birdha
32.	Khokhra	Birdha
33.	Mailwara Kalan	Birdha
34.	Mailwara Khurd	Birdha
35.	Tera	Birdha

References

Agriculture Census (2011–2012) Department of Agriculture and Cooperation. Available at: http://inputsurvey.dacnet.nic.in/RDL/districttable7.aspx. Accessed 21 Jan 17

Annual Health Survey (2011–12) Factsheet Uttar Pradesh. Vital Statistics Division, Office of the Registrar General & Census Commissioner, India New Delhi. Available at: http://www.censusindia.gov.in/vital_statistics/AHSBulletins/AHS_Factsheets_2011_12/UP_Factsheet_2011-12.pdf. Accessed 20 June 17

Annual Health Survey (2011–2012) Factsheet. Madhya Pradesh. Vital Statistics Division, Office of the Registrar General & Census Commissioner, India New Delhi: Available at: http://www.censusindia.gov.in/vital_statistics/AHSBulletins/AHS_Factsheets_2011_12/MP_Factsheet_2011-12.pdf. Accessed 20 Oct 16

Annual Health Survey (2012–13) Factsheet. Madhya Pradesh. Vital Statistics Division, Office of the Registrar General & Census Commissioner, India New Delhi: Available at: http://www.censusindia.gov.in/vital_statistics/AHSBulletins/AHS_Factsheets_2011_12/MP_Factsheet_2011-12.pdf. Accessed 20 Oct 16

Bhisht H, Nair J, Gupta R (2014) Climate resilient development in Bundelkhand region of Madhya Pradesh: vulnerability and adaptation assessment. Development Alternatives, New Delhi. Available at: http://skmcccepco.mp.gov.in/sites/default/files/resources/Climate%20Resilient%20Development%20in%20Bundelkhand%20Region%20of%20MP%20Vulnerability%20and%20Adaptation%20Assessment.pdf. Accessed 12 Jan 17

Chandramauli (2016) Ensuring food security in a scenario of unrelenting drought in Bundelkhand region of Madhya Pradesh. International Journal of Multidisciplinary Approach and Studies 3(4):182–190

Choudary M, Amarnath HK (2012) An estimate of public expenditure on health in India. National Institute of Public Finance and Policy

Development Alternatives (2007) Water resources in Tikamgarh and Jhansi Districts- A status report. New Delhi: Development Alternatives. Available at: http://www.indiawaterportal.org/sites/indiawaterportal.org/files/water_resources_in_tikamgarh_and_jhansi_districts_a_status_report_development_alternatives_2007.pdf. Accessed 10 Feb 17

Directorate of Census Operations, Madhya Pradesh (2011a) District Census Handbook, Tikamgarh, Village and Town Directory, Series 24 Part XII-A. Delhi:Government of India

Directorate of Census Operations, Madhya Pradesh (2011b) District Census Handbook, Tikamgarh, Village and Town Wise Primary Census Abstract, Series 24 Part XII-B. New Delhi: Government of India.

Directorate of Census Operations, Madhya Pradesh (2011c) District Census Handbook, Chhattarpur, Village and Town Directory, Series 24 Part XII-A. Delhi: Government of India

Directorate of Census Operations, Madhya Pradesh (2011d) District Census Handbook, Chhattarpur, Village and Town Wise Primary Census Abstract, Series 24 Part XII-B. New Delhi: Government of India

Directorate of Census Operations, Uttar Pradesh (2011a) District Census Handbook, Jhansi, Village and Town Directory, Series 10 Part XII-A. Delhi: Government of India

Directorate of Census Operations, Uttar Pradesh (2011b) District Census Handbook, Jhansi, Village and Town Wise Primary Census Abstract, Series 10 Part XII-B. New Delhi: Government of India

Directorate of Census Operations, Uttar Pradesh (2011c) District Census Handbook, Lalitpur, Village and Town Directory, Series 10 Part XII-A. Delhi: Government of India

Directorate of Census Operations, Uttar Pradesh (2011d) District Census Handbook, Lalitpur, Village and Town Wise Primary Census Abstract, Series 24 Part XII-B. New Delhi: Government of India

Madhya Pradesh Council of Science and Technology (2011) Resource atlas of Tikamgarh: a Celebration of the land and people of Tikamgarh. MPCST, Bhopal and Centre for Policy Studies, Chennai

McCarthy JJ, Canziani OF, Leary NA, Dokken DJ, White KS (2001) IPCC, 2001; climate change: impacts, adaptation, and vulnerability, contribution of Working Group II to the third assessment report. Cambridge University Press, Cambridge, UK

Ministry of Agriculture, Department of Animal Husbandry, Dairying and Fisheries (2014) 19th livestock census – 2012 All India report. Government of India, New Delhi Available at: http://dahd.nic.in/sites/default/files/Livestock 5.pdf. Accessed 25 Nov 16

Ministry of Water Resources (2013a) District ground water information booklet. Ministry of Water Resources, Tikamgarh

Ministry of Water Resources (2013b) District ground water information booklet. Ministry of Water Resources, Chhatarpur

MP State Election Commission. Panchayat Election Results 2014–2019. Available at: http://mplocalelection.gov.in/ElectionResultsP.aspx/. Accessed 22 Jan 17

National Resource Cell for Decentralised District Planning. Backward region grant fund year 2007–2012, District Tikamgarh. Tikamgarh: Zila Panchayat Tikamgarh. Available at: http://www.nrcddp.org/District_Plans/Tikamgarh_BRGF_District_Plan_2007-12.pdf. Accessed 15 Mar 17

National Rural Drinking Water Programme (2017) Format B-21- Handpump details 2014–2015. Ministry of Drinking Water and Sanitation. Available at: http://indiawater.gov.in/IMISReports/Reports/BasicInformation/Rpt_RWS_NoOfHandPump_B.aspx?Rep=2&RP=Y. Accessed 23 Feb 17

Parry ML, Canziani OF, Palutikof JP, Linden P, Hanson C (2007) IPCC, 2007; climate change 2007: impacts, adaptation and vulnerability. Contribution of Working Group II to the fourth assessment report of the Intergovernmental Panel on Climate Change. Cambridge University Press, Cambridge, UK, p 976

Samra JS (2008) Report on drought mitigation strategy for Bundelkhand region of Uttar Pradesh and Madhya Pradesh. New Delhi: Inter-ministerial team. Available at: http://nraa.gov.in/pdf/drought%20mitigation%20strategy%20for%20bundelkhand.pdf. Accessed 25 Mar 17

Songhua U (2014) District disaster management plan [DDMP] Tikamgarh (Madhya Pradesh). Kanpur: Indian institute of technology. Available at: http://tikamgarh.nic.in/docs/dmp.pdf. Accessed 12 Oct 17

Sullivan C, Meigh JR, Fediw TS (2002) Derivation and testing of the water poverty index phase 1, Final report. UK. Department for International Development

Turner BL, Kasperson RE, Matson PA, McCarthy JJ, Corell RW et al (2003) A framework for vulnerability analysis in sustainability science. Proc Natl Acad Sci 100:8074–8079

Websites

Krishi Vigyan Kendra Tikamgarh (MP). Available at: http://kvktikamgarhjnkvv.nic.in/. Accessed 7 Nov 17

Krishi Vigyan Kendra Chhatarpur (MP). Available at: http://kvkchhatarpurjnkvv.nic.in/index.html. Accessed 7 Nov 17

Krishi Vigyan Kendra Jhansi (UP). Available at: http://jhansi.kvk4.in/. Accessed 7 Nov 17

Krishi Vigyan Kendra Lalitpur (UP). Available at: http://lalitpur.kvk4.in/. Accessed 7 Nov 17

Soil and Land Use Survey of India (SLUSI). Available at: http://slusi.dacnet.nic.in/. Accessed 7 Nov 17

Strategies for Scaling Up the Adoption of Organic Farming Towards Building Climate Change Resilient Communities

4

Teodoro C. Mendoza, Roselyn Furoc-Paelmo, Hazel Anne Makahiya, and Bernadette C. Mendoza

Abstract

Adjustments and adaptive responses to diminishing resources (land, water, and energy) in agriculture due to population increase and climate change in the recent decades are varied. Proactive adaptive coping mechanisms must be instituted to avoid the onslaught of massive starvation. Organic and agroecological innovations are the logical options. But organic farming is not one-size-fits-all solution. While organic farming is considered as one of the solutions to farming in crisis, there are many barriers to its adoption. Among these constraints are (1) the nature of organic farming being difficult, laborious, and knowledge and skills intensive, the required environment (air, soil, and water), and the certification requirement and (2) the support systems from government and consumers not in place.

Scaling up the adoption of organic farming has a number of prerequisites, specifically:

1. innovation from farmers—the farmers as innovators and scientist/technologists from the academics and science and technology (S/T) institutions;
2. reengineering agri-food systems into agroecotourism as a way of attracting farm visitors and tourist-enthusiasts and attracting human interests and investment flows to the rural areas, generating rural employment, slowing down or stopping out-migration to urban areas and overseas work (OFW);
3. innovative governance-led promotion by expediting the shift from capital and resource intensive (land, water, energy, inputs) to restorative, regenerative, and vibrant agriculture and food systems and expediting this system shift by an

T. C. Mendoza (✉) · R. Furoc-Paelmo · H. A. Makahiya
Institute of Crop Science, College of Agriculture and Food Science,
University of the Philippines Los Baños (UPLB), College, Laguna, Philippines

B. C. Mendoza
Institute of Biological Sciences, College of Arts and Sciences,
University of the Philippines Los Baños (UPLB), College, Laguna, Philippines

© Springer Nature Singapore Pte Ltd. 2020

V. Venkatramanan et al. (eds.), *Global Climate Change and Environmental Policy*,
https://doi.org/10.1007/978-981-13-9570-3_4

innovative ecological carbon emission–soil erosion–water consumption tax to finance the transition and conversion process to agroecology-based organic agriculture;

4. an innovative paradigm shift from food security to health security—from financesurance to healthsurance, from financial banking to health banking, from measuring yield per acre to health per acre as the world transitions agriculture and food system from agrochemical-intensive monoculture to organic polyculture cropping systems;
5. innovative paradigm from supply chain to value chain approach in agriculture and food systems, but implementing these innovations requires 4Ps and 2 Ms (preproduction, production, processing, postproduction linkages + marketing and management);
6. a demand-led (consumer) instead of supply-led (the farmers) approach to promotion;
7. and, finally, a consumption-led greening of agroecosystems by minimizing food wastes and consuming only what we can and reducing the thermodynamic loss in food by consuming less and less meat.

Keywords
Climate resilience · Innovation · Organic farming · Demand led · Value chain approach · Agroecotourism · Health per acre · Health banking

4.1 Introduction

In the early part of the twentieth century, chemical farming was the technological *innovation*[1] which avoided the Malthusian prognosis of massive food shortage, and it was claimed to have saved more than 1 billion people from starvation. Chemical farming started from the developed-industrialized North. Nonetheless with the recognition that chemical agriculture is unsustainable brought its subsequent change into organic farming in the developed-industrial North as early as the 1930s. Organic farming became more popular in the 1960s when Rachel Carlson published her book *Silent Spring* in 1962. Meanwhile, the International Federation of Organic Agriculture Movements (IFOAM) was organized in the 1970s, which further strengthen the promotion and adoption of organic farming.

The world population shall reach 9.8 billion by 2050. In view of this, the food supply should increase by 80%. In Africa, with a population of 1.8 billion by 2050, their food production must be doubled. However, they are confronted with the challenges of declining soil fertility, decreasing farm yard manure supply as a

[1] "Innovation is improving on or makes a significant contribution to an invention or existing product, process or service. Invention is the creation of a product or introduction of a process for the first time" (Grasty 2012).

consequence of declining population of livestock, escalating fertilizer prices, and increasing risks of droughts due to climate change. These alarming scenario deemed necessary that *out-of-the-box* approaches must be in place to avoid the onslaught of the "perfect storm" of massive starvation. "Organic and agroecological solutions are of urgent necessity to feed the world" as claimed by Cook et al. (2016). Organic, agroecological, and regenerated agriculture is the way forward. It is the agricultural systems of this millennium and beyond. Organic farming as an innovation includes the following features: *Biodiverse farms promote ecological balance and protect the environment, sustainable land management, environment- and health-friendly food production systems* yielding *safe and nutrition-healthy foods, sustain food sufficiency and food sustainability, decrease carbon footprint, and produce more with fewer inputs in a vibrant agricultural systems.*

But globally, regionally, or locally (the Philippines), there is the low adoption of organic farming. The total agricultural lands under cultivation or utilization for crops and livestock are about 1.6 billion hectares. Farmed organically are 50.9 M ha in 179 countries or 3.18% of the 1.6 billion ha agricultural lands (Willer and Lernoud 2017) cultivated by more than 570 million farmers. About two-thirds of 2.4 million organic farmers are in the developing countries. More than 90% of farms are run by an individual or a family who are primarily relying on family labor. Family farms occupy a large share of the world's agricultural land and produce about 80% of the world's food (https://www.globalagriculture.org/report-topics/industrial-agriculture-and-small-scale-farming.html). In the Asian region (ASEAN, including the Philippines) where farmers are dominated by small farms ranging from 0.5 to 3.0 ha (almost 90%), farms are cultivated using externally sourced synthetic inputs usually bought from developed countries.

The comparison is odious, but genetically modified organisms (GMO) being domesticated in farms started in the 1990s. Recent reports indicate hectarage devoted to genetically engineered corn, soybean, cotton, and canola is about 181.48 million ha (11.34% of the total agricultural lands) (CBAN 2015) or approximately 3.56 times larger than the organic farms. Adoption per decade is 60 M ha on the average, while only 10.18 M ha for the organic farms. The reckoning years were the 1970s or five decades for organic farms and 1990s for GMO farms. Despite the known merits and benefits of organic farming over the agrochemical-intensive or chemical farming, there is a low rate of adoption which explains the smaller hectarage devoted to it.

As a case study, growing rice (*Oryza sativa*) through organic farming method was four times more energy efficient than the conventional method. The agrochemical input (fertilizer/pesticides) accounted for 61% of the fossil fuel–based energy inputs and 84% of the cash cost of production in the conventional system. For every 1 calorie of fossil fuel energy used in the conventional farm, only 4 calories are produced, while from the organic farm, it is 19 calories. Organic farms were less energy consuming. One ton of paddy rice utilized only 170 of fossil fuel–based inputs (FFEI), while 844 M calories in the conventional farms (Mendoza 2004). Producing rice organically was the least energy intensive (lowest energy footprint), cheapest to produce, safe to grow (non-use of chemical pesticides; farm families are hit first), and safe to eat. Moreover, it relaxed the farmers, mainly the housewives who shoulder the burden of where to get money to buy all inputs.

Many farmers are heavily indebted due to crop failure (pest, floods, super typhoons, etc.). Comprehensive crop insurance is yet to be formulated/legislated in the Philippines. Recently, chemically grown rice and rice products were detected to have a high level of arsenic[2] (U.S. FDA, 2016 http://www.fda.gov/food/foodborneillnesscontaminants/metals/ucm319870.htm). The recent surge in rice price in the Philippines clearly indicates that farming is in crisis (*price increase of 10–15 pesos/kg in some areas declared as calamity area, justifying the use of their calamity funds by the local officials*). Farmers who grow rice have no rice to eat. Severely hit by price jump are those in the rural areas. Farmers have sold already their rice even before its planting. Only 3–4 farmers out of 100 have rice to eat before the next season of harvest. But organic farmers had improved their living conditions. In fact, they have adequate rice to eat, they are no longer indebted, they are healthier, and they are able to send their children to school (Mendoza 2004).

4.2 Diagnosing the Low Adoption of Organic Farming

Organic farming can be dubbed as a game changer, but its adoption is considerably low. The two main clustered reasons that need to be addressed are as follows.

4.2.1 Organic Farming (OF) Is Difficult, Laborious, and Knowledge and Skill Intensive

More so, it requires more patience, perseverance, and highly positive attitude. Conventional/chemical farming is easy and less laborious farming. Key features of organic farming are the following:

- Use of compost/organic fertilizer as a replacement for synthetic chemical fertilizer (it is very laborious and time consuming to prepare compost).
- Weed control through manual weeding (mechanical use of rotary weeder in rice which is labor intensive and price of labor is high). It is easier, faster, and cheaper to use herbicides in chemical farming.
- As regards the pest management, cultural method and mechanical control demand cost-intensive skilled labor force. Most often, timely availability of skilled labor is a challenge.

[2] "Arsenic exerts its toxicity by inactivating up to 200 enzymes, especially those involved in cellular energy pathways and DNA synthesis and repair. Acute arsenic poisoning is associated initially with nausea, vomiting, abdominal pain, and severe diarrhea. Encephalopathy and peripheral neuropathy are reported. Chronic arsenic toxicity results in multisystem disease. Arsenic is a well-documented human carcinogen affecting numerous organs. There are no evidence based treatment regimens to treat chronic arsenic poisoning but antioxidants have been advocated, though benefit is not proven" (Ratnaike 2003).

- To avoid field contamination, irrigation water must be purified (separate/dedicated water pond for purification).
- Planting of tree barriers around the farm perimeter to prevent wind-blown pesticides, when neighboring farmers spray pesticides.

4.2.2 The Support Systems for Organic Farming Adoption Are Not in Place

Science and technology (S/T) institutions and the academic incentives and rewards scheme are built around and are supportive of the chemical industry. In a publish-or-perish environment in the academic institutions, S/T funding accessibility is more for the chemical industries who can afford to pay higher salaries and grant more research funds for chemical farming. On extension program, "extension" is the continuation of S/T or technology generation. Blame not the extension agents skillful in extending chemical farming. But the most effective/efficient/skillful extensionists are the employees or sales representatives of chemical companies who are provided with excellent logistics for mobility, plus attractive salaries. In return, they are expected to reach sales quota or they will be reprimanded or retrenched for their ineffectivity or inefficiency. In the Philippines, the extension bureau of the government was abolished, and it was replaced by a training institute, unfortunately, primarily extending more of the chemical agriculture!

4.3 Innovation as a *Game Changer* in Scaling Up the Adoption of Organic Farming

Providing answers or strategies to scale up the adoption of organic agriculture requires understanding of the main actors or sectors involved. There are at least three main groups or sectors that play important interrelated roles in the accelerated adoption of organic farming as an innovation over conventional farming.

4.3.1 The Farmers

Organic farmers are innovative and intelligent. They are fearless as they have transcended the "fear of the unknown." They can be briefly described as "philosophers." In fact, organic farming is a philosophy! Love of knowledge is a key attitude. Organic farmers are knowledge seekers! Hence, ***innovator!*** Innovation is the ***game changer***! Organic farms are an innovation power house! Innovation translated into action indeed would yield change towards development and positive growth not only in the locality as well as transcending beyond political boundaries. Successful farmers who are into organic farming are innovators across the value chain. A number of farmers are proactively innovating across the value chain to create change positively influencing growth and development. They believe in the dictum "the quantity being demanded is the quantity to be produced." Among the documented innovations in the Philippines are as follows:

- Farmers select their own seeds that are locally adapted and with target market, mostly colored rice (red, black rice) as demanded by consumers.
- Scheduling land preparation in such a manner that sufficient lead time is provided for crop/weed residues to decompose, that is, 3–4 weeks before transplanting.
- Planting at the right time so as to decrease pests and diseases infestation.
- Planting three varietal types for early, regular, and late maturing varieties for sugarcane (*Saccharum officinarum*).
- Farmers know agronomy of yield. The farmers know when and how to plant the seed (locally adapted, G X E interaction) by season (wet, dry season) and location (soil type, topography) to obtain the genetic potentials of high yield of the seed (inbred or hybrids) as shown in the agronomy of yield in equation (Eq. 4.1)
- Implementing soil quality improvement practices such as crop/weed residue recycling through its decomposition after harvesting by spreading it evenly on the farm.
- Sugarcane planters avoid burning sugarcane trash and either perform trash shredding through a mechanical shredder or incorporate into the soil by repeatedly passing tractor drawn disc harrows.
- Farmers prepare their own compost/organic fertilizer and other preparations, thus saving a lot in terms of labor and cost of production.
- Farmers in Cagayan de Oro of the Mindanao island process their organically grown black rice which is best known for its health benefits and export market.
- Cacao farmers in Kidapawan City, North Cotabato, through processing of cacao beans into tablea, increased their income as the price of cacao bean augmented from PHP 120/kg to PHP 600/kg.
- Coffee farmers in Batangas are grinding their coffee bean and selling as kapeng barako at PHP 216/kg. On the other hand, dried coffee beans are sold only at PHP 80/kg.
- Many organic farmers do direct marketing of their farm produce and some put up their own restaurant to further increase the value of their organic farm produce through "fine dining" at PHP 1200–PHP 1800 per head.

4.3.2 Researcher's Contributions to Innovation

Yield is construed as a function of technology/innovation, use of appropriate tools/ implements, and appropriate and favorable growth environment (Eq. 4.1):

$$Y = G + E + \{(G * E) * M\} \tag{4.1}$$

where Y is yield, G is genotype (variety), E is environment (climate, soil factor), and M is management (inputs applied, cultural practices, that is, land preparation, planting, cultivation and weeding, fertilizer application practices, irrigation, harvesting/milling practices).

Innovations in organic farming aim at internalizing the inputs and using less of external inputs. Organic farmers driven by their knowledge and skills endeavor to innovate in the domain of breathing, seed production, production and use of biological control agents and botanicals, and marketing of organic farm produce.

An innovation in organic rice is planting one seedling per hill and planting in double rows spaced 20 × 10 cm (the double rows) and 40 cm interval between the double rows, using only 6 kg seeds per ha or 4 kg of seeds at 30 cm × 30 cm spacing, fertilized with bokashi and liquid manure, using rotary weeder in controlling the weeds. In Bokong, Labangan, Zamboanga del Sur, Philippines, rice-growing conditions, NSIC Rc222 (an inbred) had the highest yield at 6.07 t ha^{-1} outyielding Mestizo 19 (a hybrid) at 5.37 t ha^{-1}. Furthermore, since organically grown rice sold at 20% higher price as compared to the conventional grown price, organic rice growers earn an additional income (Lao-Ay et al. 2017).

Double-row rice planting using high-yielding location adapted inbred rice cultivars can be game changer in organic rice farming, provided (1) mechanical transplanter can be innovated from the original one-row transplanter to double-row transplanter, (2) innovated motorized rotary weeder for double-row transplanted rice, and (3) innovated-solarized submersible pump to apply liquid manure fertilizer (Lao-Ay et al. 2017). In an earlier trial, as high as 8.5 t ha^{-1} was obtained. PSB Rc18 was planted in double rows at a spacing of 20 × 10 cm and 40 cm interval between rows during the dry season (fully irrigated) at one seedling per hill, keeping the field moist, weeded by using rotary weeder, and fertilized using bokashi and amplified liquid manure fertilizer three times (Fig. 4.1) (Mendoza 2016).

Our expectation is that if we can stabilize the yield of organically grown rice at 8.5 t ha^{-1} instead of the usual 4–5 t ha^{-1}, we can convince many farmers to innovate or shift to organic rice production. In Bay, Laguna, the average yield of conventionally grown rice is only 5 t ha^{-1}. It is laborious to prepare and plant one seedling per hill, prepare and apply bokashi organic fertilizer and amplified liquid cow manure, and do the rotary weeding. At 8.5 t ha^{-1} yield level, farms workers can be paid with a competitive wage at PHP 350 per day or even higher.

Furthermore, if rice can be milled and sold as brown rice (72% milling recovery and priced at PHP 80/kg) and at 8.5 t ha^{-1} and 72% milling recovery (MR), the gross income is:

$$\text{Gross Income}(\text{GI}) = 8.5\text{t}\,\text{ha}^{-1} \times 0.72 \times \text{PHP}80/\text{kg} = (6120\,\text{kg})(\text{PHP}80/\text{kg}) = \text{PHP}489{,}600/\text{ha}$$

$$\text{Net Income}(\text{NI}) = \text{GI} - \text{CP}(\text{costs of production})$$

$$\text{NI} = \text{PHP}489{,}600 - \text{PHP}128{,}184 = \text{PHP}360{,}516/\text{ha}$$

The gross income and net income are PHP 489,600/ha and PHP 360,516/ha, respectively. With net income of PHP 360,000 ha^{-1}, many rice farmers are motivated to innovate in organic rice production.

Fig. 4.1 Double-row planting pattern (20 × 10 cm to 40 cm) at various growth stages in Bay, Laguna, Philippines

As organic farming is labor intensive and difficult, innovations, as the game changer, are necessary. At the right price or competitive wage, rural labor shall be plenty and readily available. Making farm operations easier and quicker are twin goals in farming either organic or chemical farming. This is where innovations are necessary.

- Transplanting is labor intensive, and there is dire need for mechanization. Since mechanical transplanters are expensive and are unaffordable to capital-scarce small-scale farmers, they should be made available through service providers or through government support.
- Manually operated rotary weeders should be converted into motor-driven weeders like drone so as to increase the ease of operation.
- Sourcing of raw materials like molasses, poultry, hog manure, or cattle manure for organic fertilizer production must be facilitated so as to increase the adoption of organic farming. Applying amplified liquid fertilizer can be facilitated using submersible pumps.
- Each farmer must own one to two cattle/carabao for draft power and as a source of manure for bokashi fertilizer and amplified liquid manure.
- Ducklings for integrated rice + duck farming must also be provided. On-farm upgrading/mixed breeding should be done to avoid inbreeding which is considered as one of the major reasons for the decline of egg-laying capacity of ducks.
- Seeds/planting materials (vegetatively propagated fruit trees) must be provided.

Adoption of innovation by farmers must be facilitated and be fully supported. It was not them—*the farmers*—who started to shift to agrochemical-intensive agriculture (Mendoza and Villegas 2014). Why should they be the only one now to shoulder all the burden of innovation to organic agriculture? This leads to the scrutiny of the approach in making food available to the consumers.

4.3.3 The Consumers: What Is the *Legal* Push to Organic Farming as an Innovation?

In the Philippines, even the enactment of a law, R.A. 10068, "The Organic Agriculture Act" of 2010, did not accelerate its dissemination and adoption among farmers. Consumers must realize that they exist at the core of the food systems. They determine the demand! They must be involved and be conscious of the real cost of modern agriculture or how this genetically modified organism (GMO)-based Green Revolution and the food produced out of it impact our planet's ecosystem. They should recognize the need and value of nutritional and medicinal food. In view of this, food should not be solely obtained through the chemical-based production system. As stated earlier, "Let thy food be thy medicine." This should be translated into a new demand that will lead to changes in the supply side. Consumers must directly support the plight and welfare of farmers who operate in organic and agro-ecological systems. This joint consumer–farmer partnership must be fully translated into their willingness to buy organically grown products, that is, direct product users such as hotels, restaurants, caterers, food processors, and other institutional buyers investing and buying their daily/weekly food needs or meeting their raw material requirements at a fair price.

From the past up to the present, organic farming has been promoted in the country wherein the focus has been on the farmers, "the supply side." Earlier, Mendoza (2004) had proposed demand-led promotion of "organic agriculture." "The product being demanded is the product to be produced!" The "product being demanded" refers to the consumers. In Europe, the USA, and other developed countries, the consumers are the main drivers in the adoption of organic agriculture. This explains the fact that two-thirds of lands devoted to organic agriculture are in the developed economies. There is an increasing number of middle class in the developing economies who look for organic–safe–healthy foods, but their purchasing power is not yet felt by the organic farmers.

After five decades or more after the "Green Revolution," how many organic agriculture (OA) converts does the Philippines have? The Farmer–Scientist Partnership for Development, Inc. (MASIPAG), a Philippine NGO composed of farmers, scientists, and peoples' organizations, estimates that there are approximately 30,000 Filipino organic rice farmers (http://www.masipag.org/news_india.htm). The CIA Factbook (https://www.cia.gov/cia/publications/factbook/geos/rp.html#Econ) estimates that 36% of the approximately 35,790,000-strong Filipino labor force is in the agricultural sector (12,884,400). Numerically, 30,000 is a lot, but proportionally their numbers are a miniscule (0.23% or 23 out of 10,000 farmers). The theory being advanced is that the promotion of organic agriculture must be focused on the supply side of the supply and demand curve. There must be an alternative approach, the demand-led approach, to the promotion of organic agriculture. The consumers comprise the demand side of the production to postproduction linkage. Farmers follow the economic logic in production, whereby the consumers' demand are being produced by the farmers. Following this logic, if consumers demand chemical-free agricultural products, then farmers shall simply follow that signal. As proposed by Mendoza (2010), demand can be interpreted in a number of ways as follows:

1. Consumers must be willing to support farmers in the production of chemical-free products.
2. Consumers must be willing to pay a premium. Consumers' willingness to support the farmers in the production of chemical-free products as can be done in several ways as follows:

Consumers can visit and help motivate farmers to grow crops and animals the organic way. Since organic production systems are different from the agrochemical-dependent systems, consumers must also be familiar with the organic production system. As regards the "Consumers must be willing to pay a premium price for organic products," it is indeed a big issue for organic products. Consumers in the Third World countries already consider current food prices to be too high. Approximately 85% of the Philippine population lives on less than US$2.00 per day, and more than 51% of the rural population lives below the subsistence threshold as defined by the World Bank (http://www.masipag.org/news_india.htm). The government's average mandated minimum wage is PHP 400 (US$7.400) per day. The current retail price of ordinary rice in the wet markets ranges from PHP 40 to PHP 65/kg (US$0.73–US$0.94/kg). Supermarket retail prices of organic rice range from PHP 75/kg (US$1.38/kg) for ordinary varieties to PHP 150 (US$2.77/kg) for fancy varieties (red, black, glutinous, or aromatic rice). But the current high price of organic rice retards the growth in consumption and demand for organic rice. Consumption of organic rice is thus limited to those who can really afford to pay—well-off cancer patients who are advised to eat organic products; those who have undergone heart surgery; and the few environmental and health conscious sectors of the society who can afford it.

Why pay a high price? There is a need to clarify what consumers are paying for. What the consumers are simply paying if they buy chemically grown crops is the financial price. It does not truly reflect the true value of the product since all the costs of production are not included. There were no total costs accounting. The total costs should include (1) financial—the costs of purchased inputs, such as seeds, fertilizer, pesticides, fuel, machineries, cost of money, labor, storage, packaging, marketing, and distribution), and (2) ecological—soil quality deterioration due to the inputs and farming methods applied and all other environmental and ecological costs. What is paid for is simply a small fraction of the total cost. David Gould showed the true cost accounting of Big Mac to be $ 12.00 (health care—$5.69; subsidies—$0.70; environment—$0.67; cruelty—$0.38; retail price—$4.56;

total—$12.00). This means that the current market price is so low because government subsidies and the ecological costs of raising beef are not included. It means that future generations will pay dearly for these unseen costs. Even now, we are already starting to pay the price as reflected in the rise of lifestyle-related illnesses and global warming.

With pricing parity based on the true or total cost accounting (financial + ecological), there shall be a fair price in the marketing of organically grown products. Conventionally grown products are grossly underpriced or even incorrectly priced. Their price tags are way below actual costs if the true costs of production and a reasonable profit margin are included. Because of this, the market price of organic agriculture products appears to be more expensive as they are generally priced 20–30% higher. If the true price tags of conventionally grown crops and animals are to be considered, then organic agriculture products are sold at a considerably low price. But the general consuming public who are already financially hard up will not understand this logic. What they would appreciate, considering their current shrinking purchasing power, is the financially low price of products that they buy in the market. In effect, what is being presented is that the 20–30% higher price of OA products is not really high or a premium price after all. Why?

Organically grown vegetables have higher quality and higher nutritional value with more vitamins and lower water content. Thus, they keep longer (they do not wilt) even at ordinary room temperature, and they taste better, in fresh salads or in cooked form. Organically grown rice tastes better and stores longer. A common observation is that cooked organic rice does not spoil in 24 h.

An innovative policy support to organic agriculture is an ecological tax. It should be initiated by environmentalists and organic agriculture advocates. The study of Landicho et al. (2014) validated the farmers' perception of the benefits of producing and consuming organic produce. The results gathered are substantiating that awareness is being created at the grassroots level. Specifically, farmers themselves elucidated that organic farming practices are environment-friendly, minimizing air pollution, greenhouse gases emissions, thereby contributing as well to a healthy society. The healthy society is rooted in healthy soils with crops grown organically. This ensures safe, nutritious, and healthy foods both for the farmers and the consumers. Farmer-respondents also verified that organic farming resulted in increased income due to reduced costs of production and reduced risk of crop failures. From a wider perspective, organic farming practices increase the valuation of the global ecosystem services, namely, provisioning, regulating, and supporting (Fan et al. 2016).

4.3.4 Governance/Government Sector

Organic agriculture advocates/practitioners in the Philippines were happy when the law on organic farming (R.A.10068) was enacted into a law in 2010. There were two major impacts: (1) though small, there is now regular funding for R/D for organic agriculture through the General Appropriations Act (GAA). There are now many researchers/scientists doing agroecology/organic agriculture research for major crops including livestock (poultry, hogs, cattle fattening); (2) it removed the stigma, the fear of doing research in organic farming.

Having a law and having it implemented are poles apart. Our decade old law, in general, did not "scale up" or accelerate the adoption of organic farming in the Philippines considering the current number of organic farmers. Massive adoption needs "proverbial shot in the arm." Many farmers are ageing (57 years old and above). You cannot teach old dogs new tricks! As our national hero, Dr. Jose P. Rizal, said before, "the youth is the hope of the future." Adopting that statement of our national hero by the current "old" people is relinquishing their responsibility in favor of the young people. Cook et al. (2016) are correct in titling their paper *Organic and agroecological solutions! Farming for the future. It must be now!*

In relation to governance (policy support—legislative and executive), there are many policies that are antagonistic, nonsupportive, or directly prohibitive to the promotion of ecologically sound agricultural practices. However, there is much elbow room to "put the house in order" at the local level due to the local autonomy code. This flexibility could be incorporated as follows:

- A barangay/municipal ordinance could be enacted to fully enforce existing laws, especially in prohibiting the burning of crop residues and the dumping of hog/poultry manure into rivers and streams. This is rational particularly that anti-burning laws are already stipulated in the Philippines under Republic Act (RA) 9003 (Ecological Solid Waste Management) and RA 8749 (Clean Air Act).
- The municipal agricultural officer, now directly under the supervision of the town mayor, monitors compliance with existing laws and supervises the strict implementation of environmental regulations and ordinances. A barangay/municipal ordinance will promote the implementation of zero waste management. Biodegradable wastes and materials at home and in the community can be collected, segregated, and made into compost to serve as cheap biofertilizer for farmers. There are towns/cities that do these good practices, but they are still few.
- A barangay/municipal ordinance regarding stray animals (dogs, goats, hogs, cattle, and carabao) must be passed. In one village, farmers cannot plant mangoes or fruit trees in their farm because of the practice of open grazing (goats, cattle, and carabao) after the harvest of rainfed rice. Not all barangays or towns cities have ordinances prohibiting roaming animals and pets.
- Low enforcement, marginal funding, and irrelevance of existing rules and regulations add to the problem. There is an urgent need for strict enforcement of and rationalized funding for all existing agricultural and environmental laws and regulations. There is an urgent need to revisit the implementing rules and regulations of existing laws and even repeal irrelevant laws and regulations to make them attuned to the present situation and circumstances.

For example, there is a need to reform the budgeting system of the Department of Agriculture (DA) in accordance with the Agriculture and Fisheries Modernization Act of 1997 (RA 8435, as amended). Funding must be focused on well-defined and well-characterized area-based and river basin–directed agriculture and fishery development zones under agrobased industrial clustering schemes. The DA must stop the ineffective and disjointed commodity–based and function-oriented budgeting that promotes mono-cropping or mono-enterprise development and scatters rural development initiatives.

4.4 Innovative Paradigm: Supply Chain to Value Chain Approach in the Agriculture and Food Systems

An agroecology-based organic agriculture should undergo metamorphosis. Implementing *innovations* require 4Ps and 2Ms (preproduction, production, processing, postproduction linkages) + marketing and management or value chain approach (Villegas 2018). A supply chain approach must be a tinge of the past and be replaced with a value chain approach. Value adding product must be identified; farmers should be capacitated and be supported to reap the benefits of the value adding/ financially enriching part of the food systems. The challenge is helping farmers achieve the benefits accruing across the "value chain." It will spell the differences between the past (twentieth century) to the present—the twenty-first century. The twentieth-century supply chain approach made our farmers to simply produce/supply raw materials (feedstock) to the processor/manufacturer of finished/processed high-value products. It should now metamorphose into "value chain approach" or put under the control and ownerships of the farmer/producers the full interrelated, interactive, interdependent process in processing the products so any added value could accrue to them. But this also requires asset reform, particularly, land asset. Permanent structures are necessary; investment in farm tools, processing equipment, etc., necessitate that farmers have long-term occupancy right on land or even secured land tenure as the basic asset.

A value chain assessment of six-fruit tree—intercrops under coconut (*Cocos nucifera*)—was done (Mendoza et al. 2016). Income from coconut alone is very low. But coconut is providing several ecosystems services, a microclimate modifier giving "shading effect" favorable to understory fruit trees, namely, coffee (*Coffea* sp.), cacao (*Theobroma cacao*), mangosteen (*Garcinia mangostana*), durian (*Durio zibethinus*), banana (*Musa* sp.), and even rubber *(Hevea brasiliensis)* trees. Intercropping increased the net revenue using monoculture coco–copra revenue as base revenue at P 18,251/ha as follows: coconut + banana increase the revenues 12.56 X; coconut + mango (*Mangifera indica)* 9.37X; coconut + durian 7.83X; coconut + mangosteen 5.34X; coconut + cacao 2.83X; and coconut + coffee 2.36X. Income increases progressively as the product move across the value chain. Revenues multiply 10 times (coffee, cacao) and as high as 32 times in coconut + mango before the insect and disease problems set-in. However, it is imprecise to declare one crop is profitable over the others. Cacao dried beans sell higher at PHP 120/kg than coffee bean at PHP 85/kg. When Starbucks sells coffee, the gross earnings could go as high as PHP 8.4 million worth of coffee ha^{-1}. Cacao made into chocolate (too many brands, trade names are available) could yield million (more than PHP 2 million ha^{-1}). For durian and mangosteen, farm gate sales of mangosteen and/or durian are relatively low. But retailing them increases income 2.15 times (durian) to 3.5 times (mangosteen).

In addition to asset reform as mentioned above, the central feature of the value chain approach is achieving the scale of operations by organizing farmers into cooperative. Implementing a value chain approach requires many services and technology providers/group, credit institutions (banks), granting reasonable and affordable interest rates since above 6% interest rate and consequent interest expense alone will eat up the added profit due to increased yield and income. Due to risks, credit must be insured (crop insurance that pays the crop value and not simply the credit value). Promoting agriculture value chain comprises a life cycle approach. It should equally emphasize the need to produce quality raw materials (feedstock) to feed the factory machines along with processing technology setup. Research and development (R&D) institutions (academics, S/T, or R& D) institutions should help/assist the farmers on their technology needs from production, processing, packaging, up to marketing. Constructing factory needs capital (amortization + interest), skilled labor (training), professional management (commitment and dedication), tools and equipment (locally fabricated), hauling trucks (logistics), and all weather road networks that are of high priority.

4.5 Agroecotourism Is Fast-Developing Farming System Options

Agro-tourism is "the form of tourism which capitalizes on rural culture as a tourist attraction. It has gained a new dimension as a potential income and job generating activity. The symbiosis between tourism and agriculture that can be found in agrotourism is a key element of an environmentally and socially responsible tourism." Agroecotourism offers an opportunity to experience the real enchanting and authentic contact with the rural life, taste the local genuine food, and get familiar with the various farming tasks. It is perhaps *"provides a welcome escape from the daily hectic life in the peaceful rural environment and to relax and revitalize in the pure natural environment, surrounded by magnificent landscape. Agro-eco-tourism can contribute towards a green economy transition through investments leading to energy and water efficiency, waste reduction, biodiversity and cultural heritage conservation, and the strengthening of linkages with local communities"* (Barbuddhe and Singh 2014).

Agroecotourism in the Philippines (Costales Farm, Quezon; Penalosa farm in Victorias, Negros Occidental) increased the market value of their organic produce through their package tour (plus the foods and snacks)—as a training venue, as a direct marketing for their produce, when the agroecotourists go home, they buy organic produce including seedling. As regards the case examples, an entrepreneur in Victorias, Negros Occidental, earns 70% of his total income from entrance fees of the tourists who visit his farm arriving on a bus load. To accommodate overnight visits or several days' stay, he built hotel and restaurants. But the organic foods served are all harvested from the farm. Farmers practice total quality management (TQM). They make sure that their partner-workers are happy and motivated and practice benefit sharing. Farmers are encouraged to raise native chicken and pigs, providing them the breeds, raising seedlings from fruit trees, and sales are shared equally (50–50 after deducting the initial capital).

4.6 Organic Agriculture and Health Security: An Innovative Paradigm

When we talk about agriculture, we always talk about food security, farm income, the business around it, and the life industry complex built around it. Rarely, we consider agriculture and health security or extending it further into agriculture and life security.

Agriculture is life or agriculture, ultimately, is all about health! Health is wealth! No doctor can treat death. Doctor of medicine, banking, and finance are well-established academic disciplines. Our insurance is all built around money or "finansurance[3]"—it's all about money! Money becomes evil in acquiring, using, or transactions!

What is new? Health banking is nowhere or rare in our vocabulary or even in literature and published journals. (In 38 s, 38 M total the results for banking and nothing about health banking by eating healthy foods.) Senior citizens (60 years and above) talk about maintenance medicines and their skyrocketing expenditure on the said medicines! Age is getting younger now as beginning 40 years; many are taking maintenance medicines. Search literature and mostly the explanation is related to lifestyle[4] illness (diet is more meat; less or no exercise; stress, short sleep duration, etc.). But Vandana Shiva disputed lifestyle. According to her, it is "food style."

Take a "cue" from "food style" concept; it is rooted in the way food is produced and how it is eaten! Conventional agriculture promotes monoculture rather than polyculture cropping systems. Of the 7500 edible crops, only 12 species supply 70% of our food and only three species (rice, corn, wheat) supply more than 40% of our food caloric energy. Eating less diverse food (fruits and vegetables) and more meat and eggs and drinking milk with toxins from pesticide-infested crops make the life industries of mega companies super mega rich! (http://www.prostate.net/

[3] "Finansurance—a combination of financing and insurance, which is known as the most secure kind of loan, for both parties, ever offered" (http://www.payehome.org/finansurance).

"Finansurance—a convergence of finance and insurance in which households and firms optimize their overall risk positions in life-cycle and business. ... that financial technology together with information technology accelerates the trend of functional finance and will provide products to complete an incomplete system for risk optimization." (Kariya, http://www.actuaries.org/AFIR/Colloquia/Tokyo/Kariya.pdf)

[4] "A healthy lifestyle requires regular exercise/physical activity and the consumption of a fibre rich, low-calorie healthy diet. Lifestyle intervention with a combination of regular physical activity and dietary advice showed an impressive reduction in the risk of developing diabetes in all the studies" (Desai and Tandon 2009).

Table 4.1 Indicative accumulated financial costs of maintenance medicines

	Case 1[a]		Case 2[a]		Case 3[b]	
Age (Years old)	Per mo.	Per year	Per mo.	Per year	Per mo.	Per year
40–44	1000	60,000	1500	90,000		
45–49	1500	90,000	2000	120,000	[b]	1,000,000
50–54	2000	120,000	4000	240,000	10,000	600,000
55–59	3000	180,000	5000	300,000	10,000	600,000
60–64	5000	300,000	6000	360,000	10,000	600,000
65–69	6000	360,000	6000	360,000	10,000	600,000
70–74	6000	360,000	7000	420,000	10,000	600,000
75–>	6000	360,000	7000	420,000		4,000,000
Total PHP		1.83 M		2.31 M		4.0 M

[a]Did not include medical consultation and treatment for other illnesses, [b]Underwent VAS (video-assisted surgery)
Source: TC Mendoza own calculations 2018

nutrition-cancer-diet/natural-foods-for-prostate-health/pesticide-contamination-in-food/).

The accumulated toxin will soon take a beating. Calculations on the costs of maintenance medicine range from PHP 1.8 M, PHP 2.31 M, and PHP 4.0 M till the person reach the finish line (Table 4.1). If there will be 3 billion people who are taking maintenance medicine, their medical bills amount to PHP 5.4 quadrillion pesos ($1.8 \times 10^6 \times 3 \times 10^9 = 5.4 \times 10^{15}$)(Case1) to as much as 12 quadrillion pesos (Case 3) (1 USD = PHP 53). If this trend continues, it will be no surprise that we will have trillionaire people in this millennium.

Dr. Vandana Shiva questioned the true purpose of farming as she takes serious issue with industrial agriculture's obsession with yield (volume) per acre. "If food is nourishment, then health and nutrition per acre, is what we should be measuring." *With a passion for the power of small farms, she said, "Small, organic farms grow health."* Under Indian condition, Vandana Shiva and her colleagues argued that there will be more food and Indians shall be healthier if from chemical–conventional monocultures, they shift to organic and polyculture farming systems. Rather than measuring crops yields per ha, why not calculate health per acre! They are able to show that protein production per acre (124 kg/acre at 60 g protein/adult/day), India's farmland of 184 M ha, can feed 5 billion adults and calorie energy production could feed 2.4 billion at 2500 Kcal/cap/day (the population of India was 1.3 billion people in 2018) (http://eatstayfarm.com/2017/02/health-per-acre-with-dr-vandana-shiva/).

Organic production innovation must have organic consumption innovation. Foregoing beef eating (those 1.2 billion cattle and eating the food bill of 8.5 billion people) shall make Mother Earth free from the CO_2 emission and recapture back at least 30% emitted CO_2 via crop photosynthesis as we plant trees instead of growing forage and cereals to feed them and use a lot of synthetic fertilizers and pesticides. As we promote organic farming innovation (nonetheless it should be reiterated that organic farming itself is already an innovation), the production orientation (supply side) and consumption (demand side) should be given equal importance. We should not overly target the producers-farmers only! Organic production innovation must have organic consumption innovation as well! In relation to this, food preparation/recipes innovation must be promoted with equal if not greater importance now. The consuming public must be informed/educated. Innovating production systems simultaneous with the demand side shall need to focus on the following platforms:

1. On the energy aspects of production and consumption, the logistic aspect of making food available (packaging, storing, transport) is energy intensive. Data shows that traveling food (i.e., rice) beyond 50 km distance, the increase in energy bill becomes significant (20% more) (May Soe Oo and Mendoza 2018; Sem and Mendoza 2018). It will be greater if food is imported overseas involving land, sea, and air travel. The energy bill becomes enormous as food reaches the plate. It does not include energy bill or processing and cold storage. Localized food is the obvious innovation rather than food importation or food globalization. Consumers should prefer locally grown food than imported food items to minimize the "food miles" effect.
2. Nature designs crop seasonality not simply due to crop adaptation but also to supply nutrient needed by the human body to cope with the weather changes and for health reasons. Colder months require consumption of vitamin C–rich fruits. It is the citrus fruit season. We should drink more water during hot summer months. It is watermelon and sugarcane harvest time. Drink fresh sugarcane juice instead of drinking highly carbonated cola sweetened by high fructose corn syrup (HFCS), aspartame, and other alternative sweeteners (https://www.curejoy.com/content/harmful-effects-of-soft-drinks/).

   Off-season crop production (vegetable, fruits) is energy intensive (needs greenhouses/glasshouses, highly embedded energy during construction and electric power to sustain and maintain operation). To induce them to flower and to protect the flowers and fruits as in mangoes requires energy-based inputs. Hence, their energy footprint is high.
3. The thermodynamics of food must be popularized. As we convert food energy from one form to another, there is considerable food loss as shown in the data below (http://www.fao.org/3/a-a0701e.pdf; Thorpe 2008):
   - >75%–84% loss of animal protein (poultry, pork, or beef).
   - 4 kg of plant protein is reduced to 1 kg broiler, chicken (a 75% loss of food protein).
   - 6 kg plant protein is needed to produce a kilo of pork (more than 84% loss of plant protein).
   - 50 g pure animal protein from broiler is equal to 2–3-day, pork = 3–5-day, grain-fattened beef = 9–10-day effective food day (EFD) (Mendoza 1991, 1994).
   - In the transformation from grain ➔ animals ➔ man, there is a loss of 90% protein, 96% calories, 99% carbohydrates, and 100% fiber.

In terms of resource use:

- $$1\text{cattle} = \text{consume food of } 7.25\text{ persons}, 1\text{kg beef} = 16\text{ kg grain} + \text{soybean}$$
$$= 8.8\text{ l of gasoline} = 22{,}000\text{ l of water} = 77\text{ kg of eroded soil}$$
- Producing 1 mega-calorie of beef or 1000 calories requires about 150m^2 of land. Eggs or poultry require about 5m^2.
- The beef uses 1.6 cubic meters of water compared to about 1 cubic meters of water for eggs or poultry.
- Beef contributes to water pollution eight times more than eggs or poultry. https://www.pri.org/stories/2014-08-04/yes-steak-expensive-its-true-cost-even-higher-you-may-think.
- Raising beef cattle is far more environmentally costly than poultry, pork, dairy, or eggs. Per calorie, cattle requires on average 28 times more land and 11 times more water to farm. Farming cattle releases five times more greenhouse gases and uses six times as much nitrogen as the average of other animal products. When compared with staple plant foods, these ratios roughly double. So, a beef calorie requires about 50 times more land than a wheat calorie (Adam Wernick 2014. https://www.pri.org/stories/2014-08-04/yes-steak-expensive-its-true-cost-even-higher-you-may-think).
- A University of Minnesota study showed that for every 100 calories of grain fed to animals, only 40 new calories of milk, 22 calories of eggs, 12 of chicken, 10 of pork, or 3 of beef could be obtained. The UN-FAO warns that using cereals as animal feed could threaten food security by reducing the grain available for human consumption (https://www.ciwf.org.uk/media/7431690/paying-for-the-true-costs-of-our-meat-eggs-and-dairy.pdf). Furthermore, food source from animals emit more greenhouse gases (CO_2e) per kg than plant food source as shown in Table 4.2.
- Animals require more resources, land, water, and production inputs, resulting in 40% more greenhouse gas emissions than cars!
- Animals contribute 9% of total anthropogenic carbon dioxide emissions, 65% of human-related nitrous oxide from manure, 37% of all human-induced methane produced by the digestive system of ruminants, and 64% of ammonia, which contributes significantly to acid rain (http://www.fao.org/3/a-a0701e.pdf; Thorpe 2008).

4.7 Synthesis and Recommendation

Innovation is the key strategy in scaling up the adoption of agroecology-based organic agriculture and building climate change resilient communities. Innovations should aim to overcome the barriers in the metamorphosis to scale up the adoption of agroecology-based organic agriculture. Innovations are necessary across the value chain (preproduction, production, postharvest, processing, marketing inter-linkages). Various innovations, starting from seed selection, using location-tested hybrid or inbred seeds, seedling preparation (i.e., rice at 4 kg seeds ha^{-1}, sowing

Table 4.2 Carbon emission of foods (per kg) and their equivalent car miles

Food source	Kg CO_2e Per kg	Equivalent Car miles
Lamb	39.2	91
Beef	27.0	63
Cheese	13.5	31
Pork	12.1	28
Turkey	10.9	25
Chicken	6.9	16
Tuna	6.1	14
Eggs	4.8	11
Potatoes	2.9	7
Rice	2.7	6
Nuts	2.3	5
Beans/tofu	2	4.5
Vegetables	2	4.5
Milk	1.9	4
Fruit	1.1	2.5
Lentils	0.9	2

Source: Environmental Working Group's Meat Eater's Guide and the EPA's Guide to Passenger Vehicle Emissions. http://www.greeneatz.com/foods-carbon-footprint.html

seed at 1500 seeds square meter^{-1}), transplanting 20-day-old seedling at 1 seedling per hill, paying transplanters PHP 500 higher ha^{-1}, using amplified liquid manure, indigenous microorganisms (IMO) inoculated hog-biogas-generated liquid sludge, moist fields not flooded are but some pre-to-production innovations. But farmers as entrepreneurs should also practice total quality management (TQM). They should make their partner-workers happy, motivate them, and practice benefit-sharing. Partner-workers must have an additional income after deducting costs, and profit is shared at 50–50 after deducting the initial capital.

Likewise, consumers must support organic farming by paying a fair price to put an economic value to the care farmers extend to the soil and to the Mother Earth at large by rebuilding–restoring soil fertility, by not applying "easy" but destructive, health hazardous and heavy greenhouse gas-emitting oil-based inputs. Consumers patronizing on-season produced crops and on their locality buy local to avoid food miles, discouraging food imports, as it imperils food sovereignty leading to food globalization by few multinational companies/transnational corporations. The innovative change from supply chain approach to value chains approach is necessary. A supply chain approach must be a tinge of the past where our farmers simply produce/supply raw materials (feedstock) to the processor/manufacturer of finished/processed high-value products. Value-adding product must be identified and farmers should be capacitated and be supported to reap the benefits of the value adding/financially enriching part of the food systems. The challenge is how to help farmers achieve the benefits accruing across the "value chain." It will spell the differences between the past (twentieth century) to the present (twenty-first century).

Another major value adding field metamorphosis is reshaping agriculture landscape into agroecotourism. A number of farms not only in the Philippines but across Asia and the Pacific are into agroecotourism. There are farms in the Philippines whose major income are earned through the entrance fees, seeds/seedling, and fruit sales from the agrotourists. Their farm is serving as market and educational site at the same time for environment- and health-friendly farming and food consumption style.

Agriculture is primarily about food and health. This lumps together finansurance or healthsurance. Health banking refers to the growing, eating, or buying foods grown on healthy soil. Thus, agriculture should shift into biodiverse, agroecology-based organic production that innovates yield assessment into health per acre rather than monoculture yield per ha. However, organic production innovation must simultaneously have organic consumption innovation as well. For instance, mother earth can support 43 billion if we become vegetarian. By 2050, we are anticipating 9.5 to 10 billion people globally, and in Philippines, the population is projected to be 160 million. The 1.2 billion cattle consume the food equivalent of 8.5 billion people, emits more GHGs than all our cars. There must be organic food consumption innovation!

The urgent necessity to implement and scale-up the adoption of organic farming as an adaptive and mitigating mechanism to combat the looming impacts of climate change should unite us all. The shift to organic farming should not only be compelling on the supply side. But it is equally important for consumers, the demand side, to patronize organically grown produce and be supportive. The nominal premium price associated to organically grown produce compared to the conventionally chemical-based grown products has no basis if total costs accounting (financial, ecological, health costs) will be adopted in computing prices. This will undoubtedly make chemically grown crops excessively underpriced. Integrating the control of the value chain among the producers, particularly the farmers, should be institutionalized in the organic farming vis-à-vis consumer, producer, and government.

A concerted effort is necessary among the stakeholders to ensure the adoption of organic farming consequently building the resilience of the communities towards adverse impacts of climate change. In view of this, the five capitals (Morse et al. 2009; Callaghan and Colton 2008) resulting to resilient communities should be ensured as an enabling environment to realize the shift to organic farming and consumption, and they are as follows:

(a) Natural capital referring to natural resource stocks and environmental services. The natural capital shall ensure productivity with conservation of the ecological integrity. Consequently, sustainable production is assured across generations.
(b) Social capital focused on enhancing networks, social relations/affiliations, and associations including social claims. Moreover, developmental funds are usually channeled to recognized community organizations/associations thereby necessitating affiliations to official accredited community groups. As opposed

to mere construction of physical assets, social capital is founded on building trust and confidence among stakeholders towards accessing resources and sharing of benefits.

(c) Human capital is concerned about the skills, knowledge, and labor including health and physical capability. This is intertwined in the translated financial capital.

(d) Physical capital representing infrastructure, production facilities, and technologies. This captures the idea of Villegas (2018) to make available to the farmer the 4Ps and 2Ms (preproduction, production, processing, postproduction linkages + marketing and management). This provides the control of the value chain in the hands of the producer translating to a more secure economic and financial capital to be plowed back in farm development.

(e) Economic and financial capital ensuring a guaranteed capital base either sourced as cash, credit/debt, savings, or from other economic assets.

References

Barbuddhe SB, Singh NP (2014) Agro-eco-tourism: a new dimension to agriculture. Technical bulletin no. 46. Agricultural Technology and Agro-Eco-Tourism Dissemination Centre, Indian Council for Agricultural Research-ICAR Research Complex for Goa, Old Goa. http://www.ccari.res.in/Technical%20Bulletin%20No.%2046.pdf

Callaghan EG, Colton J (2008) Building sustainable & resilient communities: a balancing of community capital. Environ Dev Sustain 10:931–942. https://www.researchgate.net/publication/225653883_Building_Sustainable_Resilient_Communities_A_Balancing_of_Community_Capital. Accessed 15 Nov 2018

Canadian Biotechnology Action Network (CBAN) (2015) Where in the world are GM crops and foods? https://gmoinquiry.ca/wp-content/uploads/2015/03/where-in-the-world-gm-crops-foods.pdf

Cook CDA, Hamerschlag, Klein K (2016) Farming for the Future. Organic and Agroecological Solutions to Feed the World. 23 pp. https://foodrevolutionnow.files.wordpress.com/2017/01/farming-for-the-future_organic-agroecological-solutions-to-feed-world_cook_et-al_2016.pdf. Accessed 9 Nov 2018

Desai A, Tandon N (2009) Challenges in prevention and management of diabetes mellitus and metabolic syndrome in India. Curr Sci 97(3):356–366

Fan F, Henriksen CB, Porter J (2016) Valuation of ecosystem services in organic cereal crop production systems with different management practices in relation to organic matter input. Ecosyst Serv 22:117–127. https://ac.els-cdn.com/S2212041616304004/1-s2.0-S2212041616304004-main.pdf?_tid=71097f02-a1b2-4059-82039396648de013&acdnat=1542081393_b2bf-83b924026668ebd764c27a447db2. Accessed 13 Nov 2018

Grasty T (2012) The difference between innovation and intervention. http://mediashift.org/2012/03/the-difference-between-invention-and-innovation086/. Accessed 9 Nov 2018

Landicho LD, Paelmo RF, Cabahug RD, Visco RG, Abadillos MG (2014) Prospects and challenges in promoting organic agriculture in the upland communities in the Philippines: implications to food security and nutrition. In: 2014 International Conference on Food Security and Nutrition. IPCBEE vol.67 (2014) © (2014) IACSIT Press, Singapore. https://doi.org/10.7763/IPCBEE.2014.V67.12

Lao-Ay D, Sabellano JM Jr, Ebillo AP, Cerilles AH, Mendoza TC (2017) Double row transplanting (DRT): an innovation in organic rice production. In: Paper presented during the 6th International

Conference on Integration of Science and technology for Sustainable Development 2017 (6th ICIST 2017), held on November, 24–26, 2017 at Baguio, Philippines

May Soe Oo, Mendoza TC (2018) The energy footprints of inbred and hybrid rice genotypes grown under organic and conventional production system in Laguna, Philippines. In: Paper presented during the 7th International Conference on Integration of Science and Technology for Sustainable Development (7thICIST), 26–29 November, 2018, The Patra Bali Resort and Villas, Bali, Indonesia

Mendoza TC (1991) Coupling food and energy production. Inaugural lecture paper for the Felix Maramba, Sr. Professorial Chair on Agro-industrial Farming Systems. Department of Agronomy, College of Agriculture, UPLB, Laguna. June 25, 1991

Mendoza TC (1994) Adoption of soil benefiting agricultural practices for lowland rice production in the Philippines. Philipp Agric 77(2):235–241

Mendoza TC (2004) Evaluating the benefits of organic farming in Rice agroecosystems in the Philippines. J Sustain Agric 24(2). Published online: 23 Sep 2008. https://doi.org/10.1300/J064v24n02_09

Mendoza TC (2010) Organic agriculture: the logical sequence to chemical agriculture. Ann Trop Res 32(1):112–121

Mendoza TC (2016) Increasing SRI-organic rice yields through double rows planting pattern and using location and season adapted rice cultivar. J Agric Technol 12(4):767–790

Mendoza TC, Villegas PM (2014) The opportunities and challenges in green agriculture in the ASEAN: focus in the Philippines. Chapter 5. In: Nelles W et al (eds) ASEAN food security and sustainable agriculture in a green economy. Cross sectoral and interdisciplinary perspectives. Chulalongkorn University Press, Bangok

Mendoza TC, Acuna VM, Delos Santos D, Sandoval P (2016) Detrashing sugarcane stalks: the 1st critical step to the shift to organic farming in sugarcane production. J Agric Technol 12(6):1057–1070. http://www.ijat-aatsea.com

Morse S, McNamara N, Acholo M (2009) Sustainable livelihood approach: a critical analysis of theory and practice. In: Mannion AM (ed) Geographical paper no. 189, 68 pp. http://www.reading.ac.uk/web/files/geographyandenvironmentalscience/GP189.pdf. Accessed 13 Nov 2018

Ratnaike RN (2003) Acute and chronic arsenic toxicity. Postgrad Med J 79(933):391–396

Sem S, Mendoza TC (2018) Comparative energy footprint of Cambodian Lowland Rice Grown under different establishment methods. In: Paper presented during the 7th International Conference on Integration of Science and Technology for Sustainable Development (7thICIST) in November 26–29, 2018, at The Patra Bali Resort and Villas, Bali, Indonesia

Thorpe A (2008) Enteric fermentation and ruminant eructation: the role (and control?) of methane in the climate change debate. Clim Chang https://doi.org/10.1007/s10584-008-9506-x.; http://www.springerlink.com/content/6081j24276876123/fulltext.pdf

U.S. Food and Drug Administration (2016) Arsenic in rice and rice products risk assessment report. http://www.fda.gov/Food/FoodScienceResearch/RiskSafetyAssessment/default.htm

Villegas PM (2018) Manufacturing-based agro-industrialization: the value chain (VC) and agro-processing cluster framework and business models. Plenary paper presented during the 7th International Conference on Integration of Science and Technology for Sustainable Development (7th ICIST) in November 26–29, 2018 at The Patra Bali Resort and Villas, Bali, Indonesia

Wernick A (2014) True costs of steak. https://www.pri.org/stories/2014-08-04/yes-steak-expensive- its-true-cost-even-higher-you- may- think. Accessed Oct 2018

Willer H, Lernoud J (2017) Organic agriculture worldwide 2017: current statistics research institute of organic agriculture (FiBL), Frick, Switzerland. http://orgprints.org/31197/1/willer-lernoud-2017-global-data-biofach.pdf

Managing Climate Risk in a Major Coffee-Growing Region of Indonesia

5

Rizaldi Boer, Syamsu Dwi Jadmiko, Purnama Hidayat, Ade Wachjar, Muhammad Ardiansyah, Dewi Sulistyowati, and Anter Parulian Situmorang

Abstract

Indonesia is currently one of the top four coffee exporting countries in the world. Climate change is projected to cause significant impacts on coffee. Without proper adaptation measures, this will significantly lower the productions. Changes in rainfall and increases in temperature will affect the phenological development that would eventually influence yield and quality of crop including the potential risks of pest and disease attacks. Assessment in Toba, a major coffee-growing region of Indonesia, indicated that in the middle of this century (the 2050s), under climate scenarios of RCP4.5 and RCP8.5, suitable areas for coffee production would decrease significantly. The average yield is projected to

R. Boer (✉)
Center for Climate Risk and Opportunity Management in Southeast Asia and Pacific, Bogor Agricultural University, Bogor, Indonesia

Department of Geophysics and Meteorology, Faculty of Mathematic and Natural Sciences, Bogor Agricultural University, Bogor, Indonesia

S. D. Jadmiko · D. Sulistyowati · A. P. Situmorang
Center for Climate Risk and Opportunity Management in Southeast Asia and Pacific, Bogor Agricultural University, Bogor, Indonesia

P. Hidayat
Department of Crop Protection, Faculty of Agriculture, Bogor Agricultural University, Bogor, Indonesia

A. Wachjar
Department of Agronomy and Horticulture, Faculty of Agriculture, Bogor Agricultural University, Bogor, Indonesia

M. Ardiansyah
Center for Climate Risk and Opportunity Management in Southeast Asia and Pacific, Bogor Agricultural University, Bogor, Indonesia

Department of Soil Science and Land Resources, Faculty of Agriculture, Bogor Agricultural University, Bogor, Indonesia

© Springer Nature Singapore Pte Ltd. 2020

V. Venkatramanan et al. (eds.), *Global Climate Change and Environmental Policy*,
https://doi.org/10.1007/978-981-13-9570-3_5

decrease between 25% and 75% of the current yield. However, the highlands that are currently not suitable for coffee (>1500 m above mean sea level) is projected to become suitable with a higher yield than the current. A significant increase in rainfall during the rainy season and prolonged dry season will also affect coffee phenological development. It will shift the peak of coffee flowering and harvesting seasons in Toba. The severity of the coffee berry borer *Hypothenemus hampei* (Ferrari) attack will also increase in the future. The current crop management farming practices should be adjusted and improved to adapt to such change.

Keywords
Climate change · Climate scenarios · Indonesia · Toba region · Coffee · Climate change adaptation · Coffee berry borer

5.1 General Condition of Toba Region

5.1.1 Biophysical Conditions

Toba region is located in North Sumatra Province, Indonesia. The main coffee-growing region in Toba is located around Lake Toba, with total area of about 69,854 ha. The area is geographically spanned within 2°21′32″–2°56′28″N and 98°26′35″–99°15′40″E situated in the districts of Samosir, Toba Samosir, Dairi, Karo, Humbang Hasundutan, North Tapanuli, Simalungun and West Pak-Pak (Ministry of Environment 2014). Hilly and mountainous areas, ranging from 600 to 1800 m above mean sea level (Fig. 5.1), dominated the topographical condition of Toba, with slope varying from extra gentle (0–8%), gentle (8–15%), critical (15–25%), steep (25–45%) and extra steep (>45%). The northern part of Lake Toba (Tanah Karo) has relatively smaller water catchment areas with extra steep hillsides, whereas Samosir Island has larger catchment areas and extra gentle slope (<3%). The central part of the island is relatively extra steep mountainous area with slopes ranging from 30.5% or even in some areas more than 75% (Ministry of Environment 2014).

The soil type in the area was likely formed by Toba acidic tuff parent material under cold climatic condition. Both factors generally formed andosol. Other materials such as sedimentary rocks, bedrocks and alluvial deposits were also found forming other major soil groups like cambisol, podzol and regosol. Table 5.1 and Fig. 5.2 below provide detailed characteristics of each soil type.

In general, Cambisol (Dystrudept and Eutrudept), Andosol (Dystrandept) and Podzolic (Tropudult) soil types dominated the Toba Region. The soils have relatively low to medium level of fertility, high acidity, rough texture and low nutrient content. The area within Sinabung Mountain has a medium level of fertility. Therefore, the chemical characteristics of the soils must be improved through liming, fertiliser application and replenishment of organic materials, e.g. compost from rice straws, cacao shells or manure. These would increase the N level, as well as improve the soil quality, physical, chemical and biological characteristics.

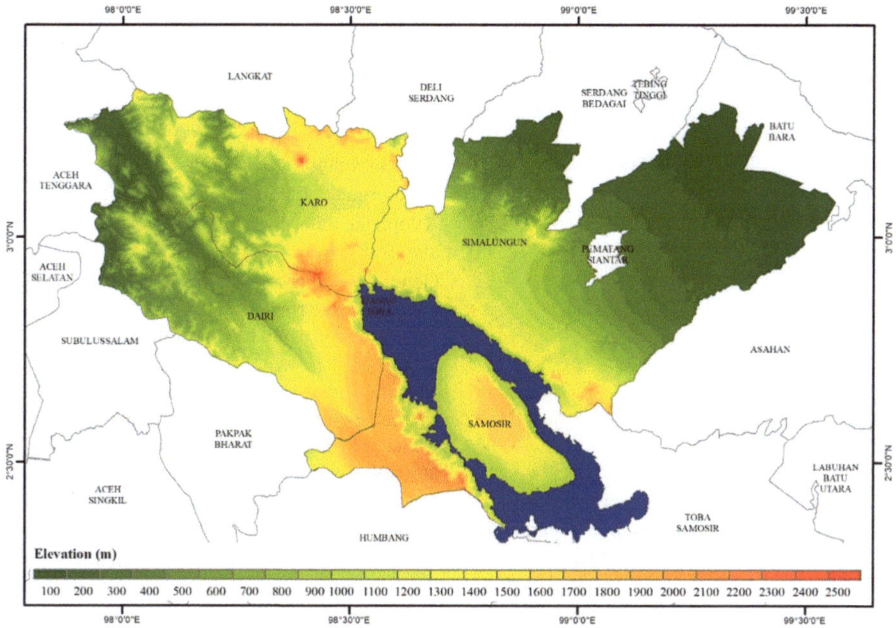

Fig. 5.1 The topography condition of the Toba Region

5.1.2 Socio-economic Conditions of the Small Holder Coffee Farming in Toba Area

The majority of Toba people engage in farming activities. Agriculture has long been their primary source of income. Agriculture, plantation and fisheries contribute to about 55% of the gross regional domestic products (Ministry of Environment 2014). Among the plantation commodities, coffee is the top of the list – well-known specialty coffee from the area including Kopi Sidikalang from Dairi and Kopi Simalungun from Simalungun. The number of coffee farm households in the four main coffee-growing districts Simalungun, Dairi, Karo and Samosir totalled to 17,036; 14,783; 6800; and 5202 families, respectively.

Statistic reports showed that Dairi District has the largest smallholder coffee plantation among the four districts, while the smallest is found in Samosir District (Table 5.2). The entire coffee plantations under study were about 29,500 ha, which accounted to more than 50% of the total coffee plantations in North Sumatra Province. Data from the years 2011–2016 have indicated that there was no significant increase in the total area of coffee plantations in Dairi, Samosir and Simalungun Districts, with an annual increase rate approximately 0.1–2.6%. On the contrary, on an average, the total plantations in Karo District have increased by 8.2% annually during these periods, indicating a relatively high expansion of coffee plantations.

Table 5.1 Soil types found in Toba area

Soil type (Center for Soil Research, 1978, 1982)	Taxonomy (USDA 1975–1990)	Characteristics
Andosol	Dystandept or Hydrandept	Has andic properties
		Content weight < 0.85 g/cm^3
		High organic material
		Allophane amorphous minerals dominate the clay fraction
		Generally categorised as fertile and has good physical properties; therefore it is common for horticulture and tea, quinine and tobacco plantations
		The andosol in the study area has relatively low fertility, rough texture, low cation exchange capacity, low base-saturation ratio and low water storage capacity
Cambisol	Dystrudept, Eutrudept, Andaquept, Tropaquept or Humitrudept	Categorised as the soil in early development phase
		Parent material varied and distributed from flat to mountainous areas, from wet to dry climate
		Cambisols found in Toba area are dystric cambisol and eutric cambisol
		In the area near water, soil colour tends to be greyer;
		Ata higher elevation with less influence from the water surface, the soil is known brownish to reddish, has a horizon of cambic and a base-saturation ratio < 50%
		The soil of this character is known as Dystrudept, and when the base saturation ratio is >50%, it is called Eutrudept
Podzolic	Tropudult	Has typically gone through the advance weathering process
		Clay translocation from upper to lower layers is obvious
		Categorised as acidic to very acidic soil
		Has low base saturation ratio and cation exchange capacity (CEC)
Regosol	Tropopsamment	Very young mineral soil
		Has a coarse texture of albic material
		Generally has low CEC, low water storage capacity and highly fast infiltration
		Commonly found in Samosir Island

(continued)

Table 5.1 (continued)

Soil type (Center for Soil Research, 1978, 1982)	Taxonomy (USDA 1975–1990)	Characteristics
Alluvial	Tropaquept or fluvaquent	Formed from the settling process of rivers, lakes and the process of colluviation at the bottom of hills
		Categorised as young soil and associated with a wet environment
		Also, this type of soil does not have a unique character/horizon, texture ranging from fine to coarse with relatively low nutrient content

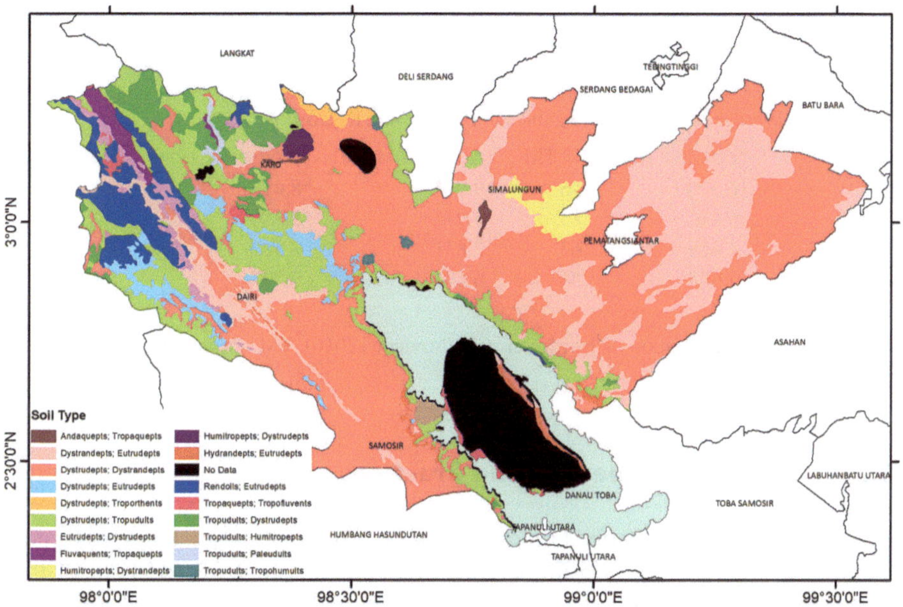

Fig. 5.2 Soil types within the Toba area

The coffee productivities in all study areas were relatively high. The average production in Simalungun Districts were 1.4 t/ha, while in Dairi was 0.85 t/ha and Samosir was 1.1 t/ha (Table 5.3). These level of yield are close to the national average (Ministry of Agriculture 2017).

Table 5.2 Average harvested area and production of coffee in 2011–2016

District	Total area (ha)	Production (tonnes)	Annual expansion rate (%)	Sub-districts with contribution >80% to the district production
Dairi	10.500	8.478	0.1	Sumbul dan Parbuluan
Karo	6.834	5.380	8.2	Merek, Simpang Empat, Barusjahe, Tiga Panah, Munte, Juhar, Payung, Naman Teran, Kutabuluh
Samosir	4.468	3.303	2.6	Rongurnihuta, Pangururan, Palipi, Simanindo, Sianjur Mulamula, Nainggolan
Simalungun	7.698	9.597	2.4	Purba, Raya, Pamatang Silimahuta, Dolok Pardamean, Dolok Silou, Silimakuta, Sidamanik, Girsang Sipangan Bolon

Source: District and sub-district statistics (BPS 2016a, b, c, d)

Table 5.3 Average production of Arabica coffee in Toba Region

District	Total area (ha)	% TM	Area of TM (ha)	Production (ton)	Yield (t/ha)
Dairi	10.500	95	9.975	8.478	0.85
Karo	6.834	85	5.975	5.380	0.93
Samosir	4.468	63	2.831	3.303	1.17
Simalungun	7.698	84	6.495	9.597	1.47
Average					1.07

Source: Office for plantation of North Sumatra Province (2016). TM: Productive Coffee (already berry producer)

5.2 Existing Coffee Farming Practices

Assessment of existing coffee farming practices in Toba area was conducted through survey and interview of 205 farmers distributed in the four districts. All of their farmlands are located in the elevation of between 600 and 1500 m MSL (Fig. 5.3). The topographical conditions of the farmlands vary from flat to very steep slopes. About 84.4% of the smallholder plantations in the four districts are located in relatively low slopes (flatlands; area with slope < 3%), 16% downhill as an area with slope ranging from 3% to 15% and 2% in very steep areas (slope > 15%).

5.2.1 Cultivation Practices

5.2.1.1 Historical Record

There were no formal records on the periods of coffee introduction and cultivation in the study area. The earliest substantiated evidence of coffee-growing as stated by the respondents appeared to be dated back before 2000, while few others indicated that coffee cultivation in the areas started after 2000. According to the respondents, the oldest plantation in the area has reached the age of 25 years. In general, coffee

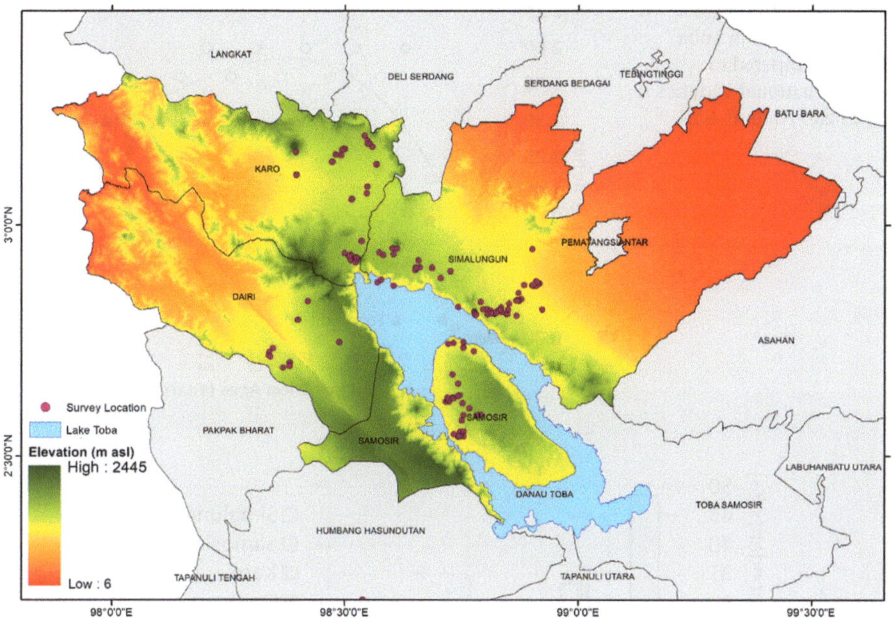

Fig. 5.3 Distribution of respondents (dot) according to elevation (m.MSL)

has been widely cultivated by the communities in the past 15 years. The age of the coffee plantations in the four districts was between 4 and 15 years. The Indonesian Coffee and Cocoa Research Center (Puslitkoka 2013) identifies that coffee is most productive between 5 and 20 years of age. Trees over 20 years of age could no longer yield fruits, and the production is 30% lower than the younger coffee trees. Morphologically, matured coffee trees have larger rod shape and tend to be porous.

Compared to the yield of the superior national coffee varieties such as S-795 and Andung Sari, the Toba coffee production is much lower. At large, the average yield of coffee in the Toba Region was found to decrease at an age above 10 years (Fig. 5.4), a similar condition found for the Andung Sari variety. On the contrary, the production of S-795 variety increases with increasing age even reaches 20 years.

5.2.1.2 Arabica Coffee

The coffee cultivars in the study areas originated from the following varieties: Katimor Sigararutang (KS), Ateng Super (AS), Katimor Andung Sari (KAS), Katimor-Ateng (KA) and Katimor-Ateng Jaluk (KAJ). Robusta is also cultivated in few areas. Figure 5.5 presents the predominant varieties cultivated in Karo and Simalungun District. In Samosir, the Ateng Super is the most dominant. Some farmers planted more than one variety.

The combination of KS and KA was identified as having the highest production at tree/plant category (Fig. 5.6), with production reaching 1 kg/tree, while other varieties like AS, KAS and KAJ could only yield <1 kg/tree. The population per ha varies from 100 to 3000 trees, with some farmers are planting more than 3000 trees per ha.

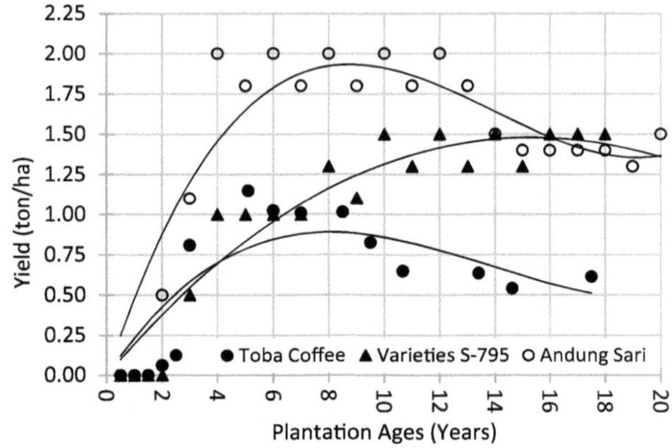

Fig. 5.4 The average yields of coffee in Toba Region compared to superior national varieties S-795 and Andung Sari

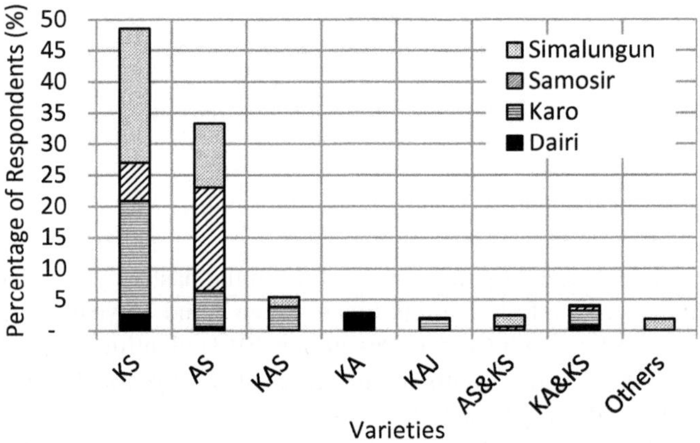

Fig. 5.5 Main coffee varieties planted in the Toba Region

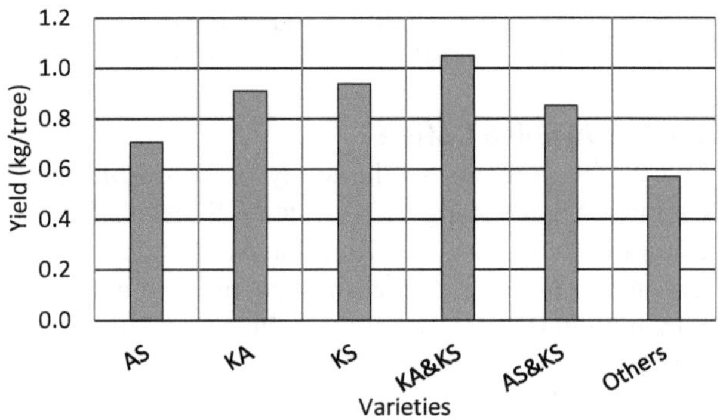

Fig. 5.6 Yield by variety

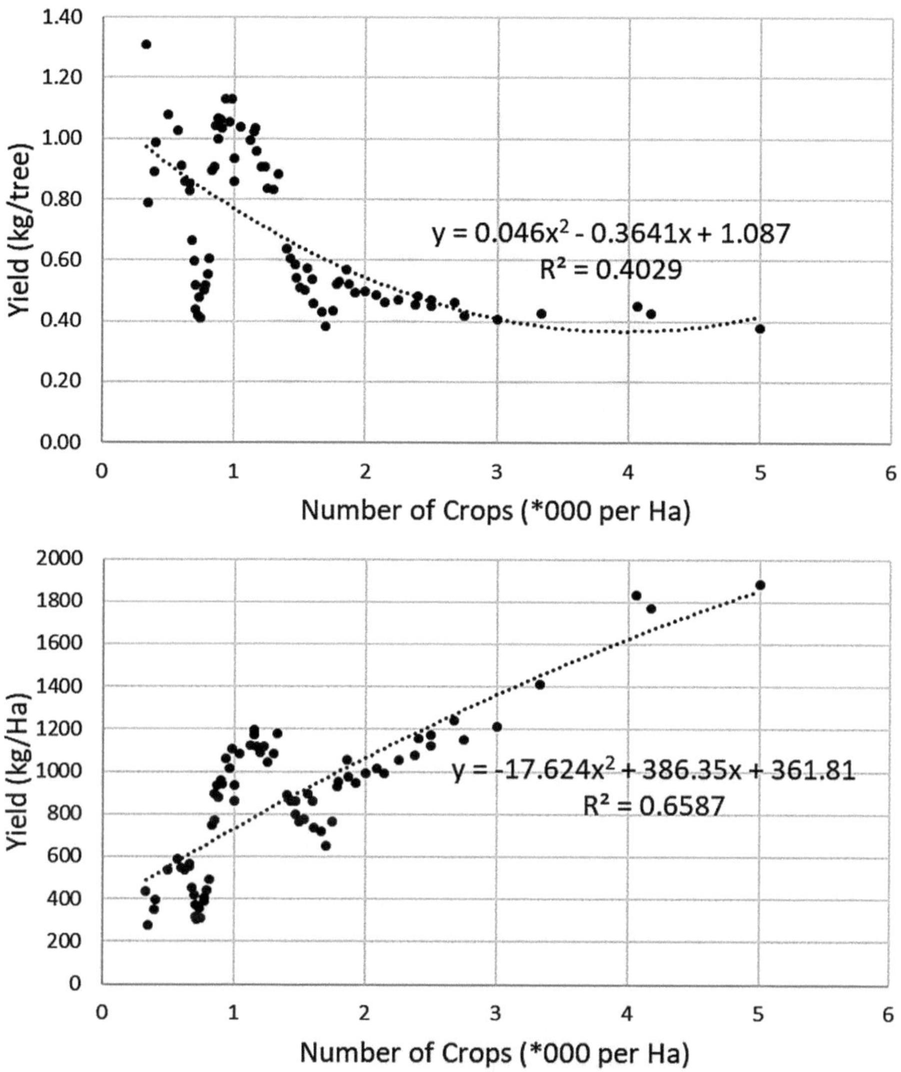

Fig. 5.7 The yield of Toba coffee vs the number of crops per hectare (production is calculated based on moving average of ten data sorted from smallest to largest land ownership)

The average land ownership area ranged between 0.06 and 4.0 ha, or on an average 0.6 ha. In general, the results showed that production lowers with an increasing number of crops per hectare (Fig. 5.7). The survey revealed that production of Arabica coffee in the four districts ranged from 0.3 t/ha to 1.9 t/ha or on an average about 0.864 t/ha. These numbers were slightly lower than those reported by the Office for plantation of North Sumatra Province.

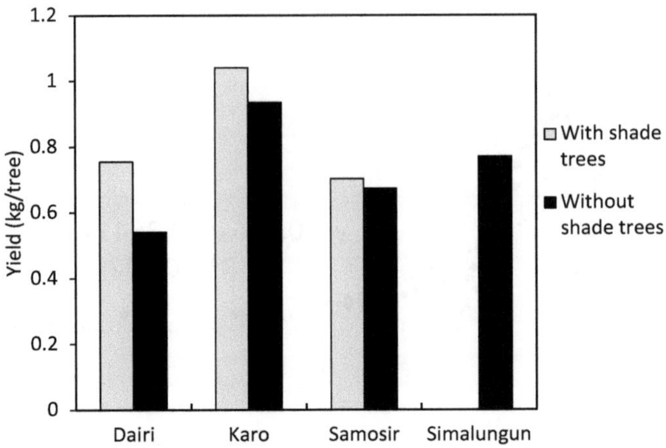

Fig. 5.8 Yield of coffee with and without shades

Seeds were obtained from farm shops, exchanges between farmers, farmers own breed or plantation office. Currently, the use of uncertified seeds has led to untraceable sources of the seeds. The respondents reported that the most productive seeds were those originated from grafting.

5.2.1.3 Cultivation Techniques

Traditional polyculture is the most common practice of coffee cultivation in the study area, with planting space ranging from 1.5 × 1.0 m to 3.0 × 4.0 m or total population per hectare ranging from 100 to 2750 trees. Few farmers are practicing monoculture with planting space ranging from 1.5 × 1.0 m to 2.0 × 4.0 m or a total of 200–2750 trees/ha.

In polyculture plantations, farmers often cultivate coffee with candlenut, palm and coconut; fruits such as avocado, mango, durian and banana; or other woody plants. Coffee planted in monoculture often experienced a higher intensity of light that leads to *biennial bearing,* a period, in which crops bear dense fruit in one period and followed by the rapidly declining intensity of flowering and fertility in the next period. Recommended protection crops include *Leucaena leucocephala* (lamtoro) and *Gliricidia maculata* (Gliricidia) and other crops from Leguminosae family. Results of the survey indicated that the production of crops was higher under specific intensity of shades, compared to crops without shades (Fig. 5.8). In a number of young or unproductive coffee plantations, farmers also planted annual seasonal horticultural crops, such as shallots and scallions.

A key aspect in coffee cultivation that plays an essential role in production, both in quality and quantity, is maintenance. Maintenance includes fertiliser application, pruning, weed control, pests and diseases control, irrigation, mulching and regulating protective plants. Most coffee farmers in the study areas have not practised this comprehensive process. The fertiliser is not regularly applied because the farmers did not have sufficient funding to cover the production costs, in addition to the long

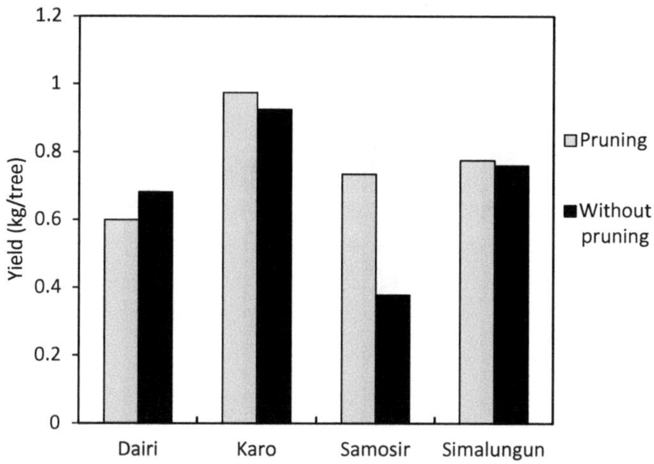

Fig. 5.9 Yield of Arabica coffee with and without pruning

distance of farm shops that sell production inputs or just because the farmers think that their plantations still produce reasonable yields without the need for additional fertiliser. Crops without additional nutrients from fertiliser would tend to grow dwarf and produce yellow crown and poor branching.

Pruning is also a vital maintenance process to ensure sustainable coffee production, which would include removing unproductive parts of the plants, e.g. shoot water, small branches that grow on the trunk, suckers and all diseased branches. The top parts of the trees were often cut to keep its height to be less than 1.8 m. Pruning also allows trees to be easier to harvest, and provides well-regulated light and air. Farmers in the study areas have not performed regular crop maintenance practices. Pruning is frequent, but not conducted on a regular basis. The farmers did not entirely understand which part of the plants that required rejuvenation to stretch the lifespan of the coffee tree and hence increase production. Unpruned trees tend to grow moult and leafless branches and produce less flowers and fruits, where, on the contrary, pruned trees would produce higher yields (Fig. 5.9).

Weed control is another central process in coffee cultivation. This maintenance process is best conducted, when the plants are in their early growing stage, has relatively open space and direct sunlight reaching the ground, thus enabling the environments to spur weed growth. The frequency of weed control can necessarily be reduced to productive/yielding plants, as the plant's canopy has grown thick and close to each other. Farmers in the study areas mainly performed weed control manually using a hoe on immature or yielding plants (Fig. 5.10).

Irrigation and mulching of coffee plantation are essential, especially during the dry season. Irrigation is not only useful for regulating water supply to support plant growth but also during flowering and fruits development. The Samosir District often experiences longer dry season, so it needs more intensive irrigation than other areas. Irrigation water is supplied from Lake Toba, and pumped up to the surrounding hilly

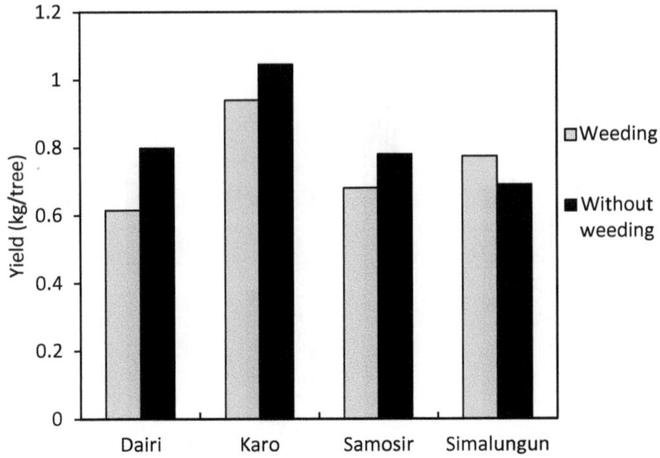

Fig. 5.10 Yield of Arabica coffee with and without weedings

Table 5.4 List of common coffee pests and diseases in the study areas

No	Coffee pests and diseases	No. of farmers	Percentage (%)
1	Coffee berry borer/CBB	166	81.0
2	Green scale	12	5.9
3	Coffee stem borer	10	4.8
4	Leaf buds/*Helopeltis* sp.	7	3.4
5	Coffee leaf rust	5	2.4
6	All plants dried and died	2	1.0
7	Branches dried	1	0.5
8	None of the above	2	1.0

areas. Mulching the grounds are necessary to reduce evaporation. Traditionally, the farmers would use weeding wastes, pruning wastes, or other materials found in the garden for mulching.

5.2.2 Pests and Diseases

5.2.2.1 Type of Common Pests and Diseases and Their Damage Rate

Plant disturbing organisms like pests and diseases, are among other factors that inhibit the growth of coffee, both in quantity and quality. In general, coffee berry borer (*Hypothenemus hampei* Ferr.) (CBB) is known as the primary pest found in the study areas. More than 80% of the respondents have also indicated other main pests, such as green lice, stem borer, leaf buds and leaf rust (Table 5.4). The highest number of coffee berry borer outbreaks was found in Simalungun and Samosir Districts while the least was found in Karo and Dairi Districts (Table 5.5). Coffee berry borer is common to attack nearly all coffee varieties including Arabica,

Table 5.5 Pests and diseases outbreaks experienced by farmers in the study areas

	No. of farmers (%)			
Coffee pests and diseases	Dairi	Karo	Samosir	Simalungun
Coffee berry borer	68.8	69.6	84.4	93.1
Green scale	9.4	8.9	2.2	4.2
Coffee stem borer	12.5	5.4	6.7	0
Leaf buds/*Helopeltis* sp.	3.1	7.1	4.4	0
Coffee leaf rust	0	7.1	0	1.4
All plants dried and died	0	0	2.2	0
Branches dried	0	1.8	0	1.4

Number of respondents in each district of Dairi, Karo, Samosir and Simalungun are, respectively, 32, 56, 45 and 72 farmers

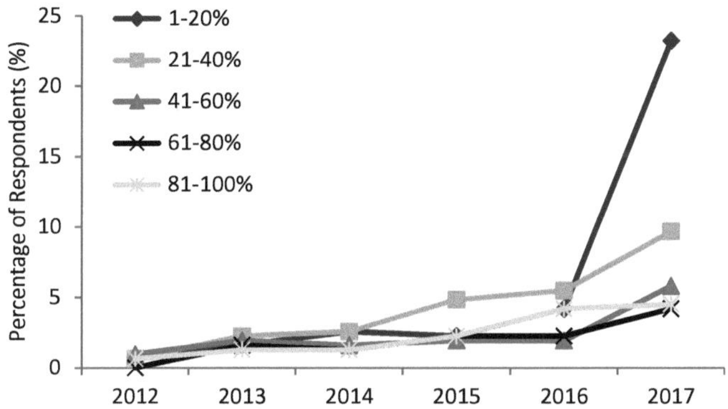

Fig. 5.11 Record of Coffee berry borer outbreaks in the study area during the periods of 2012–2017

Robusta and Liberica. Damages caused by this pest were mainly found on the apical parts of the coffee berry (Samsudin and Soesanthy 2012). The damages might cause yield decline as the young berries either dropped without ripening or showed lower quality of ripening berries.

Damage caused by coffee berry borer found to be linear with its outbreak level. In general, coffee berry borer destroyed about 80% of the total berries in the area. Farmers reported that from 2012 to 2017, the peak of coffee berry borer attacks was in 2017 (Fig. 5.11), where more than 65% of the farmers suffered crop damages with different damage levels. About 54% experienced a relatively low rate of damage (1–20%), 23% experienced medium damages (20–40%), and 10% experienced the worst damages (60–80%). The most common pest control for coffee berry borer in the study area, was using *Beauveria bassiana* and BROCAP trap. Unfortunately, only few farmers found these techniques to be effective.

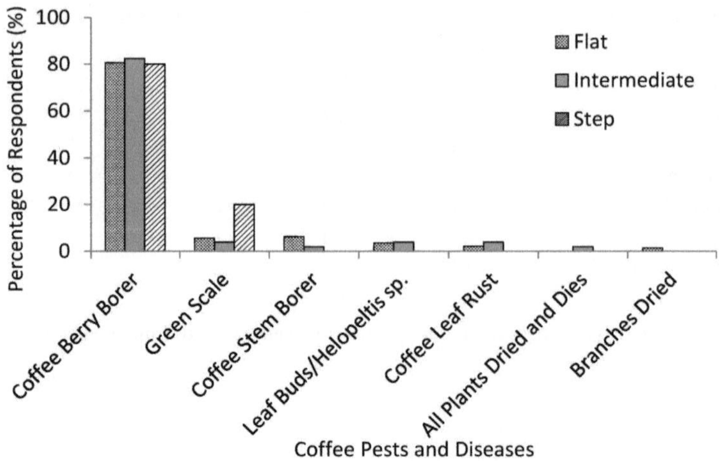

Fig. 5.12 Percentage of farmers experiencing various pests and diseases attacks by slopes

Table 5.6 Percentage of farmers experiencing various pests and diseases by cultivars type

	Cultivars					
Coffee pests and diseases	AS	KAS	KA	Aceh	KAN	KS
Coffee berry borer	90	100	62	100	100	75
Green scale	4	0	30	0	0	6
Coffee stem borer	4	0	8	0	0	6
Leaf buds/*Helopeltis* sp.	0	0	0	0	0	6
Coffee leaf rust	0	0	0	0	0	4
All plants dried and died	2	0	0	0	0	0
Branches dried	0	0	0	0	0	2

Note: *AS* Ateng Super, *KAS* Katinor Andung Sari, *KA* Katimor Ateng, *KAJ* Katimor Ateng Jaluk dan, *KS* Katimor Sigararutang

5.2.2.2 Relation of Pest and Diseases Attacks with Topography and Farming Practices

The analysis revealed no correlation between damage rate caused by coffee berry borer with plantation size, slopes, nor with coffee cultivars. Percentage of farmers experiencing coffee berry borer outbreaks on flat, intermediate and steep areas were 81%, 82% and 80% (Fig. 5.12), respectively. In general, there were no cultivars that were resistant to coffee berry borer. However, the survey revealed that KAS, KAN, AS and Aceh cultivars were relatively sensitive to coffee berry borer (Table 5.6). According to the farmers, KA variety showed the least sensitive to coffee berry borer, while KS was found to be the most vulnerable to all common pests and diseases (Table 5.6).

In the context of shade, monoculture farming without shade tent experienced the most damages from coffee berry borer (Table 5.7). Shade aims to create favourable condition for coffee plantation by reducing photorespiration and sunlight intensity. The growth of coffee plants needs a certain level of sunlight. Despite the attacks

Table 5.7 Distribution of pests and diseases attacks under different planting systems

	Farmers (%)		
Coffee pests and diseases	Unshaded monoculture	Shaded monoculture	Shaded polyculture
Coffee berry borer	85	79	80
Green scale	7	5	5
Coffee stem borer	1	0	9
Leaf buds/*Helopeltis* sp.	2	11	3
Coffee leaf rust	2	5	2
All plants dried and died	1	0.00	0
Branches dried	1	0.00	1

Table 5.8 Distribution of pests and diseases attacks under six different coffee productive ages

	Age (year)					
Coffee pests and diseases	<3	3–5	5–10	10–15	15–20	20–25
Coffee berry borer	100	80	79	90	100	0
Green scale	0	0	9	0	0	0
Coffee stem borer	0	7	3	0	0	0
Leaf buds/*Helopeltis* sp.	0	13	6	10	0	0
Coffee leaf rust	0	0	3	0	0	0
All plants dried and died	0	0	0	0	0	0
Branches dried	0	0	0	0	0	0

found on nearly all stages of productive periods, coffee plantations of 5–10 years of age were found to be the most vulnerable (Table 5.8).

5.3 Climate Variability and Its Impacts on Coffee Plantation

Two significant climatic variables that affect coffee are temperature and rainfall (ICC 2009). Both variables play critical roles in plant phenology, which eventually would determine the quality and quantity of coffee production (Haggar et al. 2011). Impacts of climate on coffee production vary among different cultivars and different growth stages. The following sections discuss the direct impacts of climate variables on coffee production, as well as indirect impacts related to the outbreaks of pests and diseases.

5.3.1 Climate Variability and Coffee Phenology

Rainfall distribution is crucial, because water is required during the vegetative to generative growth stages of coffee plants and the formation of floral primordium (Abdoellah 2016). Extreme climate conditions often hamper the development of coffee berries. Prolonged dry season and excessive rainfall would have detrimental effects on the flowering and conception processes (Susilo 2008). High rainfall

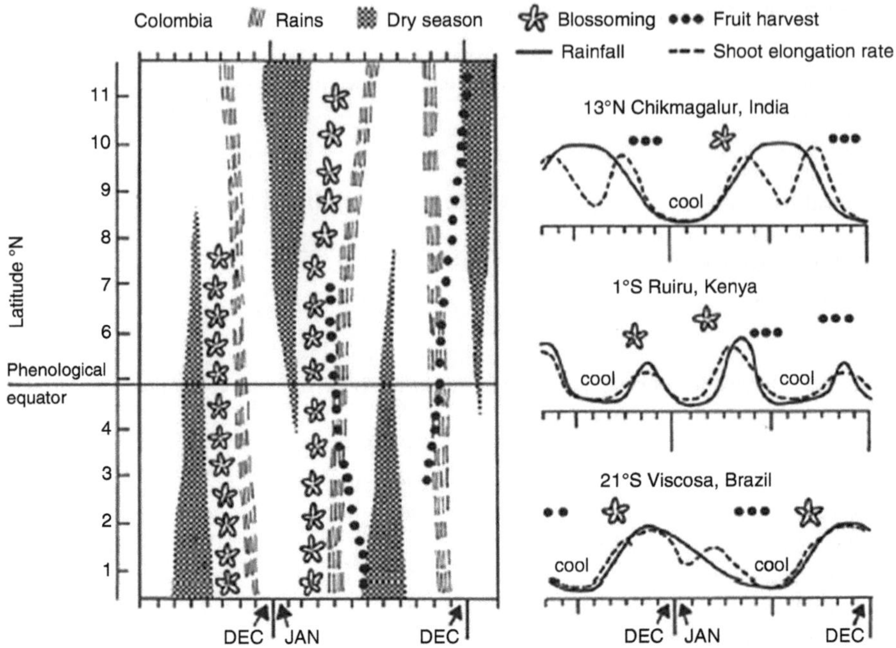

Fig. 5.13 Seasonal phenology of Arabica coffee. Times of blossoming and fruit harvesting at different latitudes (Left) and generalised seasonal changes in rainfall, shoot elongation rates and times of flowering and fruit harvesting in India, Kenya and Brazil (Cannel 1985)

would disrupt the pollination of flowers. The main driving factor for blooming is groundwater availability. Before the blooming season, flowers entered a period of dormancy for about 1–4 months and would remain dormant for an extended period under the condition of water deficit. Either the onset of rainy season or irrigation would support further development of coffee plants (Cannel 1985).

In coffee, the flowering periodicity is significantly determined by rainfall seasonality and length of seasons. Observations of coffee production centres worldwide have indicated that coffee flowering and harvesting stages are susceptible to rainfall pattern. The peak of coffee flowering and harvesting occurred twice a year in areas with bimodal rainfall seasonality near the equator (Fig. 5.13). Physiologically, the initiation and blooming of coffee flowers occur during the transition period from dry to rainy season and vice versa, because these processes were induced by changes in groundwater content. As indicated in the earlier paragraph, plants need a sufficient amount of water to leave the dormancy state and initiate blooming. However, excessive rainfall would inhibit this process, resulting in shrivelled flowers and back to the vegetative stage. Following the blooming season, plants would need at least three consecutive days without rain to start pollinating. Flowers would develop into fruits after pollination. High intensity of rainfall and short daylight during pollination would have resulted in fewer berries and abortion of flowers.

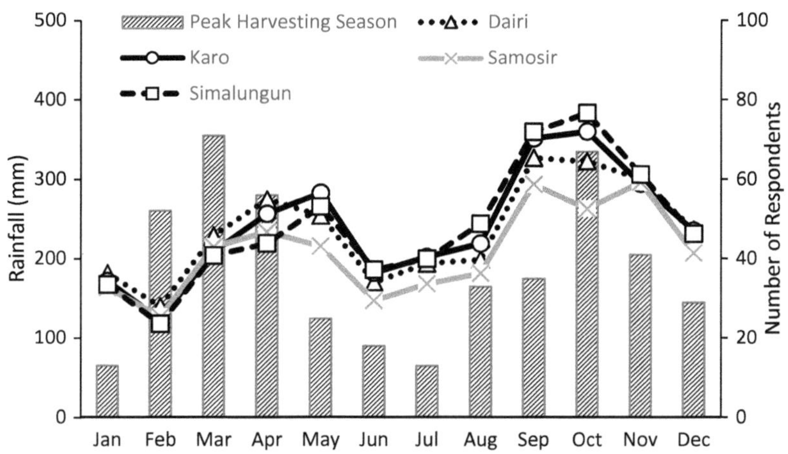

Fig. 5.14 Correlation between rainfall pattern and harvesting period of the coffee plantations in Toba areas

Once the berries are formed, it is essential to ensure an adequate amount of water to yield big and well-ripened berries. Lower rainfall might result in fewer yields due to the smaller berry size. The process from flowering to harvesting periods often takes about 9–12 months depending on the climatic conditions.

Toba is located near the equator, has bimodal rainfall seasonality, with the highest rainy season in April–May and September–October in line with the peak of coffee harvesting seasons (Fig. 5.14). Flower initiation, blooming, pollination, development and ripening of berries, followed the same pattern as the rainfall. In Toba area, flower initiation that occurs during the transition from dry to wet and wet to dry, coincided with the seasonal transition (June and December). Rainfall intensity during the transition periods played an essential role in determining the number of blooming and pollinated flowers. Erdiansyah et al. (2016) report that insufficient or irregular sunlight might create irregular fruit-set. Following the formation of fruit (berries), the plants would require further 6–7 months for ripening and development before harvesting (Fig. 5.15).

The development and growth of coffee berries require a high amount of water. Vegetative growth will be more vigorous under high intensity of rainfall, hence pruning should be scheduled more often. Super-selective pruning might necessarily be scheduled monthly in areas with higher intensity of rainfall, whereas selective pruning can only be scheduled once a year in drier areas (Yuliasmara et al. 2016), to ensure an optimum coffee-growing condition and to produce satisfying yields.

5.3.2 Climate Variability and Coffee Production

Rainfall and temperature are significant drivers for coffee production, where prolonged dry and rainy seasons will affect the flower development of coffee plants.

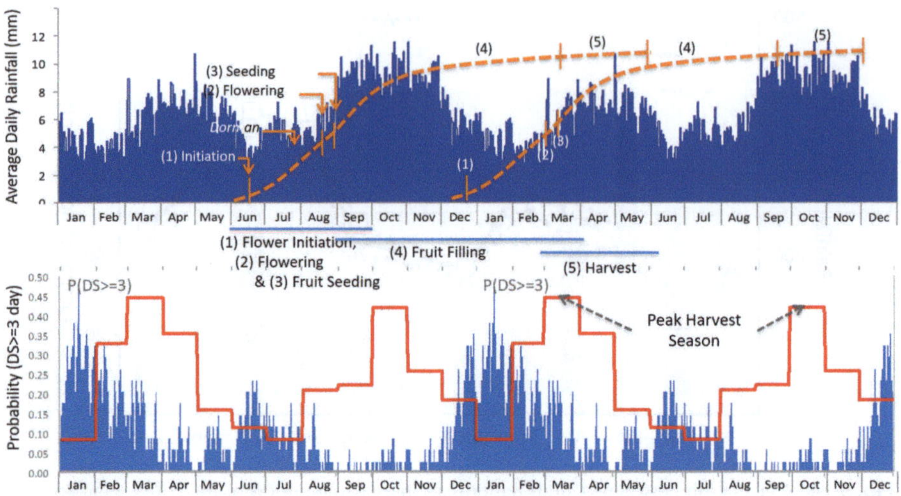

Fig. 5.15 Rainfall distribution along different phenological development of Arabica coffee (Top) and the probability of getting three consecutive days without rainfall (DS > =3 days) in Toba areas (Bottom)

Under very humid condition, the risk of shrivelling flower and fruit loss will increase (Nur 2000) and yields many empty berries (Soenarjo 1975).

Drought stress during prolonged dry season is often combined with high temperature and high solar radiation that could potentially harm coffee production (Pujiyanto 2016). Symptoms of lack of water or drought stress on coffee plantation begin with shrivelling, drying and falling of leaves. Water stress also causes stunted growth and detains buds' development and flowers' blooming, reducing production in the next year (Yahmadi 1973). Increase in temperature would affect plant metabolism such as flowering, photosynthesis and respiration that in general would affect the total coffee production (Syakir and Surmaini 2017).

Stress from rapidly increasing temperature is significant on Arabica coffee. Coffee plantation is usually located in high elevation areas, with a relatively low temperature in average. The analysis revealed that coffee growth is optimum in areas between 1300 and 1500 m MSL (Fig. 5.16). Toba coffee plantations are located within this elevation range.

Further analysis of the relationship between elevation and coffee production indicated that Super Ateng cultivar grows better at high altitudes and significantly produces higher yields on plantations located at 1600 m MSL. On the other hand, Katimor Sigararutang cultivar yields less on locations above 1400 m MSL (Fig. 5.17).

The analysis also found that coffee sensitivity to climatic conditions varied among different growth stages. The vegetative phase tends to be more sensitive to changes in temperature, while the generative phase was highly dependent on rainfall intensity and pattern. Table 5.9 summarises the impact of climate variability on coffee plantations. This information is necessary for further assessment of potential coping adaptation to climate change impacts.

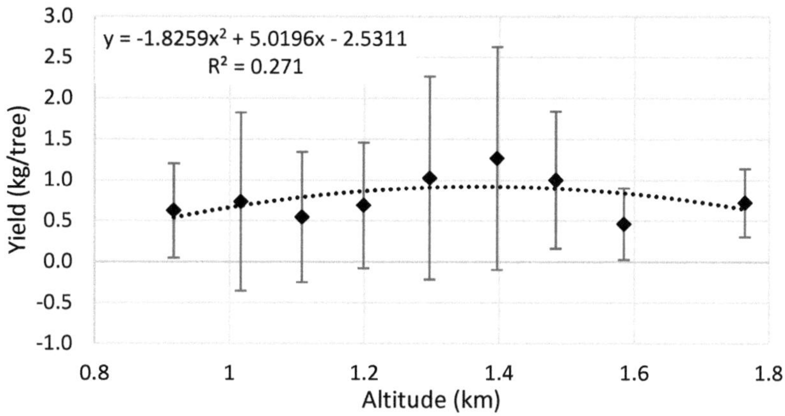

Fig. 5.16 Correlation between altitude and yield

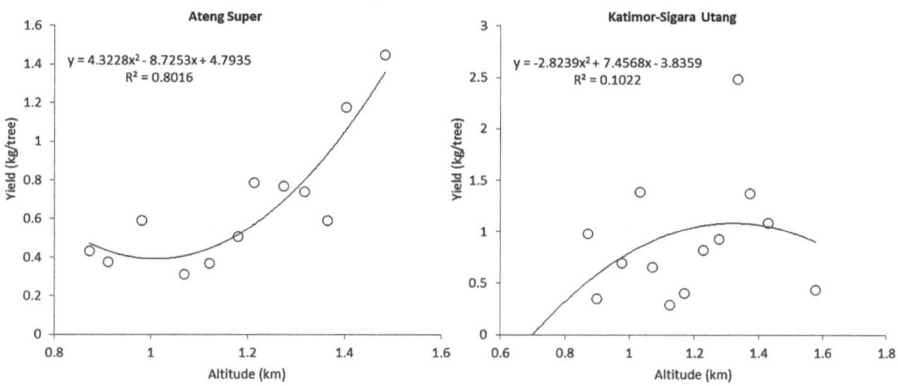

Fig. 5.17 Correlation of yields of AS and KS with location of plantations (altitude)

5.3.3 Climate Variability and Pest and Disease Occurrence

Pests and diseases have been identified as one of the critical factors in coffee production loss. Main pests and diseases include berry borer (coffee berry borer), nematode parasite (*Pratylenchus coffeae*) and leaf rust (Prastowo et al. 2010). Leaf rust is caused by *Hemileia vastatrix* Berk and Br. and nematode *Radopholus similis* (Cobb). Thorne also often harms Arabica coffee; berry borer *Hypothenemus hampei* Ferrari, *Pratylenchus coffeae* (Zimmermann) Filipjev and Schuurmans Stekhoven are commonly found to attack both Arabica and Robusta coffee. In general, common pests and diseases in Indonesia are dominated by nematode species (Wiryadiputra and Tran 2008).

Table 5.9 Summary of climate impacts on coffee production

Growth stage	Climate variable	Physiological impacts	Results	Source
Vegetative	$T > 25$ °C	Slower photosynthesis	Low production	Wilson (1985)
Vegetative (leaves development)	$T > 30$ °C (continuously)	Slower growth, yellow leaves, drying plants	Low production	Damatta and Ramalho (2006)
Generative (flowering)	Prolonged dry season	Flowers fall (direct)	Low production	Damatta and Ramalho (2006)
Generative (flowering)	High intensity of (extreme) rainfall	Flowers fall (direct)	Low production	Wintgens (2009)
Generative (seed filing/ germination)	$T > 23$ °C	Speedy growth and ripening that yield low-quality berries (indirect)	Low selling price	Damatta and Ramalho (2006)
Pest and disease	High temperature	Outbreaks	Low production	ICC (2009)

5.3.3.1 Coffee Berry Borer

Coffee berry borer is an indigenous African beetle and has been considered as the most dangerous pest for both Arabica and Robusta plants worldwide. Canephora (Robusta) coffee is the most susceptible to coffee berry borer (Lan and Wintgens 2009). Coffee berry borer has been spreading evenly in almost all coffee plantations in Indonesia including Papua, Sulawesi, Sumatra and Java (CABI 2000). The insect is predominantly found in lower altitude plantations. Berry borer has been a severe problem, not only in Indonesia but also in other coffee producing countries such as India and the Philippine. Loss of national production, both in quantity and quality, is significantly due to the attack from this pest (Sulistyowati et al. 1999; Samsudin and Soesanthy 2012).

Variation of coffee berry borer attacks depends on growth stages, land condition and cultivation system of the coffee plantations (Samsudin and Soesanthy 2012). Shades and light control may reduce the effect of coffee berry borer attack (Samsudin and Soesanthy 2012). As indicated in the earlier section, vulnerable period to coffee berry borer attack existed within the age of 5 to 10 years. The damage rate due to coffee berry borer in Toba has been categorised as high; some studies reported that production loss caused by the pest was about 25–49%. More than 60% of the Arabica coffee plantation in Toba had experienced coffee berry borer attacks within the last 10 years or equivalent to 50% of production lost (Rahayu and Wiryadiputra 2016).

The foremost climate variables that affected the level of coffee berry borer attack are temperature, rainfall and humidity. The cycle of coffee berry borer life is found to be longer at lower temperature and higher altitudes (Wiryadiputra 2012). Frequent rainfall throughout the year would also increase the possibility of coffee berry borer attacks, since under such climate, coffee flowering period is more frequent. Relative humidity and temperature also invite the spring of female insects.

Coffee Stem Borer

Coffee stem borers (*Xylotrechus quadripes* Chevrolat and *Acalolepta cervina* Hope) have been causing an extreme production loss of Arabica coffee in Thailand and China, although the effect was less harmful than berry borers (Lan and Wintgens 2009). Similarly, the climate variables influencing the intensity of stem borer are temperature and humidity. A humid environment is very favourable for the outbreaks of stem borer (Rahayu and Wiryadiputra 2016).

Green Scale

Coccus sp. is a soft green scale insect, which has also been a significant pest on Arabica and Robusta coffee plantation worldwide (Crowe 2009), although it was found to be more dominant on Arabica (Lan and Wintgens 2009). Green scales actively reproduced during the dry season, so it is necessary to wipe the eggs out before the dry season. Shading trees should also be maintained because the attacks are massive under thick set of shades.

Striped Mealybug

Striped mealybug, *Planococcus* spp., is found in almost all coffee plantations in the Philippine, Indonesia, Thailand, Malaysia, India and China (Lan and Wintgens 2009). The attack is worse during the dry season with intermittent rainfall. Random set and/or lack of shade trees worsen the damages caused by this pest. Shading should ideally be about 30% for Arabica coffee plantation and 20–25% for Robusta (Lan and Wintgens 2009).

5.3.3.2 Diseases

Microfungi, bacteria and few viruses have been the most common disease-causing organisms in a coffee plantation. The impacts on the different parts of the plant varied, from softening stem, damaged or even killed plants (Waller 1985). Climate change potentially worsens the impacts on coffee production. Increasing temperature has increased the impact of leaf rust on coffee planted in higher altitude. Lack of rainfall on low altitude plantation has caused water stress and an increasing case of Fusarium bark. On the other hand, wetter period can cause coffee berry disease (Waller et al. 2007). The following are some of the common diseases of coffee plantations in Indonesia.

Coffee Leaf Rust

Hemileia vastatrix Berk & Br is the cause of leaf rust. The fungus is originally from Africa and North America (Brown et al. 1995) and estimated to start spreading in Indonesia in 1876 as an airborne disease. Due to this disease, coffee plants usually shred leaves or even killed. Damage was about 58% or equivalent to 25% production loss (Amaria and Harni 2012). Arabica coffee is relatively more vulnerable to the disease compared to Robusta (Semangun 2000; Amaria and Harni 2012).

Leaf rust is commonly found in lower altitude plantations (<1500 m), because of the increased humidity and intensity of solar radiation (Rayner 1961; Waller et al. 2007). Therefore, Arabica planted in humid environment needs extra maintenance. Temperature is a critical climate variable in the spread of leaf rust (Waller et al. 2007).

Brown Eye Spot

Brown eye spot is caused by *Cercospora coffeicola* and *Mycosphaerella coffeicola* and commonly found in low altitude plantations, where the temperature is relatively higher. This disease is both airborne and waterborne through rainfall splash. The analysis has suggested that the current climatic condition is more favourable to the outbreaks of brown eye spots. Table 5.10 summarises the impacts of climate on pests and diseases.

5.3.3.3 Post-harvesting Infection Pests

Production and processing are susceptible processes to pests and diseases. Coffee processed in full wash, low moisture content, is relatively more resistant to pest and disease, while coffee produced or processed in humid tropical areas is more prone to pest or disease (Crowe 2009).

The major pest for stored coffee is beetle (*Araecerus fasciculatus* DeGeer). Such a case has been typical in pantropical areas. Maintaining humidity in a storage room or avoiding humid environment for storage may prevent the beans from the attack. Drying the beans under the sun can also kill this pest as the beetles will not be able to survive if the temperature is more than 37 °C (Crowe 2009).

5.4 Assessment of Historical and Future Climate

5.4.1 Historical Climate

The assessment of past climate was based on interpolated data of daily rainfall and monthly maximum and minimum temperatures. Daily rainfall was derived from CHIRPS (Climate Hazards Group InfraRed Precipitation with Station data) (Funk et al. 2014; Funk et al. 2015) which has been validated with observation data.

Toba is a part of the mountainous Bukit Barisan in the western part of Sumatra Island. Its location within equator and the topographical conditions have been the main factors determining the climatic condition. Toba has a bimodal equatorial rainfall, and the peak occurs in April/May and Sep/Oct (Fig. 5.18). Areas with an equatorial type of rainfall typically have high annual rainfalls. Analysis based on CHIRPS data (1981–2010) indicated that the annual rainfall in Toba area is high and ranges between 2250 and 3750 mm, with Samosir having the lowest annual rainfall and Simalungun with the highest annual rainfall (Fig. 5.19). The rainfall distribution pattern throughout the year is uniform in all study areas (Fig. 5.20). The lowest

Table 5.10 Summary of climate impacts on pests and diseases

Pest or disease	Climate variable	Attack level	Final impact	Region and source
Coffee berry borer (*Hypothenemus Hampei* [Ferr])	Optimum temperature 20–30 °C and humidity 90–95%	The attack is more common under humid environment. The warmer temperature may shorten the insect's life cycle so that it may increase the population	Berries are undeveloped, turned yellow, full of bores and eventually aborted	Southeast Asia, South America, Africa and China (Lan and Wintgens 2009), Indonesia (Susilo 2008; Samsudin and Soesanthy 2012; Wiryadiputra 2012)
Coffee twig borer (*Xylosandrus morigerus* and *Xylosandrus compactus*)	High humidity	The increase in temperature and humidity will increase the intensity of disease attack. The population rapidly increase in the rainy season but relatively lower in the dry season	Primary twig, seed and water shoots will shiver and die	Southeast Asia, Tropical Africa and India (Lan and Wintgens 2009), Indonesia (Rahayu and Wiryadiputra 2016)
Striped mealybug (*Ferrisia virgate*)	Humidity <70% and optimum temperature ± 30 °C	Population increase when the temperature increases	Flower buds, flowers and berries die	Southeast Asia and China (Lan and Wintgens 2009), Indonesia (Rahayu and Wiryadiputra 2016)
Coffee leaf rust (*Hemileia vastatrix*)	Humidity<70%	Increasing humidity will support the development of the disease and wind tends to spread mature spores	Yellow/ brown/black spots around the leaves and eventually leaves are falling	Nicaragua (Lan and Wintgens 2009; Brown et al. 1995; Waller 2007), Indonesia (Pratama and Aini 2016)
Brown eye spot (*Mycosphaerella coffeicola* or *Cercospora coffeicola*)	High humidity	High humidity during rainy season increases the risk to this disease	Brown to red necrosis on leaves and eventually leaves are falling	Indonesia (Pratama and Aini 2016)

(continued)

Table 5.10 (continued)

Pest or disease	Climate variable	Attack level	Final impact	Region and source
Pink diseases (*Corticium salmonicolor*) on stem, twig and berries	High humidity	The fungus spreads via wind and water/rainfall splash on humid and dark environment	Twigs are infected and killed	Indonesia (Pratama and Aini 2016)
Root diseases (*Phellinus noxius, Rosellinia bunodes* and *Rigidoporus microporus*)	Optimum temperature 25–30 °C, soil pH 3.5–7.5	Disease spreads faster in sandy areas with higher humidity	Infected the root and spread via the remaining damaged roots left under the surface	Indonesia (Wiryadiputra 2016), Kamerun (Lan and Wintgens 2009)

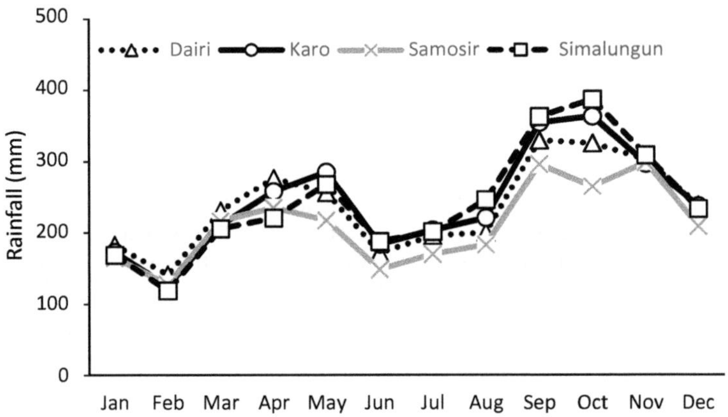

Fig. 5.18 Rainfall pattern of the four districts (study areas)

intensity of rainfall usually occurs in February, or June, July and August during the dry season or dry months. The monthly rainfall during the dry season, on an average is about 150 mm. Distribution of the dry months is in line with the rainfall pattern. The total number of dry months within a year is depicted in Fig. 5.21.

The temperatures in study areas were very much related to the topographical conditions. Higher topography experienced relatively lower temperature and vice versa. Minimum temperature ranges between 14 and 20 °C (Fig. 5.22), and maximum temperature ranges between 22 and 30 °C (Fig. 5.23). These are the ideal temperatures for coffee production. Therefore, Toba area is known as the coffee production centres. Dairi Highland is known as the best location to produce the best quality of coffee in Toba area followed by the Districs of Karo and Simalungun.

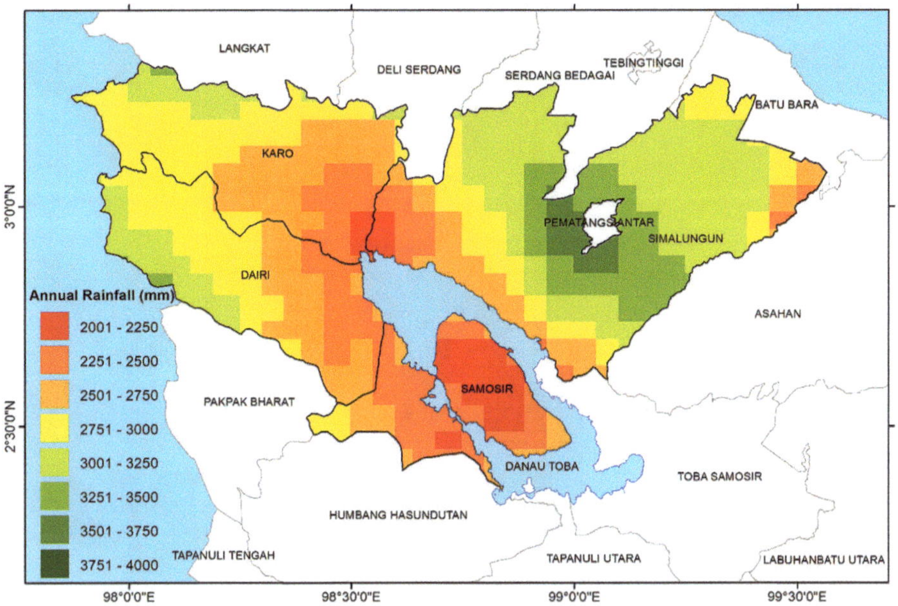

Fig. 5.19 Distribution of annual rainfall (mm) from 1981 to 2010. (Source: CHIRPS data (http://chg.geog.ucsb.edu/data/chirps/) validated with observation data)

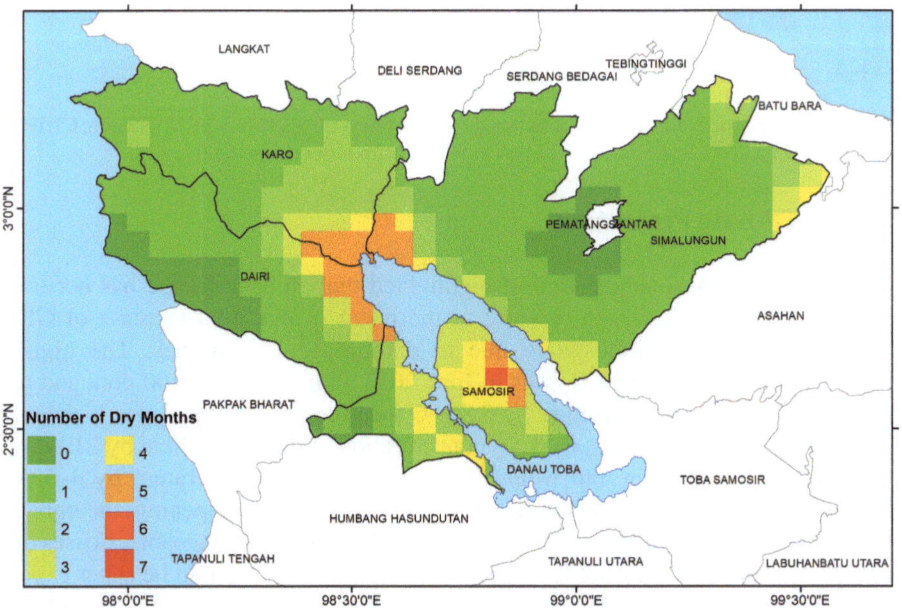

Fig. 5.20 Distribution of number of dry months in Toba area

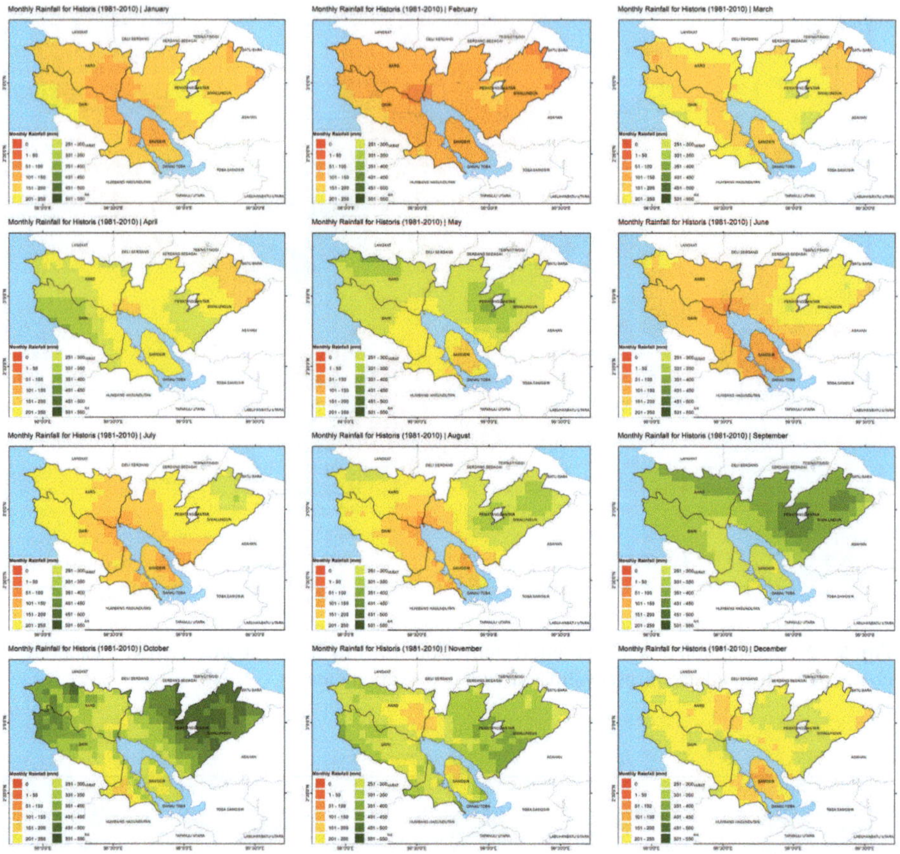

Fig. 5.21 The distribution pattern of monthly rainfall (mm) from 1981 to 2010. (Source: CHIRPS data (http://chg.geog.ucsb.edu/data/chirps/) validated with observation data)

5.4.2 Climate Projection

The IPCC Fifth Assessment Report highlighted that climate change has occurred due to increasing global temperature as the result of the rapid increase of GHG concentration in the atmosphere due to anthropogenic activities. The highest recorded concentration of CO_2 was in June 2013 with more than 400 ppm and has increased the global temperature by 0.85 °C than the pre-industrial era. If the current emission rate continues, the CO_2 concentration in the atmosphere by 2100 will be more than 1000 ppm and would increase the global temperature by 4–5 °C, higher than the pre-industrial era (IPCC 2014). Once this is happening, the impacts of climate change would be very devastating and no longer manageable. Losses due to extreme climate events are high, and costs for adaptation would be super expensive. Therefore, it is necessary to push the emission level to the lowest possible, so the future impacts would remain manageable and at low costs. Better mitigation actions would be necessary to minimise the need for adaptation.

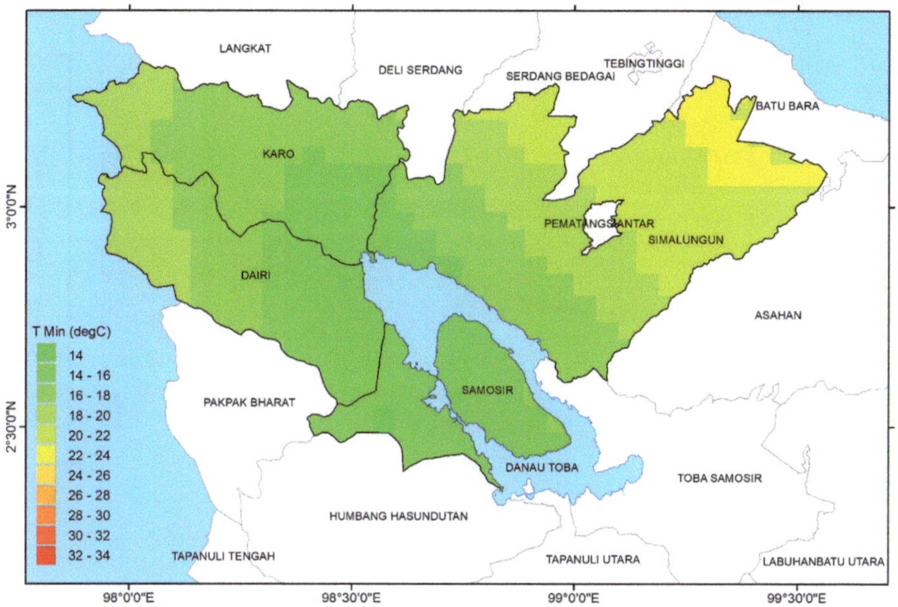

Fig. 5.22 Distribution of minimum temperature from 1981 to 2010. (Source: CRU (http://www.cru.uea.ac.uk/data) validated with observation data)

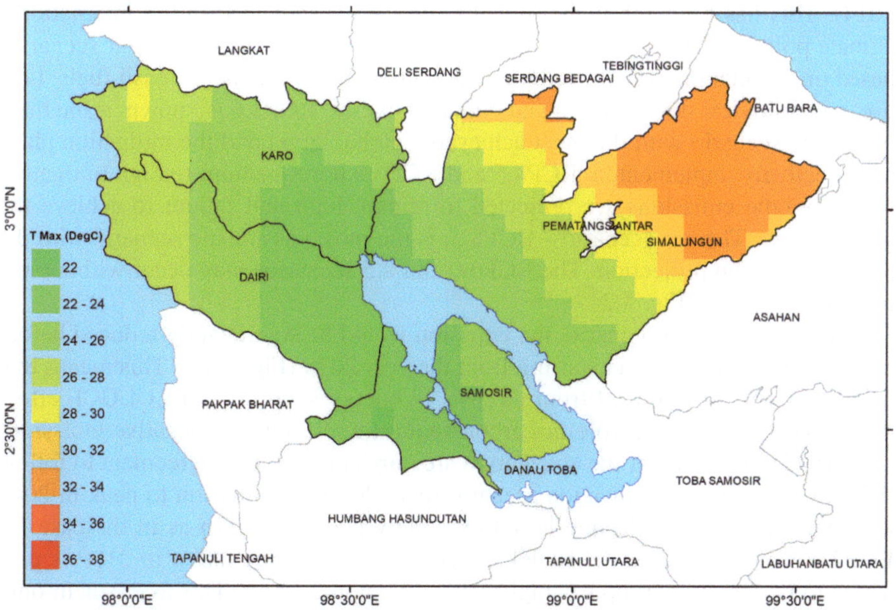

Fig. 5.23 Distribution of maximum temperature from 1981 to 2010. (Source: CRU (http://www.cru.uea.ac.uk/data) and validated with observation data)

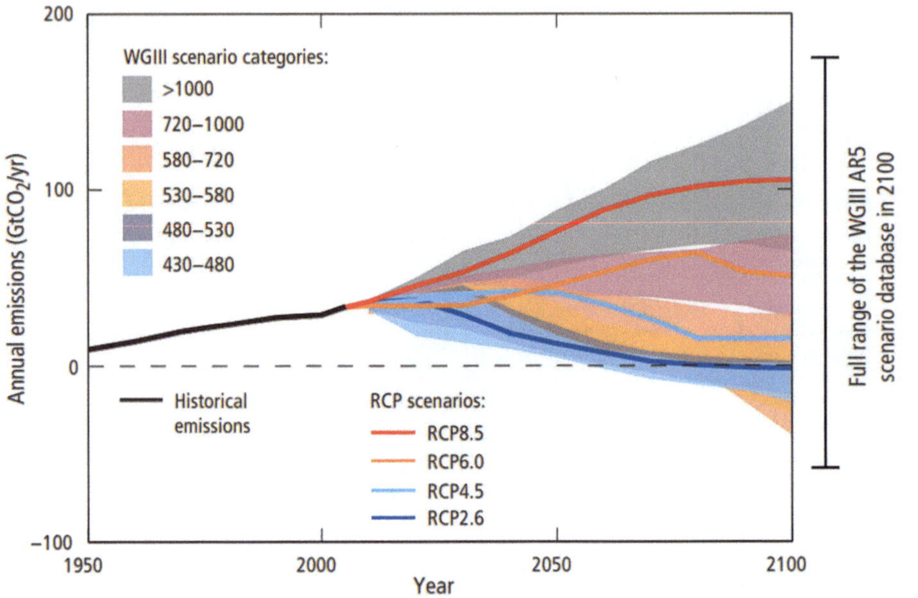

Fig. 5.24 Projection of annual CO_2 emissions under four different scenarios (IPCC 2014)

Maintaining the global temperature increase below 2.0 °C or even 1.5 °C will ensure that the impacts of climate change remain at a manageable level (IPCC 2014). This has been set as the global target and is indicated in the international climate policies. IPCC (2014) has projected future emissions under four scenarios based on population growth, economic activities, lifestyle, use of fossil fuels, land use, technological development and climate policies. RCP2.6 is known as the most optimistic scenario with the most achievable global target and the mitigation plans are effectively implemented. RCP4.5 and RCP6.0 have medium to high mitigation scenarios, the emissions are projected to remain high and failure to achieve the global target. The last scenario – RCP8.5, is the most pessimistic scenario with the highest emission projection. The RCP6.0 and 8.5 scenarios may occur without any mitigation actions (Fig. 5.24).

Under the RCP2.6 scenario, the emission would be significantly reduced before 2030 and come to a 0 level starting from the year 2070 (Fig. 5.24). This means that fossil fuel will be removed from the lists of energy sources, and LULUCF effectively plays its essential role in carbon sequestration towards negative emissions scheme. Countries under the convention are only able to set an agreement to follow RCP4.5 scenario, a stabilisation scenario that allows the emission to peak in 2040 and then decline. The decline is not extremely pushed to zero as in the case for RCP2.6 scenario. Under the RCP4.5 scenario, CO_2 concentration in 2100 is projected to be around 550 ppm – higher than the global target. This assessment only follows the emission projection under scenarios RCP4.5 and RCP 8.5 (Moss et al. 2010). Climate change under both scenarios is simulated using the regional climate model RegCM4. Scheme and simulation process in RegCM4 are explained in Faqih et al. (2016). The model has a spatial resolution of 20 × 20 km and runs ICBC data – the outputs of HadGEM2-ES global climate model (Collins et al. 2011; Martin et al. 2011). Outputs of RegCM4.4 are validated following the procedures from Jadmiko et al. (2017).

Climate change projection is specifically centred on temperature and rainfall, because both variables are the major determinant variables for plant growth. The projection period is split into (i) 2011–2040, (ii) 2041–2070 and (iii) 2071–2099. The analysis was based on general delta method and multiplier (e.g., Lenderink et al. 2007). Delta method is used to validate the projection of temperature from climate model ($X_{mp,i}$) based on the deviation of the model (U_{mb}) from observation data (U_{ob}) during the historical period of 1981–2010. The corrected temperature projection of year – i ($X_{cor,i}$) is explained below:

$$X_{cor,I} = X_{mp,i} + \left(U_{ob} - U_{mb}\right) \quad (5.1)$$

Correction method for rainfall data is as follows:

$$X_{cor,I} = X_{mp,I} * \left(U_{ob} / U_{mb}\right) \quad (5.2)$$

A more detailed description of the methodology for climate projection is available in the Indonesia Third National Communication Report (Faqih et al. 2016).

The analysis found that the minimum temperature in Toba area under the RCP8.5 scenario would increase about 2.0 °C by 2050, while under RCP4.5, the increase would only be about 1.5 °C. Later in 2099, the projected temperature increase under RCP8.5 scenario is 4.5 °C and under RCP4.5 scenario is 2 °C (Fig. 5.25). The increase varies among different areas, but in general, areas with lower altitude tend to experience higher increases of minimum temperature (Fig. 5.26). The maximum temperature is projected to increase at about 2.0 °C by 2050 under the RCP4.5 scenario and about 2.5 °C under the RCP8.5 scenario. Further in 2099, the maximum temperature is projected to increase for both RCP4.5 and 8.5 scenarios by 3 °C and 5 °C, respectively (Fig. 5.27). Similar to the minimum temperature, the increases are also relatively higher in areas of lower altitude (Fig. 5.28). In general, the RCP8.5 scenario suggested a more significant increase of temperature than RCP4.5 scenario for the post mid-twenty-first century.

Another key climate factor in coffee production is rainfall. The future rainfall in Toba area is projected to be 10–15% lower than the current condition. However, from 2071 to 2099, under the RCP8.5 scenario, it is projected to decrease by about 30%, especially within the western part of Dairi. This is equivalent to the national projection where rainfall throughout Indonesia is on an average, decreased 30% from the current condition. In contrast, Samosir area is projected to experience a 5% increase in rainfall. Analysis of the rainfall distribution in the Toba area indicated that in general, the rainfall intensity is decreasing. Changes in rainfall varied among different locations and the variations were high (Fig. 5.29). In general, the western and eastern parts of Lake Toba tend to experience lower rainfall while the northern part experienced slightly higher rainfall than the current levels (Fig. 5.30). The extreme level of rainfall decrease is projected to occur in the western part of Toba in 2099 under the RCP8.5 scenario (Fig. 5.30).

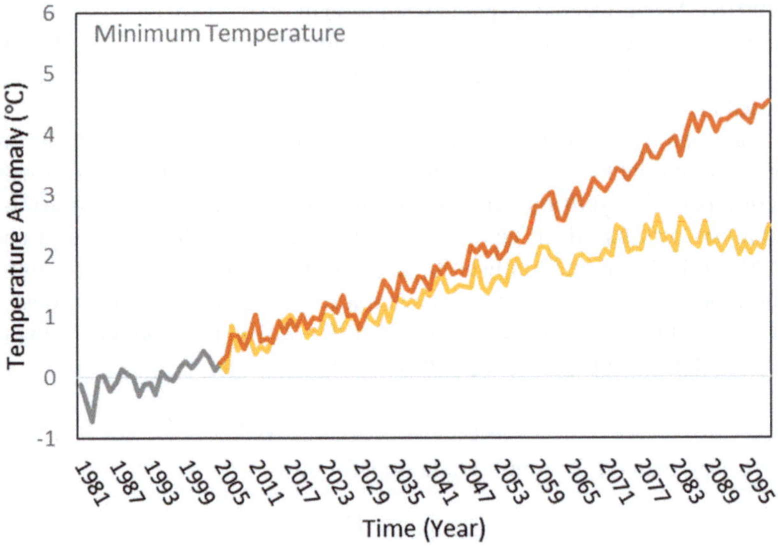

Fig. 5.25 The increasing trend of minimum temperature in Toba area (Yellow lines-RCP4.5 Scenario; Brown lines-RCP8.5 Scenario)

Projection of lower rainfall will have an impact on the length of dry season/months. In general, the total length of dry months in Toba area from 2011 to 2040 is expected to be 1–2 months longer than the existing condition (1981–2010). The prolonged dry season would be worsened by 2071–2099, especially under the RCP8.5 scenario (Fig. 5.31). Longer season is observed to be significant in the north-western part of Lake Toba.

5.5 Climate Change Impacts on Coffee Plantation

Climate change is projected to have significant impacts on yields, occurrence of outbreaks, growth and development of coffee plantation including changes in land suitability for coffee plantation. As stated in the earlier sections of this chapter, Arabica coffee is ideally grown at low-temperature areas, which is often found at an elevation of 800 m MSL. Nevertheless, under the changing climate, areas known best for growing coffee would be shifted upwards. This study focused on climate change impacts on (i) shifting of coffee-growing areas, (ii) growth and development of coffee plantation and (iii) outbreaks of major coffee pests, i.e. coffee berry borer or stem borer.

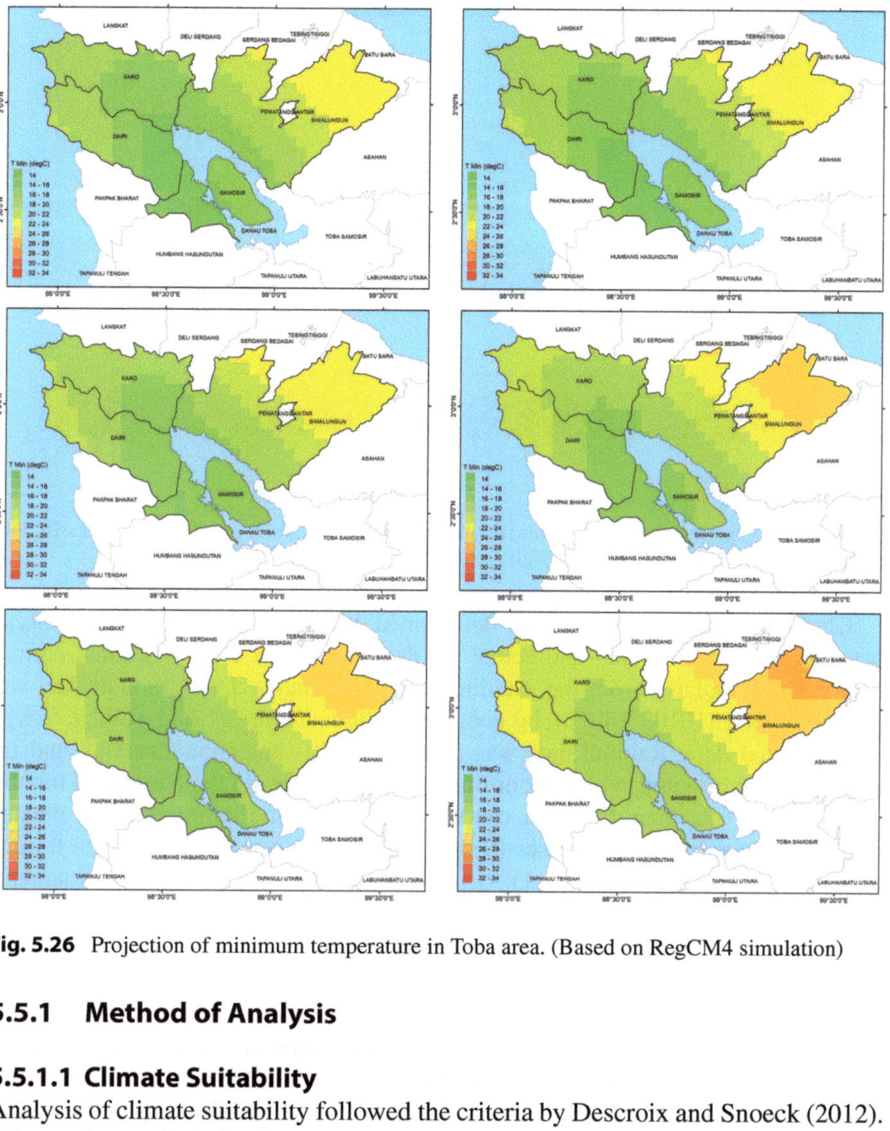

Fig. 5.26 Projection of minimum temperature in Toba area. (Based on RegCM4 simulation)

5.5.1 Method of Analysis

5.5.1.1 Climate Suitability

Analysis of climate suitability followed the criteria by Descroix and Snoeck (2012). Climate factors included in the analysis are minimum and maximum temperatures, annual rainfall and length of dry months. There are three procedures to follow for the analysis. The first procedure is the identification of suitable classes based on the correlation between each climate variable and production (survey); second, climate suitability of the area is set for each climate variable, based on suitability classes and climate suitability index (CSI); and, third, to determine the climate suitability class of the area, expressed as Area Climate Suitability Index (ACSI) based on CSI.

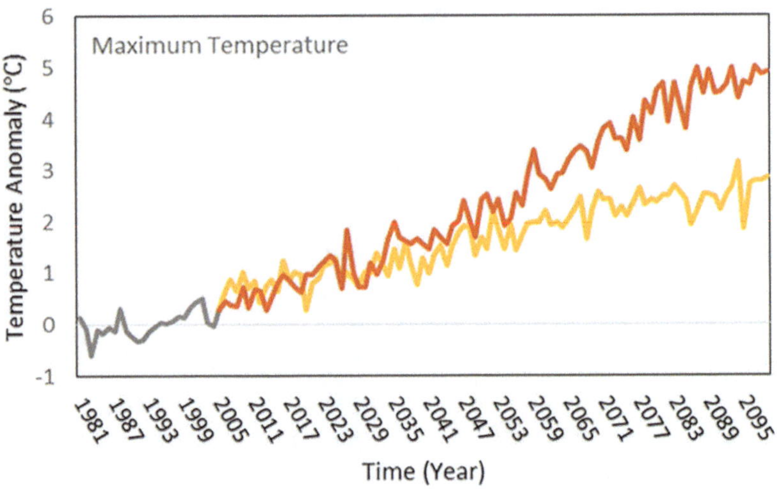

Fig. 5.27 The increasing trend of maximum temperature in Toba area (Yellow lines-RCP4.5 Scenario; Brown lines-RCP8.5 Scenario)

Climate suitability classes are determined based on the regression analysis between climate condition and actual yields from survey data. Historical climate data was interpolated from the satellite rainfall estimates. Daily rainfall data was obtained via CHIRPS (http://chg.geog.ucsb.edu/data/chirps/) and validated with observational data. Rainfall data was used to identify the annual rainfall and the number of dry months, which according to BMKG refers to the number of months with rainfall less than 150 mm (http://www.bmkg.go.id/iklim/prakiraan-musim.bmkg). Climate suitability classes (S) in this study are categorised into the following: (i) very suitable (Score = 1); (ii) suitable (Score = 2) and (iii) not suitable (Score = 3).

Climate suitability index (CSI) is calculated based on the following equation:

$$\mathrm{CSI} = 1/n * S\left(n_i * S_i\right) \tag{5.3}$$

where n is the number of recorded years, n_i is the number of years with certain climatic condition and S_i is the suitability class i, i = 1, 2 and 3. The CSI value of each climate variable is calculated for both historical and three future periods (2011–2040, 2041–2070 and 2071–2099).

Area Climate Suitability Index (ACSI) is mainly determined by CSI, using the equation below:

$$\mathrm{ACSI} = Sw_j * CSI_j \tag{5.4}$$

where w_j is a weight for CSI climate variable j. Weight is assigned to each climate variable based on its contribution in explaining the variance of coffee production. ACSI values range from 1 to 3. Based on ACSI, area suitability is categorised into (i) very suitable with two subcategories, (ii) suitable with three subcategories, (iii)

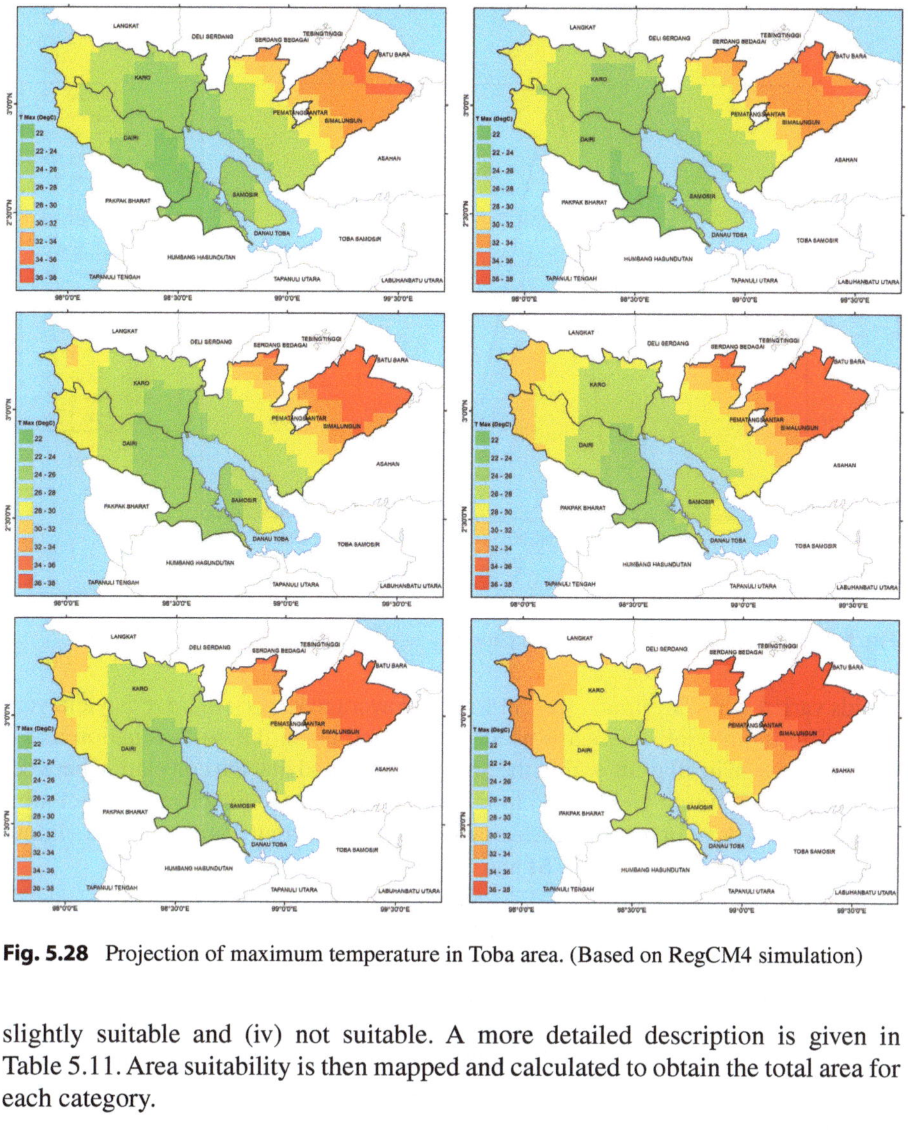

Fig. 5.28 Projection of maximum temperature in Toba area. (Based on RegCM4 simulation)

slightly suitable and (iv) not suitable. A more detailed description is given in Table 5.11. Area suitability is then mapped and calculated to obtain the total area for each category.

5.5.1.2 Plant's Growth and Development

Climate variability impacts on the growth and development of coffee plants were assessed using a qualitative approach to obtain the correlation between plants' phenological characters and climate condition. Impacts on yields are calculated using regression analysis as follows:

$$Y(\text{yield}) = f(\text{Rainfall, Number of Dry Month}, T_{\text{min}}, T_{\text{max}}) \tag{5.5}$$

Yield is measured from the average yield per elevation category (850–1650 m MSL with intervals of 50 m). Regression analysis was weighted because coffee

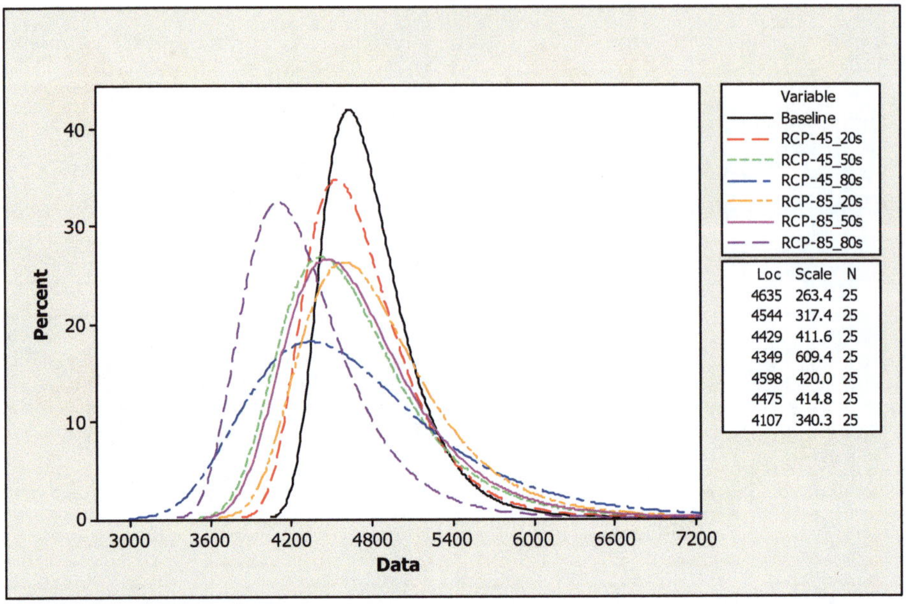

Fig. 5.29 Distribution of annual rainfall pattern

yields varied with elevations. Weight for yield average is $1/\sigma^2$ where σ^2 is yield variance. The future yield average was also estimated using the same equation as climate projection in all grids within the area. It is assumed that the correlation between the present and future yields and climate variables remained unchanged (excluding technological development in coffee cultivation). The contribution of each climate variable on yields would give CSI and ACSI.

5.5.1.3 The Level of Coffee Berry Borer Attack

Climate change impacts on the development of insects populations are approached by means of life table theory or the so-called demographic statistics. Mortality and survival of insect populations are systematically described in the demographic statistics. These are the necessary basic information to understand the changes in the density and growth rate of an insect population and the correlation with climate change.

Insects demographic statistics is defined as the quantitative analysis of insect population life cycle, the ability of female insects to produce eggs and the population growth trend (Zeng et al. 1993). Data on survival rate and egg laying ability are sorted in a life table (Table 5.12). The biological variables observed for the development of demographic statistics include (1) growth of eggs from nymph to imago, (2) growth from first to the fourth instar, (3) period of development from the fourth nymph to pupae, (4) growth from pupae to imago, (5) life as imago, (6) pre-egg laying period and (7) number of eggs laid.

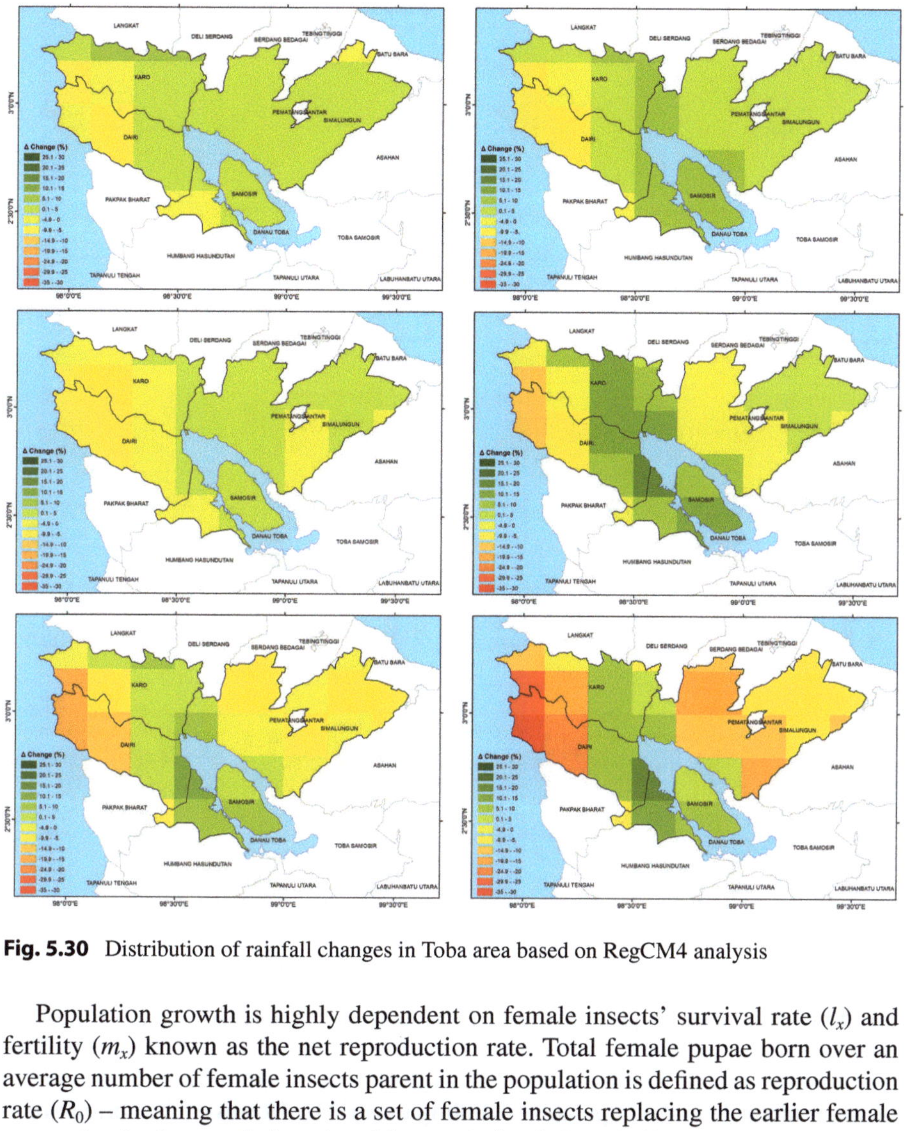

Fig. 5.30 Distribution of rainfall changes in Toba area based on RegCM4 analysis

Population growth is highly dependent on female insects' survival rate (l_x) and fertility (m_x) known as the net reproduction rate. Total female pupae born over an average number of female insects parent in the population is defined as reproduction rate (R_0) – meaning that there is a set of female insects replacing the earlier female generation in the population. A stable population is $R_0 = 0$; if $R_0 > 1$, the population tends to increase, and if $R_0 < 1$, the population tends to decrease. Once R_0 is identified, the length of a generation (T), intrinsic growth rate (r) and doubling time (DT) can be calculated. In general, DT describes the growth of population based on demographic statistics.

The major climate factor in insect demographic is temperature (Bale et al. 2002; Kiritani 2006, 2007, 2013; Gomi et al. 2007; Dingemanse and Kalkman 2008; Jaramillo et al. 2009). Research in a greenhouse indicated that temperature played

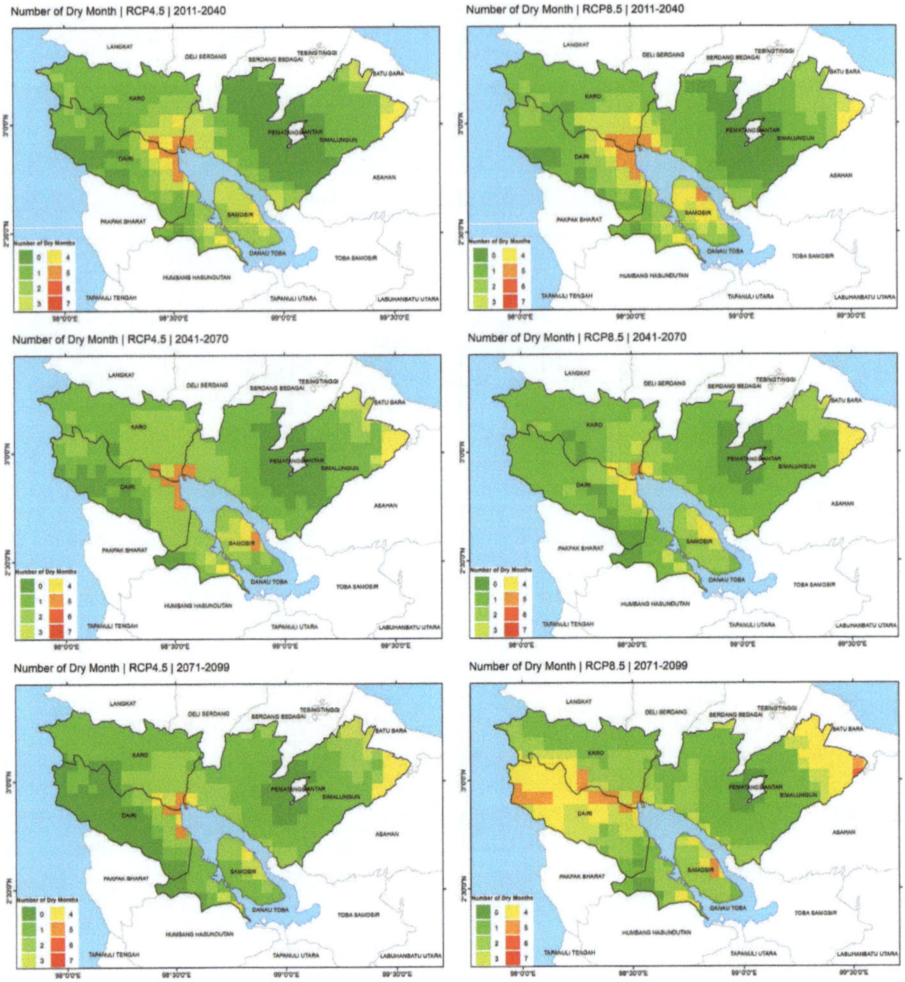

Fig. 5.31 Distribution of prolonged dry months projection in Toba area based on RegCM4 analysis

an important role in the population doubling time (DT). However, the correlation is nonlinear and follows the equation below (Fig. 5.32):

$$\mathrm{DT} = 134 - 9.93 * t + 0.191 * t^2 \tag{5.6}$$

This equation is relevant to calculate the number of days required by coffee berry borer in Toba to double.

The impact of changes in temperature on DT is expressed as the coffee berry borer population doubling index (PDI):

$$\mathrm{PDI} = \mathrm{DT}(t) / \mathrm{DT}(t_0) \tag{5.7}$$

Table 5.11 Area climate suitability classes for coffee plantations in Toba area

Area Climate Suitability Index (ACSI)	Sub-category	ACSI range
Very suitable (VS)	VS1	1.0–1.2
	VS2	1.2–1.4
Suitable (S)	S1	1.4–1.6
	S2	1.6–1.8
	S3	1.8–2.0
Moderately suitable (MS)	MS	2.0–2.2
Unsuitable (NS)	NS	>2.2

Table 5.12 Parameters in demographic statistics

Parameter	Equation
Net reproduction rate (R_0)	$R_0 = \Sigma l_x m_x$
Intrinsic growth rate (r)	$r = \ln R_0/T$
Average of life (T)	$T = \Sigma x l_x m_x/\Sigma l_x m_x$
Population multiplication (DT)	$\mathrm{DT} = \ln(2)/r$

Source: Birch (1948)

x = Cohort age class(day i); lx = Proportion of an insect living at age x; mx = Specific egg laying ability of insect in age class x

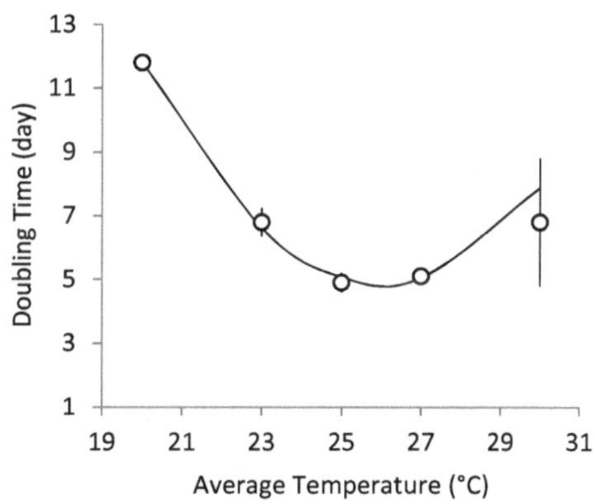

Fig. 5.32 Correlation between doubling time (DT) and temperature. (Adapted from Jaramillo et al. 2009)

where t_0 is the shortest coffee berry borer doubling time in Toba area which is set as the *base time*. DT (t_0) in Toba is 4.94 days. If PDI = 1, then DT(t) would be 4.94 days; if PDI = 2, then DT(t) would be twice longer than the base time or 9.87 days.

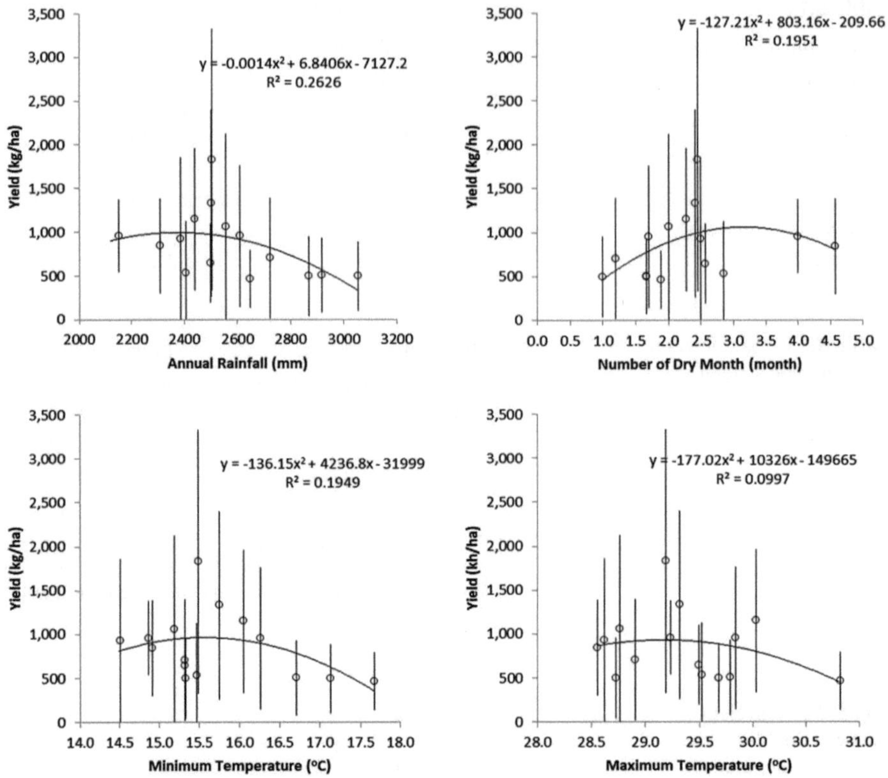

Fig. 5.33 Relationship between climate and coffee yield

5.5.2 Impacts on Coffee Plantation

5.5.2.1 Climate Suitability for Coffee Plantation Area

Regression analysis for each climate factor and coffee yield showed a nonlinear relationship (Fig. 5.33). Yield was relatively lower, if the annual rainfall is at extreme low (less than 1650 mm) or extreme high (more than 3200 mm). The highest yield is produced in areas with annual rainfall ranging from 2050 to 2840 mm (Fig. 5.33). Yield is found to be at a moderate level on areas with annual rainfall of 1650–2050 mm or 2840–3200 mm.

Another variable is the number of dry months, with an optimum range of 2.5–3.8 months. Shorter or longer dry months than the optimum range usually would result in less yield. Both low and high extremes temperatures are also unfavourable for coffee production (Fig. 5.33). Regression equation used to identify the suitability class is listed in Table 5.13. Results of the analysis were in line with existing literature that concluded Arabica coffee needed an annual rainfall of about 1250–3000 mm and showed optimum growth at 1700 mm. The ideal altitude is between 800 and 1500 m above sea level with temperature about 17 °C–23 °C. Unlike Robusta, Arabica coffee is less sensitive to the length of dry months and can survive up to 5 months.

Table 5.13 Climate suitability area for coffee plantations in Toba

	Climate suitability class (S)		
	Very suitable	Suitable	Not suitable
Climate variables	Score = 1	Score = 2	Score = 3
T_{max} (°C)	28.5–30.0	27.5–28.5 and 30.0–31.0	<27.5 and >31.0
T_{min} (°C)	14.5–16.5	13.5–14.5 and 16.5–17,5	<13.5 and >17.5
Rainfall (mm)	2050–2840	1650–2050 and 2840–3200	<1650 and >3200
Number of dry month (month)	2.5–3.8	1.0–2.5 and 3.8–5.5	<2.5 and >5.5

Table 5.14 Weight of climate suitability index (CSI)

Climate variable	Weight
Minimum temperature	0.48
Maximum temperature	0.20
Rainfall	0.30
No. dry months	0.02

Table 5.15 Total very suitable coffee plantations areas (ha) under current and future climates

		2011–2040	2041–2070	2071–2099
Elevation (m asl)	Current	RCP4.5		
700–1500	311,124	221,880	101,331	46,431
1500–2500	51,827	74,622	91,718	90,757
		RCP8.5		
700–1500	311,124	202,314	52,754	779
1500–2500	51,827	77,196	88,420	54,645

CSI is then calculated using the climate suitability class criteria (Table 5.13) for the periods of 2011–2040, 2041–2070 and 2071–2999 to obtain suitability classes of 1, 2 or 3. The analysis indicated that CSI for each climate variable is higher, meaning that climate suitability is getting less suitable. CSI determines ACSI, and in calculating the ACSI, CSI is given weights as shown in Table 5.14.

Based on the analysis, it was found that in the future, the total areas of coffee plantations located at elevations of 700–1500 m MSL that were classified as very suitable, would experience declines (Table 5.15). Very suitable areas are shifting towards higher elevations (Fig. 5.34). Nevertheless, higher elevation areas are mostly categorised as protected land. Total areas and distributions of climate suitability classes under the current and future climate are presented in Fig. 5.35.

5.5.2.2 Plant Growth and Development Index

In general, the climate change in Lake Toba was indicated by increase in temperature (Fig. 5.26 and Fig. 5.28) and changes in rainfall (Fig. 5.30). The future

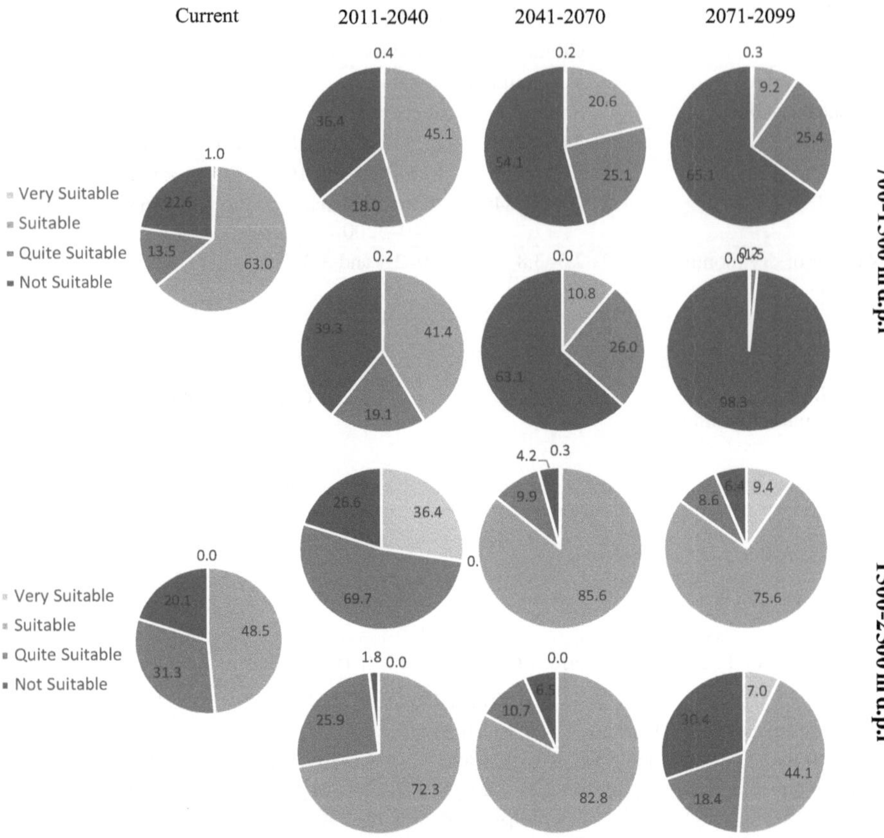

Fig. 5.34 Percentage of area of climate suitability index

occurrence of extreme low or high temperatures were projected to be more frequent and more prolonged. Change of climate as well as the impacts varied among locations. A significant increase in rainfall during the rainy season and prolonged dry season would influence the coffee phenological development. Climate change tends to shift the peaks of flowering and harvesting seasons in Toba. The high intensity of rainfall would reduce yields because it might cause the flowers to fall and increase the risks to grow vegetative branches more than the productive branches.

Analysis of the correlation between yields and climate is represented well by temperature and rainfall. The equation is as follows:

$$\begin{aligned} Y(\mathrm{kg/ha}) = {} & -28{,}174 + 6118 * T_{\min} - 173 * T_{\min}^{2} \\ & - 651 * T_{\max} - 1.89 * P - 125 * \mathrm{DRY}; R^2 = 68\% \end{aligned} \tag{5.8}$$

T_{min} and T_{max} are the minimum and maximum temperatures (°C), P is the annual rainfall (mm) and DRY is the length of dry months (month). Yield is projected to

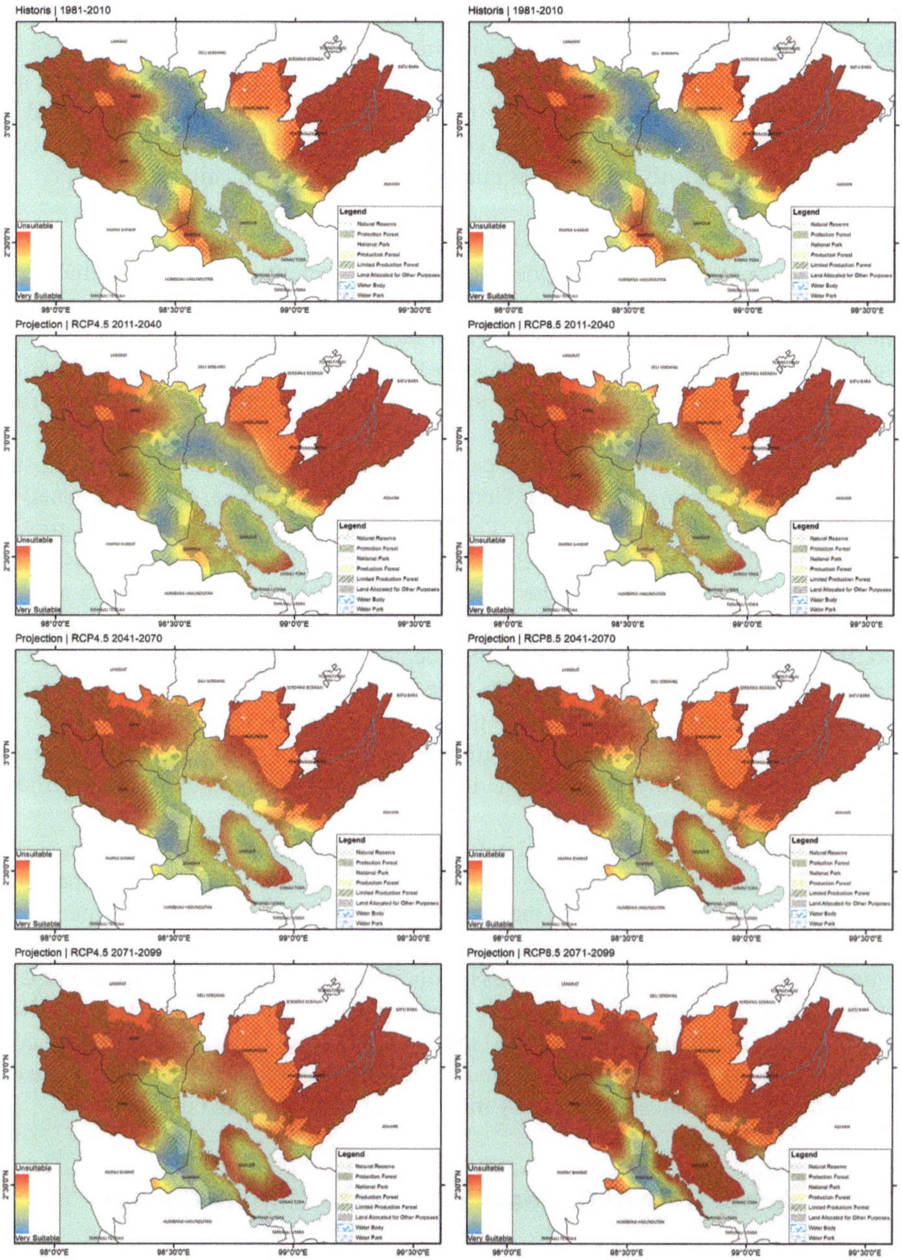

Fig. 5.35 Climate suitability index for coffee plantations in Toba area

decline with the increase of temperature and rainfall in the future. In general, yield is projected to decrease to about 25–75% in the future.

On the contrary, coffee planted on areas 1500 m MSL currently classified as not very suitable is projected to yield 1.5 to 2.0 times higher than the current yield or even twice in some areas (Fig. 5.36). This indicates a potential shift of suitable coffee production areas to higher areas currently considered not suitable.

5.5.2.3 Level of Coffee Berry Borer Attack

In addition to direct impacts on phenological development and yield of Arabica coffee, climate change is also projected to exacerbate the outbreaks of certain pests and diseases (Table 5.10). In general, future climate change is projected to increase the level of attack and modify the attack to higher altitude plantations. Pollination agents are also estimated to decline under the changing climate, hence flowering will eventually decline too.

The analysis identified that the impact of climate change on pest attack is mostly related to the doubling time (DT). Under the changing climate, DT is estimated to range around 5–14 days (Fig. 5.37). The lower PDI values indicated that the growth rate of coffee berry borer increases, increasing the potential risk of yield loss. PDI is projected to decrease or coffee berry borer population will double in a shorter period. Figure 5.37 depicted PDI projection values from 2071 to 2099 under the RCP4.5 and RCP8.5 scenarios. If RCP8.5 occurs, then the PDI values are approaching 1 in most areas. Research in South America and Africa reported that the population of coffee berry borer is increasing from 2 generations per year in the 1970s to 3.5 generations per year lately. This implies that risks of coffee berry borer attack have continuously elevated from time to time and without immediate actions, these would threaten coffee production.

The increased population of the coffee berry borer in some of the Toba areas are threatening the coffee productions, both in quantity and quality. The larger the damages caused by pests, the higher the loss of production experienced by farmers. The direct and indirect impacts of pest attacks were devastating. Based on the survey interviews, it was found that yield losses due to coffee berry borer attacks in Dairi and Simalungun districts totalled to 40%, and in Karo district about 30–40%, while the most significant loss was observed in Samosir District (50%). The declining coffee production, which currently ranged from 30% to 50%, is projected to increase in the next few decades, if necessary measures are not undertaken.

5.6 Adaptation of Coffee Plantation to Climate Change

A whole body of literature have suggested that global climate change has significant impacts on the growth and yield of coffee. As discussed in the earlier sections of this chapter, the observed changing climates occurring in Toba area (Samosir, Simalungun, Karo and Dairi) were mainly related to the decreasing intensity of rainfall and increasing temperature which were projected to have negative impacts on the growth and production of coffee. The impacts will be exacerbated in the

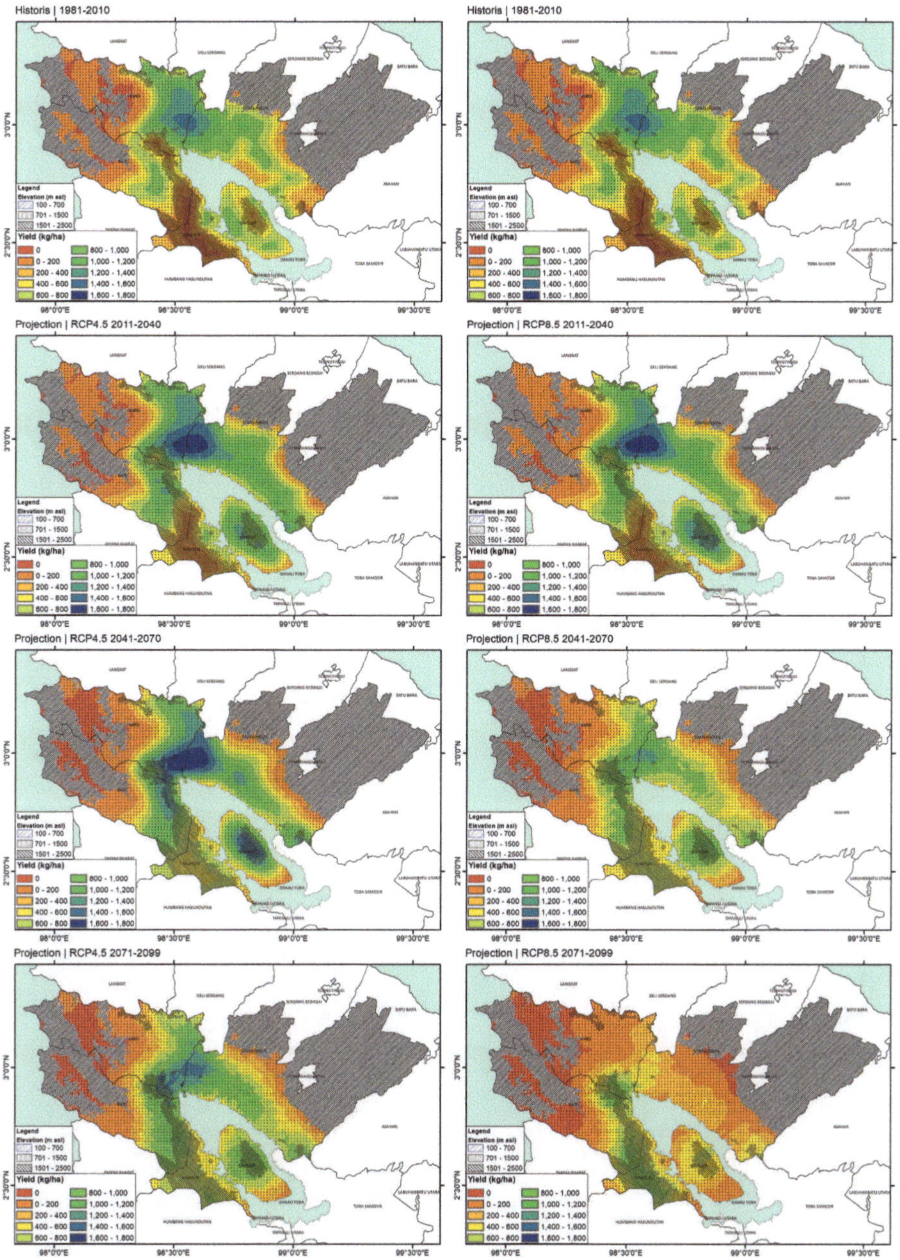

Fig. 5.36 Average yield decreases relative to current yields in Toba area

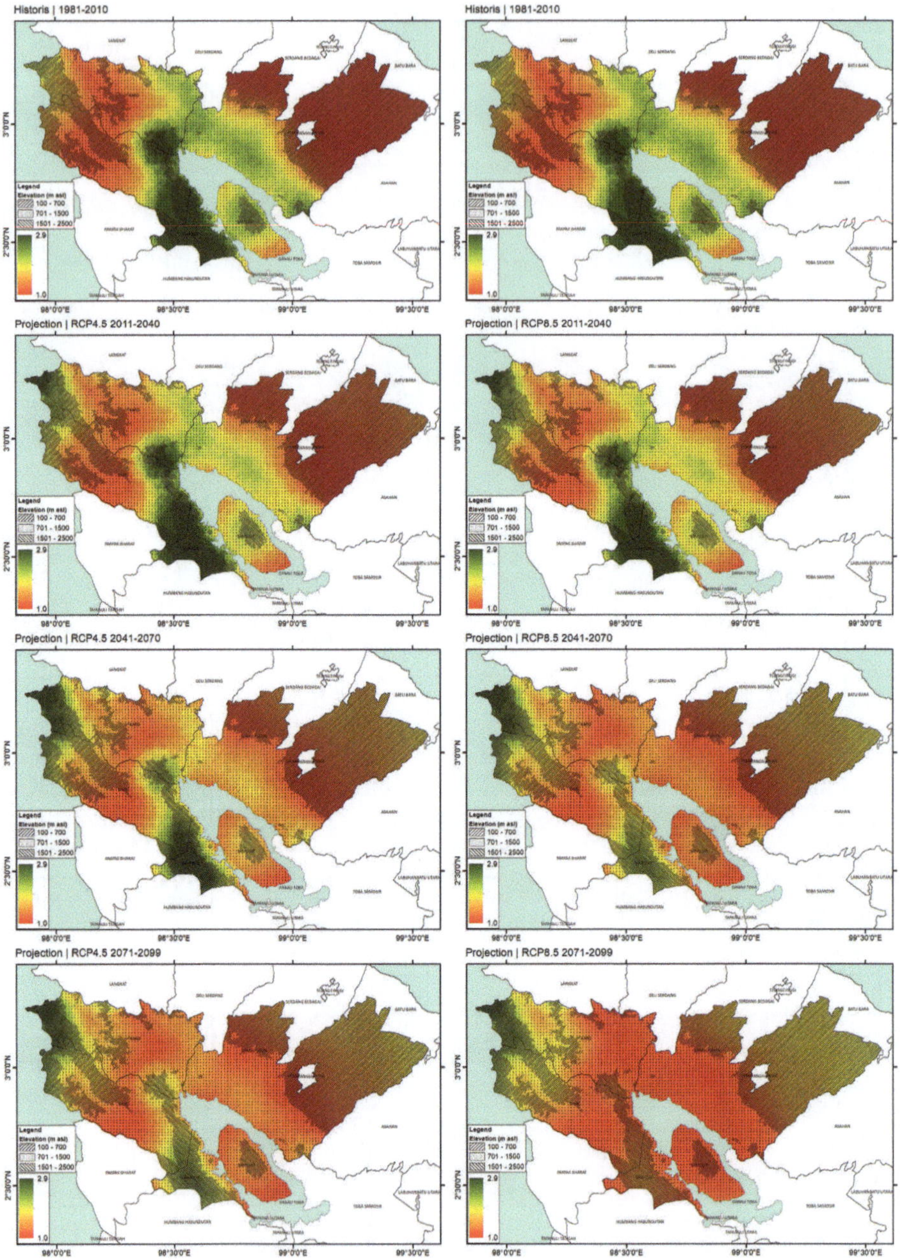

Fig. 5.37 Population doubling index (PDI) of coffee stem borer (coffee berry borer) in Toba (2011–2040; 2041–2070; and 2071–2099) under two projection scenarios ($DT(t_0) = 4.94$ days)

absence of optimum cultivation techniques. Adaptation measures should therefore focus on improving the cultivation techniques and the technology to cope with drought and warmer temperature in coffee production centres, for example, through shade control, addition and management of organic matter, balanced fertiliser application, regular pruning, harvest and post-harvest management and improvement of pest/disease control – especially on coffee berry borer.

5.6.1 Planting and Management of Shade Trees

Pujiyanto (2016) states that the micro-climatic condition of coffee plantation is not possible to be altered by cultivation techniques. Yet, shade trees are beneficial to adjust the climatic condition to a certain level. Umbrella-shaped canopies help to mitigate against excessive light, heat and humidity. Located on the equator line, Indonesia has two distinct dry and rainy seasons. During the dry season, the shade will reduce direct soil and plant exposure to sunlight, retain soil moisture and may also reduce evaporation rate. Moreover, shade trees also protect the coffee plants from the risks of strong winds and frost on higher altitude (>1500 m MSL), prevent soil erosion and loss of organic matter and suppress weeds. During the rainy season, shade trees can protect the flowers and berries from falling due to heavy rains.

In particular, shade trees are urgently needed in Samosir area as it is projected that there would be prolonged dry season in the area. Without shade trees, the soil water level will be lower, harming the rooting system and microorganisms in the soil. If the drought remains persistent, the productive age of the plantation may be shorter. As an example, some coffee plantations in Sidikalang, North Sumatra, are unexpectedly experiencing loss of production after reaching the age of 10 years although the average productive period is 25 years.

In response to the changing climate, improved cultivation techniques are urgently required, such as the addition of shade or protective plants to adjust the exposure of coffee plantation to sunlight. Nevertheless, as a consequence, farmers need to schedule extra maintenance to prune the protective plants. Pruning of protective plants is conducted based on the extra treatment required by the coffee plantation. Throughout dry season with higher intensity of sunlight, protective plants canopies are ideally thicker. In contrast, more open space or less thick canopies are useful to maintain the exposures to sunlight and humidity. Shade plants often comprised of legume family that has been acknowledged as fast-growing, wind resistant, robust and require easy maintenance and able to produce high organic matter.

There are two types of shade plants, the temporary and the permanent shades. Temporary shades are vital during the preparation of permanent shade and before the productive phase of coffee plants. Most common temporary shade plants are *Moghania macrophylla*, followed by *Crotalaria usaramoensis, C. juncea* and *C. anagyroides*. Temporary shade plants should be trimmed, reduced or even pulled

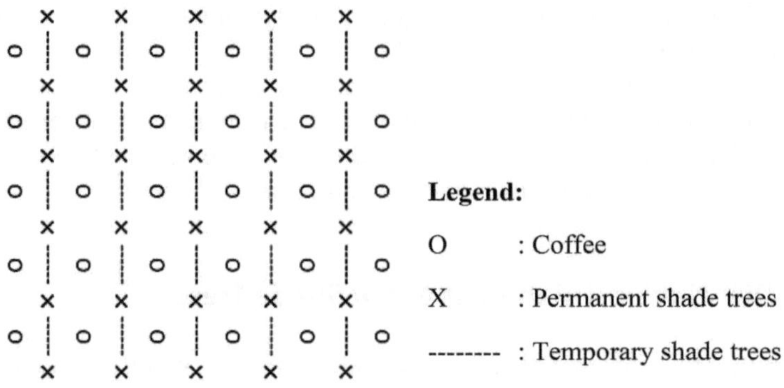

Fig. 5.38 The arrangement of temporary and permanent shades on a young coffee plantation

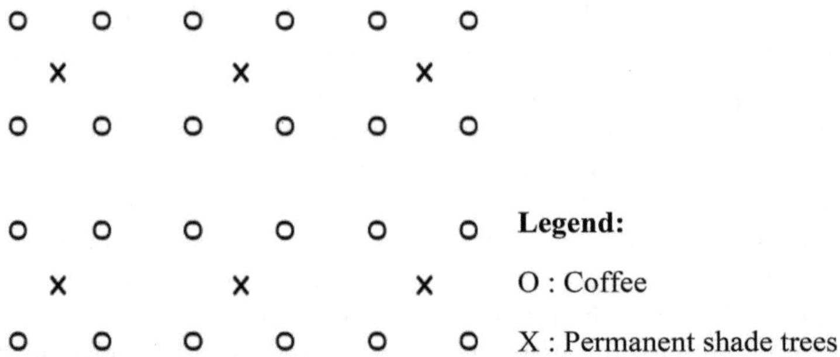

Fig. 5.39 The arrangement of permanent shade trees and coffee (has been 3 years in the productive phase)

completely once the permanent shade is ready and coffee plants come to the production phase. Litters from the temporary shade plants are often applied to the soil to add organic matters. Recently, it is suggested to plant eggplant, corn or beans as temporary shade plants so that farmers can benefit from harvesting the yield before getting benefit from the coffee plantation.

Best permanent shade plant is *Leucaena glauca* klon L2, or other common shade plants including *Gliricidia sepium, G. maculata.* In agroforestry, stands of forest trees such as *Durio zibethinus, Altingia excelsa* Noronha, *Swietenia mahagoni, Shorea* sp. *and Agathis dammara* can serve as permanent shade plants for coffee plantation. Else, productive fruit plants can also be favourable as shade trees such as *Parkia speciosa*, durian, avocado or orange.

Figure 5.38 gives an illustration of coffee and shade trees arrangement plots, both temporary and permanent shades. Moreover, Fig. 5.39 depicts permanent shade trees with coffee at the age of 6 years (has been in productive phase for 3 years). An example of a coffee plantation with *Leucaena leucocephala* is shown in Fig. 5.40.

Fig. 5.40 Example of a coffee plantation with *Leucaena leucocephala* (*lamtoro*) as shade trees

The ratio of shade trees to coffee is 1:1 for a young coffee plantation, while the ratio is 1:4 for permanent shade tree and coffee on the production phase. Once the coffee plants come into the production phase, temporary shade trees will be pulled out and applied as extra organic matter to the soil. Maintenance of shade trees include regular trimming of the branches to keep the height below 3 m and cutting small shoots that appear more frequently during the rainy season.

Most coffee plantations in Toba are monoculture, only small area practices polyculture. In general, monoculture coffee does not use shade tree to control exposure to sunlight. Extreme sunlight exposure has detained the plants to have optimum growth, yellow leaves, thicker, less chlorophyll and biennial bearing of production by year. Addition of shade trees as in Fig. 5.40 is thus relevant in this context.

Polyculture coffee plantation also often combined fruits with forest stands. Unlike monoculture, polyculture coffee shows healthier growth with green and thin leaves and fruitful vegetative phase although the production might be lower during the rainy season. The extra coverage from shade trees reduced the penetration of sunlight causing the coffee plants to receive lower amount of sunlight important for initiating flower. Regular pruning of shade trees may adjust the level of sunlight intake to 70–80%. Another option would be to trim the water shoots 1–2 times a month.

5.6.2 Addition of Organic Matter

Organic matter is a fraction of living organism, both plant and animal. Organic matter is a good source of minerals for plants once they are decomposed by bacteria,

Table 5.16 Suggested type and dosage of fertiliser for coffee plantation

	Dose (g/plant/year)				
Plants age (year)	Urea	SP36	KCl	Kieserit	Dolomite
1	40	50	30	15	30
2	100	80	80	40	50
3	150	100	100	60	80
4	200	100	140	75	110
5–10	300	160	200	100	150
>10	400	200	250	140	200

Source: Puslitkoka (2013) and PT Perkebunan Nusantara XII (Persero) (2013)

fungus or other organisms into CO_2, H_2O and mineral. Organic matter increases the ability of soil to retain water, increase cation exchange capacity (CEC), microorganism activities, retain minerals, improve soil structure and aeration. The main soil types found within the study area were entisol, inceptisol and ultisol with coarse, slightly coarse and mild textures, which are relatively low in organic matter, N and other nutrients. Such soil conditions commonly occurred in soils with high acidity (low pH). Extra applications of organic matter, compost and fertilisers are necessary to improve the soil nitrogen availability, soil quality and soil physical, chemical and biological characteristics.

Under changing climate that mainly affected by changes in rainfall pattern and temperature, an extra application of organic matter can minimise the impacts of climate change such as water and heat stresses. Both small and large coffee plantations have considerable sources of organic matters. The organic matters may originate from shade trees' pruning wastes, coffee pruning wastes, weeding wastes and coffee processing wastes (berries skin). The potential range of organic matters from these activities is about 15–20 t/ha annually. In principle, organic matters from coffee plantations should be kept and returned to the soil in the plantation.

Organic matter is applied to young and mature plants by spreading them over the surface as mulch 10–20 per kg per plant per year. Application of organic matter also helps the plants to maintain optimal humidity throughout the dry season. Another option of application during the rainy season would be to make holes or create a channel for spreading the organic matter. Burying the organic matters in holes or channel will avoid rotten shoot.

5.6.3 Fertiliser Application

Fertiliser is extra nutrients added to the soil or plants to improve the plants' nutrient level. Fertiliser application should follow six correct steps in terms of type, dosage, time, way, placement and supervision. Type and dosage of fertiliser are based on the analysis of the soil and leaves. In the case that an analysis cannot be conducted, the farmers should refer to the Guidelines from the Indonesian Coffee and Cocoa Research Center (Table 5.16).

The minimum number of fertiliser's application is twice a year – the first application should be made by the onset of the rainy season (Oct/Nov) and the second application by the end of rainy season (Mar/Apr). Urea, SP36 and KCl are mixable as long as they are applied within a short period after the mix. Kieserite (magnesium sulphate mineral) and Dolomite can be used as substitution for each other. Application of Kieserite or Dolomite is in sequence with Urea/SP36/KCl. Fertiliser channel in between coffee plants is ideally 20 cm wide and 10 cm depth. The fertiliser is spread on to the channel and covered with soil on top (buried). Fertiliser application needs serious supervision, especially on dosage, application and selection of the location to be applied.

Farmers in the study areas rarely applied extra fertiliser for their plants; few sometimes applied fertiliser without appropriate dosage and without considering the needs of their plants. On the other hand, Sigararutang which is the common cultivar in the area needs a higher amount of nutrients to produce an optimum growth. The cultivar needs at least 0.5 kg of inorganic and 10 kg of organic fertiliser. This is reasonable since the coffee plants in the area showed some unhealthy symptoms like yellowish leaves, dried branches, easily infected by diseases and less producing berries.

Farmers did not regularly apply fertiliser due to limitations in financial capital, knowledge of plant growth or human resources working on the plantation. One of the common socio-economic problems observed in the study areas was the age of farmers. Older generation were seen working as farmers, while most of the young generation prefer to work in big cities.

5.6.4 Pruning

Pruning or trimming of coffee plants are alternatives to manage the stability of production and to avoid biennial-bearing due to absorption of extreme sunlight and temperature. Biennial-bearing occurred when coffee plants are overexposed to the sun in one season, causing a high number of flowering and berry production, but followed by rapidly declining yields in the next season. Coffee grows productive branches during vegetative phase followed by berry branches. A higher number of productive branches assured higher yields and vice versa (Yuliasmara et al. 2016).

Pruning is necessary for both immature and productive plants. For immature plants, pruning aims to shape while for productive plants, pruning aims to maintain production. Shaping the immature plants would ensure the plants to grow strong, balanced and produce well. The plants are maintained to a certain height to simplify the harvesting process, branches are sturdy and well balanced, and there is good air circulation. Pruning can be conducted in two steps. The first step is toping, to adjust to the desired height, and then followed with pruning unwanted branches. On flowering and robust plants, direct cut at the height of 1.8 m is possible. Weaker plants are first cut at the height of 1.2 m and then re-cut at 1.8 m in the next year. In general, 1.8 m is ideal to allow harvesting to go smoothly without damaging the branches. The second step of pruning is to clear unwanted branches, for each 60 cm zone, to stimulate new and healthier branches growth.

Pruning of productive plants aims to maintain the height of plants at 1.7–1.8 m, to stimulate new and healthier branches that would lead to higher production, to optimise sunlight penetration and air circulation, to support primordial development and flowering, to remove unwanted branches, to control pest and disease and to obtain ideal shape.

Production/maintenance pruning consists of post-harvest pruning and removal of water shoots. Post-harvest pruning is conducted directly after the harvest, involving cutting unproductive branches (branches B3, branches that have been producing berries three times; adventive branches, wormed, turned, fan-shaped; infected branches; dried and damaged branches). Pruned plants should have a balanced proportion of strong branches of category B0 (not producing yet), B1 (producing once) and B2 (producing twice) each having equal composition of 33%. Some farmers maintain the branches to produce up to three times. In such case, the composition of B0:B1:B2:B3 would be 30%:30%:30%:10% to support sustainable production. The water shoots are removed 3 months after post-harvest pruning. Pruning is a skill-intensive operation and it is essential for the growth of coffee plants and coffee production.

Farmers in the study area did not prune their plants, few occasionally pruned parts of the plants randomly. Irregular pruning does not help plants to grow better, in fact, the height remains unfavourable for harvesting (Fig. 5.41a), drying branches, imbalance composition of B0, B1, B2 and B3. Unpruned coffee plants in the study areas showed the symptoms of having scattered new branches on top, but no new branches grown in the lower parts of the plant. Further, unpruned coffee tends to have a shorter economic period, their ability to produce berries is rapidly decreasing with time (Fig. 5.41b). In contrast, pruned plants' ability to produce berries is increasing with time (Fig. 5.41c).

5.6.5 Harvest

Harvest is the final process in crop production before the product entered post-production process. Harvesting determines the quality and quantity of the product. Therefore harvesting needs to follow specific procedures, such as selective picking (only pick red and ripe berries), ensuring no damage on stems and no falling berries around the plants. As a consequence of selective harvest, there is a picking rotation about 7–15 days continuously which requires human resources accordingly. Even with selective picking, we cannot avoid yellow or green berries picked at maximum tolerance of 10%. Blackened and dried berries should also be picked to avoid infection of pest or diseases into the plants – especially coffee berry borer.

Farmers in the study area usually performed semi-selective harvest (Fig. 5.42). Yellow and green berries were also picked, although it might lower the quality of the coffee. Harvesting system will influence further processing. Selective harvest (only red berries) will be followed by wet processing to produce high-quality green beans.

Fig. 5.41 Condition of plants with and without pruning in Samosir District (**a**) and Karo District (**b** and **c**)

Semi-selective harvest should be followed by a sorting process to separate red berries from green and yellow berries, red berries will be processed wet while others will be dry-processed. The dry process often results in lower quality beans.

5.6.6 Pest Management

Pest control in Toba is not merely about technology or technique but also socio-economic challenge related to the awareness of the farmers to remove infected plants because even damaged/low-quality berries are still saleable. Farmers are not aware of the symptoms, i.e. unhealthy plants; thus they were resistant to adopt pest control technologies.

However, reduction of risk to coffee berry borer in the future should be listed as a priority and supported by field school on integrated pest management (FSIPM). Farmers could learn pest management control from FSIPM that would allow them to have better skills to handle infected plants in the future (Table 5.17). Currently,

Fig. 5.42 Harvested berries from one of the coffee plantations in Samosir

the conventional techniques to deal with coffee berry borer are by introducing its biological control agent *B. bassiana,* use of a trap and population monitoring, but they have been limited to a single measure. Thus, an integrated coffee berry borer control is necessary.

Another option for adaptation is to plant varieties that are more resilient to climate stress. Technical control such as shade management and pruning should be conducted interactively. As indicated in earlier sections, pruning is aimed to maintain sun radiation, humidity and temperature at optimum level for coffee production. In addition, sanitation, weed control and balanced fertiliser application are also necessary. Fallen berries post-harvesting should be cleared to cut the life cycle of the coffee berry borer. Use of *B. bassiana* as a biological agent to control coffee berry borer is also beneficial.

Controlling the coffee berry borer needs a holistic and integrative approach, because there is no single solution to annihilate the pest. For the coffee berry borer control technology to be effective in removing the pest completely, it must be accompanied with regular maintenance of the plantation. Other supporting factors include farmers' commitments and knowledge to conduct integrative intervention, as well as the availability of institutional support, guidance and support from the government in FSIPM.

Some farmers relied heavily on insecticides and traps to control pests or diseases. However, the results were varied and unpredictable. Despite the increasing number of research, pest and disease remain persistent in coffee plantation across the country. Integrated Pest Management (IPM) is the major procedure to abolish coffee berry borer using an ecological approach. IPM strategies involved regular monitoring to evaluate the economical and environmental impacts from pest control. This suggests that farmers should master the basic knowledge and techniques in pest

Table 5.17 Summary of pest and disease management control in coffee production

OPT	Control
Coffee berry borer (*Hypothenemus hampei*)	Cut the life cycle
	Introduction of biological control agents such as *Beauveria bassiana* or parasitoid *Cephalonomia stephanoderis*
	Trap
Twig borer (*Xylosandrus morigerus* and *Xylosandrus compactus*)	Technical intervention: improve sanitation, cutting or burning infected stem
	Balanced fertiliser and maintenance
	Introduction of biological control agents such as parasitoid *Tetrastichus xylebororum*
Striped mealybug (*Ferrisia virgate*) harms the stem and leaves	Technical intervention, biological and chemical control
	Introduction of biological control agents such as *Nephus roepkei, Scymus apicivlavus, Cryptolaemus montrouzieri, Brumus suturalis*
	Technical culture w/ addition of shade to maintain humidity
	Chemical: application of insecticide
Coffee leaf rust (*Hemileia vastatrix*)	Technical intervention: reduce humidity (by pruning shade trees and unproductive branches of coffee plants)
	Introduction of biological control agents such as *Verticillium lecanii* or *Trichoderma* spp.
	Balance of fertiliser application to improve plants immunity to pest and disease
Brown eye spot (*Mycosphaerella coffeicola* or fungi *Cercospora coffeicola*)	Technical intervention: reduce humidity (by pruning shade trees and unproductive branches of coffee plants), improve drainage
Pink diseases (*Corticium salmonicolor*) harm the stem, branches, and berries	Technical culture: reduce humidity (by pruning shade trees and unproductive branches of coffee plants)
	Shade trees management
Root disease (*Phellinus noxius*, *Rosellinia bunodes*, and *Rigidoporus microporus*)	Pull out infected plants and its roots

control and also be able to estimate infestation. General information is available in plantation pest control guidebook published by the Directorate General of Plantation, Ministry of Agriculture, on "Technical Guidebook on Crop Pest Organism of Coffee" (Harni et al. 2015).

5.7 Conclusion

Climate change would have significant impacts on the growth and development of coffee plantations in the Toba Region. Following changes in rainfall, the coffee phenology, flowering season and peak of harvesting would experience some shifts.

Similarly, crop yields would also decline due to the changes in rainfall patterns and rising temperatures, which would ultimately increase the number of pest and disease attacks, especially of the coffee berry borer. The current technology adoption by the farmers is still inadequate. Without serious and integrated efforts to improve cultivation technology, impacts of climate change would be exacerbated.

Beyond reducing yields and amplifying pest and disease attacks, climate change has the potential to shift upwards the suitable climate area for coffee production to higher elevation and threaten the existence of protection forests. Current lands under the suitability classifications of "very suitable" and "suitable" are facing some alteration to become "less suitable" and "unsuitable" areas in the future.

5.8 Recommendation

In addition to cultivation techniques, there are some specific measures to overcome and reduce the impacts of climate change on coffee productions:

- Development of nurseries from selected parent trees. This requires specific assessment with experts and practitioner on the identification of selected parent trees.
- Development of Arabica coffee should focus on increasing yield through intensification, rehabilitation and replantation.
- Necessary steps for intensification include socialisation, maintenance of plants based on technical operational standard (SOP) that highlights weed control, fertiliser application, pruning, shade trees management and pest and disease control.
- Replantation activities must comprise of the following steps: Socialisation programme, preparation of seeds, preparation of protective/shade trees, pulling out old plants, spacing design, planting shade trees, digging planting holes, planting and maintenance.
- Rehabilitation activities must include socialisation programme, preparation of entrees, cutting stems of old plants and maintenance of water shoots which will be connected by grafting.
- Expansion programme involves socialisation, measurement of the area, preparation of seeds, preparation of shade trees, spacing, planting shade trees, digging holes, planting and maintenance.
- In order to improve dry beans quality, it is necessary to have a post-harvesting unit complete with berry skin peeler, drier and sorting machine. Roasting machine is also necessary. Cheap and locally available energy sources should be used as fuels. It is also necessary to develop coffee processing SOP to sell high-quality Arabica.
- To increase the selling point of Arabica coffee in the market, promotion by local government in collaboration with coffee exporters association (AEKI), *Disperindag* or private (CSR) is essential. This would create strong branding of Toba coffee in both domestic and international markets.

- The coffee trading system needs to be improved to enhance the bargaining position of coffee farmers and provide a fair distribution of benefits among the farmers.
- Skilled and knowledgeable human resources are necessary for coffee-growing activities. Both formal and informal educations such as training, counselling, workshops and field schools should be set as capacity building for farmers in the area. In parallel, it is necessary to initiate a pilot project that entirely adopts guidebooks on coffee cultivation as a real laboratory for farmers to learn.
- Infrastructural improvement in transportation is also necessary to support post-harvest processing.

Continuous and integrated control of coffee berry borer should at least consider the following:

- *Use of resistant varieties*: Resilience to coffee berry borer varied among some of the coffee varieties planted in Toba region. This would require research to identify resistant varieties that could provide the highest yield.
- *Pruning and clearing up of falling berries*: Pruning will ensure coffee plants are exposed to an optimum level of sunlight, humidity and temperature. Maintaining a micro-climatic condition can also keep away pest and disease. *Rampesan* or clearing falling berries aim to cut food supply for Coffee berry borer and eventually cut its life cycle. This would be useful only if all of the coffee farmers in the area practising the same strategy, otherwise, Coffee berry borer would move from one plantation to another within the neighbourhood.
- *Use of biological control agents* can naturally reduce the population of pests. Examples of biological agents are *Cephalonomia stephanoderis* (Hymenoptera: Bethylidae) or *Beauveria basiana* that are known as parasites to coffee berry borer and has been proven to be successful to minimise the population of coffee berry borer. This parasite should be introduced in the early planting years while the pest population is still low. Furthermore, training on parasitoid breeding for farmers is required to allow them to be able to breed their variety, thus do not rely upon other sources.
- *Pheromones/attractant trap*: Coffee berry borer traps have been common in the market. Use of traps is often suggested before or during harvest as the Coffee berry borer starts to find new food sources during this period. On an average, a farmer needs 18 traps/ha, set with a spacing of 24 m and a height of 1.2 m. Nevertheless, using traps is very costly and less effective, if only used by few farmers. For more effective control, all farmers in the area should set the trap at the same time and regularly reapply in combination with other pest control methods.
- *Insecticide*: In the context of outbreaks, the use of insecticide is possible, but with a rigorous procedure in a way that does not harm the surrounding environment. Field School for Integrated Pest Management (FSIPM) is an ideal programme to introduce the basic concept of integrated pest control. Knowledgeable farmers are more sustainable in maintaining their plantations in the correct way, as they are entirely aware of the consequences.

- *Support from the government*: Implementation of programmes in Indonesia is currently relying on government supports for technicalities and financial and comprehensive guidance. However, the existence of institutions for integrated pest management guidance might reduce this dependency in the near future.

Acknowledgements We thank for the financial support provided by Rainforest Alliance-UTZ. We also extend our gratitude for BITRA that provided support in conducting survey.

References

Abdoellah S (2016) Irrigation on coffee plantation. In: Wahyudi T, Pujiyanto M (eds) Coffee: history, botany, production process, processing, downstream products, and partnership systems. Gadjah Mada University Press, Yogyakarta, pp 253–258

Amaria W, Harni R (2012) Leaf rust in the planting of coffee and its control. In: Technology innovation for community coffee plantation. Indonesian Spices and Industrial Plants Research Institute, Sukabumi, pp 115–120

Bale J, Masters G, Hodkinson I, Awmack C, Bezemer T, Brown V, Butterfield J, Buse A, Coulson J, Farrar J, Good J, Harrington R, Hartley S, Jones T, Lindroth R, Press M, Symrnioudis I, Watt A, Whittaker J (2002) Herbivory in global climate change research: direct effects of rising temperature on insect herbivores. Glob Chang Biol 8:1–16. https://doi.org/10.1046/j.1365-2486.2002.00451.x

Birch L (1948) The intrinsic rate of natural increase of an insect population. J Anim Ecol 17:15. https://doi.org/10.2307/1605

BPS (2016a) Statistics of Karo Regency, Karo. ISSN:2301-8852

BPS (2016b) Statistics of Dairi Regency, Dairi. ISSN:2354-578X

BPS (2016c) Statistics of Samosir Regency, Samosir. ISSN:2301-976X

BPS (2016d) BPS-statistics of Simalungun Regency, Simalungun. ISSN:0215-2339

Brown JS, Kenny MK, Whan JH, dan Merriman PR (1995) The effect of temperature on the development of epidemics of coffee leaf rust in Papua New Guinea. J Crop Prot 14(8):671–676

CABI (2000) Crop protection compendium. CAB International, Wallingford

Cannel MGR (1985) Physiology of the coffee crop. In: Clifford MN, Wilson KC (eds) Coffee: botany, biochemistry, and production of beans and beverage. AVI Publishing Company, Connecticut, pp 108–134. https://doi.org/10.1007/978-1-4615-6657-1

Center for Soil Research (1978) National land classification system. Soil Research Center, Bogor

Center for Soil Research (1982) National land classification system. Soil Research Center, Bogor

Collins WJ, Bellouin N, Doutriaux-Boucher M, Gedney N, Halloran P, Hinton T, Hughes J, Jones CD, Joshi M, Liddicoat S, Martin G, O'Connor F, Rae J, Senior C, Sitch S, Totterdell I, Wiltshire A, Woodward S (2011) Development and evaluation of an earth-system model-HadGEM2. Geosci Model Dev 4(4):1051–1075

Crowe TJ (2009) Coffee pests in Africa. In: Wintgnes JN (ed) Coffee: growing, processing, sustainable production, 2nd edn. Willey-VCH Verlag GmbH &Cp. KGaA, Weinheim, pp 425–462. https://doi.org/10.1002/9783527619627

Damatta FM, Ramalho JDC (2006) Impacts of drought and temperature stress on coffe physiology and production: a review. Braz J Plant Physiol 18(1):55–81. https://doi.org/10.1590/S1677-04202006000100006

Descroix F, Snoeck J (2012) Environmental factors suitable for coffee cultivation. In: Wintgens JN (ed) Coffee: growing, processing, sustainable production, 2nd Rev edn. Wiley-VCH Verlag GmbH & Co. KGaA, Weinheim. ISBN:978-3-527-33253-3

Dingemanse NJ, Kalkman V (2008) Changing temperature regimes have advanced the phenology of Odonata in the Netherlands. Ecol Entomol 33:394–402. https://doi.org/10.1111/j.1365-2311.2007.00982.x

Erdiansyah NP, Soemarno D, Mawardi S (2016) Sidikalang coffee production in North Sumatra. Indonesian Coffee Cocoa Res Inst Newslett 25:10–14

Faqih A, Hidayat R, Jadmiko SD, Radini (2016) Historical climate and future climate scenarios in Indonesia: climate modelling and analysis. United Nation Development Program (UNDP), Ministry of Environment and Forestry (KLHK)

Funk C, Peterson P, Landsfeld M, Pedreros D, Verdin J, Rowland J, Romero B, Husak G, Michaelsen J, Verdin A (2014) A quasi-global precipitation time series for drought monitoring. Data Ser (832):4. https://doi.org/10.3133/ds832

Funk C, Peterson P, Landsfeld M, Pedreros D, Verdin J, Shukla S, Husak G, Rowland J, Harrison L, Hoell A, Michaelsen J (2015) The climate hazards infrared precipitation with stations-a new environmental record for monitoring extremes. Sci Data 2:150066. https://doi.org/10.1038/sdata.2015.66

Gomi T, Nagasaka M, Fukuda T, Higahara H (2007) Shifting of the life cycle and life history traits of the fall webworm in relation to climate change. Entomol Exp Appl 125:179–184. https://doi.org/10.1111/j.1570-7458.2007.00616.x

Haggar J, Barrios M, Bolaños M, Merlo M, Moraga P, Munguia R, Ponce A, Romero S, Soto G, Staver C, de M. F. Virginio E (2011) Coffee agroecosystem performance under full sun, shade, conventional and organic management regimes in Central America. Agrofor Syst 82:285–301. https://doi.org/10.1007/s10457-011-9392-5

Harni R, Samsudin WA, Indriati G, Soesanthy F, Khaerati ET, Hasibuan AM, Hapsari AD (2015) Technology of pest and disease control on coffee plantation. IAARD Press, Jakarta

ICC (2009) Climate change and coffee. International Coffee Council, 103rd session, 23–25 September 2009. http://www.ico.org/documents/icc-103-6-r1e-climate-change.pdf. Accessed 8 Jan 2016

Indonesian Coffee and Cocoa Research Center (Puslitkoka) (2013) Practical guidelines for cultivation and maintenance of coffee plants. Indonesian Coffee and Cocoa Research Center, Jember

IPCC (2014) Climate change 2014: synthesis report. Contribution of Working Groups I, II and III to the fifth assessment report of the Intergovernmental Panel on Climate Change [Core Writing Team, Pachauri RK, Meyer LA (eds)]. IPCC, Geneva, 151 pp

Jadmiko SD, Murdiyarso D, Faqih A (2017) Bias correction of regional climate model outputs for drought analysis. Indonesian Soil Clim J 41(1):25–36. https://doi.org/10.2017/jti.v41i1.5983

Jaramillo J, Chabi-Olaye A, Kamonjo C, Jaramillo A, Vega F, Poehling H, Borgemeister C (2009) Thermal tolerance of the coffee berry borer Hypothenemus hampei: predictions of climate change impact on a tropical insect pest. PLoS One 4:e6487. https://doi.org/10.1371/journal.pone.0006487

Kiritani K (2006) Predicting impacts of global warming on population dynamics and distribution of arthropods in Japan. Popul Ecol 48:5–12. https://doi.org/10.1007/s10144-005-0225-0

Kiritani K (2007) Pest status of rice and fruit bugs (Heteroptera) in Japan. Glob Chang Biol 13:1586–1595

Kiritani K (2013) Different effects of climate change on the population dynamics of insects. Appl Entomol Zool 48:97–104. https://doi.org/10.1007/s13355-012-0158-y

Lan CC, Wintgens JN (2009) Major pests of coffee in the Asia-Pacific region. In: Wintgnes JN (ed) Coffee: growing, processing, sustainable production, 2nd edn. Wiley-VCH Verlag GmbH &Cp. KGaA, Weinheim, pp 463–477

Lenderink G, Buishand A, Van Deusen W (2007) Estimate of future discharges of the river Rhine using two scenarios methodologies: direct versus delta approach. Hydrol Earth Syst Sci 11:1145–1159. https://doi.org/10.5194/hess-11-1145-2007

Martin GM, Bellouin N, Collins WJ, Culverwell ID, Halloran PR, Hardiman SC, Hinton TJ, Jones CD, McDonald RE, McLaren AJ, O'Connor FM, Roberts MJ, Rodriguez JM, Woodward S, Best MJ, Brooks ME, Brown AR, Butchart N, Dearden C, Derbyshire SH, Dharssi I, Doutriaux-Boucher M, Edwards JM, Falloon PD, Gedney N, Gray LJ, Hewitt HT, Hobson M, Huddleston MR, Hughes J, Ineson S, Ingram WJ, James PM, Johns TC, Johnson CE, Jones A, Jones CP, Joshi MM, Keen AB, Liddicoat S, Lock AP, Maidens AV, Manners JC, Milton SF, Rae JGL, Ridley JK, Sellar A, Senior CA, Totterdell IJ, Verhoef A, Vidale PL, Wiltshire

A (2011) The HadGEM2 family of met office unified model climate configurations. Geosci Model Dev 4(3):723–757
Ministry of Agriculture (2017) Agricultural statistics 2017. Center for Agricultural Data and Information System: Ministry of Agriculture, Jakarta
Ministry of Environment (2014) Rescue movement of Lake Toba (GERMADANI). Ministry of Environment of the Republic of Indonesia, Jakarta
Moss RH, Edmonds JA, Hibbard KA, Manning MR, Rose SK, Van Vuuren SP, Carter TR, Emori S, Kainuma M, Kram T, Meehl GA, Mitchel JFB, Nakicennovic N, Riahi K, Smith SJ, Stouffer RJ, Thomson AM, Weyant JP, Wilbanks TJ (2010) The next generation of scenarios for climate change research and assessment. Nature 463:747–756. https://doi.org/10.1038/nature08823
Nur AM (2000) Impact of La Nina on Robusta coffee production. Indonesian Coffee Cocoa Res Inst Newslett 16(1):50–58
Prastowo B, Karmawati E, Rubijo S, Indrawanto C, Munaro SJ (2010) Coffee cultivation and post-harvest. Indonesian Center for Estate Crops Research and Development, Bogor
Pratama SW, Aini FN (2016) Main diseases of coffee plants. In: Coffee. Gadjah Mada University Press, Yogyakarta, p 15
PT Perkebunan Nusantara XII (Persero) (2013) Guidelines for managing Arabica coffee plants. PT Perkebunan Nusantara XII (Persero), Surabaya
Pujiyanto (2016) Water stress on coffee plant. In: Wahyudi T, Pujiyanto M (eds) Coffee; history, botany, production process, processing, downstream products, and partnership systems. Gadjah Mada University Press, Yogyakarta, pp 382–401
Puslitkoka (2013) Practical guidelines for cultivation and maintenance of coffee plants. Indonesian Research Center for Coffee and Cacao, Jember
Rahayu DS, Wiryadiputra S (2016) Main insect pest and it's control. In: Coffee. Gadjah Mada University Press, Yogyakarta
Rayner RW (1961) Spore liberation and dispersal of coffee rust Hemileia vastatrix B.et Br. Nature 191:725
Samsudin, Soesanthy F (2012) Coffee berry borer and its control: technology innovation for community coffee plantation. Indonesian Spices and Industrial Plants Research Institute, Sukabumi, pp 121–130
Semangun H (2000) Plantation plant diseases in Indonesia. Gadjah Mada University Press, 835pp
Soenarjo (1975) Effect of climate anomaly on coffee production. Indonesian J Biotechnol Res Estate Crops 43(2):79–91
Sulistyowati E, Mangoendihardjo S, Wagiman FX (1999) Functional response of parasitoid *Cephalonomia stephanoderis Betr.* against coffee berry borer (*Hypothenemus hampei Ferr.*). Coffee Cocoa Res J 15(2):101–108
Susilo AW (2008) Resistance of coffee plants (*Coffea* spp.) from coffee berry borer (*Hypothenemus hampei Ferr.*). Rev Coffee Cocoa Res 24(1):1–15
Syakir M, Surmaini E (2017) Climate change in the context of the coffee production and development system in Indonesia. Agric Res J 36(2):77–90
Waller JM (1985) Control of coffee diseases. In: Clifford MN, Willson KC (eds) Coffee: botany, biochemistry and production of beans and beverage. AVI Publishing Company, Connecticut, pp 219–229
Waller JM, Bigger M, Hillocks RJ (2007) Coffee pests, diseases and their management. CAB International, Wallingford, p 400. ISBN-10: 1 84593 129 7 and ISBN-13: 978 1 84593 129 2
Wilson KC (1985) Climate and soil. In: Clifford MN, Wilson KC (eds) Coffee: botany, biochemistry, and production of beans and beverage. AVI Publishing Company, Connecticut, pp 97–107
Wintgens JN (2009) Coffee: growing, processing, sustainable production. 2. Wiley-VCH Verlag GmbH & Cp. KGaA, Weinheim
Wiryadiputra (2012) Effectiveness of Cyantraniliprole insecticides on coffee berry borer (*Hypothenemus hampei Ferr.*) in Arabica coffee. Coffee Cocoa Res J 28:103–114

Wiryadiputra S, Tran LK (2008) Indonesia and Vietnam. In: Souza RM (ed) Plant-parasitic nematodes of coffee. Springer, Dordrecht

Yahmadi M (1973) Effect of long drought on coffee plantation. Indonesian J Biotechnol Res Estate Crops 41:235–240

Yuliasmara F, Suhartono, Hulupi R (2016) Pruning on coffee plants. In: Wahyudi T, Pujiyanto M (eds) Coffee: history, botany, production process, processing, downstream products, and partnership systems. Gadjah Mada University Press, Yogyakarta, pp 195–217

Zeng F, Pederson G, Ellsbury M, Davis F (1993) Demographic statistics for the pea aphid (Homoptera: Aphididae) on resistant and susceptible red clovers. J Econ Entomol 86(6):1852–1856

Global Climate Change and Biofuels Policy: Indian Perspectives

6

Shiv Prasad, Sandeep Kumar, K. R. Sheetal, and V. Venkatramanan

Abstract

In the past few decades, unparalleled efforts have been made to reduce the dependence on fossil fuels, aiming to reduce air pollution and lessen the climate change effect. Fossil fuel burning in the energy sector is the significant contributor to GHG, global warming and climate change. These scenarios have motivated scientists and policy makers to look for eco-friendly energy supply options with enhanced energy efficiency. Globally, biofuels, considered as a substitute for fossil fuels, have become top priority due to its eco-friendly nature, and many energy policy initiatives have taken place, especially in biofuel production and its use, and various advances in technology are in progress. The biofuel policies are instrumental in improving energy security by reducing foreign oil imports by promoting renewable energy resource. The government is also focusing on advances in research and industrial development on biofuel and biomass-based economy. New policy initiatives have ample opportunities for biofuel production from agricultural residues, industrial waste, and other wastes, which exist as unused or surplus. The sustainable production of biofuel and its use as a substitute for fossil fuels and related policy initiatives would help in the mitigation of climate change.

Keywords

Biofuel policy · Climate change · Sustainability · Fossil fuels

S. Prasad (✉) · S. Kumar
Centre for Environmental Sciences and Climate Resilient Agriculture (CESCRA), ICAR-Indian Agricultural Research Institute, New Delhi, India

K. R. Sheetal
Central Arid Zone Research Institute (CAZRI), Regional Research Station, Bikaner, Rajasthan, India

V. Venkatramanan
School of Interdisciplinary and Transdisciplinary Studies, Indira Gandhi National Open University, New Delhi, Delhi, India

© Springer Nature Singapore Pte Ltd. 2020

V. Venkatramanan et al. (eds.), *Global Climate Change and Environmental Policy*,
https://doi.org/10.1007/978-981-13-9570-3_6

6.1 Introduction

The discovery and use of fossil sources like coal for energy have revolutionized the wheel as a means of transportation in the history of humanity. It continuously satisfies the need for energy of the entire world in various forms for the past few centuries. Fossils fuels were the driving force of the industrial revolution, and the modern world greatly owes its technological and mechanical progress to it. However, irrational use and burning in power generation and automotive transport have led to severe implications for the environment, especially emission of CO_2 which plays a key role in global warming and concomitant climate change (IPCC 2014; Prasad et al. 2014; Aradhey 2018; Malav et al. 2017; Kumar et al. 2016). Power generation, industries and transport sectors are primarily responsible for increasing atmospheric CO_2 emissions. Global CO_2 emission on account of combustion of fossil fuels was reported to be 32.3 Gt CO_2 in 2016. The reported increase in CO_2 emission is due to the upswing in economic activity both in the emerging and developed economies (IEA 2018).

Biofuels are fuels, primarily produced from biomass, and used as liquid fuels, i.e. bioethanol, biomethanol, biodiesel and gaseous fuels such as biohydrogen and methane (Demirbas 2008). Liquid biofuels from biomass deliver unique environmental benefits and strategic economic returns and can be considered as the eco-friendly and clean energy alternative to fossil fuels (Puppan 2002; Prasad et al. 2019b). The leading promoter for producing biofuels is global warming triggered by the continual burning of fossil fuels. Use of biofuels in compression ignition engines adding oxygen to fuel results in complete fuel combustion, supposed to cause less air pollution, and they are also biodegradable and nontoxic and has negligible sulphur content (Vasudevan and Boyi 2010; Prasad et al. 2014). The most significant difference between biofuels and petroleum feedstocks is oxygen content. Biofuels have oxygen levels from 10% to 45% while petroleum has essentially none, making the chemical properties of biofuels very different from petroleum. Ethanol when blended with petrol lowers stoichiometric air-fuel ratio that improves engine performance due to the complete fuel combustion process. Thus, it significantly reduces the emissions of CO, UHC and emission of GHG, compared to regular gasoline fuel (Prasad et al. 2014).

In absolute terms, crude oil remains a highly political energy item of trade with challenging variable pricing, and many of the long-range price estimates and crude oil market forecasting have failed miserably. Further, the uneven geographical distribution of fossil fuels and natural resources warrant countries to import crude oil from the Middle East to meet their energy needs. India relies heavily on conventional fuels to meet its 95% transportation fuel needs through petroleum products and is continually dependent on crude oil imports. Import dependency on petroleum has been steadily high at above 75% and has virtually hit 80% in the year 2014–15 (United Nations 2016). Heavy dependence on imported crude oil and its variable prices is a significant challenge in ensuring energy security. In India, the yearly energy demands upward gradually at a rate of 4.8% and is anticipated to become the world's third-largest energy end user by 2030 (IEA 2010). Ironically, India has the world's fifth largest coal reserves and still facing severe challenges to meet the growing domestic energy demand, typically due to swift urbanization and population growth, industrial relocations and expansions. On World Biofuel Day (10th August), the Prime Minister of India raised a hope that India can save about Rs. 12,000 crores in foreign exchange in the next 4 years due to ethanol blending of petrol. The Union Road Transport Minister has also been harping on the value of alternative fuels (Down To Earth 2003).

As concerns grow over variable pricing, uncertain climate change and the imminent depletion of non-renewable resources, different factors in society have placed increasing emphasis on the search for new, sustainable and renewable energy sources (Prasad et al. 2012, 2019a). As a result of this endeavour, policy makers focus increasingly on advancing liquid biofuels as an energy source for transportation. The Government of India has initiated several programmes and adopted a series of policies on biofuels due to its potential contributions towards a wide range of uses to achieve a reduction in air pollution, lowering greenhouse gas (GHG) emissions and climate change mitigation. This would also help to achieve sustainable development goals like energy security, foreign exchange savings, health and well-being, environmental safety and socio-economic issues related to rural areas (Faaij and Domac 2006; Basavaraj et al. 2012; Prasad et al. 2018).

6.2 Ethical Principles and Biofuels Policy

Scarcity of finite fossil fuels, changing climate, increasing demand for transportation fuel and mandatory biofuel blending policy have positioned biofuel as a significant energy source in the global arena. Scientific studies have been focused on sustainable production of biofuels which can lessen the impacts of climate change (Prasad et al. 2019a). Nevertheless, the increased focus on biofuel production has brought to the forefront the ethical dimensions of biofuel production. Buyx and Tait (2011) have distilled the following ethical principles from the prevailing ethical values. These principles shall guide "biofuel policy making" and also objectively analyse the sustainability of biofuel technologies.

- Environmentally sound and sustainable biofuel production should not purge the "existential rights" of citizens.
- Biofuel production in addition to contributing a "net decrease of GHG emissions" should adopt fair trade policies that also as the connotation of equity and benefit sharing.

Biofuel development policy making should entertain following queries so as to build a sustainable and ecological valuable biofuel policy. They include:

- Is there a danger that the costs of the development be out of proportion to the benefits, linked to other major (public) spending priorities?

- Are there any competing energy resources that might be even better, for example, at decreasing GHG emissions while still meeting all the expected ethical principles?
- Is there an alternative and better use of the lignocellulosic biomass feedstock in question?
- Has due attention been paid to the voices of those directly affected by the implementation of technology?

Careful reflection of these questions should be an integral part of a comprehensive, comparative analysis of all several future energy and climate change abatement options, including the comparison of energy portfolios with a different mix of technologies. A biofuel policy must incorporate ethical principles which focus not only on people's rights and environmental security but also imbibe the virtues of sharing of benefits and net negative emissions of GHG (Buyx and Tait 2011).

6.3 Biofuel Policy and Programme in India

Biofuel policies in India began over a decade ago to promote energy security and self-sufficiency and reduce foreign oil imports. A large number of initiatives and programs have been taken by the Government of India to support research and development in the biofuels sector and also encourage the biomass-based economy in India. Policy frameworks are in place with important visions to develop ease of doing business, set up start-ups, SMEs (small and medium enterprises) and large enterprises for the production of biofuels and chemicals from agro-residues and industrial waste, mostly which are available as surplus.

6.3.1 National Biofuel Mission

In India, the "National Biofuel Mission" was started in 2003 with the establishment of the Planning Commission, now called NITI Aayog, Government of India (GoI). The mission aim was to develop various policies and projects to accelerate the growth and advances in biofuel crop cultivation, biofuels production and its use and also replace or moderate the reliance on imported crude oil. The Government of India decided to launch the Ethanol Blended Petrol (EBP) Programme in January 2003. Soon the effort led to compulsory blending in states. The Ministry of Petroleum and Natural Gas (GoI) has made 5% ethanol blending compulsory in petrol in nine states and five union territories (Planning Commission 2003). In the second phase of the Ethanol Blended Petrol Programme (EBPP) in 2006, the blending mandate was further extended to cover 20 states and 8 union territories. However, there were shortcomings as a result of which the established targets could not be achieved. As a result, the commencement of large-scale production units and the implementation of modern technologies to achieve the target were sought after. In September 2007, the Cabinet Committee on Economic Affairs (CCEA) recommended 5% ethanol blending across the country except for Jammu and Kashmir, the North East and island territories (Prasad et al. 2007).

6.3.2 National Biofuel Policy 2009

The "National Policy on Biofuels" was passed on 24 December 2009 by the "Ministry of New and Renewable Energy" (MNRE), Government of India. The policy had set an indicative target of 20% blending of both ethanol and biodiesel by 2017. However, it was achieved partly because of the insufficiency of a fair amount of ethanol. India's "National Biofuel Policy" continues to focus on the use of non-food items, including molasses for bioethanol production and non-edible oils for biodiesel production (Aradhey 2010). The National Policy on Biofuels 2009 has the National Registry on Feedstock Availability and also a periodical analysis of biodiesel-blending targets. The Government of India contemplated to create a "National Biofuel Fund" (NBF) so as to meet the objectives of the national policy and to provide financial support and other possible incentives to innovate and materialize the practical applications of "second-generation biofuel production technologies." Further, the biofuel technologies and projects were allowed 100% foreign equity by automatic approval plans to bring foreign direct investment (FDI) on condition that biofuel is for domestic use only and not for export (USDA GAIN reports 2017).

India's National Biodiesel Policy is driven by the fact that energy is a critical input for the socio-economic improvement of our country. The energy strategy aims at efficiency and security to give access while being environment friendly and achieve the optimum mix of basic resources for energy production. In the past, the National Biodiesel Mission (NBM) identified jatropha (*Jatropha curcas*) as the most suitable inedible oilseed to help achieve a proposed biodiesel blend of 20% with conventional diesel by 2017. However, using jatropha proved untenable due to a host of agronomic and economic constraints.

6.3.3 India's National Policy on Biofuels (2018)

Worldwide, biofuels have grabbed the attention in the last few decades, and it is crucial to carry on with the pace of developments in the biofuels field. In order to promote biofuels in India, the central government came out with the National Policy on Biofuels 2018 to reduce import dependency on petroleum and natural gas and to move towards renewable clean energy and mitigating climate change. As per the government's targets, biofuels would contribute 10 gigawatts (GW) of power by 2022. The policy seeks to achieve 20% blending of ethanol with gasoline and 5% blending of biodiesel with diesel by 2030.

The National Policy on Biofuel 2018 seeks to not only help farmers economically dispose of their surplus stock but also reduce India's oil import dependence. Biofuels in India is of strategic significance as it augurs well with the ongoing initiatives of the Government of India such as "Make in India," "Skill Development" and "Swachh Bharat Abhiyan" and offers excellent opportunity to integrate with the driving targets of "doubling of farmers' income," "import reduction," "employment generation" and "waste to wealth creation." The biofuels programme in India has been primarily impacted due to the non-availability of domestic feedstock for biofuel generation which needs utmost attention so as to upscale biofuel production (PIB 2018).

6.3.3.1 Salient Features of National Policy on Biofuels

1. The National Policy on Biofuels (2018) describes biofuels as Basic biofuels viz. 1G (first-generation) bioethanol and biodiesel; and Advanced biofuels -2G (second-generation) ethanol, municipal solid waste (MSW) to drop-in fuels, 3G (third-generation) biofuels and bio-CNG, to enable the extension of relevant financial and fiscal incentives under each category.
2. The policy increases the scope of raw material for ethanol production by allowing the use of sugarcane juice; sugar-containing feedstocks like sugar beet and sweet sorghum; starch-containing feedstocks like corn and cassava; damaged food grains like wheat, rice and rotten potatoes unfit for human consumption for ethanol production.
3. The new biofuel policy allows the use of surplus food grains for ethanol production for blending with petrol/gasoline with the approval of the National Biofuel Coordination Committee.
4. With a thrust on advanced biofuels, the policy indicates a viability gap funding scheme for 2G ethanol biorefineries of Rs. 5000 crores in 6 years in addition to additional tax incentives, a higher purchase price as compared to 1G biofuels.
5. The policy supports setting up of supply chain mechanisms for biodiesel making from non-edible oilseeds, waste cooking oil and short gestation crops.
6. Roles and responsibilities of all the concerned authorities for biofuels have been captured in the policy document to synergize efforts.

6.3.3.2 Expected Benefits

- *Reduce import dependency*: One crore litre of E10 (ethanol mix) saves Rs. 28 crores of foreign currency exchange at current rates. The ethanol supply year (2017–18) is expected to mark a supply of around 150 crore litres of ethanol which results in savings of over Rs. 4000 crores of foreign exchange.
- *Cleaner environment*: One crore liter of E10 saves around 20,000 tons of CO_2 emissions. For the ethanol supply year (2017–18), there are secondary emissions of CO_2 to the tune of 30 lakh tons. Conversion of agricultural residue to biofuels would help reduce crop burning to reduce greenhouse gas emissions.
- *Health benefits*: Consumption of repeatedly heated cooking oil (RHCO), particularly in deep frying, is a potential health hazard and can lead to multiple diseases. Used cooking oil is a potential source for making biodiesel. Its use is environmentally safe and prevents the diversion of used cooking oil to small restaurants/roadside vendors.

- *MSW management*: India generates 62 million tons of waste every year. There are technologies available which can convert MSW to drop-in fuels. It is estimated that 1 ton of MSW has the potential to produce around 20% of drop-in fuels.
- *Infrastructural investment in rural areas*: It is expected that one 100 klpd biorefinery will require around Rs. 800 crores capital investments. At present, oil marketing companies (OMCs) are in the process of setting up 12 second-generation (2G) biorefineries with an investment of nearly Rs.10,000 crores. Further, addition of 2G biorefineries across the country will spur infrastructural investment in rural areas.
- *Employment generation*: One 100 klpd 2G biorefinery can contribute 1200 jobs in plant operations, village level entrepreneurs and supply chain management.
- *Additional income to farmers*: By adopting 2G technologies, agricultural waste/crop residues which otherwise are burnt by the farmers can be used to produce ethanol and can fetch a price for these wastes if a market is developed for the same. Also, farmers are not certain of getting the relative price for their produce during the surplus production phase. Thus conversion of surplus grains and agricultural biomass can help in price stabilization.

6.3.4 The Draft Auto Policy 2018

According to the draft National Auto Policy, February 2018, from the Ministry of Heavy Industries and Public Enterprises, the Government of India seeks to promote clean, safe, efficient and comfortable mobility for every person with a focus on emission control, environmental protection and affordability. The new policy includes:

- *"A 10-year strategy (until 2028) for emission standards;*
- *Adopting reductions in CO_2 through the Corporate Average Fuel Economy (CAFE) regulations;*
- *Introducing criteria for vehicle length and CO_2 emissions to classify vehicles for taxation;*
- *Defining a list of target technologies in the areas of green mobility, emission control, and safety with components and equipment that will be eligible for import duty reductions;*
- *Finalizing a green mobility roadmap including emission and fuel consumption standards, along with incentives and related infrastructure investments;*
- *Conducting a detailed study on the requirements of people, infrastructure for green vehicles to manage the quantity, density and mix of mobility infrastructure requirements, and*
- *Including standards for green vehicle infrastructure regarding power supply, connectors, and refueling."*

Moreover, the growing concern about environmental pollution through carbon emission has led the Government of India transport policy to target Euro III and IV vehicle norms (Aradhey 2013). Intending to conserve natural resources as well as to reduce pollution, the Government of India has also decided to implement Bharat Stage VI or BS VI emission norms in Delhi NCR from 1 April 2018. In the rest of the country, these will be implemented from 1 April 2020. Also, these standards permit to increase ethanol blending in petrol and biodiesel in diesel, which will assist in natural resources conservation and reduction in import of crude oil (The Economic Times 2018).

6.4 India's Dependence on Fossil Fuel Energy Imports

Although energy consumption per capita in India is estimated to be one-third of the global average, strong growth prospects in the world's fastest-growing economy will drive demand for energy across different sectors. Hence, access to sufficient and reliable energy sources becomes vital, mainly when one-quarter of the people lack access to electrical energy and reliance on coal, crude oil and natural gas continues to rise. Fossil fuels supply about three-quarters of India's energy demand, and India is the third-biggest importer of crude oil after China and the USA and remains to rely frequently on imports (mostly from Russia and Algeria).

In the last 5 years, import volumes of petroleum and petroleum products have risen by 25% to 307 billion litres. However, the associated cost fell by $155 billion in the Indian fiscal year 2013–2014 to $81 billion until the fiscal year 2016–2017. It grew by more than 25% in the last fiscal year to $101 billion (Fig. 6.1). India's total installed power capacity is just under 344 thousand megawatts, of which significant portions are produced from coal (57%), followed by renewables (20%),

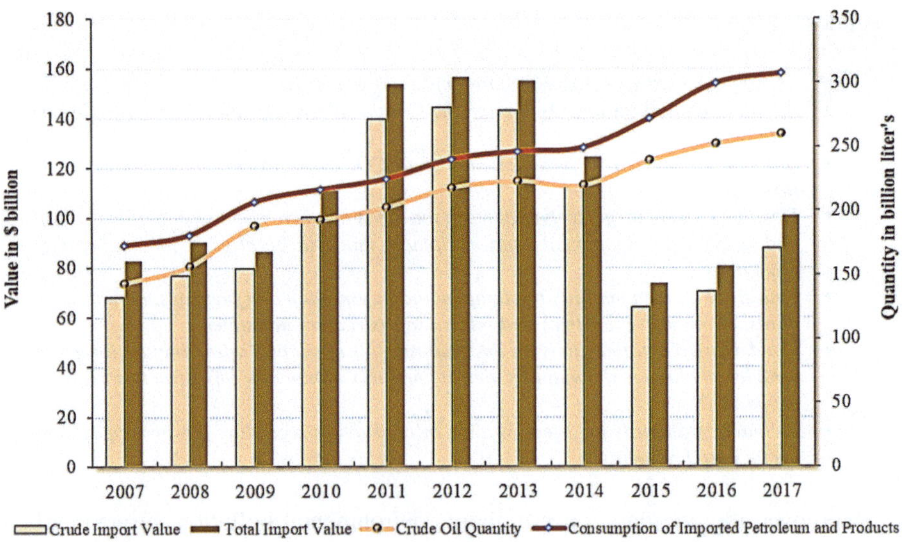

Fig. 6.1 India: crude oil import, petroleum products and consumption. (Source: Petroleum Planning and Analysis Cell, Gov. of India (GOI), timescale in the Indian fiscal year)

hydroelectricity (13.2%), natural gas (7.2%), nuclear source (1.9%) and diesel (0.24%). Among renewable energy sources, 49.3% is contributed by wind energy, 31.3% by solar energy, 12.8% by bioenergy and the rest is from small hydropower generation (Aradhey 2017, 2018).

In the economy of any country, crude oil prices play a crucial role. Changes in crude oil price indirectly impact the local currency owing to its effects on fiscal deficit and current account deficit. However, crude oil prices are dependent on several factors, the impact being political situation in and around major oil-producing countries. As far as India is concerned, crude oil price rise has a significant impact on various segments of the Indian economy; falling price is undoubtedly a blessing for the economy as it helps macro-economic management of inflation, fiscal deficit and current account deficit. In the long run, the only sustainable technology and viable policy to deal with high international oil prices are to rationalize the tax burden on oil products over time, pass on some increase in oil prices to consumers and protect especially those who are below the poverty line.

6.5 The Transport Sector and Commitment of Government of India to Reduce GHG Emissions

Continued economic growth coupled with urbanization and the fast lifestyle of Indian people pushes them to go in favour of road transport. The total registered vehicles have recorded a trend growth rate of 9.8% during the last 10 years (2005–2015), and as on 31 March 2015, there were 210 million registered vehicles in India (Transport Year Book 2013–2014 and 2014–2015). The Indian transport sector is the largest and the fastest-growing consumer of petroleum, with 39% of petroleum products consumed (TERI 2013). Transportation is the fastest growing and a significant contributor to GHG emissions. Growth in fuel use is higher for the transport sector than any other end-use sector. The primary drivers of global transport energy growth are land transport, mostly light-duty vehicles, such as cars, as well as freight transport. Transport sector accounted for nearly 23% of global CO_2 emissions in 2010 and 27% of end-use energy emissions with urban transport accounting for around 40% of end-use fuel consumption. CO_2 persists in the environment for over a century, with long-term warming effects (IPCC 2014).

According to the Fifth Assessment Report of IPCC, human-induced activity has caused an imbalance in the natural phenomenon of the greenhouse effect and related processes. Retention of excess heat by the increasing concentration of GHGs over the years (Table 6.1) traps more heat, resulting in changes in climatic processes as evidenced by an increase in temperatures, changes in rainfall patterns, rising sea levels, increased ocean acidity, and melting of glaciers IPCC 2014; Blasing 2016). The IPCC (2014) report reveals that total anthropogenic GHG emissions have continued to grow, with more significant increases towards the end of 2010. Despite the large number of mitigation policies, annual GHG emissions rose by 1.0 Gt carbon dioxide equivalent (Gt CO_2eq) (2.2%) per year from 2000 to 2010 compared to 0.4 Gt CO_2eq (1.3%) per year from 1970 to 2000, with highest recorded emission of 49

Table 6.1 Recent tropospheric greenhouse gas (GHG) concentrations

GHGs	Pre-1750 level	Recent level	GWP (100-year time horizon)	Increased radiative forcing (W/m^2)
Carbon dioxide (CO_2)	280	399.5 ppm	1	1.94
Methane (CH_4)	722	1834 ppb	28	0.50
Nitrous oxide (N_2O)	270	328 ppb	265	0.20
Tropospheric ozone (O_3)	237	337 ppb	n.a.	0.40
CFC-11(CCl_3F)	zero	232 ppt	4660	0.060

Source: Blasing (2016)

(±4.5) Gt CO_2eq/year in 2010 (IPCC 2014). India's greenhouse gas (GHG) emissions rose by an alarming 4.7% in 2016, compared to the previous year, the report released by the Netherlands Environmental Assessment Agency. The USA saw a decline of 2% and even China reported a decrease of 0.3%. The good news is that a global carbon dioxide emission has remained flat in the past 2 years registering only marginal increases.

India is the sixth largest greenhouse gas emitter, contributing almost 3% of the world's total emissions. India's per capita CO_2 emission is projected to increase to 1.6 tons by 2030. India's huge population, however, aggravates the net emissions into the atmosphere (Francis et al. 2005). Experts say India is expected to be hit severely by global warming. It is already one of the most disaster-prone nations in the world, and many of its 1.2 billion people live in regions vulnerable to hazards such as floods, cyclones, and droughts. Significant GHG reduction *is* necessary in order to achieve the prescribed levels (Prasad et al. 2019b).

The GOI is committed to reducing GHG by 20–25% by 2020 and 33–35% by 2030 over 2005 levels. Moving towards green mobility and low carbon economy, a dedicated transport sector will likely support GHG reduction goals (Aradhey 2018). Through upgradation of fuel efficiency standards, the GHG emissions can be reduced. Additionally, there is a proposal to upgrade the vehicles from current Bharat Stage (BS)-IV standard to BS-VI fuel compatible by 2020. The new fuel engine standards are likely to reduce harmful emissions and increase fuel efficiency. The GOI was committed to rolling out BS-VI norms by 2023, but it revised the deadline to 2020.

6.6 Biofuel Consumption, Production, Trades, and Projections

Over the next few decades, the most certain increase in demand for biofuels is going to concentrate on displacing liquid fuels for transportation, particularly ethanol which currently supplies over 95% of the biofuels for transport (Fulton et al. 2004). At present, ethanol production is mostly based on sugarcane and maize sources. At

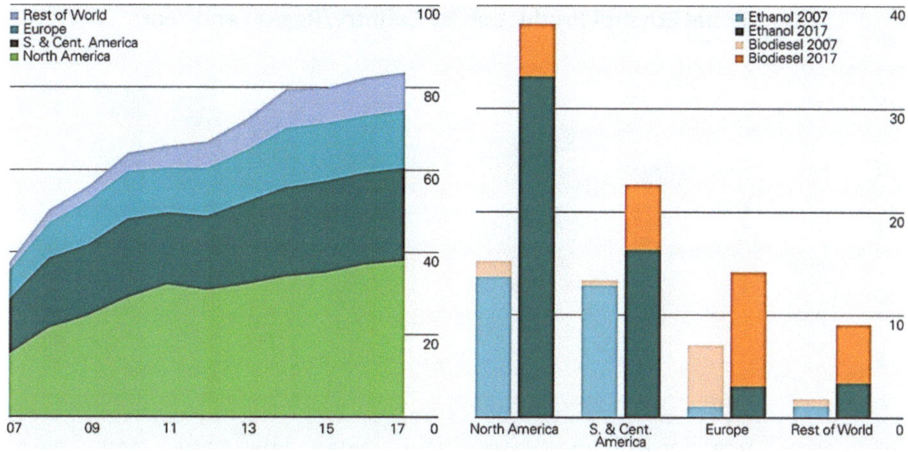

Fig. 6.2 World biofuels production by region and type (million tonnes oil equivalent). (Sources: Includes data from F.O. Lichts; Strategie grains; US-EIA (March 2018))

the same time, these crops likely have the most significant impact on food supply and demand systems. That is particularly true, if the production occurs on prime agricultural lands as is likely given the need to decrease transport costs of both the feedstocks and fuel products to and from larger, centralized ethanol production facilities.

As regards the global biofuels scenario, the production increased only by 3.5% (2017) and the USA has contributed significantly for the total production. While the growth of global production of ethanol was 3.3%, the growth rate of biodiesel production was 4% (Fig. 6.2). The USA is the world's largest producer of ethanol, and in 2017, the US production was about 16 billion gallons. The USA and Brazil together contribute 85% of the global ethanol production (Fig. 6.3). The vast majority of US ethanol is produced from corn, while Brazil primarily uses sugarcane (Renewable Fuels Association 2018).

With respect to biodiesel production, the USA and Brazil were among the biggest biodiesel producers in the world, totalling 6.0 and 4.3 billion litres, respectively, and Germany is ranked third with a production volume of around 3.5 billion litres in 2017 (Fig. 6.4). After the adoption of the Energy Policy Act, 2005, which provided tax incentives for certain types of energy, biodiesel production in the USA begins to expand. Currently, volumetric ethanol excise tax credit is one of the chief financial supports for biofuels production in the USA. The USA exported around 85 million gallons of its biodiesel products in 2010. Comparatively, Argentina accounted for over half of the world's total exports.

Currently, in India, almost 330 distilleries are producing more than 4.6 billion litres rectified spirits (alcohol) annually. Of this total, about 165 distilleries can distil over 2.2 billion litres of ethanol (denatured and undenatured) used as fuel, industrial chemicals, and beverage. India's EBP was based more on sugarcane molasses, not directly from sugarcane juice, corn, or any other potential domestic raw material

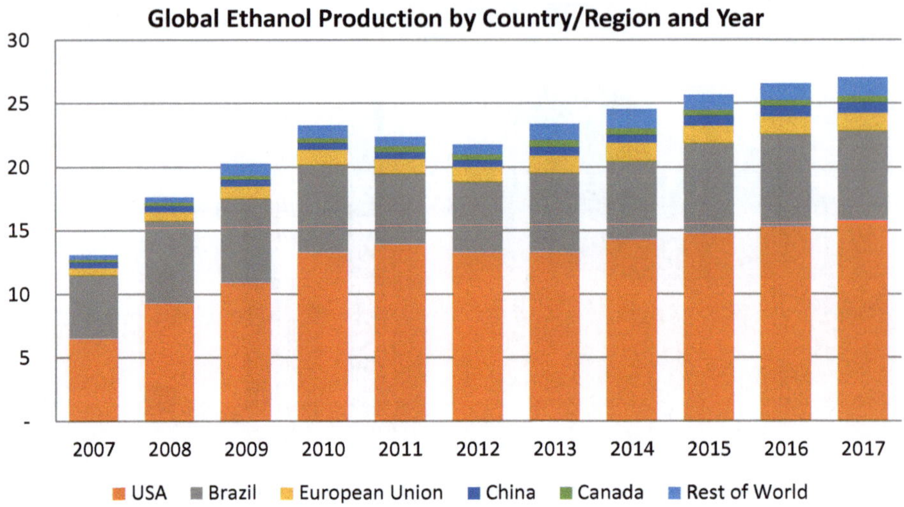

Fig. 6.3 Global ethanol production by Country/Region and Year. (Sources: www.afdc.energy.gov/data)

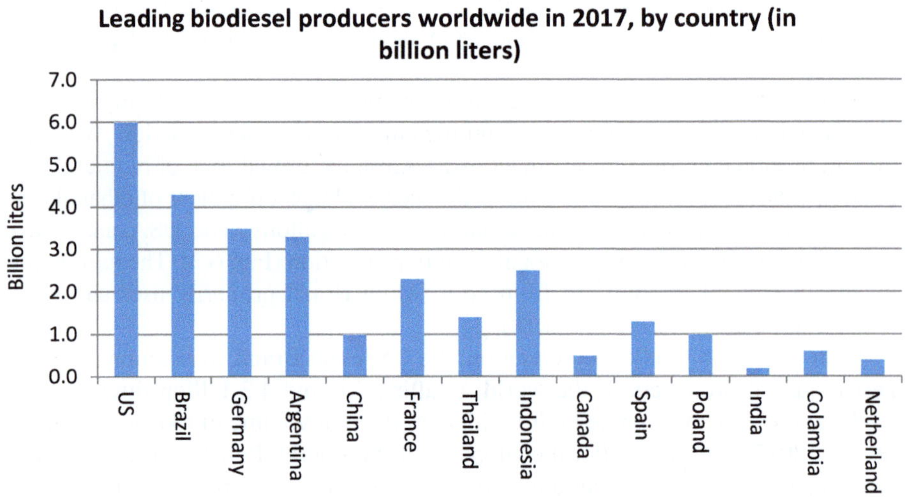

Fig. 6.4 Global biodiesel producers worldwide in 2017. (Sources: www.afdc.energy.gov/data)

sources available in the country. However, the current biofuel policy for 2018 has increased the scope for including other raw materials, in addition to encouraging optimal capacity utilization of grain-based distilleries (Aradhey 2018).

India's total ethanol consumption will outgrow production for the fourth consecutive year due to an uptick in fuel ethanol purchases by industry and a consistent rise in demand from the industrial and potable sectors (Fig. 6.5). The ethanol

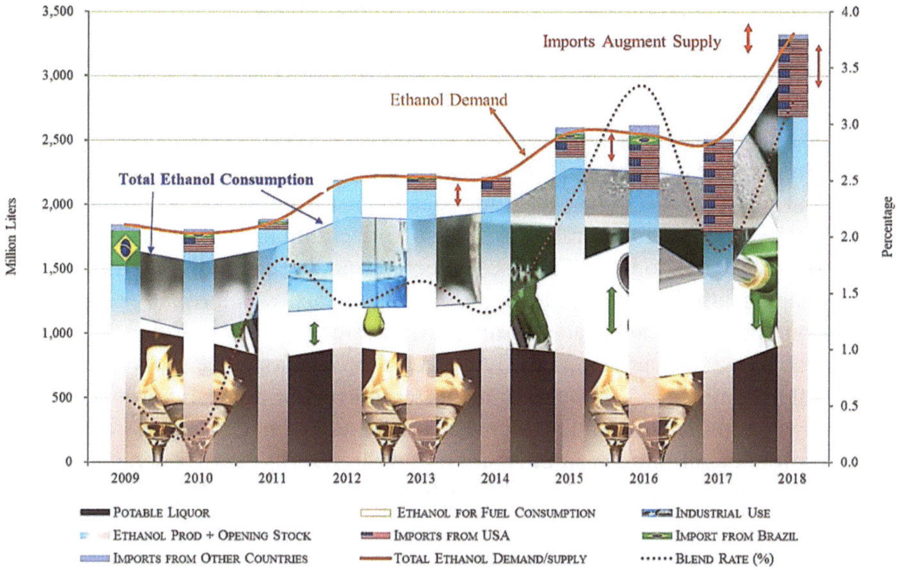

Fig. 6.5 India: Ethanol production, Supply, and Consumption. (Source: Aradhey 2018, FAS/USDA Data)

consumption demand growth (8% annual, 5 year average) is rather stiff compared to production growth although both have risen, but in response to different demand drivers: the rise in fuel prices have contributed to growth in ethanol consumption and a strong recovery in sugarcane production this year has contributed to production growth.

Total ethanol consumption is expected to reach a record high of 3.1 billion litres in 2018. (India achieved its highest ethanol market penetration at 3.3% (national-level blend) in 2016). The GOI mandatory use of 'indigenous ethanol only' for fuel under the EBP is expected to rise from 675 million litres in 2017 to a record 1.25 billion litres this year; this is 85% above last year and marginally above the 1.1 billion litres blended in 2016. The remaining 1.9 billion litres will be the industrial and potable alcohol sectors (which are exempted from GST). Since the quantity of ethanol demanded at higher prices may be less, the industrial uses and the potable sector will need to augment some of its supply from grain-based distilleries, partly from raw material imports or by directly importing the finished products.

An estimated 2.55 billion litres (record) of ethanol will be produced in 2018, 52% above last year. An anticipated rise in sugarcane production in 2018 and consequent increase in availability of molasses will bring an additional 875 million litres of ethanol into the supply chain compared to last year's supply. Hypothetically, if all the ethanol produced in 2018 is made available for EBP, then it will meet a 6.5% blend target. However, demand rationing from potable and industrial sectors will limit the national blend average to 3.2%. Industry sources indicated that the OMCs might be able to procure upwards of 1.3 billion litres in 2018. With around

166 refineries, the production capacity of combined plants in the last 10 years has risen by 800 million litres to 2.3 billion litres in 2018. The capacity utilization in 2018 will be 111% after 2 years of successive underutilization on supply concerns.

In 2018, India will continue to be a net importer of ethanol (all end uses) despite the rise in its domestic production. The USA has become the near-sole supplier of India's imports, and exports will rise on growing demand from African nations and neighbouring countries. The preceding statement assumes current market conditions, and the subsequent narrative is based on prevailing market trends. Imports are allowed only for the non-fuel purpose subject to actual user condition. They backfill supply gaps in ethanol demand for industrial chemicals, which usually gets smaller during bumper sugarcane harvests and surplus sugar production thereby leaving additional feedstock (molasses) for fuel ethanol production. The supply deficit will be slightly narrower in 2018 due to a substantial rise in domestic production, but imports will continue to augment increasing demand not met through local supplies. As a result, ethanol imports in 2018 are 640 million litres (mostly denatured), second highest in a decade. The share of US ethanol in the total import basket has grown by 22% in the last 5 years to 96% in 2017. China, South Korea, Pakistan and Bhutan filled in the remaining 4% market share.

Ethanol exports[1] in 2018 are expected to rise to 164 million litres (mostly undenatured), 16% above last year. After its peak export sales in 2013 (233 million litres), Indian exports of ethanol have declined by an average of 15% per year on tighter supply and strong local demand. That trend seems to have reversed itself in the last 2 years as export sales have recovered (close to the five-year average sales figure), as local supply is enough to match the consistent rise in demand from African nations and neighbouring countries. In 2017, Nigeria, Ghana, Angola, Nepal, and Kenya were the top five export destinations for Indian ethanol. However, India faces stiff competition from other major ethanol suppliers from the USA, South Africa, the UK and Canada.

The marketing of biodiesel is still nascent. Presently, India has an annual capacity to produce biodiesel around 650 million litres. Estimated annual biodiesel production upwards of 185 million litres in 2018, an additional 18 million litres above last year, consumption of biodiesel is growing steadily at 2–3%. However, producing close to 29% of the installed capacity, with most of the feedstock sourced from the food processing industry and restaurants. It is estimated that transport by road and rail account for roughly half of all biodiesel use, and the other half is consumed by off-road farm transport and various stationary applications. The national average blend rate for on-road transport and stationary applications are each estimated at one-seventh of one percent (0.14%) today or slightly higher than the estimated 0.04% blend rate 10 years ago (Table 6.2).

Beginning 2014 and running through 2016, a little more than a quarter of total biodiesel production was exported before declining through 2018 as growing domestic demand left a little exportable surplus. Compared to average export sales

[1] Biofuel (e.g. fuel grade) exports are not allowed when domestic supply is lower than the country's requirement.

Table 6.2 India's total biodiesel production, trades, and projections

Calendar year	2009	2010	2011	2012	2013	2014	2015	2016	2017	2018
Beginning stocks	0	45	15	13	14	14	9	12	12	18
Production	75	100	111	126	132	138	152	158	170	187
Imports	0	0	0	0	0.3	1.7	0.8	2.7	7.1	9.9
Exports	0	0	0	0	3.9	41.5	33.1	41.7	7.6	7.4
Consumption	30	131	113	125	125	81	106	108	163	187
Ending stocks	45	15	13	14	18	31	23	23	18	21
Production capacity (million litres)										
Number of biorefineries	5	5	5	5	6	6	6	6	6	6
Nameplate capacity	450	450	450	460	465	480	500	550	600	650
Capacity use (%)	16.7	22.2	24.7	27.4	28.4	28.8	30.4	28.7	28.3	28.8
Feedstock Use for Fuel (1000MT)										
Non-edible industrial	30	50	58	65	70	75	85	90	100	110
Used cooking oil	35	38	42	48	49	50	55	55	55	60
Animal fats and Tallow's	3	6	6	7	7	6	5	6	6	8
Market penetration (million litres)										
Biodiesel, on road use	15	36	31	42	49	32	41	48	72	83
Diesel, on-road use	39.8	42.6	45.5	49.3	49.4	49.6	52.2	55.2	57.0	61.2
Blend rate (%)	0.04	0.09	0.07	0.08	0.10	0.06	0.08	0.09	0.13	0.14
Diesel total use	66.4	71.0	75.9	82.3	82.3	82.7	87.1	92.0	95.0	102.1

Source: Aradhey (2018)

of 39 million litres made from 2014 to 2016, the export forecast for 2018 is just 7.4 million litres (Table 6.2). However, during the same period, imports grew five-fold to 10 million litres in 2018, indicating small but steady growth. The EU, China, Malaysia, and Indonesia are significant suppliers of biodiesel to India while India's major export destinations are the Malaysia and EU. Additionally, for sustainable biodiesel production to grow there is a need for a strong commercially viable strategy, as capacity utilization is less than 30%, due to multiple constraints, including limited feedstock availability, lack of integrated and dedicated supply chain, and restrictions on imports. The government of India and the private sector are working together in production and distribution. Financial institutions such as IREDA, NABARD, and SIDBI are supporting for biodiesel production infrastructure especially cultivation, extraction, processing, storage, and distribution. The Biodiesel Association of India, working as a coordinating body for supporting Research & Development and marketing of biodiesel in the country.

6.7 Global Climate Change and Biofuels

Climate change has been a major global concern in recent years. That has led to various initiatives, such as the UNFCCC and Kyoto Protocol (KP). The energy sector being a significant emitter of GHGs has also motivated the political progress towards improved energy supplies with enhanced efficiency as far as their eco-friendly nature is concerned (Prasad et al. 2014). According to Tilman et al. (2009), balancing needs and seeking solutions to energy, environment, and food security challenges, society cannot afford to miss out on greenhouse gas emission reductions and multiple benefits, when biofuels are done right. However, the world also cannot accept the undesirable impacts of biofuels done wrong. Biofuels done right can be produced in large quantities (NAE 2009). However, they must be derived from materials produced with much lower life-cycle greenhouse gas emissions than traditional fossil fuels and with slight or no competition with food production. The feedstocks materials that can be included in this category are (1) perennial plants grown on degraded lands/waste or abandoned from agricultural use, (2) crop residues, (3) sustainably harvested wood and forest residues, (4) double crops and mixed cropping systems, and (5) conversion of municipal and industrial wastes into liquid fuels (Puppan 2002; Kammen et al. 2007; Prasad et al. 2019a).

In its fourth climate change assessment report, biofuels were identified as a "key mitigation strategy" (IPCC 2007). However, the debate surrounding biomass in the food versus fuel competition, and growing concerns about land use, water, and replacement of forests has acted as incentives for the development and implementation of sustainability criteria and frameworks (IPCC 2011). Furthermore, the support for advanced biorefinery and next-generation biofuel options is driving biofuel to be more sustainable (IPCC 2014). Conversion of biomass feedstock to biofuels and its use as a supplement to petrol-fuels is more environmentally friendly than petrol-fuels alone (Prasad et al. 2019b). When we use ethanol instead of petroleum fuel, we help reduce atmospheric CO_2 by avoiding the emissions associated with the use of petrol; not releasing CO_2 stored in the fossil fuels and contributing a mechanism for CO_2 absorption by growing new biomass. Because of their adaptability with the natural carbon cycle, biofuels offer the most beneficial alternative for reducing greenhouse gases from the transportation sector (US DoE 1999).

Advanced conversion technologies including the use of biomass residues and long-lived plantations can lead to 80–90% reduction in GHG emissions compared to the fossil fuel use levels (IPCC 2011). Figure 6.6 shows a range in reductions of GHG emissions per vehicle-km (v-km) obtained from various studies. Larson (2006) assessed GHG emissions from transportation fuels, petroleum fuels, first-generation biofuels (sugar and starch-based ethanol, oilseed-based biodiesel), and selected second-generation biofuels derived from lignocellulosic biomass (ethanol and Fischer-Tropsch diesel) on a well-to-wheel basis.

Many studies have been found that the application of 1st-generation biofuels results in emission reductions of 20–60% of CO_2eq relative to fossil fuels. Expected reductions for future commercialized 2nd generation biofuels are in the range of 70–90% of CO2eq relative to fossil fuels (FAO 2008). The broad range of emission

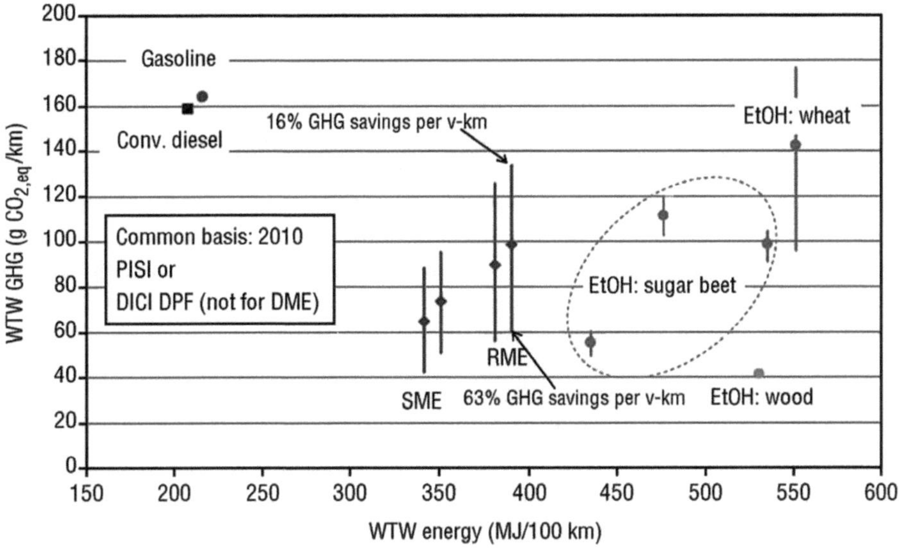

Fig. 6.6 Well-to-wheels energy requirements and GHG emissions for conventional biofuel pathways compared with gasoline and diesel pathways, assuming 2010 vehicle technology. (Note: *EtOH* ethanol, *SME* soy methyl ester, *RME* rape methyl ester, *PISI* port injection spark ignition, *DICI DPF* direct injection compression ignition with diesel particulate filter. (Source: Larson 2006)

reductions for the 1st-generation biofuels is due to several types of feedstock and conversion processes and to the different sites of production and consumption. Substituting biofuels for petroleum-based fuels was found to have the potential to reduce lifecycle GHG emissions directly associated with the fuel supply chain. Second generation biofuels (with lifecycle GHG emissions between 10 and 38 g CO_2eq/MJ) were reported to provide greater mitigation potential over first-generation biofuels (with lifecycle GHG emissions between 19 and 77 g CO_2eq/MJ) compared to 85–109 g CO_2eq/MJ for petroleum fuels (Larson 2006). Estimates of GHG emissions are variable for both biofuels and petroleum fuels, primarily due to assumptions about technological issues and where and how the feedstock is produced. Adopting a consistent 10% blending regimen for gasoline and a 5% biodiesel regimen for diesel could potentially reduce CO_2 emissions by 3–4%, creating a total abatement of 10–12 million tons of CO_2. A compelling blend regime could also deliver annual economic savings of $1.2–1.5 billion in a country like India (Wang et al. 2011).

6.8 Policy Initiative by Government of India to Combat Biofuel Trilemma

In India, National Policy on Biofuels 2009 and the recently approved the national policy on biofuel 2018 have a clear-cut guidelines to combat biofuel trilemma on food, energy, and environment. The Government of India is focusing on utilizing waste and degraded forest and non-forest lands only for the cultivation of shrubs and trees bearing non-edible oilseeds to produce biodiesel. In the country, bioethanol is produced mainly from molasses, a by-product of the sugar industry. In the future, it would be ensured that the next generation of technologies is based on non-food feedstocks. Therefore, the issue of fuel vs. food security is not relevant in the Indian context (MNRE 2010). The Government of India is also committed to reducing GHG by 33–35% by 2030 over 2005 levels and moving towards green biofuels and low carbon economy to combat harmful emissions and its impact on the environment by increasing fuel efficiency standards, upgrading vehicles from BS-IV standard to BS -VI fuel compatible norms by 2023. Looking forward to the IPCC AR6 cycle, key emerging concerns are likely to be (i) trade-offs between the use of land for bioenergy generation, food and fibre production and conservation of ecosystem integrity and (ii) the co-delivery of bioenergy based climate change mitigation (with or without CCS and the UN Sustainable Development Goals) (Smith and Porter 2018).

6.9 Conclusion

Biofuels production has increased significantly in the past few years to reduce fossil fuel dependence. The use and cautious biofuel production from renewable sources can help to satisfy not only the energy demand but also cut carbon emissions as committed by India through various multilateral agreements and at International fora. The biofuel production and multi-functional biofuel benefits can also help to sustain biofuel producers and boost rural employment and strengthen the bio-based economy in India. There is an urgent need to develop a strategy for efficient and resilient biofuel production system under changing climatic scenario. Hence, our effort should bring in a concrete change for the society and ensure the protection of natural resources and ecosystem.

References

Aradhey A (2010) India-biofuels Annual Tech Rep GAIN report. USDA foreign agricultural service, New Delhi, India

Aradhey A (2013) India biofuels annual, In: GAIN-report no IN3073, Global agricultural information network (GAIN), US Department of Agriculture-Foreign Agricultural Services (FAS). http://gain.fas.usda.gov/Recent%20GAIN%20Publications/Biofuels%20Annual_New%20Delhi_India_8-13-2013.pdf

Aradhey A (2017) India biofuels annual 2017. Global Agricultural Information Network (GAIN) Report Number IN-7075, USDA Foreign Agricultural Service, New Delhi. Available from: https://gain.fas.usda.gov/Recent%20GAIN%20Publications/Biofuels%20Annual_New0De%21hi_India_6-27-2017.pdf

Aradhey A (2018) India biofuels annual, In: GAIN-report no IN8085, Global agricultural information network (GAIN), US Department of Agriculture-Foreign Agricultural Services (FAS). http://gain.fas.usda.gov/Recent%20GAIN%20Publications/Biofuels%20Annual_New%20Delhi_India_9-28-2018.pdf

Basavaraj G, Rao PP, Reddy CR, Kumar AA, Rao PS, Reddy BVS (2012) A review of the national biofuel policy in India-a critique of the need to promote alternative feedstocks, Working paper series no. 34. ICRISAT. Patancheru, Andhra Pradesh

Blasing TJ (2016) Recent greenhouse gas concentrations, updated April 2016, Carbon Dioxide Information Analysis Center (CDIAC). https://doi.org/10.3334/CDIAC/atg.032. Available electronically at: https://cdiac.ess-dive.lbl.gov/pns/current_ghg.html

Buyx AM, Tait J (2011) Biofuels: ethics and policy-making. Biofuels Bioprod Biorefin 5:631–639

Demirbas A (2008) Biofuels sources, biofuel policy, biofuel economy and global biofuel projections. Energy Convers Manag 49(8):2106–2116

Down To Earth (2003) India's ambitious plan to boost biofuel likely to fail without a roadmap. https://www.downtoearth.org.in/news/energy/india-s-ambitious-plan-to-boost-biofuel-likely-to-fail-without-roadmap-61892

Draft Auto Policy (2018) Govt. of India. https://dhi.nic.in/writereaddata/UploadFile/DHI-NAB-Auto%20Policy%20Draft%20Document_vDRAFT.pdf

Faaij AP, Domac J (2006) Emerging international bio-energy markets and opportunities for socio-economic development. Energy Sustain Dev 10(1):7–19

FAO (2008) The state of food and agriculture. Biofuels: prospects, risks, and opportunities. ISBN:978-92-5-105980-7, Rome, 138 p

Francis G, Edinger R, Becker K (2005) A concept for simultaneous wasteland reclamation, fuel production and socio-economic development in degraded areas in India: need, potential, and perspectives of Jatropha plantations. Nat Res Forum 29:12–24

Fulton L, Howes T, Hardy J (2004) Biofuels for transport: an international perspective. International Energy Agency, Paris

IEA (2010) International energy agency, world energy outlook- 2010. Available online http://www.worldenergyoutlook.org/media/weo2010.pdf

IEA (2018) CO_2 Emissions from fuel combustion: an essential tool for analysts and policy makers. https://www.iea.org/statistics/co2emissions/

IPCC (2007) Climate change 2007: synthesis report. Contribution of Working Groups I, II and III to the fourth assessment report of the Intergovernmental Panel on Climate Change [Core Writing Team, Pachauri RK, Reisinger A (eds)]. IPCC, Geneva, Switzerland, 104 pp

IPCC (2011) Summary for policymakers. In: Edenhofer O, Pichs-Madruga R, Sokona Y, Seyboth K, Matschoss P, Kadner S, Zwickel T, Eickemeier P, Hansen G, Schlömer S, von Stechow C (eds) IPCC special report on renewable energy sources and climate change mitigation. Cambridge University Press, Cambridge/New York

IPCC (2014). In: Edenhofer O, Pichs-Madruga R, Sokona Y, Farahani E, Kadner S, Seyboth K, Adler A, Baum I, Brunner S, Eickemeier P, Kriemann B, Savolainen J, Schlömer S, von Stechow C, Zwickel T, Minx JC (eds) Climate change 2014: mitigation of climate change. Contribution of Working Group III to the fifth assessment report of the Intergovernmental Panel on Climate Change. Cambridge University Press, Cambridge/New York

Kammen DM, Farrell AE, Plevin RJ, Jones AD, Nemet GF, Delucchi MA (2007) Energy and greenhouse impacts of biofuels: a framework for analysis. OECD Roundtable on Biofuels, Paris

Kumar S, Prasad S, Malav MK, Gupta N (2016) Characterization and pretreatment of wheat straw using microwave assisted dilute acid to produce fermentable sugars for hydrogen production. Green Farming 7(1):95–98

Larson ED (2006) A review of life-cycle analysis studies on liquid biofuel systems for the transport sector. Energy Sustain Dev 10(2):109–126

Malav LC, Khan SA, Kumar S, Gupta N, Chaudhary P (2017) Effect of air pollutants on growth and yield of rice (*Oryza sativa*) and wheat (*Triticum aestivum*) crops around the coal based thermal power plant. Int J Curr Microbiol App Sci 6(10):3151–3165

MNRE (2009) National policy on biofuels, Govt of India. Available at: http://mnre.gov.in/file-manager/UserFiles/biofuel_policy.pdf

MNRE (2010) Bioenergy India. Magazine on Biomass Energy. https://mnre.gov.in/file-manager/bioenergy-india/bio4-0610.pdf

NAE (2009) National Academy of Engineering (NE), National Research Council (NRC), Liquid transportation fuels from coal and biomass: technological status, costs, and environmental impacts. National Academy of Sciences, Washington, DC, 2009

PIB (2018) Press Information Bureau of India Government Cabinet, Cabinet approves National Policy on Biofuels – 2018. http://pib.nic.in/new site/PrintRelease.aspx?relid=179313

Planning Commission (2003) GOI, Report of the committee on the development of biofuels. Retrieved from http://planningcommission.nic.in/reports/genrep/cmtt_bio.pdf

Prasad S, Singh A, Joshi HC (2007) Ethanol as an alternative fuel from agricultural, industrial and urban residues. Resour Conserv Recycl 50(1):1–39

Prasad S, Dhanya MS, Gupta N, Kumar A (2012) Biofuels from biomass: a sustainable alternative to energy and environment. Biochem Cell Arch 12(2):255–260

Prasad S, Amit K, Muralikrishna KS (2014) Biofuels production: a sustainable solution to combat climate change. Indian J Agric Sci 84(12):1443–1452

Prasad S, Malav MK, Kumar S, Singh A, Pant D, Radhakrishnan S (2018) Enhancement of bio-ethanol production potential of wheat straw by reducing furfural and 5-hydroxymethylfurfural (HMF). Bioresour Technol Rep 4:50–56

Prasad S, Sheetal KR, Venkatramanan V, Kumar S, Kannojia S (2019a) Sustainable energy: challenges and perspectives. In: Shah S et al (eds) Sustainable green technologies for environmental management. Springer, Singapore, pp 175–197. https://doi.org/10.1007/978-981-13-2772-8_9

Prasad S, Venkatramanan V, Kumar S, Sheetal KR (2019b) Biofuels: a clean technology for environment management. In: Shah S et al (eds) Sustainable green technologies for environmental management. Springer, Singapore, pp 219–240. https://doi.org/10.1007/978-981-13-2772-8_11

Puppan D (2002) Environmental evaluation of biofuels. Periodica Polytechnica Ser Soc Man Sci 10:95–116

Renewable Fuels Association (2018) Global ethanol production by country/region and year. The Alternative Fuels Data Center (AFDC), https://afdc.energy.gov/data/10331

Smith P, Porter JR (2018) Bioenergy in the IPCC assessments. GCB Bioenergy 10(7):428–431

TERI Energy Data Directory and Yearbook (2012/2013) Indian energy sector: an overview. TERI, New Delhi, pp 1–14.

The economic times (2018) BSVI norms: India revises standards specifications for petrol and diesel. https://energy.economictimes.indiatimes.com/news/oil-and-gas/bsvi-norms-india-revises-standards-specifications-for-petrol-and-diesel/62553578

Tilman D, Socolow R, Jonathan AF, Hill J, Larson E, Lynd SP, John R, Tim S, Chris S, Robert W (2009) Beneficial Biofuels: the food, energy, and environment trilemma. Science 325:270–271

Transport Year Book (2013–14 and 2014–2015) Ministry of Road Transport and Highways, Delhi, Government of India: 2015.116p. https://data.gov.in/catalog/road-transport-year-book-2013-14-and-2014-15

United Nations (2016) World economic situation and prospects 2016. http://www.un.org/en/development/desa/policy/wesp/wesp_current/2016wesp_ch1_en.pdf

US DoE (1999) Biofuels: a solution for climate change. U.S. Department of Energy Office of Energy Efficiency and Renewable Energy Office of Transportation Technologies DOE/GO-10098-580 Revised September 1999

USDA GAIN Report (2017) India Biofuels Annual. USDA Foreign Agricultural Service. https://gain.fas.usda.gov/Recent%20GAIN%20Publications/Biofuels%20Annual_New%20Delhi_India_6-27-2017.pdf

Vasudevan PT, Boyi F (2010) Environment sustainable biofuels: advances in biodiesel research. Waste Biomass Valori 1:47–63

Wang H, Srinivasan R, Yu F et al (2011) Effect of acid, alkali, and steam explosion pretreatments on characteristics of bio-oil produced from pinewood. Energy Fuel 25:3758–3764

Climate Change, Water Resources, and Agriculture: Impacts and Adaptation Measures

7

Durba Kashyap and Tripti Agarwal

Abstract

Agriculture is one of the key domains that is significantly affected by climate change. The chapter presents the observed and projected impact of climate change on freshwater resources globally. In addition to this, case studies of successful implementation of adaptation measures adopted to tackle climate change-induced water stress in agriculture have been discussed with a special focus on high-altitude farming systems particularly vulnerable to increasing climate risk. As one of the potential adaptation measures, the relevance of water footprint as a tool to optimize water use and strategize cropping patterns with respect to crop water use efficiency and prevailing climatic conditions has also been discussed.

Keywords

Climate change · Freshwater resources · Water footprint · Agriculture · Adaptation · Irrigation water policy

7.1 Introduction

The water cycle is dynamic and intimately connected with atmospheric temperature and radiation balance. "*Changes in the large-scale hydrological cycle like increasing atmospheric water vapor content; changing precipitation patterns, intensity, and extremes; reduced snow cover and widespread melting of ice; and changes in soil moisture and runoff have been linked to the observed warming over the past several decades*" (Bates et al. 2008). Such climate-driven changes in the hydrological cycle along with non-climatic drivers of change such as population growth,

D. Kashyap · T. Agarwal (✉)
National Institute of Food Technology Entrepreneurship and Management, Kundli, Sonipat, Haryana, India

© Springer Nature Singapore Pte Ltd. 2020

V. Venkatramanan et al. (eds.), *Global Climate Change and Environmental Policy*, https://doi.org/10.1007/978-981-13-9570-3_7

economic development, urbanization, land use changes and water management responses can lead to diverse impacts and risks (Cisneros et al. 2014). Agriculture is one of the important sectors where changes in the hydrological cycle will have greater impact. It is among the most vulnerable sector to the risks of climate change (Smit and Skinner 2002). Both rainfed and irrigated agriculture are affected by climate-induced changes in precipitation and the availability of water. Climatic drivers are conditioned by and interact with non-climatic drivers which challenge the sustainability of resources. The significant role of non-climatic drivers in altering water supply and demand indicates that adaptation practices hold potential to improve the availability of water in the future (Cisneros et al. 2014).

Precipitation, temperature and evaporative demand are the most dominant climate drivers for water availability. Of these, temperature is predominantly significant in snow-covered basins and coastal areas where sea level rise due to high-temperature-induced expansion of water poses a threat. Groundwater tables and recharge rates are also affected by climate change through hydrological drivers that charge and recharge groundwater. Changes in permafrost thaw, surface water flow and precipitation variability also affect groundwater levels (IPCC 2007b). Table 7.1 presents the highlights of climate change-induced risks for freshwater resources and projections based on IPCC (Intergovernmental Panel on Climate Change) AR 4 (4th Assessment report) and AR 5 (5th Assessment report). Table 7.2 presents a few examples of the observed effects on freshwater water resources mainly attributable to climate change. The next section describes the effects of climate change on different hydrological variables that serve as sources of freshwater.

7.2 Impact of Climate Change on Freshwater Resources

7.2.1 Cryosphere

The cryosphere, derived from the Greek word "krios" meaning cold, is the frozen water part of the Earth system. The cryosphere consists of (a) ice and snow on land and (b) ice on water. Ice and snow on land include continental ice shelves, ice caps, glaciers and areas of snow and permafrost and stores around 75% of world's freshwater whereas ice on water includes frozen parts of oceans, rivers and lakes. The general trend in cryosphere shows a significant decrease in ice storage due to warming (IPCC 2007a). Snow cover is dependent on both air temperature and precipitation. Climate change is said to be responsible for the increased summer thaw which is found to be more than the snowfall during winters. There are projections of a widespread reduction in snow cover, increase in thaw depths of permafrost region and mass loss of glaciers and ice caps (IPCC 2007a). When in equilibrium, glaciers store water during cold or wet years and release them in warmer years, thus reducing the inter-annual variability of water resources (Viviroli et al. 2011). Shrinking glaciers due to global warming might reduce their capacity to store water thus making water supply less dependable (Cisneros et al. 2014). In high altitudes, summer

Table 7.1 Highlights of AR 4 and AR 5 with respect to climate change-induced freshwater-related risks

Freshwater-related risks at the global scale (AR 5)	Degree of uncertainty (qualitative)
Freshwater-related risks of climate change increase significantly with increasing greenhouse gas (GHG) concentrations	*Robust evidence, high agreement*
Climate change is projected to reduce renewable surface water and groundwater resources significantly in most dry subtropical regions	*Robust evidence, high agreement*
Climate change is likely to increase the frequency of meteorological and agricultural drought in presently dry regions by 2100 under the RCP8.5 scenario	*Medium evidence, medium agreement*
In regions with snowfall, climate change has altered observed streamflow seasonality, and increasing alterations due to climate change are projected	*Robust evidence, high agreement*
Total meltwater yields from stored glacial ice in glacial-fed rivers will increase in many regions during the next decades, but decrease thereafter	*Robust evidence, high agreement*
Some GHG emission mitigation measures like bioenergy crops imply risks for freshwater systems, if they demand more water than other measures	*Medium evidence, high agreement*
Freshwater-related risks at the global scale (AR 4)	**Degree of uncertainty (quantitative)**
Globally, the negative impacts of future climate change on freshwater systems are expected to outweigh the benefits	*High confidence*
Water supplies stored in glaciers and snow cover are projected to decline during the twenty-first century	*High confidence*
Annual average river runoff and water availability are projected to increase at high latitudes and in some wet tropical areas and decrease over some dry regions at mid-latitudes and in the dry tropics by 2050	*High confidence*
Increased precipitation intensity and variability are projected to increase the risks of flooding and drought in many areas	*Very likely(for flooding)* *Likely(for droughts)*

Source: Bates et al. (2008) and Cisneros et al. (2014)

irrigation depends on annual snowfall, which the large glaciers and snowfields help to trap and maintain. With changing climate and receding glaciers, the summer irrigation reserves might continue to shrink (Rizvi 1998). Consequently, high-altitude agriculture, which is dependent on snowmelt and glacial water, is already facing a crisis situation (discussed in Box 7.1).

An increased melting and consequent mass loss of glaciers and ice caps worldwide and glacial retreat have also led to the formation of moraine-dammed glacial lakes. These lakes are a threat for glacial lake outburst floods particularly in the mountain ranges that are steep (e.g., the Himalayas, the Andes and the Alps) (IPCC 2007b). In the western Himalayas, there are small and stable lakes in Nepal; while Bhutan has large and numerous glacial lakes, most of which are growing and causing a continued increase in hazard (Gardelle et al. 2011). Also, the enhanced

Table 7.2 Few examples of observed changes in freshwater resources mostly attributable to climate change

Hydrological variable	Observed change	Region (Period)	References
Cryosphere	Decreases in the extent of permafrost	Arctic and Eurasia	IPCC (2013) and Rabassa (2009)
		Andes	
	Mass loss of glaciers	Global	Gardner et al. (2013), Rabassa 2009, Rabatel et al. 2013, and Bolch et al. 2012
		Andes	
		Himalayas	
	Consistent decrease in snow cover	Northern Hemisphere (1966–2005)	IPCC (2007a)
	Decreased glacier meltwater yield	Europe (Alps) (1910–1940 vs. 1980–2000)	Collins (2008)
	Disappearance of Chacaltaya Glacier	Bolivia (2009)	Rosenzweig et al. (2007)
	Reduction of snow water equivalent	Norway	Skaugen et al. (2012)
	Shorter snowfall season with earlier snowmelt	Most of Northern Hemisphere	Takala et al. (2009)
Runoff and discharge	Changed runoff	Global (1960–1994)	Gerten et al. (2008), Piao et al.(2007), and Alkama et al. (2011)
	Reduced runoff	China (Yellow River)	Piao et al. (2010)
	Earlier annual peak discharge	The Russian Arctic (1960–2001)	Shiklomanov et al. (2007) and Hidalgo et al. (2009)
		W. USA, Columbia River (1950–1999)	
	Decreased dry-season discharge	Peru (the 1950s–1990s)	Baraer et al. (2012)
Extreme events	More intense extremes of precipitation	Northern tropics and mid-latitudes (1951–1999)	Min et al. (2011)
	The fraction of the risk of flooding	England and Wales (2000)	Pall et al. (2011)
	More severe, frequent and longer soil moisture droughts over 37% of land area	China (1950–2006)	Wang et al. (2011)
Groundwater	Decreased recharge of karst aquifers	Spain (20th century)	Aguilera and Murillo (2009)
	Decreased groundwater recharge	Kashmir, India (1985–2005)	Jeelani (2008) and Government of Western Australia (2003)
		Southwestern Australia	

Source: Adapted from Cisneros et al. (2014)

Box 7.1 Water-Related Adaptations for High-Altitude Agriculture

Mountains have been referred to as the water towers of the world (Viviroli 2007). They are a major source of stored water as they contain vast reserves of water in the form of snow, ice and glaciers. Changes in temperature and precipitation are expected to seriously affect the snow and glacial melt characteristics, which are important hydrological processes in the mountains (Immerzeel et al. 2010). The water towers of Asia, the Hindu Kush Himalayas, are also known as the third pole as these are the largest areas outside the poles that are covered by glaciers and permafrost. The huge amounts of snow and ice are crucial to water, energy, food and ecological security for a large part of Asia.

Increased variation in precipitation is observed and strongly projected for the future (discussed previously in Sect. 7.2). The precipitation variation triggers floods in monsoons and scarcity in dry seasons. Thus to buffer and manage the seasonal shocks of water availability, appropriate infrastructure and institutional capacities are required in the Himalayan region (Molden et al. 2014). Water infrastructure, both physical (e.g., reservoirs, water retention ponds, etc.) and natural (e.g., watersheds, forested land, floodplains, wetlands, etc.), can moderate the variability and improve water security, agricultural productivity and adaptive capacity, if properly managed (Clements et al. 2010; Molden et al. 2014).

Among the basic approaches of adaptation to water scarcity, developing water storage facilities had been regarded as particularly relevant for the Himalayan Hindu Kush mountain regions which, despite the sufficient precipitation in most parts, face seasonal water scarcity due to intra-annual variability in precipitation, short duration of rainfall and losses due to runoff (Vaidya 2015). Water storage has been suggested as a key strategy in climate change adaptation as it can ensure water availability throughout the dry year (ICIMOD 2009; McCartney and Smakhtin 2010).

Traditional systems of water harvesting from glaciers and streams that tap glacier water are common in certain regions of the Himalayas like Chitral and Hunza districts of Pakistan and Spiti Valley of Himachal Pradesh, India; while in certain other regions in Himachal Pradesh diversion channels called *kuhls* are used to tap mountain stream for irrigation water (Agarwal and Narain 1997). The participation of the local community institutions played a fundamental role in the allocation of water and service provisions in traditional water-harvesting systems, thus also creating an appropriate "institutional environment" for making and enforcing rules (Vaidya 2015).

In Nepal, community-driven water storage structures have proven to be successful adaptation measure for seasonal water scarcity. Concrete ponds fed by streams or springs that capture excess water during periods of high flow are used to irrigate arable crops in Nepal. Low-cost plastic-lined ponds that

(continued)

Box 7.1 (continued)

reduce seepage are an effective way to store water, particularly for vegetable crops. However, it is also necessary that water storage development with appropriate storage infrastructure be part of an integrated approach that includes improvement of land cover, soil moisture retention and facilitating aquifer recharge (Sugden et al. 2014). Often overlooked by government strategies, the approach of building decentralized small-scale and very small-scale water storage systems at a local level can ensure water availability at times of high requirement and help build community resilience to changing climate (Vaidya 2015).

In Ladakh, ice reservoir structures like artificial glaciers and ice stupas which evolved from local knowledge guided with engineering design principles have emerged as an innovative solution of water storage in the cold desert area. Ladakh is a high altitude (over 3000 m) cold desert situated in the western Himalayas, India. In the past, glaciers and snowfields have adequately supported subsistence agriculture in the Ladakh region, the sustainability of which in recent years is threatened by changing weather patterns (Mingle 2015). Unlike rain or seasonal snowmelt that reaches peak flow in a short duration, glaciers act as long-term storage units of water in the form of ice and play a central role in irrigation (Malone 2010). There is a significant reduction in water reserves in the region, which is likely to continue as the glaciers and snowfields retreat due to climate change (Grossman 2015; Mingle 2015; Rizvi 1998).

In Ladakh, winter snowfall bears a strong correlation to the amount of water available for irrigation in summer, and a low snowfall in winter indicates a summer drought (Gutschow and Mankelow 2001). Further, until seasonal glacier meltwater becomes available, typically there is a water shortage of about 2 months just before the onset of sowing due to low temperature and high variability of seasonal snow cover. Artificial glaciers are engineered ice reservoirs at lower altitudes which melt earlier and supply the necessary water for agriculture season. Artificial glacier is an intricate network of water channels and check dams that capture water for seasonal storage by freezing thin layers of water, creating superimposed sheets of ice along the upper slopes of the village valley (Fig. 7.2a, b) (Shaheen 2016; Nusser et al. 2018). These are intentional variations of traditional agriculture systems of water management like *kuhls* (channels) and zings (reservoirs located above cultivated fields) which, while retaining the conventional walls and snow barrier bands, incorporate "engineered design thinking" to manage and manipulate scarce water resource for irrigation (Clouse et al. 2017). Apart from allowing villagers to have timely and controlled access to water for irrigation water throughout the year (Vince 2010), such structures have led to a decline in water losses through

(continued)

Box 7.1 (continued)
seepage and an extension of summer cropping season to grow cash crops like potatoes and green peas (Shaheen 2016).

Ice stupas are another recently constructed (in 2015) variation of artificial glaciers which do away with the limitation of appropriate site associated with artificial glaciers. Ice stupas use underground pipes to divert water from upper streams to preferred locations where the water is frozen as it falls through sprinkler fountainhead into a conical structure, thus holding water that would flow downstream during winters to be made available in time of spring sowing (Fig. 7.2a, b). Diversion of water, however, can create a problem in downstream areas (Shaheen 2016; Nusser et al. 2018). While artificial glaciers are site-specific and cater to previously cultivated lands, ice stupas can be constructed in barren lands, thus leading to an expansion of irrigated areas.

Climate-adaptive design interventions could become necessary to ensure agricultural stability in the face of climate change-induced diminished surface meltwater and out of season precipitation and agriculture irrigated by snowmelt (Clouse et al. 2017). Moreover, the innovative adaptation strategies (as in Ladakh) could guide similar initiatives in regions facing similar issues due to changing environmental conditions (Mingle 2009). Another noteworthy aspect of such water storage-related adaptations in high altitudes (both Nepal and Ladakh) is the community involvement in planning and maintaining such structures, besides the use of local knowledge and adaptation to community needs.

Fig. 7.2 (**a**) Artificial glacier in summer time at Igloo, Ladakh. (Source: Clouse et al. 2017; Taylor & Francis Ltd. www.tandfonline.com). (**b**) The artificial glacier at Igloo in winter turns into a solid water reservoir. (Source: Leh Nutrition Project)

melting and increased duration of melting of glaciers led to increased river run-off and discharge peaks in shorter time frame (Jansson et al. 2003). This has been observed in the Andes and the Alps (IPCC 2007a, b).

7.2.2 Runoff and Streamflow

Runoff is the part of the excess water that flows over land surface under gravity instead of getting absorbed. It is caused by both natural process and human activities. Glaciers, snow and rain contribute to natural runoff. Runoff from glacier melt serves as an important water source and is occasionally the primary water source for the arid lowlands in mountainous regions. Runoff from glaciers accounts for the most important water supply source in central Asia (Chen et al. 2016). Retreat of glaciers that sustain rivers draining the glaciated regions during warm and dry periods (particularly in the high mountain ranges in Asia and the Andes) are projected to lead to intensified short-term river flows due to an increased glacier melt induced by warming, followed by a gradual fall in the longer term (IPCC 2007b). Further, the annual runoff peaks from glacierized catchments are strongly expected to shift towards spring from summers due to climate change and shrinking glaciers (Huss 2011). This has implications for downstream water flow, assuming greater significance during drought and heat waves (Koboltschnig et al. 2007).

A component of runoff which enters waterbodies is called streamflow or discharge and is expressed as the volume of water per unit time. Climate-driven changes in river flows depend largely on changes in volume, time and form (rain or snow) of precipitation and changes in evaporation (Bates et al. 2008). Changes in streamflow are also driven by non-climatic confounding factors like land use change, irrigation and urbanization (Cisneros et al. 2014). Trends in streamflow have been found to be consistent with observed trends in regional precipitation and temperature since the 1950s (Cisneros et al. 2014). At a global scale, the annual runoff increased in some regions (particularly higher latitudes in the USA) and decreased in others (parts of southern parts of Europe and South America) (Bates et al. 2008). However, the number of streamflow showing significant decreasing trends is higher than that showing a significant increasing trend in discharge in a global analysis of simulated streamflow (1948–2004) (Dai et al. 2009). Projections of runoff indicate a notable reduction in average annual runoff in dry tropical regions and southern Europe while an increase is projected for wet tropics, south-east Asia and high latitudes. "*Flows in high-latitude rivers are expected to increase, while flows from major rivers in the Middle East, Europe, and Central America might decrease*" (IPCC 2007a; Cisneros et al. 2014). In total, an increase in annual runoff is projected over the entire land surface, which in order to be fully made use of would require adequate infrastructure to capture and store the excess water (Bates et al. 2008).

Significant changes in river flows have been observed in many regions where snow falls in the winter (Bates et al. 2008). In regions with seasonal snow storage, warming has led to increased winter flows because a greater proportion of precipitation in winter falls as rain instead of snow (Clow 2010; Korhonen and Kuusisto

2010; Tan et al. 2011). Also reduced or earlier snowmelt leading to decreasing spring flows and increasing winter flows is strongly projected to alter seasonality of river flows in regions where snowfall is the predominant form of winter precipitation. Such occurrences are observed in the Himalayas, the Alps, the Scandinavian region, etc. (IPCC 2007b).

7.2.3 Ground Water

A decreasing trend in groundwater level of aquifers has been observed in the last few decades mostly due to groundwater abstractions (IPCC 2007b). However, the extent to which climate change is responsible for groundwater withdrawals is not known (Cisneros et al. 2014). Projected changes in groundwater levels are linked to changes in other hydrological variables due to warming. Increased groundwater abstractions and lower groundwater levels are expected due to projected increases in precipitation and streamflow variability (Taylor et al. 2013) and decreasing snowfall in sites where snowmelt provides most of the groundwater recharge (e.g., the southwestern USA) (Earman et al. 2006). With each degree rise in global mean temperatures between 0 °C and 3 °C, land areas affected by groundwater decrease are expected to increase linearly (Portmann et al. 2013). Surface water recharges and groundwater recharges are interlinked and vary with precipitation and type of soil. In humid areas, an increased precipitation variability might decrease groundwater recharge as heavy precipitation results in exceeding of infiltration capacity of the soil; while in semi-arid and arid areas intense rainfall leads to faster infiltration and, therefore, increase in recharge (IPCC 2007b).

7.2.4 Precipitation and Extreme Events

Precipitation varies largely in both temporal and spatial scales and is strongly influenced by "*large-scale patterns of natural variability*". This makes it a difficult parameter to monitor trends for over a global scale and attribute changes to with certainty (IPCC 2007a). On an average, global mean precipitation trends during 1901–2005 have been found to be statistically insignificant. However, heavy precipitation events have increased extensively even where total precipitation has decreased. A notable decrease in precipitation has been observed in the tropic and sub-tropic zones (10°S to 30°N) in the past 30–40 years (Bates et al. 2008).

According to theoretical and climate model studies, increasing warming and a consequent rise in atmospheric water vapor are expected to lead to extreme precipitation events as compared to mean precipitation. Unlike extreme precipitation events that are controlled by the availability of water vapor in the atmosphere, average rainfall is influenced by the atmospheric ability to radiate long-wave energy which is, in turn, restricted by increasing greenhouse gases (GHGs) (Bates et al. 2008).

Heavy precipitation events are projected to occur more frequently with an increasing intensity at higher latitudes. Flood hazards are expected to increase over half of the world in the south and Southeast Asia, tropical Africa, northeast Eurasia and South America (Hirabayashi et al. 2013). A projected increase in precipitation intensity is also accompanied by an increase in the number of consecutive dry days in many regions due to a decrease in even spread of precipitation into concentrated intense events interspersed with longer dry periods and increased evapotranspiration, particularly, in the sub-tropical region (Bates et al. 2008). Future tropical cyclones are also more likely to get more intense with more heavy precipitation (IPCC 2007c).

Drought is another kind of extreme event that is generally classified into three types: (a) "meteorological drought" (low precipitation), (b) "hydrological drought" (low river flows and groundwater levels) and (c) "agricultural drought" (low soil moisture). Decrease in precipitation and soil moisture reduction (including diminishing snowpack in cold areas) along with increasing temperatures and the associated increase in evapotranspiration are the important factors that have contributed to the expansion of regions suffering from drought (Dai et al. 2004). Changes in sea surface temperatures also contribute to droughts in the tropics where droughts have become more common since the 1970s. Further, excessive water withdrawals can aggravate the impact of drought (IPCC 2007b). Meteorological droughts and agricultural droughts are projected to become more frequent and/or last longer in some regions and some seasons by the end of twenty-first century. There are projections of increased severity of droughts in southern Europe and the Mediterranean region, central Europe, central and southern North America, central America, northeast Brazil and southern Africa (Cisneros et al. 2014).

7.3 Projections of Change in Freshwater Resources and Impacts on Agriculture

Projections of freshwater-related impacts are generally evaluated by comparison to historical data. Such projections are useful in making adaptation decisions with regard to climate change (Cisneros et al. 2014). Projections also help to quantify the likely outcomes under current management practices apart from indicating the necessary actions to avoid unfavourable outcomes in future (Oki and Kanae 2006). Projections are based on different scenarios simulated using different climate models, the outcome to which might vary widely. Scenarios represent possible future emission pathways and are used as a basis for exploring a realistic set of future projections of climate change. Previously IPCC assessments used a set of scenarios (A1, A2, B1 and B2 family) called SRES (Special Report on Emissions Scenarios) which were driven by a range of factors like demographic development, socio-economic development and technological change. Scenario A1 assumed a globalized future with rapid economic and technological growth. A2 scenario assumes more regional emphasis while scenarios B1 and B2 assumed more sustainable practices with a global and regional focus, respectively (IPCC 2000). SRES was replaced by RCP (representative concentration pathways (RCPs) in the IPCC fifth assessment report (AR 5) which takes into account climate change mitigation policies to limit emissions. The four RCP scenarios (RCP 2.6, RCP4.5, RCP6, RCP8.5) are named after the approximate radiative forcing values (2.6 W/m^2, 4.5 W/m^2, 6.0 W/m^2, 8.5 W/m^2, respectively) in year 2100 relative to the pre-industrial values. Radiative forcing or climate forcing is the difference between solar irradiation absorbed by the Earth and energy radiated back to space. A positive radiative forcing indicates a net gain in energy and causes warming. As of 2016, the radiative forcing was 3.027 W/m^2. The RCP2.6 scenario is the lowest emission range pathway and requires robust mitigation of GHG concentrations in the twenty-first century. The RCP4.5 and RCP6.0 scenarios are medium emission range scenarios. While the RCP8.5 scenario is a high range emission scenario with no mitigation, it is closest to a 'business as usual' scenario of fossil fuel use (Mann and Gaudet 2018). Few projections of future changes in freshwater resources under different climate scenarios are presented in Table 7.3.

A general trend in projections of water resources due to global warming in the future indicate decreased renewable water resources (Portmann et al. 2013; Schewe et al. 2014), aggravated water scarcity (Gerten et al. 2013) and increased exposure to floods (Hirabayashi et al. 2013). Precipitation events are expected to be more variable and more intense with increased frequency of extreme events like floods and droughts. Higher temperatures and higher variability in precipitations will lead to an increase in irrigation water demand irrespective of total precipitation in the growing season (IPCC 2007b). The increased variability of river flow, due to variability in precipitation and decreased snow and ice storage, is expected to decrease reliable surface water supply in the future. This would further complicate the situation in areas where the groundwater levels are projected to decline, as well as groundwater resources cannot be used to ease stress caused by freshwater unavailability (Kundzewicz and Döll 2009). Overall by 2050, under A2 and B2 scenarios of SRES, water stress is projected to increase over 62–76% of the area globally, primarily due to increased water withdrawals and decrease in 20–29% area due to increased precipitation (IPCC 2007b). Another study projected a 10- to 30-fold increase in the proportion of land facing extreme drought and a 2-fold and 6-fold increase in the frequency of 'extreme drought events' and 'mean drought duration', respectively, for A2 scenario by 2090 (Burke et al. 2006). However, a change in population is anticipated to have a greater impact on water availability and water demand than climate change in the future (Cisneros et al. 2014; Gerten et al. 2011).

Climate change can threaten water security by altering the availability of water. Crops depend on a balance between soil water moisture and atmospheric moisture deficit apart from crop management practices. Future crop water demand would be affected by variations in precipitation, and changes in radiation and temperature in both irrigated and rainfed systems (Cisneros et al. 2014). However, rainfed agriculture, which accounts for 80% of the total cultivated land area, is more vulnerable to variability in precipitation (projected to affect yields) (Finger et al. 2011). In areas where rainfall is projected to increase in the monsoon season, less irrigation may be required for paddy rice cultivation (Yoo et al. 2013). Irrigation, on the other hand, accounts for 70% of global water withdrawals and greater than 90% of consumptive water use (IPCC 2007b). Without considering climate change effects, the FAO

Table 7.3 Projected hydrological changes under different climate change scenarios

Projected hydrological change	Region/scale	Change in different emission scenarios/degrees of global warming	References
The decrease of renewable water resource by more than 20%	Global	–	Schewe et al. (2014)
The decrease of more than 10% in groundwater resource by the 2080s as compared to the the 1980s	Global	–	Portmann et al. (2013)
Increase in irrigation water demand	Global	Mean change of required irrigation water withdrawals by the 2080s as compared to the 1980s	Hanasaki et al. (2013)
		RCP2.6: −0.2 to 1.6%	
		RCP4.5: 1.9–2.8%	
		RCP8.5: 6.7–10.0%	
Shifts in river flow regime from perennial to intermittent and vice-versa	Global	Percent of the global land area affected by regime shifts between the 1970s and the 2050s	Döll and Müller Schmied (2012)
		SRES B2: 5.4–6.7%	
		SRES A2: 6.3–7.0%	
Decrease in groundwater recharge	Australia	Probability that groundwater recharge decreases to less than 50% of the 1990s value by 2050: GW 1.4°C: Close to 0 almost everywhere GW 2.8°C: In western Australia 0.2–0.6, in central Australia 0.2–0.3, elsewhere close to 1	Crosbie et al. (2013) Holman et al. (2009)
	East Anglia, UK	Percent change between baseline and future (by 2050s) groundwater recharge	
		SRES B1: −22%	
		SRES A1f: −26%	
Increase in agricultural (soil moisture) droughts	France	Smaller increase for SRES B1 as comapred to A2 and A1B	Vidal et al. (2012)
9–17% reduction in the annual mean snow coverage by 2100.	Northern Hemisphere	B2 scenario	IPCC (2007a)
20–35% likely decrease in permafrost area by 2050	Northern Hemisphere	–	IPCC (2007b)

Source: Adapted from Cisneros et al. (2014)

projected a likely expansion of irrigated area by 0.6% per year till 2030 in the developing countries (water-stressed areas including southern Asia, northern China, northern Africa, etc.) and a cropping intensity increase from 1.27 to 1.41 crops per year with slight increase in irrigation water use efficiency (Bruinsma 2003). Under climate change, future irrigation water demand is projected to surpass water availability in various regions (Wada et al. 2013). Irrigation water demand is projected to increase significantly in Europe, USA and parts of Asia. Different models using different scenarios have projected varying levels of changes in irrigation water demand (Cisneros et al. 2014). The reversal of 20–60 million hectares of irrigated cropland to rainfed management by 2100 due to freshwater limitations is projected in some irrigated regions (western USA; China; and West, South and Central Asia) while in other parts (northern/eastern USA, parts of South America, Europe and South East Asia) increase in water supply is expected to increase net irrigation, provided substantial investments are made in irrigation infrastructure (Elliott et al. 2011).

7.4 Adaptation Options in Agriculture

Adaptation, in general, means the process of changing to suit a new situation. With respect to climate change, adaptation is defined by Burton (1996) as "all those responses to climate change that may be used to reduce vulnerability or to actions designed to take advantage of new opportunities that may arise as a result of climate change". It can involve both building adaptive capacity and implementing adaption decisions. Building adaptive capacity implicates increasing the ability of individuals, groups or organizations to adapt to changes while implementing decisions transform capacity into action (Adger et al. 2005). Adaptation has been recognized as a response measure to address the impacts of climate change and reduce vulnerability. Parties to the United Nations Framework Convention on Climate Change (UNFCCC) are required to promote and facilitate adaptation to address climate change (UNFCCC 1992, 1998).

Several studies have classified adaptation measures differently. However, with regard to agri-water sector, autonomous adaptation options are particularly important as well as widespread (Iglesias and Garrote 2015; Bates et al. 2008). Autonomous adjustments are independently developed and implemented mostly short-term adjustments that result from changes to meet "*altered demands, objectives, and expectations*" and are "*not a conscious response*" to cope with climate change (Bates et al. 2008). Farmers have traditionally and naturally taken action in response to changes; this is in contrast to policy-driven or planned adaptation where the action is taken after policy level decision (Stern 2006).

This section discusses some of the important and frequently implemented adaptation measures in the agriculture sector that address water scarcity issue. An important element of water scarcity responses in relation to the agriculture sector is to ensure that water demand is managed at a sustainable level given local supply and availability (Anderson et al. 2008). On the other hand, the adaptation measures related to floods usually include infrastructural reforms of surface storage (dams,

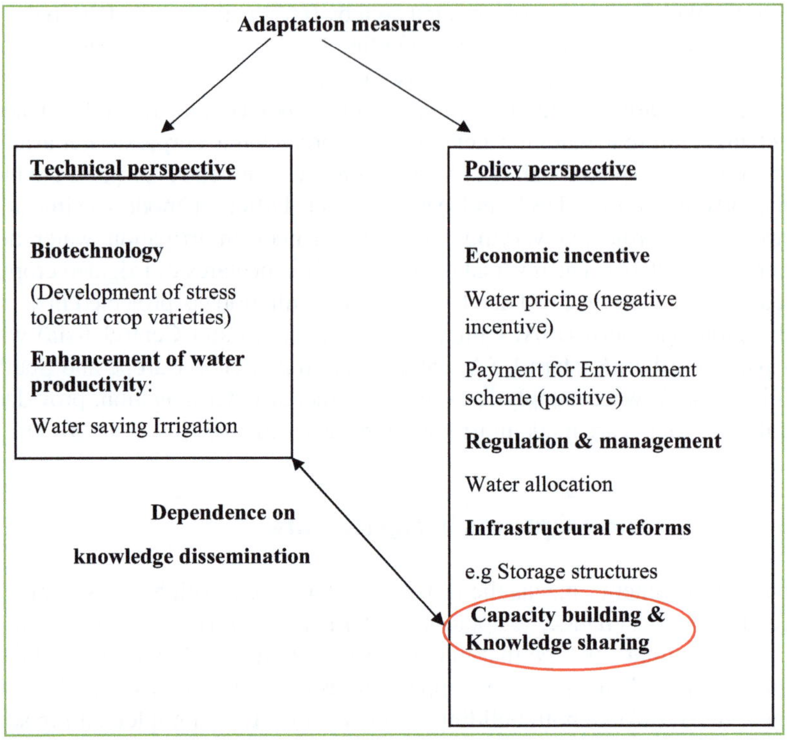

Fig. 7.1 Adaptation measures. (Source: Adapted from Gil and Kamanda 2015; Levidow et al. 2014)

reservoirs) and prevention of flows through embankments and dykes. However, specific "concrete" actions solely in response to climate change are very rare as climate change is one of the many factors affecting investment plans and strategies, and there is some degree of uncertainty associated with projections of future hydrological changes (Bates et al. 2008). The adaptation measures in the water sector can be broadly classified into technical and policy perspectives (Fig. 7.1).

7.4.1 Technical Perspective

Global studies on potential impacts of climate change indicate a reduction in water availability for irrigation purposes across all regions (Iglesias and Garrote 2015). Projected future water scarcity alongside an escalating water demand with an ever-increasing population and expansion of irrigated farming necessitates sustainable management of present water resources. Presently, irrigation efficiency is very low with only 55% crop water usage (Chartzoulakis and Bertaki 2015). Improving irrigation water productivity involves the reduction of water loss in the crop production system while maintaining crop yield at a certain level (Kang et al. 2017).

Climate-smart agricultural technologies (Venkatramanan and Shah 2019) and advanced irrigation technology like water-saving irrigation (WSI) and deficit irrigation techniques can promote sustainable use of limited water resource and minimise the impact of water scarcity (Chartzoulakis and Bertaki 2015). However, adaptation as a whole and adoption of sustainable measures are local processes, which apart from technology are also dependent upon factors like socio-economic conditions, and legal and institutional frameworks (Chartzoulakis and Bertaki 2015).

7.4.1.1 Water-Saving Irrigation

Water-saving irrigation has been noted to play a positive role in coping with the impact of climate change. For instance, in China the implemented WSI techniques are expected to save 30 billion cubic meter water per year (Zou et al. 2012). Few widely implemented water-saving irrigation techniques to address the impacts of climate change have been discussed.

Localized irrigation (drip/trickle, micro-sprayers) is a widely recognized efficient means of irrigating individual plants using pipes laid on the ground surface (Chartzoulakis and Bertaki 2015). Drip irrigation comprises of frequent (usually every 1–3 days) application of water close to plant roots through a system of small diameter plastic pipes with outlets called emitters or drippers at very low rates (2–20 l/h). Sprinkler irrigation involves the spraying of water pumped through a pipe system using rotating sprinkler heads. In canal-lining irrigation, the addition of an impermeable layer to the bed and sides of the canal significantly reduces water losses in a canal through seepage and water consumption by weeds. Low-pressure irrigation uses a pressurized piped system and uses pressure (2–3 bars) to convey and distribute the irrigation water from source to the irrigable area in closed pipes. Small flow rates cover large areas regardless of slope and topography. The quantity of water saved per acre can be determined by the water-saving rate. Table 7.4 provides a summary of water-saving rates for WSI techniques (Zou et al. 2012).

The high installation cost remains a major barrier to the expansion of the use of micro-irrigation techniques which represent only 6% of the total irrigated area despite a 50-fold increase in the last 20 years (Chartzoulakis and Bertaki 2015). On the other hand, propagation of use of drip irrigation by government subsidies has not borne the desired results in some cases. Government subsidies in drip irrigation in Madhya Pradesh, India, have had negative consequences and hindered access to the technology to smallholder farmers because of "*technical requirements, highly bureaucratic process*" and agents emerging as middlemen, thus increasing prices up to 40% (Malik et al. 2016).

Table 7.4 Summary of water-saving rates for WSI techniques

Technique	Range(%)	Average(%)
Micro-irrigation	35–70	52.7
Sprinkler irrigation	30–65	45.3
Low-pressure irrigation	20–38	29.4
Canal lining	16.2–36.2	24.5

In a study on the economic feasibility of water-saving irrigation (WSI) techniques in China, micro-irrigation (including drip irrigation, mini-sprinkler irrigation and bubbler irrigation) was found to be the most effective measure, in terms of both adaptation and mitigation to climate change. Mitigation here means lesser energy consumption and reduction in GHG emissions. Using micro-irrigation techniques led to the least amount of CO_2 eq. emissions per unit cubic meter of water saved as well as per kilogram increase in grain yield among the other WSI. From an economic perspective, canal-lining irrigation, which involves adding an impermeable layer, was found to be the least cost option in the long term due to energy and water cost savings. While sprinkler irrigation and low-pressure pipe irrigation have a larger potential for water saving and increasing grain yield, these techniques also incur additional energy needs to attain a certain water pressure, therefore, leading to higher GHG emissions. Hence considering both adaptation and mitigation, they were the least preferred (Zou et al. 2013).

7.4.1.2 Deficit Irrigation

Unlike the water-saving irrigation technique which leads to water saving through prevention of loss but meets full crop water requirements, deficit irrigation techniques maximize water use efficiency by meeting partial crop water requirements. Techniques like regular deficit irrigation (RDI) and alternate partial root drying and subsurface drip irrigation are optimizing strategies mostly used in arid and semi-arid regions for more effective use of limited water resources. Since the yield is directly proportional to crop water consumption (Perry et al. 2017), deficit irrigation techniques have a trade-off of slightly reduced yield as compared to WSI production. However, deficit irrigation prevents the reduction in irrigated area. Thus crops which are less sensitive to water stress can adapt well to various deficit irrigation practices. As water can be applied precisely to individual plants, RDI and alternate partial root zone irrigation (APRI) strategies are suitable for fruit trees. For cereal crops like maize, wheat and sorghum, research is ongoing and is required with regard to "*the timing of RDI and the degree of soil water deficit and the timing and method of APRI in different climates, cereal varieties and planting conditions*" (Kang et al. 2017).

In regular deficit irrigation (RDI), plants are exposed to a regulated level of water stress during particular stages of growth in order to control excessive vegetative growth, and irrigation is applied in the whole root zone (Goodwin and Jerie 1992; Boland et al. 1993). Introduced in the 1970s for pome and stone fruit orchards, it has been successfully used for field crops, fruit trees, and greenhouse vegetables, grape, apple, jujube, maize, sorghum, wheat, potato, cotton, etc. (Kang et al. 2017). High-yielding varieties are more sensitive to water stress and, therefore, not suitable to RDI. Crops or varieties with a short-growing season are more suitable for RDI (Chartzoulakis and Bertaki 2015).

Alternate partial root zone irrigation (APRI) makes use of alternate irrigations usually in a 7–14 day cycle in different parts of root zone such that one half of the root system remains dry while the other half is irrigated. A balance between vegetative and reproductive growth is achieved by induction of a biochemical response from plants primarily through abscisic acid (ABA) hormone, which leads to reduced transpiration by partial stomatal closure and improves water use efficiency (Loveys et al. 1999). It has been used in sugar beet, sugarcane, winter wheat, sorghum, maize, cotton, potato, tomato, beans, apple, grapes, pear, etc. (Sepaskhah and Ahmadi 2010). Even though the implementation cost of alternate partial root zone drying irrigation (APRD) is high, it is economical in places where the price of water is higher in terms of availability or irrigation cost (Chartzoulakis and Bertaki 2015).

The use of RDI necessitates implied knowledge on a number of parameters including crop response to water deficits, identification of critical crop growth stages, water retention capacity of soil, economic impact of yield reduction and most importantly a certain degree of technological development to assess crop water requirement in order to support an optimum irrigation water scheduling. To that end, the use of models like FAO CROPWAT has been proven to be very useful in predicting and disseminating practical recommendations on deficit irrigation scheduling under various conditions of water supply, soil and crop management (Smith and Kivumbi 2000). Such pre-requisite knowledge relies on an effective and continuous effort on capacity building and communication between extension workers and farmers.

Subsurface irrigation is a low-pressure highly efficient irrigation system that uses buried pipes to deliver water thus minimizing losses due to evaporation and incidences of weed and diseases. Subsurface irrigation is suitable for nearly all crops, particularly for high-value fruit and vegetables. High initial investment and the possibility of clogging due to poor water quality are the major disadvantages (Chartzoulakis and Bertaki 2015).

7.4.1.3 Agricultural Practices

Both properly managed irrigation technology and better agricultural practices determine greater water use efficiency (Levidow et al. 2014). Apart from irrigation management, good agriculture practices based on farm-level decisions for reducing exposure to climate risks have been recognized and recommended for achieving greater synergies (Molden et al. 2010). Such practices include retention of soil moisture and increasing the infiltration of rainwater into the soil through conservation tillage and mulching, soil surface tillage, bed surface profiling and increasing the amount of organic matter; diversification of crop types/change in timing and change in cropping patterns through a switch to crop types and varieties which use lesser water (e.g., varieties tolerant to abiotic and biotic stress); and decreasing runoff, retaining moisture and improving water uptake by changing land topography through contouring and terracing and building small-scale storage structures and recharge areas to capture rainwater (Smit and Skinner 2002; Anderson et al. 2008; Chartzoulakis and Bertaki 2015). The concept of water footprint as a tool to make decisions on cropping patterns has been discussed in Box 7.2.

Box 7.2 Water Footprint as a Tool to Optimize Cropping Patterns

The importance of change in cropping patterns as an adaptation measure noted by several studies is summarized by Levidow et al. (2014) as "If agricultural water demand is inelastic, then policies which encourage changes in cropping patterns can be more effective than higher prices." Water footprinting can be used as a tool to optimize cropping patterns. The water footprint is an indicator of direct and indirect freshwater consumption of a consumer or product. The water footprint of a product is the volume of freshwater used to produce that product over the entire supply chain. It is a comprehensive indicator of freshwater appropriation which is spatially and temporally explicit. The volumetric water footprint comprises three components: *blue water footprint* refers to water consumed (lost water that does not immediately return to the same water body/catchment) from surface and groundwater sources; *green water footprint* refers to consumption of rainwater while *grey water footprint* is an indicator of the volume of water polluted, i.e., the freshwater volume required to assimilate the pollutant load to bring it to natural conditions/ambient standards (Hoekstra et al. 2011). The volume of blue water used depends on the crop type, crop tolerance to water deficits, irrigation efficiency and on green water. Blue water is used whenever green water is insufficient to meet the crop water requirement (Lovarelli et al. 2016). Green water particularly refers to the rainfall evaporated from the soil, absorbed and transpired by the crop, and the water incorporated in the harvested crop (Hoekstra et al. 2011). The total water footprint (WF) of the process of growing a crop is equal to the sum of all three components, i.e., green, blue and grey. The WF is commonly expressed as the water volume used to produce a unit of product (m^3/tonne), the blue and green components of which are calculated by calculating the accumulation of evapotranspiration of rainwater and irrigation water, respectively, over the complete growing period divided by the yield.

Crop evapotranspiration (ET) is defined as the combination of separate processes of water evaporation from the soil surface and transpiration from the crop (Allen et al. 1998). It can be measured directly in the field or estimated based on models. FAO CROPWAT is one of the most commonly used models for estimating crop ET based on climate data of a place. Irrigation schedules for deficit irrigation practices are based on crop evapotranspiration rates. Evapotranspiration rates depend on soil and weather parameters like sunshine hours, radiation, wind speed, precipitation and temperature which vary from place to place resulting in a large variation in water footprint irrespective of crop variety. Hence, for a particular crop, blue WF might be higher in some areas and lower in others, within the same basin or region. Allocation of crops from high to low WF can result in significant savings in surface and groundwater resources (Mali et al. 2018). Optimizing cropping patterns is a key to sustainable water use in areas facing water crisis (Zeng et al. 2012).

(continued)

Box 7.2 (continued)

Evaluation of regional water footprint to influence decisions on cropping pattern can lead to considerable water saving. A study on assessments of water footprint in the Gomti river basin of the Indo-Gangetic plains, India, found that under the present cropping pattern, a considerable part of blue water was used to grow low-value crops like chickpea, mustard and pearl millet, which can yield well even under rainfed conditions, and confining such crops to rainfed areas would lead to savings in blue water. Reallocation of cropping patterns through policy reforms that limit farmers' choice of crops and areas to minimize blue water use within a water scarce basin; and improving the efficiency of rainwater use, can potentially reduce blue water use in the basins (Mali et al. 2018). Cropping patterns that focus on high-value low water-intensive crops rather than higher net returns per unit area contribute to higher economic water productivity (Mali et al. 2018). This is particularly relevant in water-scarce places where water allocation and water pricing practices are implemented.

As a part of the integrative approach (discussed in Sect. 7.4.2.3), successful reduction in irrigation water use while maintaining farmer's income was achieved in the Shiyang river basin, China, by changing cropping patterns. Replacement of cereal crops (spring wheat and maize) by cash crops (grape and cotton) led to lesser water consumption as grapevine and cotton (350 and 320 mm, respectively) have lesser evapotranspiration rates compared to wheat and maize (450 and 500 mm, respectively). However, economic sustainability of farmers is a criterion to consider while planning a switch in cropping patterns (Kang et al. 2017).

Another approach of water footprinting is the stress weighed water footprint which measures the impact of water consumption on regional water stress by multiplying the volume of blue water with water stress index (WSI). It takes into account only blue water as green water does not contribute to water scarcity (Ridoutt and Pfister 2010). WSI is a sustainability indicator, defined as the relationship between water use and availability. Stress-weighted WF could be used to quantitatively compare production systems/products in terms of their contribution to water scarcity. Considering the spatial variability in stress-weighted WF, allocation of water-intensive crops to regions having lower stress-weighted WF would reduce pressure on blue water. Thus, both volumetric WF and stress-weighted WF can be used to make decisions on ways to increase crop water productivity and reduce water stress from a national perspective (Wang et al. 2015).

Further, it is shown that international trade via virtual water trade of agriculture products, i.e., export of a product from water-efficient region (relatively low virtual water content of the product) to water-inefficient region

(continued)

Box 7.2 (continued)
(relatively high virtual water content of the product) saves water globally (Chapagain et al. 2006). Closely linked with water footprint concept, the virtual water content is the volume of water embedded in a food or other products. The redistribution of food trade has been considered as a potential climate change adaptation measure (Nelson et al. 2009). Staple food trade-related water savings are projected to increase under climate change particularly for wheat trade where large volumes of wheat would be traded from relatively water-efficient exporters to less-efficient importers (Konar et al. 2013). However, optimization of global water resources through enhanced virtual water trade while relieving pressure on water-scarce regions might also create additional pressure on export-oriented countries that produce the water-intensive commodities (Chapagain et al. 2006).

7.4.2 Policy Perspectives

7.4.2.1 Jevon's Paradox

There is a postulation in economics called Jevon's paradox or rebound effect which states that a technological progress that enhances the efficiency of use of a natural resource tends to lead to an increase in its consumption due to an increase in demand. Lately, Jevon's paradox has been discussed and observed with regard to agricultural water use and irrigation technologies that have improved water use efficiency. The very few studies on the impact of the use of irrigation technology indicate that technologies that improve irrigation efficiency lead to increased consumptive water use and groundwater extractions (Grafton et al. 2018; Perry et al. 2017; Pfeiffer and Lin Lawell 2014). This is because the "saved" water, which results from the efficient use of water, is diverted to gain higher production (since yields increase with a marginal increase in water supply), increasing acreage or changing cropping patterns to more water-intensive (and probably income-generating) crops, thus leading to an increase in on-farm water consumption and groundwater extraction (Perry et al. 2017; Pfeiffer and Lin Lawell 2014). Increase in the irrigated area leads to increased crop transpiration and reduced percolation for aquifer recharge or downstream flows, thus increasing local water consumption (Perry et al. 2009).

Subsidies in technologies that promote irrigation efficiency have also led to decreased groundwater levels in several cases. Subsidies to drip irrigation have been found to increase water extractions due to increase in irrigated area in New Mexico (Ward and Pulid 2008) and Rajasthan, India (Birkenholtz 2017), where the volume of water applied also increased. Reduced recoverable return flows to aquifers and overexploitation of groundwater and aquifer due to cropping intensification after the adoption of drip irrigation in Souss and Tensift basins, Morocco, has been reported (Molle and Tanouti 2017). In Snake River, Idaho, USA, growth in irrigation efficiency has led to reduced groundwater recharge and drop in the Eastern Snake Plain

Aquifer level by 30% since the mid-1970s (McVeigh and Wyllie 2018). Like these cases, irrigation-efficient technologies in general, like drip irrigation, are applied in arid and semi-arid areas already facing acute water scarcity. Increased water withdrawal due to the expansion of irrigated area or increase in water consumption per unit area exacerbates the water scarcity problem. Therefore, water allocation/regulated access to water has been strongly suggested as a measure for decreasing water consumption and recommended as a crucial step to be taken prior the introduction of hi-tech irrigation (Perry et al. 2017). Introduction of water-efficient irrigation technology is not an isolated process as remarked by Grafton et al. (2018) that:

"Increases in irrigation efficiency must be accompanied by robust water accounting and measurements, a cap on extractions, an assessment of uncertainties, the valuation of trade-offs, and a better understanding of the incentives and behavior of irrigators."

7.4.2.2 Water Pricing and Water Allocation

The price of water is generally low, and correct water pricing is essential for water pricing to be an effective measure (Perry et al. 2009). There are few positive references on water pricing policy. Higher irrigation water prices have encouraged farmers in Europe to adopt IS-DSS (irrigation scheduling decision support system) (discussed in Sect. 7.4.2.4) (Giannakis et al. 2016). Volumetric water metering and water accounting procedures have been recommended for proper water pricing (Chartzoulakis and Bertaki 2015). A study of farmer's responses to changes in water pricing practices indicated that solely increasing water prices is not a viable option and an integrated approach is necessary (Mamitimin et al. 2015). The efficiency of water pricing mechanisms is further hampered by low water price elasticity, large differences in prices within and outside country and political issues related to enforcement (Giannakis et al. 2016). Largely, the economic instrument of water pricing to reduce water demand has been found to be ineffective (Molle and Berkoff 2007; Molle 2008). Also, crop-specific differences in irrigation efficiency are not accounted for in this measure, causing a heavy penalty for certain crop growers like rice, and in a few cases it has led to abusive water utilization by the wealthier stakeholders (Gil and Kamanda 2015); besides farmers view it as an extra penalty (Molden et al. 2010).

Water allocation measures, on the other hand, are more effective in protecting available water resources (Gil and Kamanda 2015), leading farmers to adopt water-efficient practices (Molden et al. 2010). It is one of the key adaptation measures to tackle water scarcity (Iglesias and Garrote 2015), and a crucial supplement to irrigation technologies (Perry et al. 2017; Grafton et al. 2018). Even though quotas can be subject to arbitrariness and adapt to changing economic conditions; they were found to be consistently preferred over economic regulations, as they were more equitable, transparent and efficient in aligning demand and supply with limited overall income loss as compared to price-based regulation (Molle 2008). However, practically determining a "socially optimal distribution" for water allocation is difficult, as is meeting the high costs associated with the assignment, administration and monitoring (Gil and Kamanda 2015). Because the potential solutions to allocation problems like changes in infrastructure, land use or limitations of irrigation could lead to conflict among stakeholders, this necessitates the incorporation of interests of different stakeholders while decision-making (Iglesias and Garrote 2015). Additionally, the sustainable threshold of water demanded by each region needs to identified and considered in decisions related to the distribution of regional water resources in different sectors (Kang et al. 2017).

7.4.2.3 Integrated Measures

Integrated measures combining multiple approaches have been successfully implemented in the arid inland area of Shiyang River Basin, China, in order to address water scarcity and improve overall water productivity. The approach included a combination of measures like closing over 3000 wells, reducing planting and irrigation area through changing cropping patterns and building solar greenhouses for high value crops, adjusting planting systems, "developing comprehensive water-saving technologies and their integrated models such as adopting deficit irrigation, drip irrigation under mulch, field levelling, etc., implementing water right management through strict irrigation quota and step-price system", and demonstrations and farmers' trainings and workshops on water-saving technologies, water resources management and water-saving in agriculture. These measures worked synergistically to not only improve irrigation efficiency, water productivity and groundwater levels but also maintained farmers' income (Kang et al. 2017).

7.4.2.4 Capacity Building

Compared to measures that place obligations and forceful implementations for water saving, positive incentives which provide compensations and tax exemptions for adopters of water management like the payment for environment services schemes (PES) are becoming popular particularly in Latin America (Gil and Kamanda 2015).

A bottleneck over the use of technical advice on irrigation schedule is the lack of adequate assistance and knowledge gap between farmers' actual practice and potential crop water use (Levidow et al. 2014). Irrigation scheduling decision support systems (IS-DSS) is a computer-based tool which provides advice to farmers on real-time irrigation scheduling and ensures water-efficient agriculture. It has been largely applied in southern parts of Europe and Italy (Mannini et al. 2013). However, not all technologies are equally widely adopted in all regions and the reasons also help to identify bottlenecks. For instance, despite the large potential for water saving by the use of IS-DSS in the Mediterranean region, it was not widely adopted due to lack of proper dissemination and the actual transfer of knowledge in the field. This highlights the need for technology to be user-friendly and the importance of a participatory approach which encompasses the stakeholder's perspective in the technology design or implementation (Giannakis et al. 2016).

A farmer's general perspective on improving water efficiency is on revenue maximization rather than water saving (Knox et al. 2012). Even while policymakers and water managers limit allocation of water as part of reducing water use, water saving is not a priority for farmers (Luquet et al. 2005). The farmer exhibits a tendency to avert risks and wants to derive the maximum of the limited resource, in this case water available to him, by consuming as much as possible. This is also the reason why subsidies and technology have not worked in the intended ways as discussed previously.

Analysis of farmers' adaptation to water scarcity in Al-Bayda canal in Nile Delta, Egypt, revealed there are a number of factors beyond water scarcity and profit maximization that shape the response of farmers' which include collective dimension of individual crop choice, farm fragmentation, risk aversion, social capital and history of farmers (Ghazouani et al. 2014). The way farmers react to a change in policy also depend on a number of factors, for instance, a study on change in water pricing in Tarim Basin, China, found that small farmers and farmers growing perennial fruit trees tend to opt for drill wells when water prices increased while large farmers and farmers with annual crop production opted for options other than drill wells (like adopting improved irrigation technology, improved crop production or do nothing). Further, it was found that farmers who suffered slight water shortages in the past were more likely to use water more wisely (Mamitimin et al. 2015). The behaviour of irrigators is an important factor that should be taken into account while implementing policies (Grafton et al. 2018). The significance of understanding stakeholder's perception on climate change and adaptation policies and their participation in decision-making have been highlighted by several studies (Iglesias and Garrote 2015; Gil and Kamanda 2015; Kang et al. 2017; Giannakis et al. 2016; Levidow et al. 2014).

In a study by Levidow et al. (2014), it has been demonstrated that farmers in spite of making investments in irrigation technologies cannot reap the full benefits because of "*inadequate knowledge regarding crops' water use, soil-moisture conditions, irrigation scheduling techniques, crops' yield response to different irrigation management strategies, etc.*" While farmers have a responsibility for increasing water use efficiency, it is getting displaced due to lack of knowledge exchange systems that can identify scope for improvements and uphold the shared responsibility of relevant stakeholders across the entire water supply chain. Only a greater institutional responsibility for water policies and strategies can support an adequate knowledge system (Levidow et al. 2014). Adopting optimized and integrated technologies requires cooperation from different agencies like scientists, technicians, local agencies, financial incentives and farmers themselves (Kang et al. 2017), including a stronger extension service (Mamitimin et al. 2015).

7.5 Conclusion

There are evidences that climate change is associated with changes in the hydrological cycle. Freshwater resources are already vulnerable to climate change and have the potential to be strongly influenced by it in the future. Future projections in general indicate a reduction in water availability at the global scale and risk of frequent extreme events. This would have a huge impact on the agriculture sector which relies heavily on water. The situation is further complicated by confounding non-climatic drivers like population growth which would continue to increase the demand for freshwater and put tremendous pressure on available water resources. In this context, adaptation can increase preparedness to face the uncertainties posed by climate change and reduce vulnerability. The adaptation measures in agriculture to address freshwater scarcity have been discussed. In high altitude mountains, the observed effects of climate change on water resources are profoundly impacting agriculture. Communal water harvesting through storage structures is the primary means of adaptation in high altitudes. Innovative adaption measures in Ladakh which amalgamate local knowledge, community participation and technical expertise are particularly noteworthy; such measures can guide similar initiatives in other regions. It is important that adaptation measures should be evaluated for their long-term feasibility and effect on water scarcity. A rebound effect/Jevon's paradox has been observed in places where increases in irrigation efficiency were achieved. Policymakers have to be cautious of such effects before promoting and subsiding efficient irrigation technology; integrating it with other measures like water allocation has been strongly recommended. Integrated approaches work synergistically and, therefore, can be much more effective than measures implemented in isolation. Technology interventions also necessitate knowledge sharing and capacity building so that it remains effective. Lastly, adaptation is a local process, and every region should actively involve the stakeholders in decision-making to develop an appropriate adaptation strategy that is in harmony with its institutional and socio-economic capacities and geographic limitations.

References

Adger N, Arnell N, Tompkins E (2005) Successful adaptation to climate change across scales. Glob Environ Chang 15(2):77–86

Agarwal A, Narain S (eds) (1997) Dying wisdom: rise, fall and potential of India's traditional water harvesting systems. Center for Science and Environment, New Delhi, 404 pp

Aguilera H, Murillo JM (2009) The effect of possible climate change on natural groundwater recharge based on a simple model: a study of four karstic aquifers in SE Spain. Environ Geol 57(5):963–974

Alkama R, Decharme B, Douville H, Ribes A (2011) Trends in global and basin scale runoff over the late twentieth century: methodological issues and sources of uncertainty. J Clim 24(12):3000–3014

Allen RG, Pereira LS, Raes D, Smith M (1998) Crop evapotranspiration: guidelines for computing crop water requirements. FAO Irrigation and Drainage Paper 56, 300 pp

Anderson J, Arblaster K, Bartley J (2008) Climate change-induced water stress and its impacts on natural and managed ecosystems. European Parliament, Brussels

Baraer M, Mark BG, McKenzie JM, Condom T, Bury J, Huh K, Portocarrero C, Gomez J, Rathay S (2012) Glacier recession and water resources in Peru's Cordillera Blanca. J Glaciol 58(207):134–150

Bates BC, Kundzewicz ZW, Wu S, Palutikof JP (eds) (2008) Climate change and water. Technical Paper of the Intergovernmental Panel on Climate Change. IPCC Secretariat, Geneva

Birkenholtz T (2017) Assessing India's drip-irrigation boom: efficiency, climate change and groundwater policy. Water Int 42:663–677

Boland AM, Mitchell PD, Jerie PH, Goodwin I (1993) Effect of regulated deficit irrigation on tree water use and growth of peach. J Hortic Sci 68:261–274

Bolch T, Kulkarni A, Kaab A, Huggel C, Paul F, Cogley JG, Frey H, Kargel JS, Fujita K, Scheel M, Bajracharya S, Stoffel M (2012) The state and fate of Himalayan glaciers. Science 336(6079):310–314

Bruinsma J (2003) World agriculture: towards 2015/2030. An FAO perspective. Earthscan, London, 444 pp

Burke EJ, Brown SJ, Christidis N (2006) Modelling the recent evolution of global drought and projections for the 21 st century with the Hadley Centre climate model. J Hydrometeorol 7:1113–1125

Burton I (1996) The growth of adaptation capacity: practice and policy. In: Smith JB et al (eds) Adapting to climate change: an international perspective. Springer, New York, pp 55–67

Chapagain AK, Hoekstra AY, Savenije HHG (2006) Water saving through international trade of agricultural products. Hydrol Earth Syst Sci 10:455–468

Chartzoulakis K, Bertaki M (2015) Sustainable water management in agriculture under climate change, 4 agriculture and agricultural. Sci Procedia 4:88–98

Chen Y, Li W, Deng H, Fang G, Li Z (2016) Changes in Central Asia's water tower: past, present and future. Sci Report 6:35458

Cisneros JBE, Oki T, Arnell NW, Benito G, Cogley JG, Döll P, Jiang T, Mwakalila SS, (2014) Freshwater resources. In: Climate change 2014: impacts, adaptation, and vulnerability. Part A: Global and sectoral aspects. Contribution of Working Group II to the Fifth Assessment Report of the Intergovernmental Panel on Climate Change. Cambridge University Press, Cambridge/New York, pp 229–269

Clements T, D'Amato V, Taylor T (2010) Integrating water infrastructure in a new paradigm for sustainable, resilient communities. In: Proceedings of the Water Environment Federation, Cities of the Future/Urban River Restoration, pp 801–825

Clouse C, Anderson N, Shippling T (2017) Ladakh's artificial glaciers: climate-adaptive design for water scarcity. Clim Dev 9(5):428–438. https://doi.org/10.1080/17565529.2016.1167664

Clow DW (2010) Changes in the timing of snowmelt and streamflow in Colorado: a response to recent warming. J Clim 23(9):2293–2306

Collins DN (2008) Climatic warming, glacier recession and runoff from Alpine basins after the Little Ice Age maximum. Ann Glaciol 48(1):119–124

Crosbie RS, Pickett T, Mpelasoka FS, Hodgson G, Charles SP, Barron OV (2013) An assessment of the climate change impacts on groundwater recharge at a continental scale using a probabilistic approach with an ensemble of GCMs. Clim Chang 117(1–2):41–53

Dai A, Trenberth KE, Qian T (2004) A global data set of Palmer Drought Severity Index for 1870–2002: relationship with soil moisture and effects of surface warming. J Hydro meteorol 5:1117–1130

Dai A, Qian T, Trenberth KE, Milliman JD (2009) Changes in continental freshwater discharge from 1948 to 2004. J Clim 22(10):2773–2792

Döll P, Müller Schmied H (2012) How is the impact of climate change on river flow regimes related to the impact on mean annual runoff? A global-scale analysis. Environ Res Lett 7:014037

Earman S, Campbell AR, Phillips FM, Newman BD (2006) Isotopic exchange between snow and atmospheric water vapor: estimation of the snowmelt component of groundwater recharge in the southwestern United States. J Geophys Res Atmos 111(D9):D09302

Elliott M, Armstrong A, Lobuglio J, Bartram J (2011) In: De Lopez T (ed) Technologies for climate change adaptation: the water sector, TNA guidebook series. UNEP Risø Centre, Roskilde, 114 pp

Finger R, Hediger W, Schmid S (2011) Irrigation as adaptation strategy to climate change – a biophysical and economic appraisal for Swiss maize production. Clim Chang 105(3–4):509–528

Gardelle J, Arnaud Y, Berthier E (2011) Contrasted evolution of glacial lakes along the Hindu Kush Himalaya mountain range between 1990 and 2009. Glob Planet Chang 75(1–2):47–55

Gardner AS, Moholdt G, Cogley JG, Wouters B, Arendt AA, Wahr J, Berthier E, Hock R, Pfeffer WT, Kaser G, Ligtenberg SRM, Bolch SMJ, Hagen JO, Van den Broeke MR, Paul F (2013) A reconciled estimate of glacier contributions to sea level rise: 2003 to 2009. Science 340(6134):852–857

Gerten D, Rost S, von Bloh W, Lucht W (2008) Causes of change in 20th century global river discharge. Geophys Res Lett 35(20):L20405

Gerten D, Heinke J, Hoff H, Biemans H, Fader M, Waha K (2011) Global water availability and requirements for future food production. J Hydrometeorol 12(5):885–899

Gerten D, Lucht W, Ostberg S, Heinke J, Kowarsch M, Kreft H, Kundzewicz ZW, Rastgooy J, Warren R, Schellnhuber HJ (2013) Asynchronous exposure to global warming: freshwater resources and terrestrial ecosystems. Environ Res Lett 8(3):034032

Ghazouani W, Molle F, Swelam A, Rap E, Abdo A (2014) Understanding farmers' adaptation to water scarcity: a case study from the western Nile Delta, Egypt, IWMI Research Report 160. International Water Management Institute (IWMI), Colombo. 31p

Giannakis E, Bruggeman A, Djuma H, Kozyra J, Hammer J (2016) Water pricing and irrigation across Europe: opportunities and constraints for adopting irrigation scheduling decision support systems. Water Sci Technol Water Supply 16(1):245–252

Gil J, Kamanda J (2015) Policies and initiatives related to water and climate change in agriculture: case studies from Brazil and Africa. In: Zolin CA, Rodrigues R (eds) Impact of climate change on water resources in agriculture. CRC Press, Taylor and Francis Group, Boca Raton, pp 39–61

Goodwin I, Jerie P (1992) Regulated deficit irrigation: from concept to practice. Aust N Z Wine Ind J 5:131–133

Government of Western Australia (2003) Securing our water future: a state water strategy for Western Australia. Government of WA, 64 pp. Available http://dows.lincdigital.com.au/files/State_Water_Strategy_complete_001.pdf

Grafton RQ, Williams J, Perry CJ, Molle F, Ringler C, Steduto P et al (2018) The paradox of irrigation efficiency. Science 361:748–750

Grossman D (2015) As Himalayan glaciers melt, two towns face the fallout. Yale Environment 360. http://e360.yale.edu/feature/as_himalayan_glaciers_melt_two_towns_face_the_fallout/2858/. Accessed 8 Oct 2018

Gutschow K, Mankelow S (2001) Dry winters, dry summers: water shortages in Zanskar. Ladakh Studies 15:28–32

Hanasaki N, Fujimori S, Yamamoto T, Yoshikawa S, Masaki Y, Hijioka Y, Kainuma M, Kanamori Y, Masui T, Takahashi K, Kanae S (2013) A global water scarcity assessment under shared socio-economic pathways – part 2: water availability and scarcity. Hydrol Earth Syst Sci 17:2393–2413

Hidalgo HG, Das T, Dettinger MD, Cayan DR, Pierce DW, Barnett TP, Bala G, Mirin A, Wood AW, Bonfils C, Santer BD, Nozawa T (2009) Detection and attribution of streamflow timing changes to climate change in the western United States. J Clim 22(13):3838–3855

Hirabayashi Y, Mahendran R, Koirala S, Konoshima L, Yamazaki D, Watanabe S, Kanae S (2013) Global flood risk under climate change. Nat Clim Chang 3:816–821

Hoekstra AY, Chapagain AK, Aldaya MM, Mekonnen MM (2011) The water footprint assessment manual: setting the global standard. Earthscan, London

Holman IP, Tascone D, Hess TM (2009) A comparison of stochastic and deterministic downscaling methods for modelling potential groundwater recharge under climate change in East Anglia, UK: implications for groundwater resource management. Hydrogeol J 17(7):1629–1641

Huss M (2011) Present and future contribution of glacier storage change to runoff from macroscale drainage basins in Europe. Water Resour Res 47:W07511

ICIMOD (2009) Water storage: a strategy for climate change adaptation in the Himalayas, sustainable mountain development no. 56, special issue. International Centre for Integrated Mountain Development, Kathmandu

Iglesias A, Garrote L (2015) Adaptation strategies for agricultural water management under climate change in Europe. Agric Water Manag 155:113–124

Immerzeel WW, Beek LPH, Bierkens MFP (2010) Climate change will affect the Asian water towers. Science 328:1382–1385

IPCC (2000) Summary for policy makers: emission scenarios. A special report of IPCC Working Group III. Available https://ipcc.ch/pdf/special-reports/spm/sres-en.pdf. Accessed 25 Oct 2018

IPCC (2007a) Climate change 2007: the physical science basis. Contribution of Working Group I to the Fourth Assessment Report of the Intergovernmental Panel on Climate Change. Cambridge University Press, Cambridge/New York, 996 pp

IPCC (2007b) Climate change 2007: impacts, adaptation and vulnerability. Contribution of Working Group II to the Fourth Assessment Report of the Intergovernmental Panel on Climate Change [Parry ML, Canziani OF, Palutikof JP, van der Linden PJ, Hanson CE, eds]. Cambridge University Press, Cambridge, 976pp

IPCC (2007c) Climate change 2007: synthesis report. Contribution of working groups I, II and III to the Fourth Assessment Report of the Intergovernmental Panel on Climate Change. IPCC, Geneva, 104 pp

IPCC (2013) Climate change 2013: the physical science basis. Contribution of Working Group I to the Fifth Assessment Report of the Intergovernmental Panel on Climate Change. Cambridge University Press, Cambridge/New York, 1535 pp

Jansson P, Hock R, Schneider T (2003) The concept of glacier storage: a review. J Hydrol 282:116–129

Jeelani G (2008) Aquifer response to regional climate variability in a part of Kashmir Himalaya in India. Hydrogeol J 16(8):1625–1633

Kang S, Hao X, Du T, Tong L, Su X, Lu H, Li X, Huo Z, Li S, Ding R (2017) Improving agricultural water productivity to ensure food security in China under changing environment: from research to practice. Agric Water Manag 179:5–17. https://doi.org/10.1016/j.agwat.2016.05.007

Knox JW, Kay MG, Weatherhead EK (2012) Water regulation, crop production and agricultural water management—understanding farmer perspectives on irrigation efficiency. Agric Water Manag 108(1):3–8

Koboltschnig GR, Schöner W, Zappa M, Holzmann H (2007) Contribution of glacier melt to stream runoff: if the climatically extreme summer of 2003 had happened in 1979. Ann Glaciol 46(1):303–308

Konar M, Hussein Z, Hanasaki N, Mauzerall DL, Rodriguez-Iturbe I (2013) Virtual water trade flows and savings under climate change. Hydrol Earth Syst Sci 17(8):3219–3234

Korhonen J, Kuusisto E (2010) Long-term changes in the discharge regime in Finland. Hydrol Res 41(3–4):253–268

Kundzewicz ZW, Döll P (2009) Will groundwater ease freshwater stress under climate change? Hydrol Sci J 54(4):665–675

Levidow L, Zaccaria D, Maia R, Vivas E, Todorovic M, Scardigno A (2014) Improving water-efficient irrigation: prospects and difficulties of innovative practices. Agric Water Manag 146:84–94

Lovarelli D, Bacenetti J, Fiala M (2016) Science of the total environment water footprint of crop productions: a review. Sci Total Environ 548–549:236–251

Loveys BR, Dry PR, McCarthy MG (1999) Using plant physiology to improve water use efficiency of horticultural crops. Acta Hortic 537:187–199

Luquet D, Vidal A, Smith M, Dauzatd J (2005) More crop per drop: how to make it acceptable for farmers? Agric Water Manag 76(2):108–119

Mali S, Singh DK, Sarangi A, Parihar SS (2018) Assessing water footprints and virtual water flows in Gomti river basin of India. Curr Sci 115(4):721–728

Malik RPS, Giordano M, Rathore MS (2016) The negative impact of subsidies on the adoption of drip irrigation in India: evidence from Madhya Pradesh. Int J Water Resour Dev 0627:1–12

Malone EL (2010) Changing glaciers and hydrology in Asia: addressing vulnerabilities to glacier melt impacts. United States Agency for International Development (USAID) Report November 2010, pp 1–113

Mamitimin Y, Feike T, Seifert I, Doluschitz R (2015) Irrigation in the Tarim Basin, China: farmers' response to changes in water pricing practices. Environ Earth Sci 73(2):559–569

Mann M, Gaudet B (2018) SRES' scenarios and 'RCP' pathways. In: Meteorology to mitigation: understanding global warming. Available https://www.e-education.psu.edu/meteo469/node/145. Accessed 25 Oct 2018

Mannini P, Genovesi R, Letterio T (2013) IRRINET: large scale DSS application for on-farm irrigation scheduling. Procedia Environ Sci 19:823–829

McCartney M, Smakhtin V (2010) Water storage in an era of climate change: addressing the challenges of increasing rainfall variability, IWMI Blue Paper. International Water Management Institute, Colombo

McVeigh M, Wyllie A (2018) Memo on irrigation efficiency and ESPA storage changes, State of Idaho Department of Water Resources

Min S, Zhang X, Zwiers FW, Hegerl GC (2011) Human contribution to more intense precipitation extremes. Nature 470(7334):378–381

Mingle J (2009) When the glacier left. Boston Globe. Available http://archive.boston.com/bostonglobe/ideas/articles/2009/11/29/when_the_glacier_left/. Accessed 20 Oct 2018

Mingle J (2015) Fire and ice: soot, solidarity and survival on the roof of the world. St. Martin's Press, New York

Molden D, Oweis T, Steduto P, Bindraban P, Hanjra MA, Kijne J (2010) Improving agricultural water productivity: between optimism and caution. Agric Water Manag 97:528–535

Molden DJ, Vaidya RA, Shrestha AB, Rasul G, Shrestha MS (2014) Governance and management of local water storage in the Hindu Kush Himalayas. Int J Water Resour Dev. Taylor & Francis

Molle F, Berkoff J (2007) Water pricing in irrigation: the lifetime of an idea. In: Molle F, Berkoff J (eds) Irrigation water pricing: the gap between theory and practice. CABI, Wallingford, pp 1–20

Molle F (2008) Can water pricing policies regulate irrigation use? In: 13th World Water Congress, September, Montpellier, France

Molle F, Tanouti O (2017) Squaring the circle: agricultural intensification vs. water conservation in Morocco. Agric Water Manag 192:170–179

Nelson G, Palazzo A, Ringler C, Sulser T, Batka M (2009) The role of international trade in climate change adaptation, ICTSDIPC platform on climate change, Agriculture and Trade Series, Issue Brief 4

Nüsser M, Dame J, Kraus B, Baghel R, Schmidt S (2018) Socio-hydrology of "artificial glaciers" in Ladakh, India: assessing adaptive strategies in a changing cryosphere. Reg Environ Chang:1–11

Oki T, Kanae S (2006) Global hydrological cycles and world water resources. Science 313(5790):1068–1072

Pall P, Aina T, Stone DA, Stott PA, Nozawa T, Hilberts AGJ, Lohmann D, Allen MR (2011) Anthropogenic greenhouse gas contribution to flood risk in England and Wales in autumn 2000. Nature 470(7334):382–385

Perry C, Steduto P, Allen RG, Burt CM (2009) Increasing productivity in irrigated agriculture: agronomic constraints and hydrological realities. Agric Water Manag 96:1517–1524

Perry C, Steduto P, Karajeh F (2017) Does improved irrigation technology save water? A review of the evidence. Food and Agriculture Organization of the United Nations, Cairo

Pfeiffer L, Lin Lawell C-YC (2014) Does efficient irrigation technology lead to reduced groundwater extraction? Empirical evidence. J Environ Econ Manag 67(2):189–208. https://doi.org/10.1016/j.jeem.2013.12.002

Piao S, Friedlingstein P, Ciais P, de Noblet-Ducoudre N, Labat D, Zaehle S (2007) Changes in climate and land use have a larger direct impact than rising CO2 on global river runoff trends. Proc Natl Acad Sci U S A 104(39):15242–15247

Piao S, Ciais P, Huang Y, Shen Z, Peng S, Li J, Zhou L, Liu H, Ma Y, Ding Y, Friedlingstein P, Liu C, Tan K, Yu Y, Zhang T, Fang J (2010) The impacts of climate change on water resources and agriculture in China. Nature 467(7311):43–51

Portmann FT, Döll P, Eisner S, Flörke M (2013) Impact of climate change on renewable groundwater resources: assessing the benefits of avoided greenhouse gas emissions using selected CMIP5 climate projections. Environ Res Lett 8(2):024023

Rabassa J (2009) Impact of global climate change on glaciers and permafrost of South America, with emphasis on Patagonia, Tierra del Fuego, and the Antarctic Peninsula. Dev Earth Surf Process 13:415–438

Rabatel A et al (2013) Current state of glaciers in the tropical Andes: a multi-century perspective on glacier evolution and climate change. Cryosphere 7(1):81–102

Ridoutt BG, Pfister S (2010) A revised approach to water footprinting to make transparent the impacts of consumption and production on global freshwater scarcity. Glob Environ Chang 20(1):113–120

Rizvi J (1998) Ladakh: crossroads of high Asia. Oxford University Press, Delhi

Rosenzweig C, Casassa G, Karoly DJ, Imeson A, Liu C, Menzel A, Rawlins S, Root TL, Seguin B, Tryjanowski P (2007) Assessment of observed changes and responses in natural and managed systems. In: Climate change 2007: impacts, adaptation and vulnerability. Contribution of Working Group II to the Fourth Assessment Report of the Intergovernmental Panel on Climate Change [Parry ML, Canziani OF, Palutikof JP, van der Linden PJ, Hanson CE (eds)]

Schewe J, Heinke J, Gerten D, Haddeland I, Arnell N, Clark D, Dankers R, Eisner S, Fekete B, Colón-González F, Gosling S, Kim H, Liu X, Masaki Y, Portmann F, Satoh Y, Stacke T, Tang Q, Wada Y, Wisser D, Albrecht T, Frieler K, Piontek F, Warszawski L, Kabat P (2014) Multimodel assessment of water scarcity under climate change. Proc Natl Acad Sci 111:3245–3250. https://doi.org/10.1073/pnas.1222460110

Sepaskhah AR, Ahmadi SH (2010) A review on partial root-zone drying irrigation. Int J Plant Prod 4(4)

Shaheen FA (2016) The art of glacier grafting: innovative water harvesting techniques in Ladakh. IWMI-Tata Water Policy Research Highlight, 8. 8p

Shiklomanov AI, Lammers RB, Rawlins MA, Smith LC, Pavelsky TM (2007) Temporal and spatial variations in maximum river discharge from a new Russian data set. J Geophys Res Biogeosci 112(G4):G04S53

Skaugen T, Stranden HB, Saloranta T (2012) Trends in snow water equivalent in Norway (1931–2009). Hydrol Res 43(4):489–499

Smit B, Skinner MW (2002) Adaptation options in agriculture to climate change: a typology. Mitig Adapt Strateg Glob Chang 7:85–114

Smith M, Kivumbi D (2000) Use of the FAO CROPWAT model in deficit irrigation studies. In: Deficit irrigation practices. Food and Agriculture Organization, Rome. Available http://www.fao.org/docrep/004/Y3655E/y3655e05.htm. Accessed 3 Oct 2018

Stern N (2006) The economics of climate change: the stern review. Cambridge University Press, Cambridge

Sugden F, Shrestha L, Bharati L, Gurung P, Maharjan L, Janmaat J, Price J, Sherpa TYC, Bhattarai U, Koirala S (2014) Climate change, out-migration and agrarian stress: the potential for upscaling small-scale water storage in Nepal (research report). International Water Management Institute (IWMI), Colombo. https://doi.org/10.5337/2014.210

Takala M, Pulliainen J, Metsamaki SJ, Koskinen JT (2009) Detection of snowmelt using spaceborne microwave radiometer data in Eurasia from 1979 to 2007. IEEE Trans Geosci Remote Sens 47(9):2996–3007

Tan A, Adam JC, Lettenmaier DP (2011) Change in spring snowmelt timing in Eurasian Arctic rivers. J Geophys Res Atmos 116:D03101

Taylor IH, Burke E, McColl L, Falloon PD, Harris GR, McNeall D (2013) The impact of climate mitigation on projections of future drought. Hydrol Earth Syst Sci 17(6):2339–2358

United Nations Framework Convention on Climate Change (UNFCCC) (1992) United nations framework convention on climate change: text, Geneva, World Meteorological Organization and United Nations Environment Program

United Nations Framework Convention on Climate Change (UNFCCC) (1998) The Kyoto Protocol to the UNFCCC. In: UNFCCC, Report of the Conference of the Parties Third Session, Kyoto, UNFCCC, pp 4–29

Vaidya RA (2015) Governance and management of local water storage in the Hindu Kush Himalayas. Int J Water Resour Dev 31(2):253–268

Venkatramanan V, Shah S (2019) Climate smart agriculture technologies for environmental management: the intersection of sustainability, resilience, wellbeing and development. In: Shah S et al (eds) Sustainable green technologies for environmental management. Springer Nature Singapore Pte Ltd., Singapore, pp 29–51. https://doi.org/10.1007/978-981-13-2772-8_2

Vidal JP, Martin E, Kitova N, Najac J, Soubeyroux J-M (2012) Evolution of spatio-temporal drought characteristics: validation, projections and effect of adaptation scenarios. Hydrol Earth Syst Sci 16(8):2935–2955

Vince G (2010) A Himalayan village builds artificial glaciers to survive global warming. Sci Am. https://www.scientificamerican.com/article/artificial-glaciers-to-survive-global-warming/. Accessed 21 Oct 2018

Viviroli D, Dürr HH, Messerli B, Meybeck M, Weingartner R (2007) Mountains of the world, water towers for humanity: typology, mapping, and global significance. Water ResWater Res 43:W07447

Viviroli D, Archer DR, Buytaert W, Fowler HJ, Greenwood GB, Hamlet AF, Huang Y, Koboltschnig G, Litaor MI, López-Moreno JI, Lorentz S, Schädler B, Schreier H, Schwaiger K, Vuille M, Woods R (2011) Climate change and mountain water resources: overview and recommendations for research, management and policy. Hydrol Earth Syst Sci 15(2):471–504

Wada Y, Wisser D, Eisner S, Flörke M, Gerten D, Haddeland I, Hanasaki N, Masaki Y, Portmann FT, Stacke T, Tessler Z, Schewe J (2013) Multi-model projections and uncertainties of irrigation water demand under climate change. Geophys Res Lett 40(17):4626–4632

Wang A, Lettenmaier DP, Sheffield J (2011) Soil moisture drought in China, 1950–2006. J Clim 24(13):3257–3271

Wang YB, Wu PT, Engel BA, Sun SK (2015) Comparison of volumetric and stress-weighted water footprint of grain products in China. Ecol Indic 48:324–333

Ward FA, Pulido-Velazquez M (2008) Water conservation in irrigation can increase water use. Proc Natl Acad Sci U S A 105:18215–18220

Yoo S-H, Choi J-Y, Lee S-H, Oh Y-G, Yun DK (2013) Climate change impacts on water storage requirements of an agricultural reservoir considering changes in land use and rice growing season in Korea. Agric Water Manag 117:43–54

Zeng Z, Liu J, Koeneman PH, Zarate E, Hoekstra AY (2012) Assessing water footprint at river basin level: a case study for the Heihe River Basin in Northwest China. Hydrol Earth Syst Sci 16(8):2771–2781

Zou X, Li Y, Gao Q, Wan Y (2012) How water saving irrigation contributes to climate change resilience – a case study of practices in China. Mitig Adapt Strateg Glob Chang 17:111–132

Zou X, Cremades R, Gao Q, Wan Y, Qin X (2013) Cost-effectiveness analysis of water-saving irrigation technologies based on climate change response : a case study of China. Agric Water Manag 129:9–20

Mitigating Enteric Methane Emission from Livestock Through Farmer-Friendly Practices

8

C. Valli

Abstract

Enteric methane emission in ruminants in addition to being an environmental pollutant causes a loss of 10–11% of the total gross energy intake of the animal. The various strategies available to mitigate enteric methane emission include management strategies, feeding strategies, rumen manipulation and advanced strategies. Feeding strategies are practical approaches to mitigate enteric methane emission and can be practiced with ease by farmers under field conditions. Various approaches that cause rumen manipulation and thereby reduce enteric methane emission are supplementation of bacteriocins, ionophores, fats, oils, organic acids, probiotics, prebiotics, sulphate, halogenated methane analogues, nitroxy compounds, fungal metabolite, secondary plant metabolites, microalgae and exogenous enzymes. In the Indian context, ruminant livestock are grazed in wastelands or fed with poor-quality agricultural waste, whose digestibility is low, and the nutritional requirement of the animals is not met resulting in poor productivity. Improving per animal productivity is a potent tool to reduce enteric methane emission per unit of product produced, and this can be achieved through ration balancing. As regards the greenhouse gas abatement opportunities, measures to reduce enteric methane production have immense economic and ecological benefits to the farmers.

Keywords

Ruminants · Enteric methane emission · Mitigation · Livestock management · Manure management · Pasture management · Feed management · Rumen manipulation · Secondary metabolites

C. Valli (✉)
Institute of Animal Nutrition, Tamil Nadu Veterinary and Animal Sciences University, Chennai, Tamil Nadu, India

© Springer Nature Singapore Pte Ltd. 2020

V. Venkatramanan et al. (eds.), *Global Climate Change and Environmental Policy*,
https://doi.org/10.1007/978-981-13-9570-3_8

8.1 Introduction

The “agriculture, forestry and other land use (AFOLU)” sector releases 10–12 Gt CO_2e per annum. Greenhouse gas emissions from agricultural activities (Fig. 8.1) include “land use changes”; enteric methane emission from ruminants; lowland rice cultivation; N_2O emissions from nitrogenous fertilizer use; and crop residue burning (Lipper et al. 2014; Smith et al. 2014; Venkatramanan and Shah 2019). Methane is a very significant greenhouse gas in view of its global warming potential and atmospheric lifetime. The causes for “increases in CH_4 concentration include natural wetlands emissions (177–284 Tg CH_4 yr^{-1}), agriculture and waste (187–224 Tg CH_4 yr^{-1}), fossil fuel related emissions (85–105 Tg CH_4 yr^{-1}), other natural emissions (61–200 Tg CH_4 yr^{-1}), and biomass and biofuel burning (32–39 Tg CH_4 yr^{-1})” (Venkatramanan and Shah 2019). Livestock are one of the largest single sources of anthropogenic methane emission in the world. Amongst livestock, ruminants (cattle, buffaloes, sheep and goats) are the principal sources of enteric methane emission.

About 30% of methane emissions at the global level are released from the ruminant activity. In other words, globally, 2.7 Gt of CO_2e of enteric CH_4 per annum are released from the ruminants. As regards the livestock methane emissions, 2.1 Gt (77%) of CO_2e is contributed by cattle, 0.37 Gt (14%) of CO_2e by buffaloes and 0.26 Gt CO_2e by small ruminants like sheep and goats (FAO 2017).

In India, 49.1% of enteric methane was contributed by cattle, 42.8% by buffaloes, 5.38% by goats and 2.59% by sheep. Importantly, during 1961–2010, the increase in methane emissions (70.6%) from livestock population of India is much

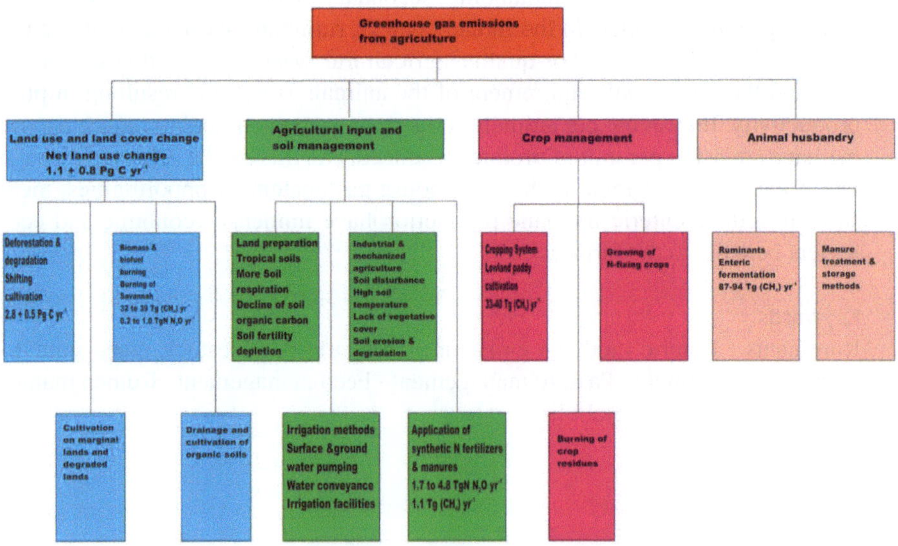

Fig. 8.1 Greenhouse gas emissions from agriculture. The GHG values mentioned represent the annual anthropogenic fluxes in Tg/year averaged over 2000–2009 time period. (Source: Venkatramanan and Shah 2019; GHG values from Ciais et al. 2013)

greater than the increase in methane emissions from livestock population worldwide (54.3%). It is reported that by 2050, about 15.7% of enteric CH_4 emission at the global level will be contributed by the Livestock population of India (Patra 2014). One of the reasons for high enteric methane emission in India is the larger bovine population which emits more methane than any other livestock species. On an average, cattle and buffaloes aggregately emit more than 90% of the total enteric methane emission of the country. The contribution from small ruminants is relatively small and constitutes only 7.7%. Rest of the methane emissions arise from the species such as yak and mithun, which are scattered to specific states only. Enteric methane emission from crossbred cattle is comparatively much more than the emissions from indigenous cattle (46 versus 25 kg/animal/year). Enteric methane emission from livestock is not uniform across the states and varies considerably according to the livestock numbers, species, type of feed and fodders, etc. (Malik et al. 2015).

The National Institute of Animal Nutrition and Physiology, India, has developed a statewise inventory for enteric methane emission from Indian livestock. Their estimates revealed that Uttar Pradesh was the largest enteric methane emitting state and other major methane emitting states were Rajasthan, Madhya Pradesh, Bihar, West Bengal, Maharashtra, Karnataka and Andhra Pradesh. These states altogether have 66% of the total livestock population and are responsible for 68% enteric methane emissions (Bhata et al. 2016).

Enteric methane emission in livestock is a consequence of ruminal/hindgut fermentation. In the reticulo rumen or hindgut of livestock, simple and complex carbohydrates are hydrolyzed to 5 or 6 carbon sugars by enzymes secreted by microbes. These sugars are then fermented to volatile fatty acids through metabolic pathways that produce metabolic hydrogen. The metabolic hydrogen is converted to hydrogen by hydrogenase expressing bacterial species (Methanogens), and the hydrogen is converted to methane by archaea. In fact, the dissolved hydrogen through a negative feedback restrains fermentation pathways occurring in the rumen. Incidentally, the non-removal of dissolved hydrogen can lessen the degradation of carbohydrate, growth rate of microbes, and also the production of microbial protein (Wolin 1974; McAllister and Newbold 2008). Methanogenesis can be regarded as an evolutionary adaptation that enables the rumen ecosystem to dispose of hydrogen, which may otherwise accumulate and inhibit fibre degradation.

Enteric methane emission in ruminants in addition to being an environmental pollutant causes a loss of 10–11% of the total gross energy intake of the animals. Methane mitigation options are therefore associated with improved efficiency of animal production. Thus enteric methane mitigation measures would be advantageous both environmentally and nutritionally.

8.2 Strategies to Mitigate Enteric Methane Emission

There are various strategies available to mitigate enteric methane emission. The strategies may be broadly grouped as management strategies, feeding strategies, rumen manipulation and advanced strategies.

8.2.1 Management Strategies

Management strategies that could be adopted to control enteric methane emission are reducing the ruminant livestock population, breeding management and manure management.

8.2.1.1 Reducing the Ruminant Livestock Population

In underdeveloped and developing nations, agriculture and livestock rearing is a way of life and indeed contributes greatly to the income of the farmers. Nevertheless, unable to maintain and rear high producing ruminants, the small-scale farmers primarily meet their demands by maintaining a large number of livestock which are low producing ruminants (Tarawali et al. 2011). This can increase the total enteric methane emission in a given area. Thus the most effective methane mitigating strategy will be to increase livestock productivity, which may permit reduction in livestock numbers to provide the same product output at a reduced enteric methane emission. Higher animal performance also reduces methane emission because this can lead to a reduction in the number of animals in the production system.

8.2.1.2 Breeding Management

Adopting a superior breeding program to enhance per animal productivity is one of the best options to reduce enteric methane emission as the production of enteric methane is greatly reduced on account of improvement in the productivity of livestock. It is thus important to consider the enteric methane production per unit of the livestock product produced (kg of milk, meat, or wool).

Another school of thought is that genetics influences enteric methane emission in ruminants (Robertson and Waghorn 2002; Pinares-Patino et al. 2007). Different amounts of methane emission per unit intake of the same quality ration at the same level of performance were recorded, and accordingly livestock were categorized as "high" and "low" methane emitters (Pinares-Patino et al. 2007). Genetic selection and breeding of such animals that consume less feed and produce less methane per unit of feed may be one of the management strategies for reduction in methanogenesis.

8.2.1.3 Manure Management

Animal manure from both ruminant and non-ruminant livestock contributes to methane emission. Stored manure is an important source of methane emission (Priano et al. 2014). Manures are organic compounds which on decomposition by anaerobic bacteria transform their carbon skeleton into methane. Production of methane from manure from dairy cattle, beef cattle and dairy ewe was reported to be 33.2, 2.0 and 0.3 kg/head/year, respectively (Merino et al. 2011).

Many factors indeed influence the emission of methane from manure. Physical properties of manure like density, humidity, amount of digestible material and temperature influence methane emission (Priano et al. 2014; Saggar et al. 2004). The type of ration fed to the animal influences the manure quality and the quantity of methane emitted from it. Methane emissions from manure storage arise only when animals are intensively reared. When animals are managed on grazing this scenario does not arise. Methane emitted from manures can be greatly reduced by minimizing the duration of storage of manure and giving lesser time for the "microbial fermentation" to take place. Anaerobic digesters are recommended as a mitigation strategy for methane, and are a source of renewable energy as well and provide sanitation opportunities for developing countries.

8.2.2 Feeding Strategies

Feeding strategies are practical approaches to mitigate enteric methane emission and can be practiced with ease by farmers under field conditions. These may include pasture improvement, feed processing, increasing concentrate in ration and strategic supplementation.

8.2.2.1 Pasture Improvement

Enteric methane emission is influenced by the type of forage the animal feeds upon. Animals grazing on mixed pastures, having leguminous forages, emit less methane than animals grazing on grass alone pastures. This effect can be explained by the presence of condensed tannins in legumes (Waghorn 2007).

There is also variation in enteric methane emission of ruminants grazing on tropical or temperate pastures. C_4 grasses in tropical pastures can produce more methane per kg of dry matter ingested than temperate grasses with C_3 photosynthesis. This is due to the poor digestibility of C_4 grasses compared to C_3 grasses (Archimède et al. 2011).

In temperate pastures itself, Boadi et al. (2002) reported less energy loss as methane in steers grazing on pastures during the early period of the grazing season than the mid- and late-grazing period. Pastures rich in grasses with high sugar content are suitable for animal grazing for a simple reason that high sugar-rich grasses promote the growth of rumen microbes resulting in the increased synthesis of microbial protein and decreasing the production of methane. Hence improving pasture quality is another important way by which enteric methane emission can be reduced.

8.2.2.2 Feed Processing

The processing of feeds especially roughages can improve the feeding value by increasing its digestibility and/or by increasing feed intake. Any attempt to increase feed digestibility/intake has potential to reduce enteric methane emission. Improving the nutritive value of low-quality forages/feeds that are fed to ruminants could increase their productivity and consequently reduce methane emissions.

Feed processing techniques may include chaffing and grinding of straws, alkali/ammonia treatment of straws and feed residues and urea-molasses blocks. These processing techniques are reported to check the methane emission from rumen by 10%. The reduction of forage particle size has significant benefits like increasing feed intake and digestibility, increasing microbial activity due to increase in the surface area of the substrate and consequent improvement in the productivity of livestock (Gerber et al. 2013). Ammoniation or protein supplementation of low-quality forages or crop residues increased methane losses in proportion to digestibility; however methane production per unit of product produced was reduced (Johnson and Johnson 1995). Beauchemin et al. (2008), had stated that methanogenesis was lower on feeding silage compared to feeding hay and was also lower in finely ground feed or pellets compared to roughly processed feed.

8.2.2.3 Increasing Concentrate in Ration

Altering the roughage concentrate ratio of ruminant ration by increasing concentrates at the expense of roughage causes reduction in methane emission. Methane production can be reduced to 3% from 6.5% by feeding ruminants more concentrates (Beauchemin and McGinn 2005).

Higher concentrate feeding decreases fibre (cellulose and hemicellulose) and increases starch which results in significant modifications in both the physiochemical conditions in the rumen and microbial populations. Increase of amylolytic bacteria results in a change in short chain fatty acid production, promoting a proportional increase of propionate and a reduction of acetate which reduces the hydrogen availability in the rumen, causing reduced methane emission (Pereira et al. 2015). Beauchemin and McGinn (2005) reported that beef cattle, fed diets containing mainly corn grain produce about 30% less methane than when fed diets containing mainly barley grain. Methane emission can be reduced by enhancing the nutritive value and digestibility of feed by balancing the ration with addition of concentrate for grazing animals.

This strategy of increasing concentrate in ration to reduce methane emission has its own limitations. In high concentrate rations, increased production of lactic acid could trigger ruminal acidosis, reduce milk fat and shorten the productive life span of animals. Moreover, its economic viability is also questionable. Increasing the concentrate proportion will have negative impact on fibre digestibility which could cause potential loss in production and will also result in increased concentration of fermentable organic matter in manure and is likely to increase methane emissions from stored manure (Lee et al. 2012).

8.2.2.4 Strategic Supplementation

In tropical countries, where ruminant livestock are grazed in wastelands or on very poor-quality pastures or fed with poor-quality agricultural waste, the digestibility of the diet is low and the nutritional requirement of the animals is not met. In this context protein supplementation in the diets increases the nutrient digestibility and significantly decreases methane production in the rumen (Mehra et al. 2006). Balanced diet with inclusion of protein supplements when fed to the lactating cows and buffaloes caused a marked reduction in the methane emission and an observable increase in both the production of milk and fat content in milk (Kannan and Garg 2009). Similar findings were observed when lactating crossbred cows are fed with balanced diet with respect to calcium, phosphorus and protein (Kannan et al. 2011).

8.2.3 Rumen Manipulation

Enteric methane emission from ruminant livestock can be reduced by various approaches that cause rumen manipulation viz., supplementation of bacteriocins, ionophores, fats, oils (Beauchemin et al. 2008), organic acids (Newbold et al. 2005; Wallace et al. 2006), probiotics, prebiotics, sulphate, halogenated methane analogues, nitroxy compounds (Morgavi et al. 2010), fungal metabolites and secondary plant metabolites (Patra et al. 2006; Wallace et al. 2002; Calsamiglia et al. 2007), microalgae, exogenous enzymes, etc.

8.2.3.1 Bacteriocins

Bacteriocins are naturally occurring bacterial products with a bactericidal activity. They are effective as they directly inhibit methanogens and redirect hydrogen to other reductive rumen bacteria such as propionate producer or acetogens. Callaway et al. (1997) have shown that nisin, a food additive, reduces methanogenesis by 36%. However, the organisms developed resistance quickly. This problem can be overcome by the use of bacteriocin of rumen origin. Lee et al. (2002) found that semi-purified bacteriocin inhibits 50% of methane production *in vitro* and methanogens did not show any adaptation to these bacteriocins.

8.2.3.2 Ionophores

Ionophores are generally used as feed additives in order to improve the efficiency of digestion in ruminants. Ionophores include tetronasin, monensin, lasalocid, salinomycin, narasin, lysocellin, etc. These ionophore antibiotics are carboxylic polyether compounds produced by various strains of *Streptomyces*, e.g. monensin by *S. cinnamonensis* and lasalocid by *S. lasoliensis*. Monensin is moderately active against gram-positive bacteria, certain mycobacteria and coccidia, while lasalocid is specifically active against hydrogen producing bacteria and results in higher propionate production which in turn is related to low methane production (Kobayashi et al. 1992).

Ionophores increase the proportion of gram-positive bacteria in the rumen, resulting in a shift in fermentation acids from acetate and butyrate to propionate, and hence decrease the methane production. Several studies were conducted to know the efficacy of monensin in rumen modification. Few studies found the positive mitigating potential of monensin in enteric methane emission. Nevertheless, the ionophore effects on enteric methane emission are ephemeral as the microbes adapt themselves (Kataria 2015).

8.2.3.3 Fats/Oils

Adding fats to the diet can reduce methane emission by lowering ruminal fermentability and also through hydrogenation of unsaturated fats (Johnson and Johnson 1995). As methane is produced in the rumen to act as a hydrogen sink during the fermentation of carbohydrate, polyunsaturated fatty acids (PUFA) having double or triple bonds have potential to be used as hydrogen sinks, because these bonds will get saturated by hydrogen and less hydrogen will be available for methane production. Adding 4.6% canola oil, a source of unsaturated fat, to a high-forage diet was an effective suppressant of methane, with daily methane emissions decreasing by 32% and methane emissions as a percentage of GE intake decreasing by 21% (Beauchemin and McGinn 2006).

Fat feeding has also been shown to reduce methane through a reduction in protozoal numbers. Methanogenic bacteria are metabolically associated with ciliate protozoa (Newbold et al. 1995), and feeding oil can cause substantial decreases in protozoan populations (Ivan et al. 2004). Rumen protozoans were found to be responsible for higher hydrogen production and also acting as a habitat for the methanogens present in the rumen (Newbold et al. 1995). In defaunated ruminants, the methanogenic bacteria do not get the symbiotic partner and methane synthesis is partially inhibited. On defaunation, the methane production is reduced by 20–50% (Van Nevel and Demeyer 1996) depending on the various factors in the diet of the animal.

8.2.3.4 Organic Acids

Within the rumen, methane represents a terminal hydrogen sink. Propionate production represents an alternative hydrogen sink in normal rumen fermentation, provided sufficient precursors are available. The precursors to propionate production are "pyruvate", "oxaloacetate", "malate", "fumarate" and "succinate". The propionate can also be produced from pyruvate through an important pathway called acrylate pathway. Organic acids mentioned above have potential to jettison reducing power and thereby methane production can be reduced. Dietary supplementation of dicarboxylic organic acids such as malate, fumarate, aspartate etc. reduces methane production (Martin 1998). These organic acids are converted to succinate or propionate by reduction process, and less hydrogen will be available for methane production.

Addition of fumaric acid decreased methane emissions *in vitro* (Asanuma et al. 1999) and in vivo (Bayaru et al. 2001) but not in all studies (McGinn et al. 2004), possibly because of the level of supplementation used. Malate, a potent methane inhibitor, is present in animal feeds like alfalfa (2.9–7.5% of DM) and Bermuda grass (1.9–4.5%), but its level varies with variety and stage of maturity. *In vitro*, continuous rumen culture experiments have shown that adding graded levels of fumarate to fermentations of rye grass pasture substrate linearly depressed methane production while linearly increased propionate production (Kolver et al. 2004). The long-term effects of organic acids on methane mitigation have not been demonstrated. Moreover, the high price of organic acids makes its use uneconomical.

8.2.3.5 Probiotics

Probiotics such as yeast cultures are used to stimulate bacterial activity in the rumen. The probiotics have been shown to stabilize the rumen pH, increase propionate levels and decrease the amount of acetate, methane and ammonia production. Probiotics cause the redirection of hydrogen ions away from methanogenesis and also decrease production of hydrogen during feed fermentation. Addition of *Saccharomyces cerevisiae* reduced methane production *in vitro* (Mutsvangwa et al.

1992). Lactic acid utilizing bacteria like *Megasphaera elsdenii* and *Propionibacterium* spp. and yeast like *S. cerevisiae* have major effects on methanogenesis by decreasing methane (Seo et al. 2010). Chaucheyras et al. (2012) suggested that live yeast cells can stimulate the use of hydrogen by acetogenic strains of ruminal bacteria thereby enhancing the formation of acetate and decreasing the formation of methane.

8.2.3.6 Prebiotics

Prebiotic compounds such as mannan-oligosaccharide (MOS), fructo-oligosaccharide (FOS), galacto-oligosaccharide (Mwenya et al. 2004) enhances propionate production by stimulating *Selenomonas, Succinomonas* and *Megasphera* with inhibition of acetate producers such as *Ruminococcus* and *Butyrivibrio.* The administration of galacto-oligosaccharides has brought about reduction of methane production up to 11%.

8.2.3.7 Sulphate

Sulphate reducing bacteria have the highest affinity to utilize hydrogen in the rumen, even better than methanogens, but the availability of sulphate in the rumen appears to be a limitation. Hegarty (1999) showed that to reduce methane emissions by 50%, about 0.75 moles of sulphate ingestion per day is required. Kamra et al. (2004) observed that sulphate supplementation helps in increasing the production of fibre degrading enzymes and fibre degradation in the rumen. Therefore, for a high fibre ration, sulphate/sulphite supplementation improves fibre degradability and inhibits methanogensis. However, a proper dose needs to be optimized, as toxic levels of sulphide can be generated on sulphate reduction.

8.2.3.8 Halogenated Methane Analogues

Various halogenated methane analogues have been tried as methane inhibitors. They include carbon tetrachloride, chloral hydrate, trichloroacetamide, DDT, trichloroacetaldehyde, bromochloromethane, chloroform, methylene chloride, methylene bromide, nitrapyrin, hemiacetal of chloral and starch (Haque 2001). Favourable effects on supplementation of these have been reported only in those animals fed on high roughage diets. Bromochloromethane inhibits methane production by reacting with reduced form of vitamin B_{12} which inhibits methanogenesis.

8.2.3.9 Nitrooxy Compounds

The “3-nitrooxypropanol (NOP)” and “ethyl-3-NOP” were found to possess an important anti-methanogenic property which indeed aids in reducing methane production. In fact, “3-nitrooxypropanol (NOP)” through interference with a co-enzyme (methyl-coenzyme M reductase) of methanogens checks the methanogen growth and thereby reduces methane production. Also, 3-NOP has inhibitory effects against methanogenic archaea without inducing any effects on growth of non-methanogenic bacteria in the rumen (Duin et al. 2016).

8.2.3.10 Fungal Metabolites

Lovastatin which is a secondary fungal metabolite is found to limit the activity of "3-hydroxy-3-methyl glutaryl coenzyme A reductase". HMG-CoA reductase is an important enzyme involved in the biosynthesis of cholesterol (Jahromi et al. 2013). Rice straw fermented with *Aspergillus terreus* was found to possess lovastatin which was observed to reduce the population of methanogens and production of methane. Saprophytic fungal strains *Mortierella wolfii* were found to be a suitable candidate for the inhibition of the methanogenesis process (Cosgrove et al. 2012). The fungal metabolites, viz. "pravastatin" and "mevastatin", were observed for reducing the ratio of ruminal acetate to propionate and also reducing the methanogen population (Morgavi et al. 2013).

8.2.3.11 Secondary Plant Metabolites

Plants produce chemicals collectively called as secondary plant metabolites that help in the self-defence of the plant species concerned. Most of these secondary plant metabolites have anti-microbial and anti-methanogenic activity. There are numerous plant extracts each with varying effects on ruminal fermentation and feed digestion (Kamra et al. 2005).

Garlic oil extracted from *Allium sativum* is a complex mixture of many secondary plant products including allicin, diallyl sulfide, diallyl disulfide and allyl mercaptan and a decrease in methane production was observed on supplementing garlic oil (Busquet et al. 2005). Ethanol and water extracts of the bulb of *Allium sativum* (garlic) reduced methane production significantly in the rumen liquor of buffaloes (Patra et al. 2006).

Methanol extract of seed pulp of *Terminalia chebula* (kadukka) reduced methane production significantly in the rumen liquor of buffaloes (Patra et al. 2006). The presence of tannins in *Terminalia chebula* might be responsible for reduction in methane emission. Phenolic acids such as p-coumaric acids, ferulic acids, cinnamic acids and phloretic acids and some monomeric phenolics have been found to decrease methane, acetate and propionate production (Asiegbu et al. 1995). Tannin, an important secondary plant metabolite, was found to be toxic against rumen organisms like protozoans, methanogens and other microbes that degrade the fibrous material. There are reports indicating a decrease in methane emission with dietary addition of condensed tannins or inclusion of condensed tannin containing forage (Carulla et al. 2005). Negative effects on ruminal fibre digestion, which may relate to decreased number of cellulolytic bacteria, formation of condensed tannin complexes that are resistant to enzymatic digestion and impaired substrate adhesion by fibrinolytic microbes, would reduce hydrogen availability to lessen methanogenesis (Carulla et al. 2005). Patra et al. (2006) showed that adding water extracts of neem seeds decreased total rumen volatile fatty acid (VFA) digestibility and also reported antiprotozoal activity.

Essential oils beneficially affect ruminal fermentation, causing an increase in VFA (Castillejos et al. 2005) and decrease in the rate of amino acid deamination (McIntosh et al. 2003). The antimicrobial activity of essential oils and secondary plant metabolites is highly specific, which raises the possibility that these compounds can be used to target methanogens. McIntosh et al. (2003) reported that growth of the methanogen *Methanobrevibacter smithii* was inhibited when the concentration of essential oil in supplemental product exceeded 1000 ppm.

Saponins have been found to have anti-protozoal and anti-methanogenic activities (Sliwinski et al. 2002). There are plants, which contain saponins such as *Alfalfa* (3–5%), *Sapindus rarak*, *Sapindus mokorossi*, *Yucca schidigera* (4%), *Quillaja saponaria* (10%), etc. that cause a decrease in methane production by 20–60%. Saponins reduce the protozoal population which reduces the inter species hydrogen transfer to the methanogenic bacteria attached to the protozoa, thereby decreases the hydrogen availability to the methanogens.

8.2.3.12 Microalgae

Asparagopsis, a genus of microalgae, even at 2% of total substrate organic matter, was reported to reduce methane production by 99% in, *in vitro* condition. Similarly, *Oedogonium,* a member of filamentous algae, was found to decrease methane production (Machado et al. 2014). Further, *Cystoseira trinodis* and *Dictyota bartayresii*, members of brown algae, are found under *in vitro* conditions to inhibit methane production (Dubois et al. 2013).

8.2.3.13 Exogenous Enzymes

The role of exogenous enzymes in mitigating enteric methane emission has been reviewed with contradictory results. Exogenous enzymes increase fibre digestibility, improves feed efficiency and decreases fermentable organic matter in manure, thus reducing overall methane emission from ruminant production system. However, Chung et al. (2012), reported increased enteric methane emission per unit dry matter intake or milk yield by 10–11% on exogenous enzyme supplementation with endoglucanase and xylanase activity.

8.2.3.14 Defaunation

Ten per cent decrease in enteric methane emission has been reported on defaunation. However, defaunation may have a negative impact on feed digestibility, animal productivity and milk fat content. Practically, it is not possible to maintain defaunated livestock as possibility of the animals getting refaunated always exists.

8.2.4 Advanced Strategies

Advanced strategies can be adopted in technologically advanced livestock farms to mitigate enteric methane emission. Some of the advanced strategies are precision feeding, vaccines that target methanogens, genetic transformation of rumen bacteria and transferring rumen microbiome of low methane emitting ruminants.

8.2.4.1 Precision Feeding

Precision feeding is feeding the livestock with the right feed at the right time, so that it satisfies its nutritional requirement at that point of time. Precision feeding combines the genetics of the animal with feed management. Precision feeding will improve feed digestibility, feed efficiency and livestock productivity and hence indirectly reduce enteric methane emission. Precision feeding can be adopted only in high value livestock farms that are highly advanced technologically.

8.2.4.2 Vaccines that Target Methanogens

Vaccines have been developed that specifically target methanogens in the rumen. There are immunogenic fractions on the surface of methanogens and antibodies developed against these could curtail methanogenic activity in ruminants (Wedlock et al. 2013). However, the majority of the vaccines have failed due to poor efficacy and low cross-reactivity. The reverse vaccinology technology has provided solution for identifying conserved vaccines. Moreover sequencing the genome of *Methanobacterium ruminantium* has provided data for identifying conserved vaccine targets.

8.2.4.3 Genetic Transformation of Rumen Bacteria

Genetic modification of rumen microorganisms is aimed to alter their fermentation characteristics. Extensive characterization of the molecular genetics of rumen bacteria with high precision is needed for the success of this genetic modification. Even then the viability of the altered organisms in the natural environment is doubtful. Moreover, the product obtained/organism used has to be approved by both national and international regulatory standards.

8.2.4.4 Transferring Rumen Microbiome of Low Methane Emitting Ruminants

Another possibility of mitigating enteric methane emission is through transferring the microbiome of low methane emitters to the rumen of high methane producing ruminants. Such intervention when made in the early stage of life of the calf may have greater stability. Better understanding of host microbiome relations is needed for the success of such a venture.

8.2.4.5 Farmer-Friendly Approaches to Mitigate Enteric Methane Emission

As discussed earlier in this chapter, feeding strategies such as pasture improvement, feed processing, increasing concentrate in ration and strategic supplementation are practical approaches to mitigate enteric methane emission. Farmers can be advised to chaff the forage before feeding it to their livestock. Chaff cutters are available with the farmers, or farmers can purchase them at subsidized cost. Increasing concentrate in ration and strategic supplementation can be achieved through ration balancing. Ration balancing is considered as the practical approach to mitigate enteric methane in the Indian context.

Indian livestock are grazed on very poor-quality tropical pastures or fed on crop residues having a very low nutritive value. Hence the productivity of animals is far below their genetic potential. Improving per animal productivity is a potent tool in reducing enteric methane emission per unit of product produced. Feeding balanced diets reduced enteric methane emission in terms of g/d and g/kg milk yield by 10.1 and 13.5% respectively, which further reduced the part of dietary gross energy loss as methane by 10.3%. Intestinal flow of microbial nitrogen increased by 51.4 g/d whereas, faecal nitrogen excretion reduced by 19.4% (Garg et al. 2013). Ration balancing shifts the rumen fermentation more towards propionate production and microbial protein production; thus enteric methane emission is reduced. Rumen fermentation efficiency can be improved through critical nutrient supplementation.

8.3 Conclusion

Mitigating enteric methane emissions in ruminants will assist not only in the achievement of international commitments that benefit climate change but also in the improvement of energy utilization efficiency and the performance of livestock. Increasing livestock productivity across production systems increases food security and improves livelihood of farmers. In the perspective of GHG emission management, enteric methane emission reduction through significant gain in productivity is construed as a "lowest-cost option", and it has positive economic repercussions to the farmers dependent on livestock (Gerber et al. 2013). A country like India cannot afford the energy loss as methane in livestock, as it demands additional feed resources to compensate the loss (Malik et al. 2016).

Livestock farming in India is a low productivity venture. But the demand for livestock products is ever rising. In order to meet the consumer demand for livestock products, farmers rear a large number of livestock, larger than the number that would be economically viable with the available feed resources. More pressure is thus created on availability of feed, and livestock productivity declines further. In this context any increase in productivity resulting in an increased livestock product will lead to increase in employment in livestock enterprises for marketing and distribution of livestock products.

References

Archimede H, Eugène M, Marie-magdeleine C, Boval M, Martin C, Morgavi D, Lecomte P, Doreau M (2011) Comparison of methane production between C3 and C4 grasses and legumes. Fuel Energy Abstr 166:59–64

Asanuma N, Iwamoto M, Hino T (1999) Effect of the addition of fumarate on methane production by ruminal microorganism *in vitro*. J Dairy Sci 82:780–787

Asiegbu FO, Paterson A, Morrison IM, Smith JE (1995) Effect of cell wall phenolics and fungal metabolites on methane and acetate production under in vitro conditions. J Gen Appl Microbiol 41:475–485

Bayaru E, Kanda S, Kamada T, Itabashi H, Andoh S, Nishida T, Ishida M, Itoh T, Nagara K, Isobe Y (2001) Effect of fumaric acid on methane production, rumen fermentation and digestibility of cattle fed roughage alone. J Anim Sci 72:139–146

Beauchemin KA, McGinn SM (2005) Methane emissions from feedlot cattle fed barley or corn diet. J Anim Sci 83:6

Beauchemin KA, McGinn SM (2006) Methane emissions from beef cattle: effects of fumaric acid, essential oil, and canola oil. J Anim Sci 84:1489–1496

Beauchemin KA, Kreuzer M, O'Mara F, McAllister TA (2008) Nutritional management for enteric methane abatement: a review. Aust J Exp Agric 48:21–27

Bhata R, Malik PK, Kolte AP, and Gupta R (2016) Annual progress report on outreach project on methane. NIANP, Bangalore

Boadi D, Wittenberg KM, McCaughey WP (2002) Effects of grain supplementation on methane production of grazing steers using the sulphur (SF6) tracer gas technique. Can J Anim Sci 82:151–157

Busquet M, Calsamiglia S, Ferret A, Carro MD, Kamel C (2005) Effect of garlic oil and four of its compounds on rumen microbial fermentation. J Dairy Sci 88:4393–4404

Callaway TR, Carneiro De Melo AMS, Russell JB (1997) The effect of nisin and monensin on ruminal fermentations *in vitro*. Curr Microbiol 35:90–96

Calsamiglia M, Busquet P, Cardozo W, Castillejos L, Ferret A (2007) Invited review: essential oils as modifiers of rumen microbial fermentation. J Dairy Sci 90:2580–2595

Carulla JE, Kreuzer M, Machmuller A, Hess HD (2005) Supplementation of *Acacia mearnsii* tannins decreases methanogenesis and urinary nitrogen in forage-fed sheep. Aust J Agric Res 56:961–970

Castillejos L, Calsamiglia S, Ferret A, Losa R (2005) Effects of a specific blend of essential oil compounds and the type of diet on rumen microbial fermentation and nutrient flow from a continuous culture system. Anim Feed Sci Technol 119:29–41

Chaucheyras-Durand F, Chevaux E, Martin C, Forano E (2012) Use of yeast probiotics in ruminants: effects and mechanisms of action on rumen pH, fibre degradation, and microbiota according to the diet. Chapter 7. In: Rigobelo EC (ed) Probiotic in animals. Intech, Rijeka. https://doi.org/10.5772/50192:119-152

Chung YH, Zhou M, Holtshausen L, Alexander TW, McAllister TA (2012) A fibrolytic enzyme additive for lactating Holstein cow diets: ruminal fermentation, rumen microbial populations, and enteric methane emissions. J Dairy Sci 95:1419–1427

Ciais P, Sabine C, Bala G et al (2013) Carbon and other biogeochemical cycles. In: Stocker TF, Qin D, Plattner G-K, Tignor M, Allen SK, Boschung J, Nauels A, Xia Y, Bex V, Midgley PM (eds) Climate change 2013: The physical science basis. Contribution of Working Group I to the Fifth Assessment Report of the Intergovernmental Panel on Climate Change. Cambridge University Press, Cambridge/New York

Cosgrove GP, Muetzel S, Skipp R, Mace WJ (2012) Effects of endophytic and saprophytic fungi on in vitro methanogenesis. N Z J Agric Res 55:293–307

Dubois B, Tomkins NW, Kinley RD, Bai M, Seymour S, Paul NA (2013) Effect of tropical algae additives on rumen in vitro gas production and fermentation characteristics. Am J Plant Sci 4:34–43

Duin EC, Wagner T, Shima S, Prakash D, Cronin B, Yanez-Ruiz DR, Duval S, Rumbeli R, Stemmler RT, Thauer RK, Kindermann M (2016) Mode of action uncovered for the specific reduction of methane emissions from ruminants by the small molecule 3-nitrooxypropanol. Proc Natl Acad Sci 113:6172–6177

FAO (2017) Reducing Enteric Methane for improving food security and livelihoods. www.fao.org/in-action/enteric-methane/background/why-is-enteric-methane.../en/

Garg MR, Sherasia PL, Phondba BT, Sheike SK, Patel CT (2013) Effect of feeding balanced ration on milk production enteric methane emission and metabolic profile in cross bred cows under field conditions. Indian J Dairy Sci 66:113–119

Gerber PJ, Steinfeld H, Henderson B, Mottet A, Opio C, Dijkman J, Falcucci A, Tempio G (2013) Tackling climate change through livestock–A global assessment of emissions and mitigation opportunities. Food and Agriculture Organization of the United Nations (FAO), Rome

Haque N (2001) Environmental implication of methane production: diet and rumen ecology. Short course, CAS in Animal Nutrition, IVRI, Izatnagar

Hegarty RS (1999) Mechanisms for competitively reducing ruminal methanogenesis. Aust J Agric Res 50:1299–1305

Ivan M, Koenig KM, Teferedegne B, Newbold CJ, Entz T, Rode LM, Ibrahim M (2004) Effects of the dietary *Enterolobium cyclocarpum* foliage on the population dynamics of rumen ciliate protozoa in sheep. Small Rumin Res 52(1–2):81–91

Jahromi FM, Liang JB, Ho YW, Mohamad R, Goh YM, Shokryazdan P (2013) Lovastatin in *Aspergillus terreus*: fermented rice straw extracts interferes with methane production and gene expression in *Methanobrevibacter smithii*. Biomed Res Int

Johnson KA, Johnson DE (1995) Methane emissions from cattle. J Anim Sci 73:2483–2492

Kamra DN, Agarwal N, Yadav MP (2004) Methanogenesis in the rumen and the Greenhouse effect on the environment. Livest Int 8(9):2–8

Kamra DN, Agarwal N, Chaudhary LC (2005) Inhibition of ruminal methanogenesis by tropical plants containing secondary plant compounds. In: Soliva CR, Takahash J, Kreuzer M (eds) Proceeding of the 2nd international conference of greenhouse gases and animal agriculture. ETH Zurich, Zurich, pp 102–111

Kannan A, Garg MR (2009) Effect of ration balancing on methane emission reduction in lactating animals under field conditions. Indian J Dairy Sci 62:292–296

Kannan A, Garg MR, Mahesh Kumar BV (2011) Effect of ration balancing on milk production, microbial protein synthesis and methane emission in crossbred cows under field conditions in Chittoor district of Andhra Pradesh. Indian J Anim Nutr 28:117–132

Kataria RP (2015) Use of feed additives for reducing green house gas emissions from dairy farms. Microbiol Res 6:6120

Kobayashi Y, Wakita M, Hoshino S (1992) Effects of ionophore salinomycin on nitrogen and long-chain fatty acid profiles of digest in the rumen and the duodenum of sheep. Anim Feed Sci Technol 36:67–76

Kolver ES, Aspin PW, Jarvis GN, Elborough KM, Roche JR (2004) Fumarate reduces methane production from pasture fermented in continuous culture. Proc N Z Soc Anim Prod 64:155–159

Lee SS, Hsu JT, Mantovani HC, Russell JB (2002) The effect of bovicin HC5, a bacteriocin from *Streptococcus bovis* HC5, on ruminal methane production in vitro. FEMS Microbiol Lett 217:51–55

Lee C, Hristov AN, Dell CJ, Feyereisen GW, Kaye J, Beegle D (2012) Effect of dietary protein concentration on ammonia and greenhouse gas emissions from dairy manure. J Dairy Sci 95:1930–1941

Lipper L, Thornton P, Campbell B et al (2014) Climate-smart agriculture for food security. Nat Clim Chang 4:1068–1072. https://doi.org/10.1038/nclimate2437

Machado L, Magnusson M, Paul NA, De Nys R, Tomkins N (2014) Effects of marine and freshwater macroalgae on *in vitro* total gas and methane production. PLoS One 9:e85289. https://doi.org/10.1371/journal.pone.0085289

Malik PK, Bhatta R, Soren NM, Sejian V, Mech A, Prasad KS, Prasad CS (2015) Feed based approaches in enteric Methane Amelioration. livestock production and climate change. CAB International, Wallingford, pp 336–359

Malik PK, Kolte AP, Dhali A, Seijan V, Thirumalisamy G, Gupta R, Bhatta R (2016) GHG emissions from livestock challenges and ameliorative measures to counter adversity. Intech Open. https://doi.org/10.5772/6488J

Martin SA (1998) Manipulation of ruminal fermentation with organic acids: a review. J Anim Sci 76:3123–3132

McAllister A, Newbold CJ (2008) Redirecting rumen fermentation to reduce methanogenesis. Aust J Exp Agric 48:7–13

McGinn SM, Beauchemin KA, Coates T, Colombatto D (2004) Methane emissions from beef cattle: effects of monensin, sunflower oil, enzymes, yeast, and fumaric acid. J Anim Sci 82:3346–3356

McIntosh FM, Williams P, Losa R, Wallace RJ, Beever DA, Newbold CJ (2003) Effects of essential oils on ruminal microorganisms and their protein metabolism. Appl Environ Microbiol 69:5011–5014

Mehra UR, Khan MY, Lal M, Hasan QZ, Das A, Har R, Verma AK, Dass RS, Singh P (2006) Effect of sources of supplementary protein on intake, digestion and efficiency of energy utilization in buffaloes fed wheat straw based diets. Asian-Aust J Anim Sci 19(5):638–644

Merino P, Ramirez-Fanlo E, Arriaga H, Del Hierro O, Artetxe A (2011) Regional inventory of methane and nitrous oxide emission from ruminant livestock in the Basque country. Anim Feed Sci Technol 166–167:628–640

Morgavi DP, Forano E, Martin C, Newbold CJ (2010) Microbial ecosystem and methanogenesis in ruminants. Animal 4:1024–1036

Morgavi DP, Martin C, Boudra H (2013) Fungal Secondary metabolites from *Monascus* spp reduce rumen methane production *in vitro* and *in vivo*. J Anim Sci 91:848–860

Mutsvangwa T, Edward IE, Topp JH, Peterson GF (1992) Anim Prod 55:35–40

Mwenya B, Santanso B, Sar C, Gano Y, Kobayashi T (2004) Effects of including beta 1-4 galacto oligosacharides, lactic acid bacteria or yeast culture on methanogenesis as well as energy and nitrogen metabolism in sheep. Anim Feed Sci Technol 118:19–20

Newbold CJ, Lassalas B, Jouany JP (1995) The importance of methanogens associated with ciliate protozoa in ruminal methane production *in vitro*. Lett Appl Microbiol 21:230–234

Newbold CJ, López S, Nelson N, Ouda JO, Wallace RJ (2005) Propionate precursors and other metabolic intermediates as possible alternative electron acceptors to methanogenesis in ruminal fermentation *in vitro*. Br J Nutr 94(1):27–35

Patra AK (2014) Trends and projected estimates of GHG emissions from Indian livestock in comparison with GHG emissions from world and developing countries. Asian Australas J Anim Sci 27(4):592–599

Patra AK, Kamra D, Agarwal N (2006) Effect of plant extract on *in vitro* methanogenesis, enzyme activities and fermentation of feed in rumen liquor of buffalo. Anim Feed Sci Technol 128:276–291

Pinares-Patiño CS, D'Hour P, Jovana JP, Martin C (2007) Effect of stocking rate on methane and carbon dioxide emissions from grazing cattle. Agric Ecosyst Environ 121:30–46

Priano ME, Fuse VS, Ger JI, Berkovic AM, Williams KE (2014) Strong differences in the CH_4 emission from feces of grazing steers submitted to different feeding schedules. Anim Feed Sci Technol 194:145–150

Ribeiro Pereira LG, Machado FS, Campos MM, Guimaraes Júnior R, Tomich TR, Reis LG, Coombs C (2015) Enteric methane mitigation strategies in ruminants: a review. Revista Colombiana de Ciencias Pecuarias *28*(2):124–143

Robertson LJ, Waghorn GC (2002) Dairy industry perspectives on methane emissions and production from cattle fed pasture or total mixed rations in New Zealand. Proc N Z Soc Anim Prod 62:213–218

Saggar S, Bolan NS, Bhandral R, Hedley CB, Luo J (2004) A review of emissions of methane, ammonia, and nitrous oxide from animal excreta deposition and farm effluent application in grazed pastures. N Z J Agric Res 47(4):513–544

Seo JK, Kim SW, Kim MH, Upadhaya SD, Kam DK (2010) Direct fed microbials for ruminant animals. Asian-Australas J Anim Sci 23(12):1657–1667

Sliwinski BJ, Soliva CR, Machmuller A, Kreuzer M (2002) Efficacy of plant extracts rich in secondary constituents to modify rumen fermentation. Anim Feed Sci Technol 101:101–114

Smith P, Bustamante M, Ahammad H et al (2014) Agriculture, forestry and other land use (AFOLU). In: Edenhofer O, Pichs-Madruga R, Sokona Y, Farahani E, Kadner S, Seyboth K, Adler A, Baum I, Brunner S, Eickemeier P, Kriemann B, Savolainen J, Schlömer S, von Stechow C, Zwickel T, Minx JC (eds) Climate change 2014: Mitigation of climate change. Contribution of working group III to the fifth assessment report of the intergovernmental panel on climate change. Cambridge University Press, Cambridge/New York

Tarawali S, Herrero M, Descheemaeker K, Grings E, Blümmel M (2011) Pathways for sustainable development of mixed crop livestock systems: taking a livestock and pro-poor approach. Livest Sci 139:11–21

Van Nevel CJ, Demeyer DI (1996) Control of rumen methanogenesis. Environ Monit Assess 42:73–97

Venkatramanan V, Shah S (2019) Climate smart agriculture technologies for environmental management: the intersection of sustainability, resilience, wellbeing and development. In: Shah S et al (eds) Sustainable green technologies for environmental management. Springer Nature Singapore Pte Ltd., Singapore, pp 29–51. https://doi.org/10.1007/978-981-13-2772-8_2

Waghorn GC (2007) Beneficial and detrimental effects of dietary condensed tannins for sustainable sheep and goat production: progress and challenges. Anim Feed Sci Technol 147:116–139

Wallace RJ, McEwan NR, McIntosh FM, Teferedegne B, Newbold CJ (2002) Natural products as manipulators of rumen fermentation. Asian-Australas J Anim Sci 10:1458–1468

Wallace RJ, Wood TA, Rowe A, Price J, Yanez DR (2006) Encapsulated fumaric acid as a means of decreasing ruminal methane emissions. Int Congr Ser 1293:148–151

Wedlock DN, Janssen PH, Leahy SC, Shu D, Buddle BM (2013) Progress in the development of vaccines against rumen methanogens. Animal 7(S2):244–252

Wolin MJ (1974) Metabolic interactions among intestinal microorganisms. Am J Clin Nutr 27:1320–1328

Timber-Based Mixed Farming/ Agroforestry Benefits: A Case Study of Smallholder Farmers in Limpopo Province, South Africa

9

Phokele Maponya, Sonja L. Venter, Christiaan Philippus Du Plooy, Gerhard R. Backeberg, Sylvester Mpandeli, and Edward Nesamvuni

Abstract

Agroforestry is considered as "a sustainable land use system that includes the use of woody perennial, agricultural crops and animals in combination to achieve beneficial ecological and economical interactions for food, fiber and livestock production". However, limited understanding, incorrect information, insufficient awareness and a negative mindset could hinder the benefits of this practice. The survey was conducted in Limpopo Province by the Agricultural Research Council, University of Venda and Water Research Commission (WRC). The research is fully funded by the WRC. The aim of the survey was to identify and describe the farmer's benefits from timber-based mixed farming/ agroforestry cultivation in Limpopo Province. A total of 65 smallholder farmers participated in the study and were spread in districts as follows: Vhembe (40), Capricorn (21) and Mopani (4). Sixty-five potential smallholder agroforestry farmers were selected through a "purposive sampling technique" from the list of farmers' provided by the Department of Agriculture, Forestry and Fisheries (DAFF) and Forestry South Africa Limpopo. Quantitative and qualitative designs were adopted along with the use of questionnaire, stakeholder's discussion and field observations. Data was coded, captured and analysed using SPSS. The results indicated that some farmers in Limpopo Province were generating income through renting of farms for grazing and selling trees to the communities to build

P. Maponya (✉) · S. L. Venter · C. P. Du Plooy
Agricultural Research Council – Vegetable and Ornamental Plants, Pretoria, South Africa
e-mail: maponyap@arc.agric.za

G. R. Backeberg · S. Mpandeli
Water Research Commission, Pretoria, South Africa

E. Nesamvuni
University of Venda, Thohoyandou, South Africa

© Springer Nature Singapore Pte Ltd. 2020

V. Venkatramanan et al. (eds.), *Global Climate Change and Environmental Policy*,
https://doi.org/10.1007/978-981-13-9570-3_9

shelter, kraals, medicinal purposes, fuelwood, etc. Those farmers with access to water were able to grow crops and sell their produce at local communities, local municipality and international market. The majority of farmers also indicated that they were also benefiting from nitrogen fixation, increased crop production, economic gain, soil conservation and improved soil quality and sequestration of atmospheric carbon as a result of timber-based mixed farming/agroforestry practice. The identified farmers' benefits were in line with some of the researchers' field observations. It is thus recommended that stakeholders should take note of the benefits identified by farmers in an attempt to increase agroforestry farmers' participation.

Keywords
Agroforestry · Mixed farming · Limpopo Province

9.1 Introduction

Results from several studies have indicated that agroforestry practices are perceived in different ways. According to Lundgren and Raintree (1982), agroforestry is viewed as the set of land-use practices, which involves the combination of trees, agricultural crops and/or animals on the same land management unit. Nair (1993) emphasised that although cultivating trees in combination with crops and livestock is considered an ancient practice, factors such as the deteriorating economic situation in many parts of the developing world, increased tropical deforestation, incorrect agricultural practices, degradation and scarcity of land because of population pressures and growing interest in farming systems, intercropping and the environment have contributed to a rising interest in agroforestry since the 1970s. Based on the above-mentioned factors, Mercer and Miller (1998) further acknowledged that most research on agroforestry has been conducted from the biophysical perspective, but socio-economic aspects in relation to perception of farmers should be given more attention.

Combe (1982) classified agroforestry systems into three broad groups, namely agrosilvicultural (mixing trees and crops), silvopastoralism (mixing trees, pastures and animals) and agrosilvopastoralism (mixing trees with crops and animals). According to Rethman et al. (2007) these groups can further be subdivided as either simultaneous (where trees and crops are grown simultaneously) or sequential (where trees and crops are grown separately, temporally, over a number of seasons, as with improved fallows). According to DAFF (2017), agroforestry is viewed as "a land use system that includes the use of woody perennial and agricultural crops and animals in combination to achieve beneficial ecological and economical interactions for food, fiber and livestock production". Sustainable management of agroforestry system contributes immensely to the ecosystem health, and the agroforestry system greatly empowers and improves the livelihood of farmers and other stakeholders through revenue generation (DAFF 2017).

According to Hassan (2013), South Africa is a rain-starved country having semi-arid climatic conditions and is often vulnerable to water stress. As regards the Limpopo province, the mean precipitation for the province is about 600 mm. The threshold for rainfed agriculture is placed at 250 mm per annum (LDA 2010). Limpopo Province is characterised by extreme climatic events, high variability and change. From a forestry perspective, plantation forests in South Africa consume only 3% of total water resources available in the country, and so the rainfall requirement for sustaining plantation forests is more than 750 mm per year. In the previous 18 months, South Africa has experienced one of the worst droughts in history where some provinces were declared disaster areas, for example, Western, Eastern and North West Provinces. Poolman (2017) predicted an increased likelihood for the development of El Niño during the next summer, which is often associated with drought and water scarcity, a phenomenon that has been widely experienced in South Africa over the past 2 years.

It is against this background that a comprehensive survey was conducted in Limpopo Province to identify the following factors: (a) social factors (agroforestry potential benefits, perceptions of agroforestry, knowledge and skills in agroforestry management practices), (b) land-related factors (shortage of land, operations within communal areas, security on land tenure), (c) technical factors (management and integrating crops into forest systems, the ecology of the system, the need for common understanding of agroforestry, quality materials availability), (d) economic factors (return on investments, labour requirements, etc.) and (e) policy-related factors (agriculture and forestry production systems, coordination between sectors and regulation of forestry aspects, etc.). The potential benefits from properly designed and managed agroforestry practices include nitrogen fixation, increased crop production and economic gain, soil conservation and improved soil quality, sequestration of atmospheric carbon, etc.

9.2 Objectives

The aim of the survey was to identify and describe the farmer's benefits from timber-based mixed farming/agroforestry cultivation in Limpopo Province, South Africa. The following objectives were established:

- To describe the situational analysis of timber-based mixed farming/agroforestry in Limpopo Province
- To describe the socio-economic characteristics of the farmers in Limpopo Province

9.3 Methodology

All research done so far with partners is focused on achieving or working towards a participatory research approach since the researcher, collaborators, extension officers, farmers and funders were actively involved in all phases to achieve eight deliverables. According to Backeberg and Sanewe (2010), the method of participatory action research is most appropriate since people, especially farmers, benefit while the research is ongoing. The participatory action approach was also recommended by various researchers who emphasised that the participatory action approach is a good alternative to the traditional "transfer of technology" or "top – down approach" to agricultural research and extension. The research used both quantitative and qualitative methods. A detailed questionnaire written in English was developed as a quantitative data collection method and data was collected from 04th to 15th September 2017 in Vhembe, Capricorn and Mopani districts. The questionnaire used both open- and closed-ended questions. Qualitative data collection method included focus group discussions and field observations. The two methods were also used to identify the smallholder farmers and agroforestry systems/farms in Limpopo Province. A purposive sampling technique was used to select 65 potential productive agroforestry systems/farms from the list provided by the DAFF and Forestry South Africa Limpopo. As indicated in Figs. 9.18, 9.19 and 9.20, Vhembe district is dominant as a location for smallholder farmers with interest in agroforestry systems establishment. The same trend is observed at a local municipality level where Thulamela local municipality in Vhembe district had the most agroforestry systems followed by Molemole Local Municipality in Capricorn District. Data collected was analysed quantitatively using the Statistical Package for Social Sciences (SPSS).

The average monthly rainfall was determined by adopting the approach given by Malherbe and Tackrah (2003). "Decadal (ten-day period) 1km ×1km surfaces were created from the rainfall data (1920–1999) that was downloaded from the AgroMet databank at the Agricultural Research Council -Institute for Soil, Climate and Water (ARC-ISCW) (South African Weather Service and ISCW weather stations) from stations with a recording period of 10 years or more" (Malherbe and Tackrah 2003). The surface was developed using topographic indices (altitude, slope, aspect, etc.) through regression analysis and spatial modelling. Similarly, average monthly temperature was determined by adopting the approach given by Malherbe and Tackrah (2003) using the temperature data (1920–1999). Soil types were determined using the method given by ARC-ISCW (2017). Digital land type information, viz. soil depth and the spatial component, were used (Table 9.1). Soil depth is recorded as a range for each soil entry. A weighted average was calculated for each land type unit (Land Type Survey Staff 1972–2006; Land types of South Africa: Digital map 1:250000 scale). Soil drainage was determined using the method given by ARC-ISCW (2017).

Table 9.1 Digital land type information – soil type and the spatial component

Class	Soil drainage	Qualifying soil forms	Percentage qualifying soil in land type
1	Poor	Ch, Rg, Wo, Ka, Kd, Es	> 40
2	Impeded	Ch, Rg, Wo, Ka, Kd, Es, Lo, Wa, Cf, La	> 40
3	Somewhat impeded	Ch, Rg, Wo, Ka, Kd, Es, Lo, Wa, Cf, La, Bo Fw, Vf, We, Av, Gc, Pn, Bv, Ss, Va, Sw	> 40
4	Other		

9.4 Results and Discussion

9.4.1 Situational Analysis

9.4.1.1 Rainfall

With respect to the mean precipitation for the month of September, in most parts of the Limpopo Province, the mean precipitation was remarkably low between 0 and 25 mm (Fig. 9.1). Nevertheless, the mean precipitation for the month of September in parts of Vhembe, Capricorn and Mopani districts was comparatively higher (20–50 mm and 51–75 mm) (Fig. 9.1). The most potential agroforestry systems are found in Vhembe, Mopani and some parts of Capricorn districts.

As shown in Fig. 9.2, the mean precipitation for the month of October was found to increase (75–100 mm; 101–125 mm and 126–145 mm) in the study areas (Vhembe, Mopani and some of Capricorn respectively) so much so that the Forestry South Africa and DAFF had identified these districts for establishment and expansion of agroforestry.

As shown in Figs. 9.3 and 9.4, the long-term rainfall for the months November and December in the study areas, viz. Vhembe, Mopani and some parts of Capricorn, was found to be 126–150 mm, 151–175 mm, 176–200 mm and 201–220 mm, respectively. The increased rainfall during the months November and December enable expansion of agroforestry in the study areas as the rainwater is no more a constraint. The results are also in line with Rethman et al. 2007 (WRC Report

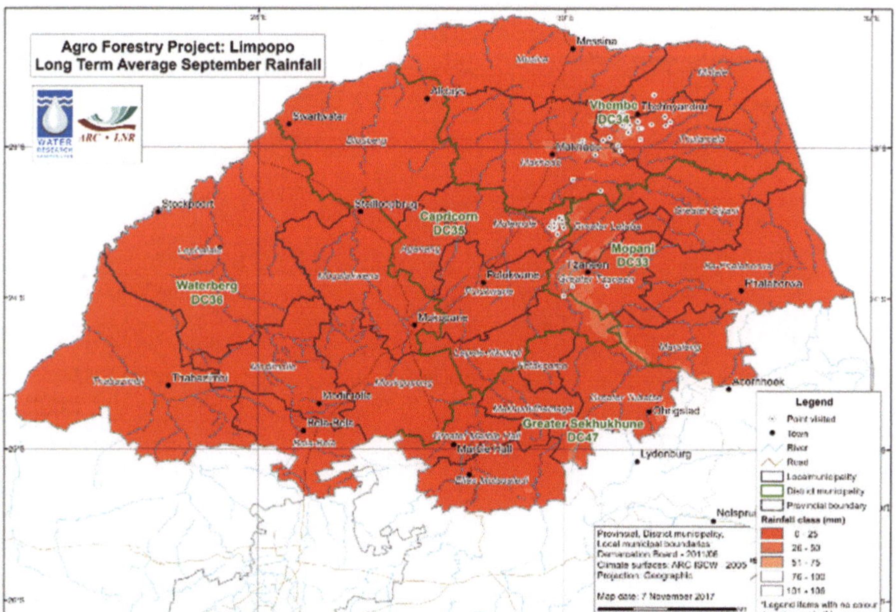

Fig. 9.1 Limpopo Province long-term average September rainfall. (ARC ISCW 2017)

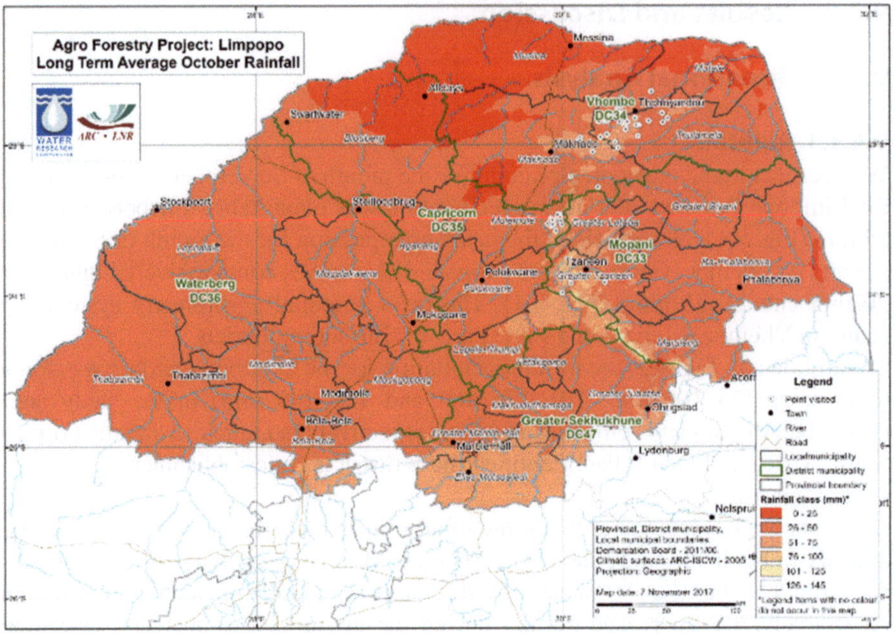

Fig. 9.2 Limpopo Province long-term average October rainfall. (ARC- ISCW 2017)

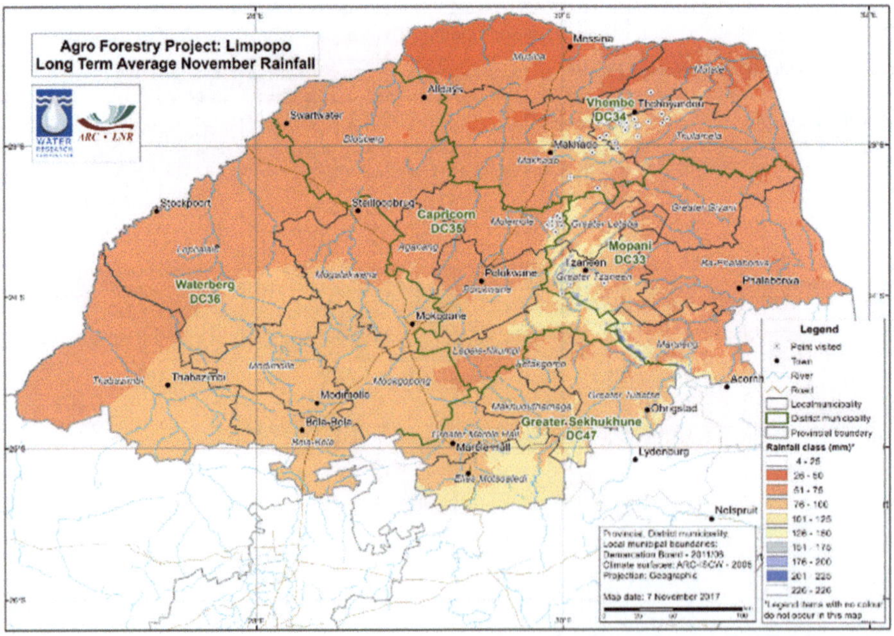

Fig. 9.3 Limpopo Province long-term average November rainfall. (ARC- ISCW 2017)

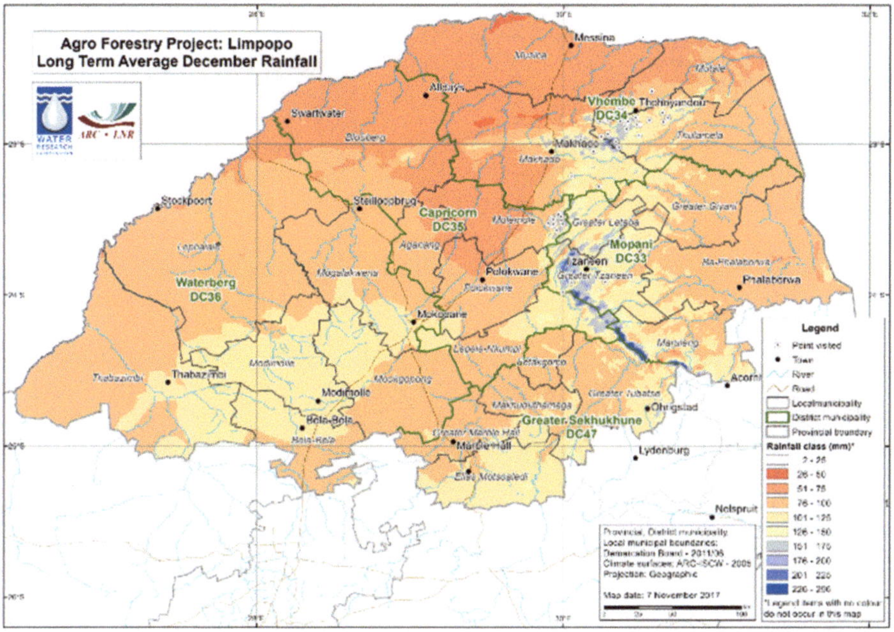

Fig. 9.4 Limpopo Province long-term average December rainfall. (ARC- ISCW 2017)

1047/1/07) who found that water resource remains a serious key problem in the establishment of agroforestry in some parts of Limpopo province. The researchers found that in Sikhakhane village, almost 97% of rural communities recognised water resource as a key problem, while in Chuene Maja village, 71% of rural communities recognised water resource as a key problem. The participatory action approach of the research project helped to identify the potential constraints of water resource in agroforestry in Limpopo province.

The annual rainfalls were estimated for 33rd, 50th and 66th percentiles for the future years, and it was found to fall between 601–700 mm and 1001–2216 mm (Figs. 9.5, 9.6 and 9.7). The estimated rainfall trend was found to be higher than the annual mean rainfall (600 mm) for the Limpopo province, and the inference that can be drawn from this trend is that rainwater is not a constraint from the perspective of establishment of agroforestry production system in the Vhembe, Mopani and some parts of Capricorn.

According to ARC-ISCW (2017), the Vhembe District Municipality covers part of the Limpopo as well as parts of Luvuvhu and Letaba water management area. With respect to Luvuvhu and Letaba, it occupies the Luvuvhu/Mutale sub-area and parts of Shingwedzi and Klein Letaba sub-areas. In the Limpopo water management area, it occupies the Nwanedi and Nzhelele sub-areas. Apart from the Luvuvhu/Mutale area, the new Nandoni Dam resulted in a temporary water surplus. As indicated in Figs. 9.8 and 9.9, the visited potential agroforestry systems fall near the water bodies, and this offers a good opportunity for establishment and expansion of

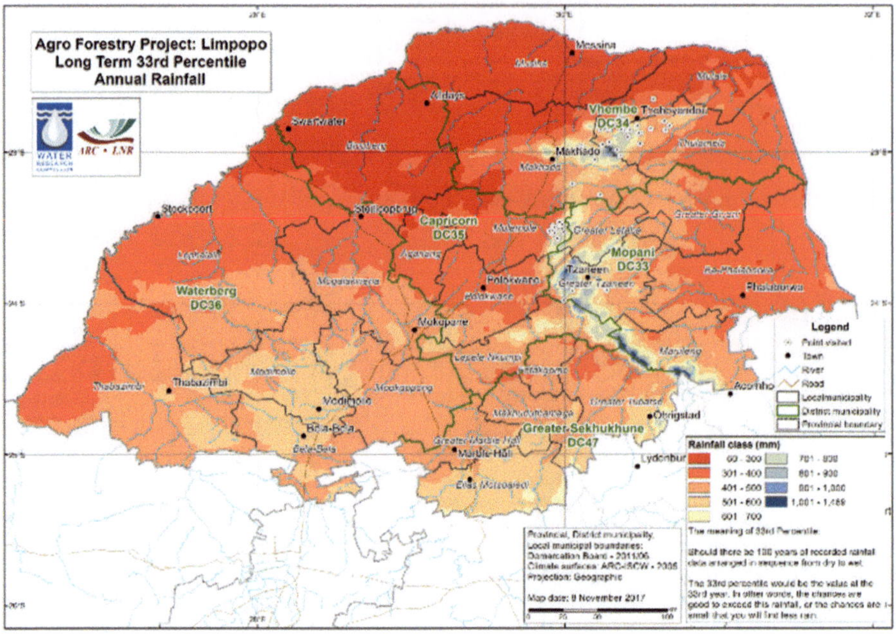

Fig. 9.5 Limpopo Province long-term 33rd percentile annual rainfall. (ARC- ISCW 2017)

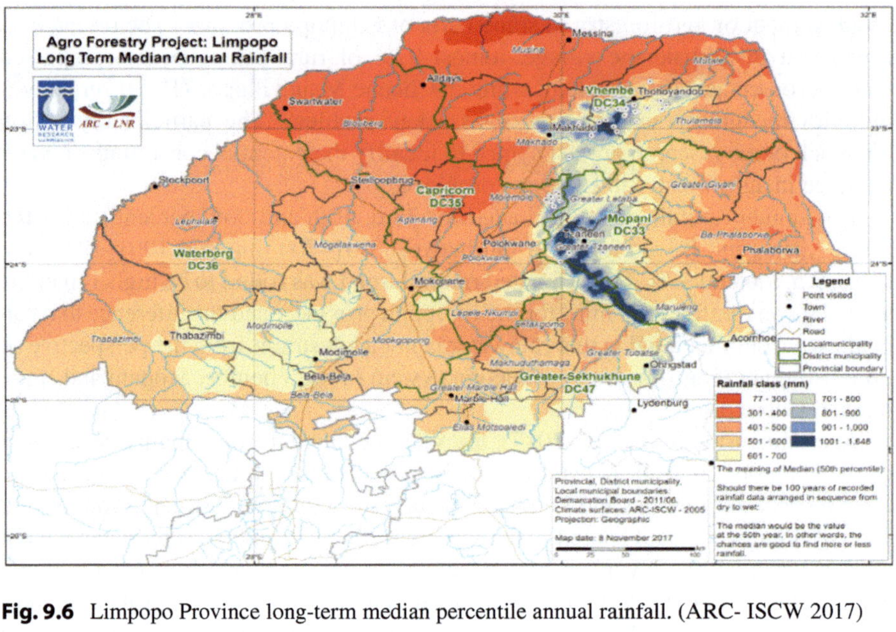

Fig. 9.6 Limpopo Province long-term median percentile annual rainfall. (ARC- ISCW 2017)

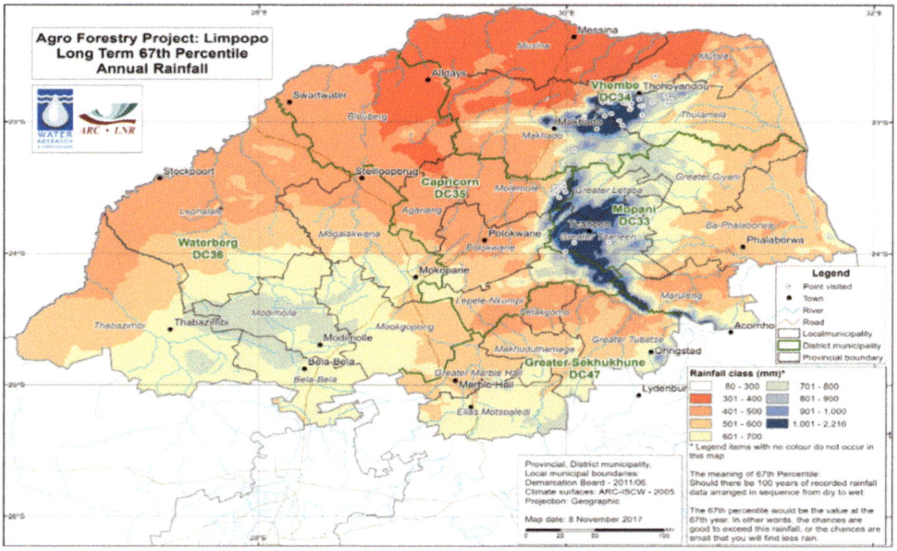

Fig. 9.7 Limpopo Province long-term 67th percentile annual rainfall. (ARC- ISCW 2017)

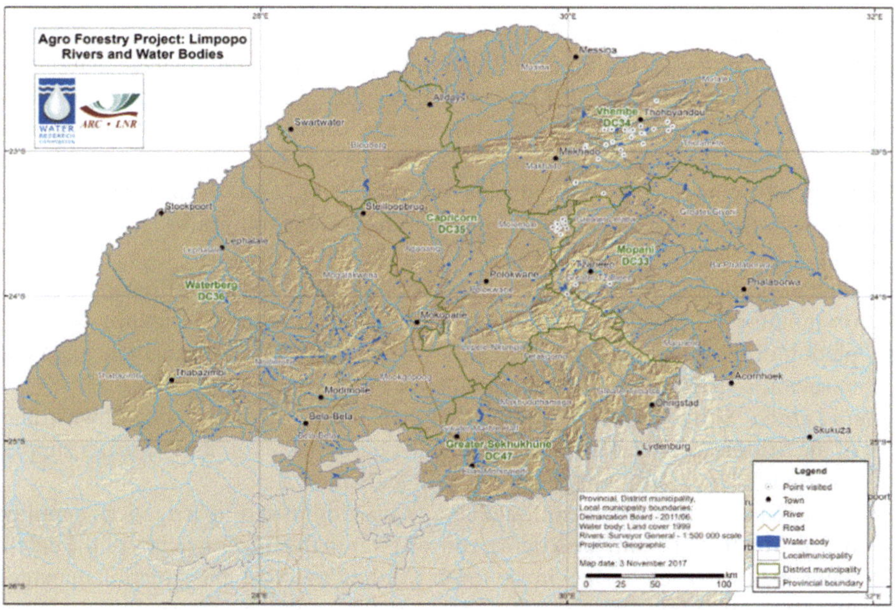

Fig. 9.8 Limpopo Province rivers and water bodies. (ARC-ISCW 2017)

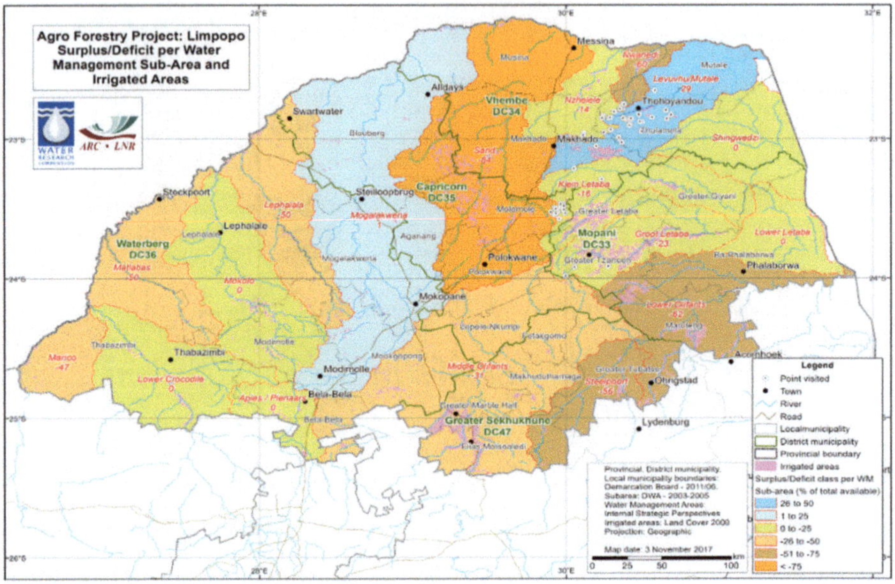

Fig. 9.9 Limpopo Province surplus/deficit. (ARC-ISCW 2017)

different agroforestry systems. The same trend is seen in Mopani district where the agroforestry systems fall within the Luvuvhu and Letaba water management areas. According to ARC-ISCW (2017), Luvuvhu/Mutale and Groot Letaba constitute important irrigation areas with high value crops. Apart from water from the new Nandoni Dam, the surface water resources are overextended, and water for irrigation is being augmented by groundwater. As indicated in Figs. 9.8 and 9.9, the water bodies with its surplus and deficit can be summarised as follows:

Luvuvhu/Mutale: 12400 ha irrigated; yield of Alabasini Dam insufficient; high but un-monitored groundwater use; short term surplus available following the completion of the Nandoni Dam; allocations being made for domestic water use and to revitalise irrigation schemes which have fallen into disuse

Klein Letaba: Surface water overextended; 5100 ha irrigated; irrigation downstream of the Middle Letaba Dam in disuse due to decreasing assurance of water supply; targeted for revitalisation, despite insufficient water supply

Nzhelele: Surface water overexploited; water use dominated by irrigation; supplied by the Mutshedzi Dam, farm dams, run-of-river in the upper reaches of the catchment and the Nzhelele Dam in the lower reaches; much of the irrigation managed by smallholder farmers

Nwanedi: Surface water overexploited; without major dams in the catchment, the surface water resource is limited; ample groundwater resources, although use of it is limited; substantial irrigation, much of it managed by smallholder farmers

Groot Letaba: Surface water over extended; 19,100 ha irrigated; groundwater supplementing irrigation supplies; irrigators upstream of Tzaneen Dam experience relatively high level of assurance; users downstream experience shortages; irrigation highly efficient and well managed; scope for further improvements limited

9.4.1.2 Soil

According to ARC-ISCW (2017), fairly large tracts of moderately deep to deep, well-drained loam or clay loam soils are found in the Tzaneen area, the alluvial valleys of the major rivers, a belt between Tzaneen and Phalaborwa and in areas between Phalaborwa and the Kruger National Park. Although the bulk of the lowveld area is dominated by shallow soils, sporadic occurrences of deeper soils occur. There is thus an overabundance of good soils in Mopani district. As indicated in Figs. 9.10 and 9.11, the potential agroforestry systems visited in Mopani district falls in well-drained soil with soil depth of 901–1200 mm. Moderately deep to deep, well-drained loamy or clay loam soils occur commonly in the escarpment and Soutpansberg areas as well as sporadically across the Vhembe district. Good soils are thus in overabundance. As seen in Figs. 9.10 and 9.11, the potential agroforestry systems falls in well-drained soils with soil depth of 601–900 mm and 901–1200 mm, respectively.

Moderately deep to deep, well-drained, sandy or loamy soils occur fairly widespread to the west and south of the Polokwane plateau, while good soils are rare on

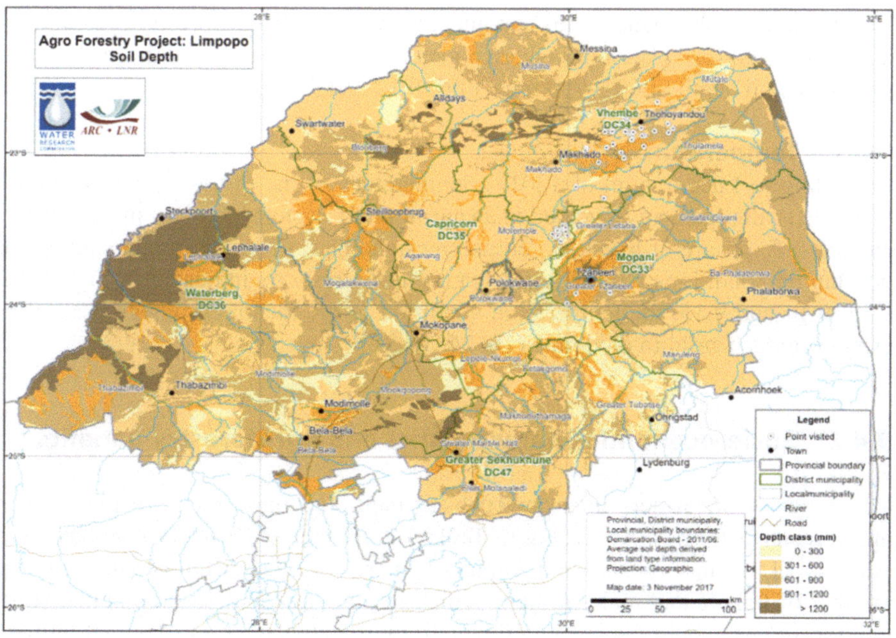

Fig. 9.10 Limpopo Province soil depth. (ARC- ISCW 2017)

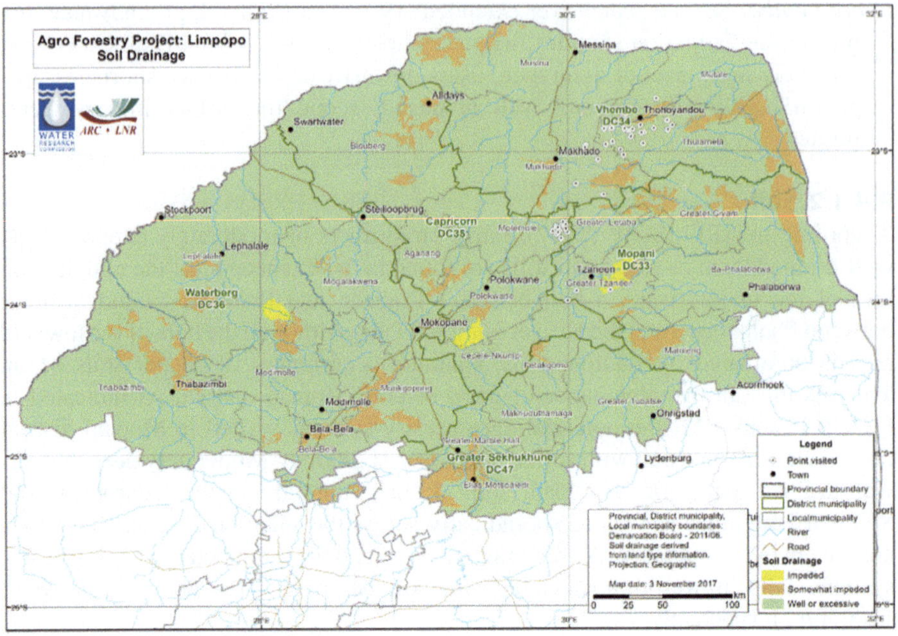

Fig. 9.11 Limpopo Province soil drainage. (ARC- ISCW 2017)

the Polokwane plateau itself. As is the case almost everywhere in the province, good soils are in over abundance. This statement is in line with agroforestry systems visited in some parts of Capricorn where potential systems were found in well-drained soils with soil depth of 601–900 mm (Figs. 9.10 and 9.11).

9.4.1.3 Temperature

According to Luedeling et al. (2013), temperature is one of the climate variables that affects all organisms involved in an agroforestry system, possibly in very different ways. Exposure of livestock, crops, trees and bees to strong, direct sunlight or hot, humid conditions will definitely cause heat stress. As indicated in Figs. 9.12, 9.13, 9.14 and 9.15, the temperature condition in the identified potential agroforestry systems is mild to moderate (18–24 °C).

9.4.2 Socio-economic Characteristics and Agroforestry Benefits

The majority of farmers interviewed were men, and these men are involved in agroforestry activities. According to Table 9.2, 51 men were interviewed as compared to 14 females. As shown in Table 9.2, there is a huge difference between the number of female and male farmers, which implied that any developmental strategy for the farmers in the areas will benefit more males than females. All farmers interviewed were black South African citizens. In terms of educational attainment

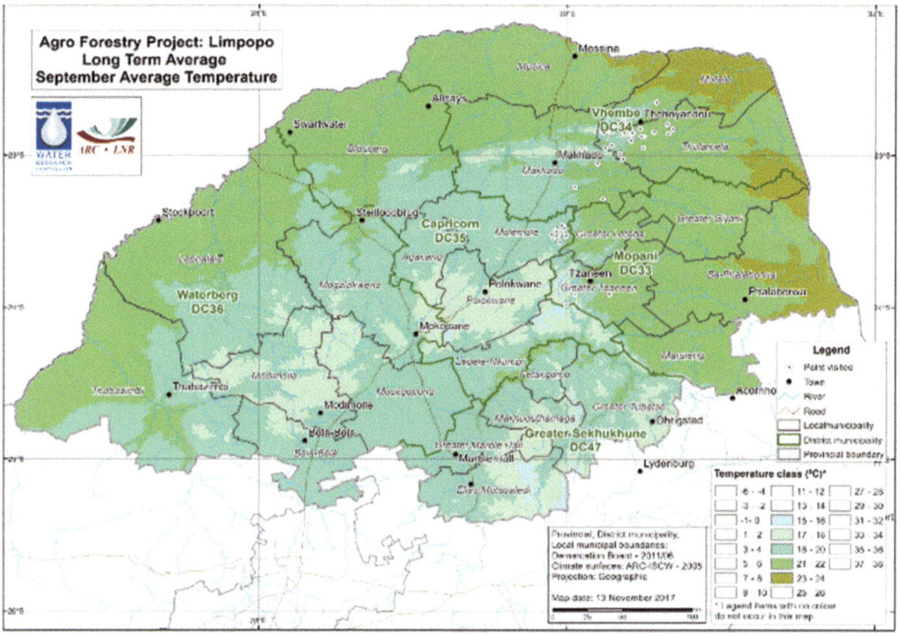

Fig. 9.12 Limpopo Province long-term average temperature in September. (ARC- ISCW 2017)

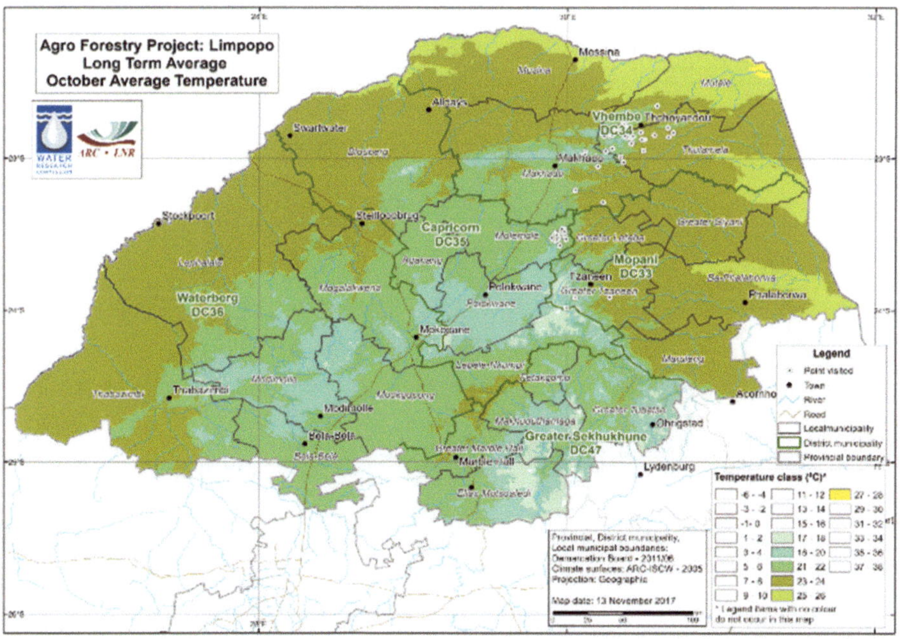

Fig. 9.13 Limpopo Province long-term average temperature in October. (ARC- ISCW 2017)

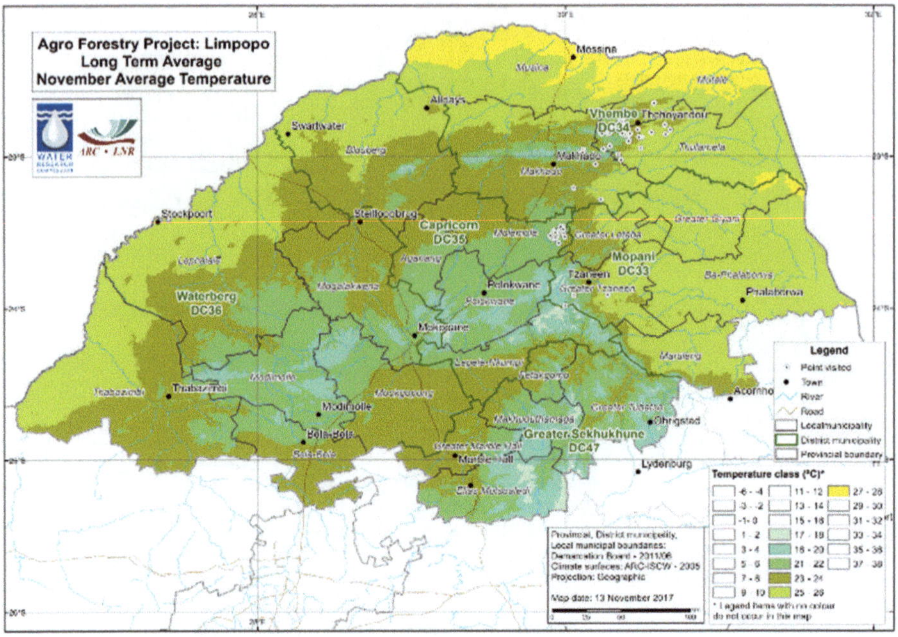

Fig. 9.14 Limpopo Province long-term average temperature in November. (ARC- ISCW 2017)

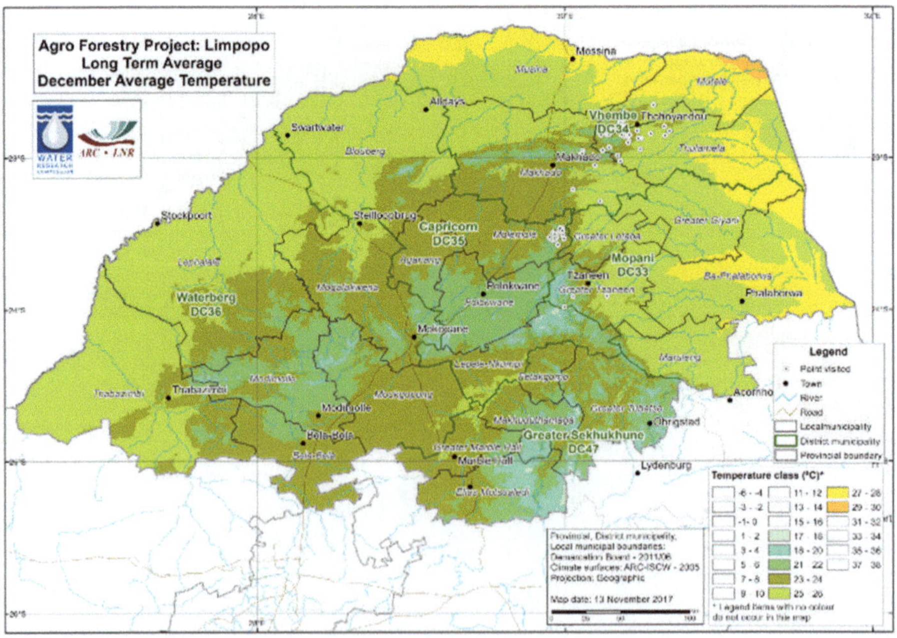

Fig. 9.15 Limpopo Province long-term average temperature in December. (ARC- ISCW 2017)

Table 9.2 Limpopo Province farmers' demographics

Variables	Respondents	% of Respondents
Gender		
Female	14	78.5
Male	51	21.5
Total	65	100
Level of education		
No education	9	13.8
Primary education	20	30.8
Secondary education	27	41.5
Tertiary education	9	13.8
Total	65	100
Training skills acquired		
Yes	24	40
No	41	60
Total	65	100
Agroforestry experience		
Less than 5 Years	21	32
6–10 Years	14	22
More than 10 Years	30	46
Total	65	100
Land acquisition		
Bought	9	14
Leased	11	17
Inherited	2	3
Government	14	22
Permission to occupy	23	35
Renting	6	9
Total	65	100

(Table 9.2), 9 farmers had no education, 20 of the farmers completed primary education, 27 secondary education and 9 tertiary education. The educational levels of the farmers were generally good. Almost all farmers attended secondary and tertiary education levels. These results indicated that educational level of the selected farmers is generally adequate to enable interpretation and understanding of basic farming activities. It is expected of farmers with tertiary education to at least interpret and understand different farming principles to make informed decisions on general farming operations and to be able to negotiable contracts better than beneficiaries with lower education levels. The research done previously shows that farmers who have basic education are far better compared to farmers that do not have education at all. The farmers can be able to make strategic or informed decisions based on their understanding of agroforestry setups and also the situations that they find themselves in (Maponya and Mpandeli 2013). The lower educational levels among the farmers implied that written information might be of minimal benefit to such farmers, and alternatives will have to be developed.

According to Maponya et al. (2016), training and education plays an important role in smallholder farmer development. Failure to address some of the training needs has led to constrained agricultural growth in some districts in South Africa (Maponya et al. 2014, 2015). As indicated in Table 9.2, only 24 farmers received training, whereas the majority of farmers did not receive any training. The results showed farmers' variation in terms of agroforestry experience acquired over time (Table 9.2). The majority of farmers (30) had more than 10 years' experience, while 21 and 14 farmers had less than 5 years and 6–10 years' experience, respectively. Results on land acquisition (Table 9.2) indicated that the majority of farmers got land through Permission to Occupy (PTO) (25) while others received land through the following: government (14), leased (11), renting (6) and inheritance (2). The role of traditional leaders must be applauded as the majority of farmers had PTO.

As indicated in Figs. 9.16, 9.17 and 9.18, the following potential agroforestry systems were identified and visited in Limpopo province: (1) agrosilvoculture, (2) silvipasture, (3) agrosilvipasture, (4) apiculture and (5) combination of agrosilvipasture and apiculture. The most popular systems identified were silvipasture (32) and agrosilvipasture (23). The systems were popular because farmers were generating good income by allowing community livestock to graze in their forests throughout the year at a cost. As indicated in Figs. 9.16, 9.17 and 9.18, Vhembe district is dominant as a location for agroforestry systems establishment. The same trend is observed at a local municipality level where Thulamela local municipality in Vhembe had the most agroforestry systems followed by Molemole Local Municipality in Capricorn.

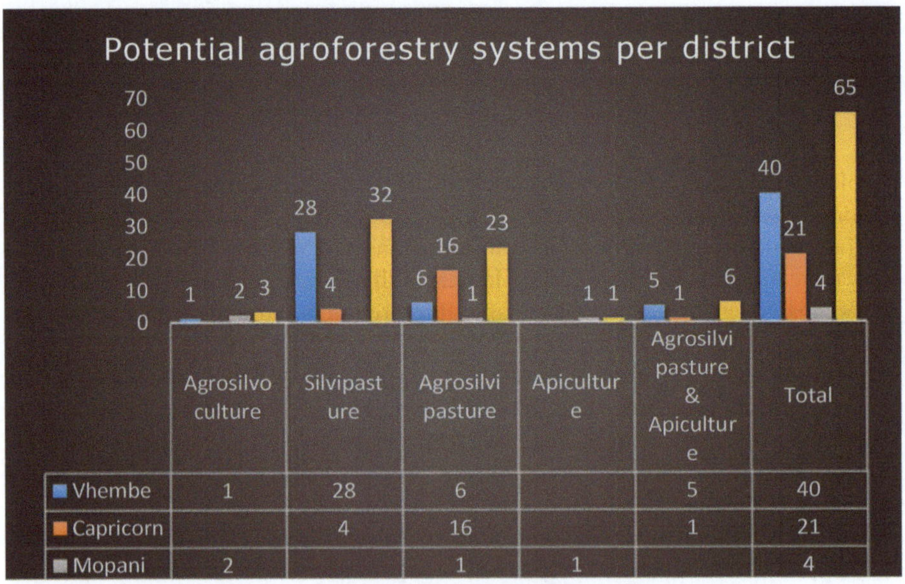

Fig. 9.16 Limpopo Province potential agroforestry systems per district

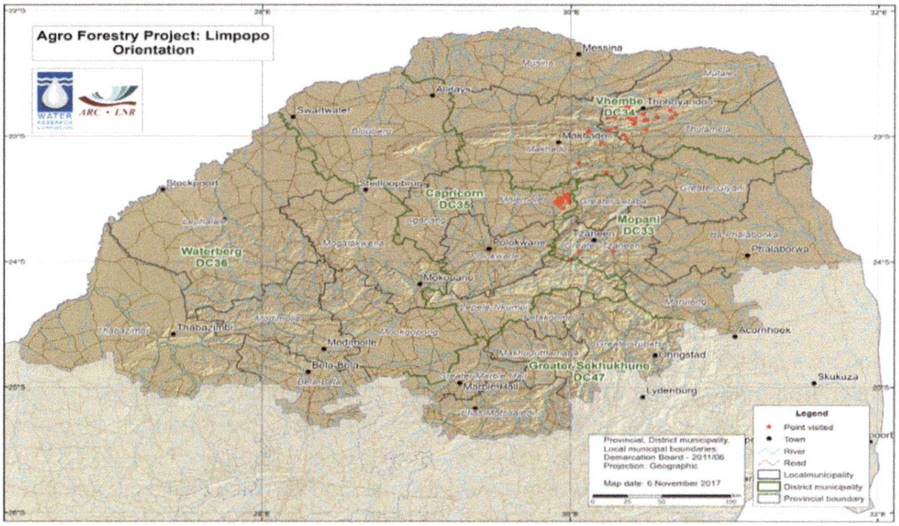

Fig. 9.17 Limpopo Province orientation. (ARC-ISCW 2017)

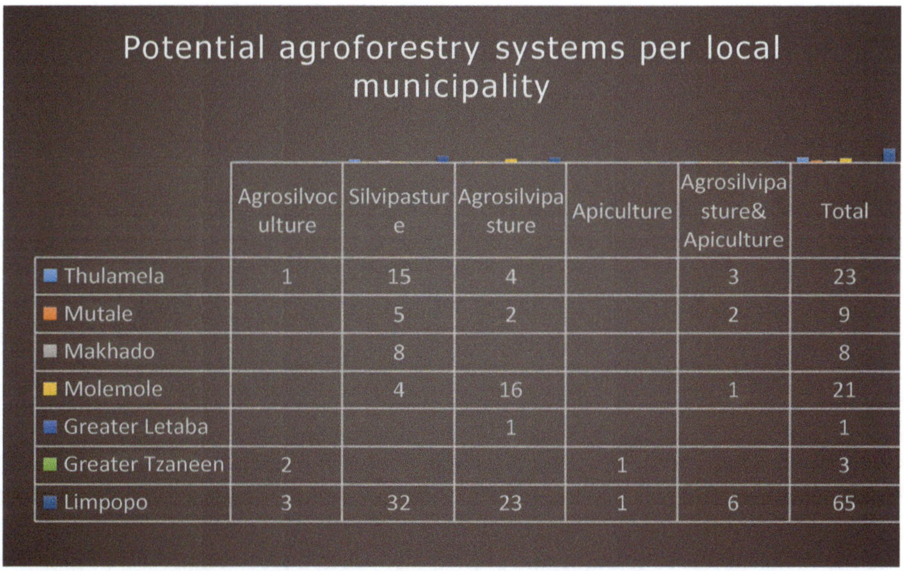

Potential agroforestry systems per local municipality

	Agrosilvoculture	Silvipasture	Agrosilvipasture	Apiculture	Agrosilvipasture& Apiculture	Total
Thulamela	1	15	4		3	23
Mutale		5	2		2	9
Makhado		8				8
Molemole		4	16		1	21
Greater Letaba			1			1
Greater Tzaneen	2			1		3
Limpopo	3	32	23	1	6	65

Fig. 9.18 Limpopo Province potential agroforestry systems per district

Table 9.3 shows the potential agroforestry systems, total land area and land used per enterprise. The results indicated that the total land available for agrosilvoculture system in the identified areas was 431 ha and the usage of land by other enterprises were as follows: (1) trees (230 ha) and crops (212 ha). The results also showed total

Table 9.3 Potential agroforestry systems (AF), total land and land use per enterprise

AF systems	Total land (ha)	Trees (ha)	Crops (ha)	Livestock (ha)	Bees (ha)
Agrosilvipasture	7069	5552	208	5703	–
Agrosilvoculture	431	220	212	–	–
Silvipasture	3528	3512	–	3528	–
Apiculture	100	80	–	–	100
Agrosilvipasture and apiculture	165	128	44	165	165
Total	11,293	9492	464	9396	265

land and land usage per agrosilvoculture system. The results showed a balanced integration of crop and tree enterprises in the agrosilvoculture system.

The total land available for the apiculture system in the identified areas was 100 ha, and the usage of land by other enterprises were as follows: (1) bees (100 ha) and trees (100 ha). The results also showed a total land and land usage per apiculture system. The results indicated that the total land available for agrosilvipasture and apiculture system in the identified areas was 165 ha and the usage of land by other enterprises were as follows: (1) trees (128 ha), crops (44 ha), livestock (165 ha) and bees (165 ha). The results showed a balanced integration of enterprises except for crops. The total land available for the silvipasture system in the identified areas was 3528 ha and the usage of land by other enterprises were as follows: (1) trees (3512 ha) and livestock (3528 ha). The results also showed a total land and land usage per silvipasture system. The results showed a very good integration of enterprises in the silvipasture system.

According to DAFF (2017), centres of excellence should be identified and established as lead research agents in agroforestry systems. These may include universities, agricultural colleges, forestry colleges, research stations and state research agencies (e.g. ARC). The department of agriculture, forestry and fisheries further suggested that a number of centres be established and provided with funding for agroforestry research. The geographic location of these centres should reflect the different agro-ecological zones that occur in South Africa, and the research focus should be on systems best suited to the region in which the centre is located (DAFF 2017). The state should provide seed funding for research and assist the centres with securing funding from other sources (e.g. Water Research Commission, SADC funding mechanisms, etc.) (DAFF 2017). The centres of excellence should conduct technical, social, environmental and economic elements of agroforestry, with a particular focus on shared learning and participatory action research (DAFF 2017). The statements from the Department of Agriculture, Forestry and Fisheries are true as indicated in Fig. 9.19, whereby majority of farmers (46) in different potential agroforestry systems in identified areas in Limpopo province had no access to research. It is against this background that the current research is being conducted to enable full agroforestry benefits in Limpopo Province.

Figure 9.20 indicated that 29 beneficiaries received extension services, while only 36 did not. Most of the farmers were receiving extension services through

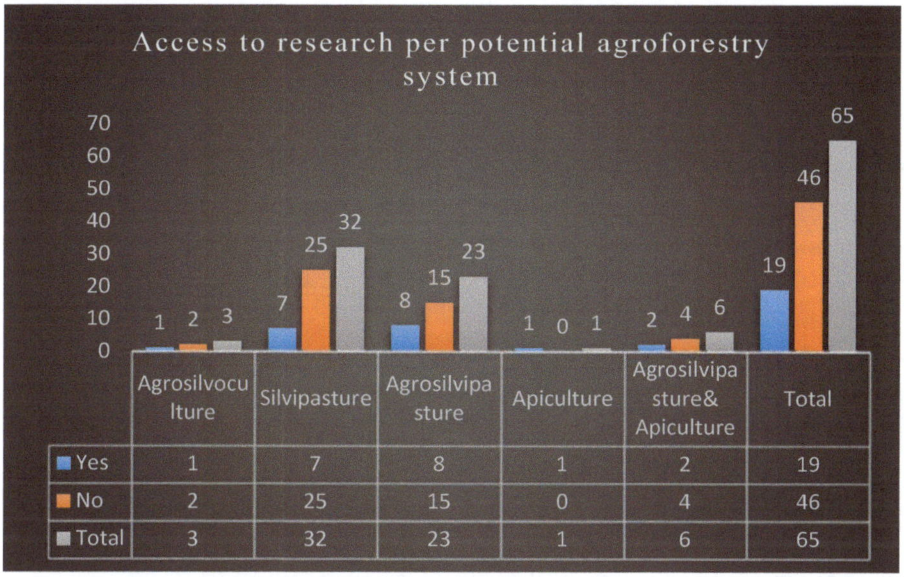

Fig. 9.19 Access to research per potential agroforestry system

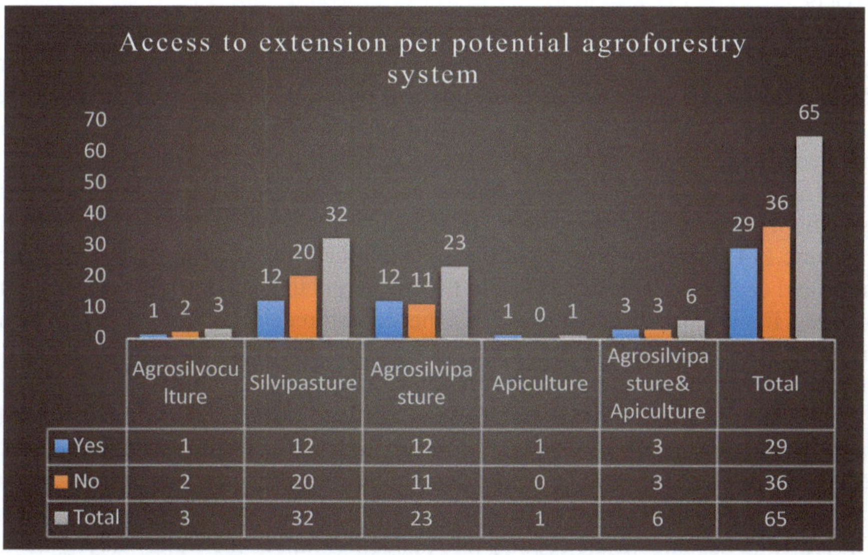

Fig. 9.20 Access to extension per potential agroforestry system

formal extension service, i.e., National, Provincial and Municipal Departments of Agriculture. This situation needs improvement especially for those who are not accessing extension service but encouraging as Mmbengwa (2009) and Maponya and Mpandeli (2013) emphasised that extension service has an important role in

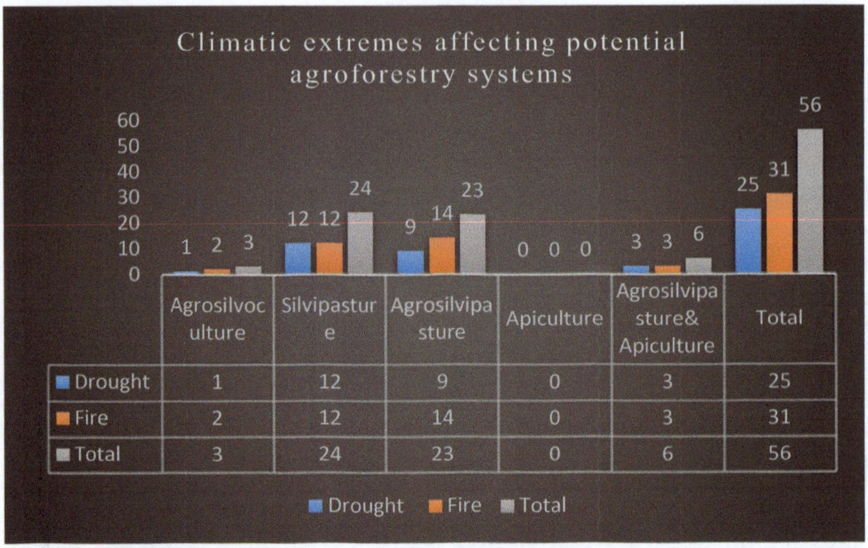

Fig. 9.21 Climatic extremes affecting potential agroforestry system

assisting farmers to acquire new technology, skills, innovation and production advice.

Climatic extremes have an adverse effect on food production and agroforestry systems as indicated in Fig. 9.21. The majority of potential agroforestry systems (56) in the identified areas has been affected by drought and fire, and this posed threats to the livelihoods of the majority of households in the province. The most affected systems were silvipasture (24) and agrosilvipature (23).

As indicated in Fig. 9.22, fire is one of the disasters that affected the agroforestry systems. The most fire happened in 2015 and affected silvipasture (12) and agrosilvipasture (14) systems the most. Farmers explained that the fire burned trees and crops and destroyed infrastructure and livestock, and the damage ranked in millions of rands. It must also be emphasised that the support received was from N1 firefighters in stopping the fire. No preparedness plan and fire belts were available to the affected farmers as mandated by the National Disaster Management Act 57 of 2002.

According to the South African Weather Service, SAWS (2016), the 2015/2016 drought has been one of the worst in the history of Southern Africa with South Africa receiving the lowest rainfall since 1904. Eight South African provinces have been or have had some of their areas declared as disaster areas, except for Gauteng Province. The situation has thrown the country into pandemonium as the water scarcity debate has taken centre stage with every sector looking for ways of conserving water. As indicated in Fig. 9.23, drought had affected most potential agroforestry systems in 2016 and had affected silvipasture (12) and agrosilvipasture (9) systems the most. These situations affected grazing and destroyed crops and plantations and resulted in low production in the affected areas. The farmers indicated that no/little support was received from the government, and they do not have any preparedness plan.

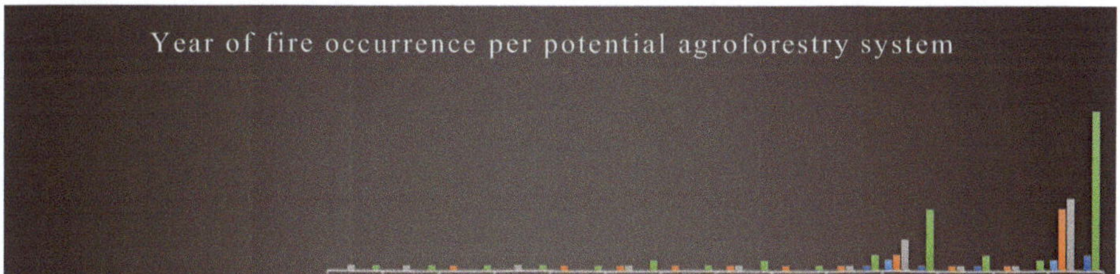

	2000	2001	2007	2008	2009	2010	2011	2012	2013	2014	2015	2016	2017	Total
Agrosilvoculture											2			2
Silvipasture			1		1	1	1	1	1	1	3	1	1	12
Agrosilvipasture	1	1		1		1		1		1	6	1	1	14
Apiculture	0	0	0	0	0	0	0	0	0	0	0	0	0	0
Agrosilvipasture& Apiculture										1	1	1		3
Total	1	1	1	1	1	2	1	2	1	3	12	3	2	31

Fig. 9.22 Year of fire occurrence per potential agroforestry system

	2001	2002	2007	2008	2009	2010	2011	2012	2013	2014	2015	2016	2017	Total
Agrosilvoculture												1		1
Silvipasture		1					1		1		1	8		12
Agrosilvipasture												9		9
Apiculture	0		0	0	0	0	0	0	0	0	0	0	0	0
Agrosilvipasture& Apiculture												3		3
Total	0	1	0	0	0	0	1	0	1	0	1	21	0	25

Fig. 9.23 Year of drought occurrence per potential agroforestry system

As indicated in Fig. 9.24, different prices per ton were received per agroforestry systems in Limpopo Province. The highest price per ton was R900 per ton (17), while the others were as follows: R300 per ton (1), R400 per ton (3), R500 per ton (9), R600 per ton (4), R700 per ton (6), R750 per ton (1), R800 per ton (12) and R850 per ton (1). This result indicated that trees serve as an important source of income and communities were using some of the trees to build shelter, kraals, medicinal purposes, fuelwood, etc. It must also be emphasised that few farmers were reluctant to disclose their farm income as they thought that it will jeopardise their chances of receiving support from the government.

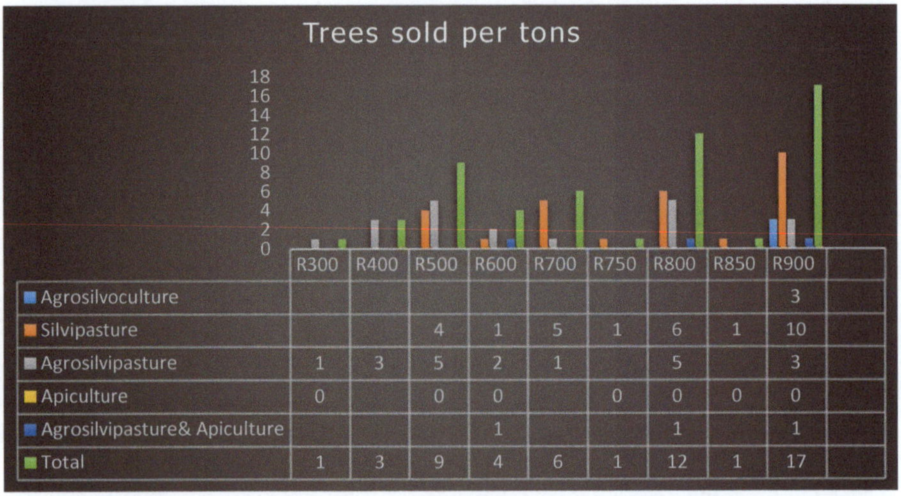

Fig. 9.24 Trees sold per tons

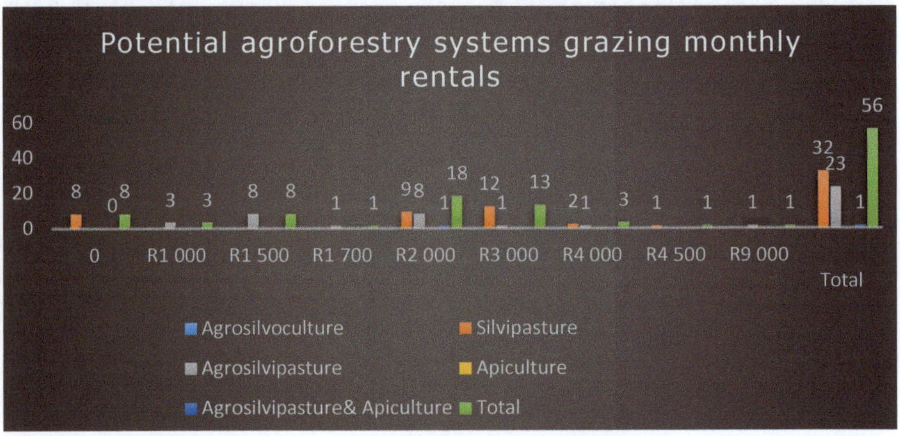

Fig. 9.25 Potential agroforestry systems grazing monthly rentals

As indicated in Fig. 9.25, it is very clear that silvipasture and agrosilvipasture remains the most popular systems in the identified areas in Limpopo Province. The reason is farmers were generating income through renting of farms for grazing. Quite a number of farmers were generating income ranging from R1000 to R9000 per month as indicated in Fig. 9.25.

Farmers used different marketing channels to market their produce for various reasons. For example, many smallholder farmers' opt for informal markets for reasons such as inability to satisfy the quality standards in the formal markets. On the other hand, most commercial farmers prefer formal markets for reasons such as secured market with better returns. Figure 9.26 indicated that almost all farmers

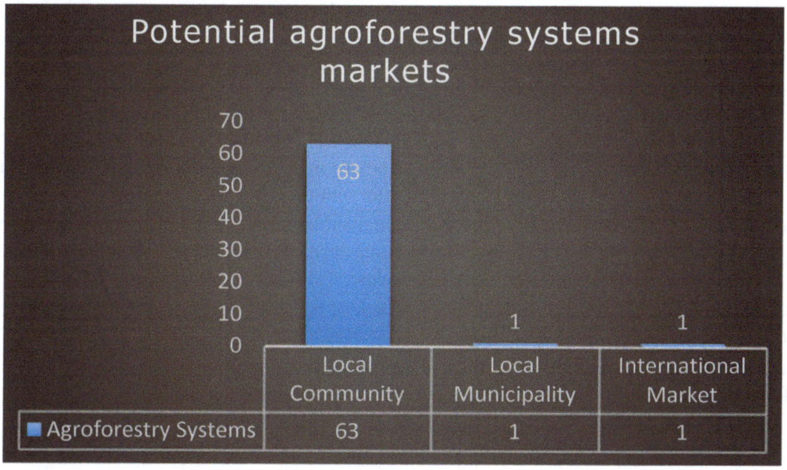

Fig. 9.26 Potential agroforestry systems' markets

were selling their produce at local communities (63), local (1) and international market (1). Due to poor record keeping of farmers, it was difficult to quantify their market information.

As indicated in Table 9.4, strength, weaknesses, opportunities and threats (SWOT) analysis was conducted to provide an accurate and relevant overview of what should be considered in taking agroforestry systems forward in the identified areas. Some of the SWOT analyses are in line with SWOT analysis conducted by the Department of Agriculture, Forestry and Fisheries.

9.4.3 Correlations and Univariate Analysis

Correlation is a bivariate analysis that measures the strengths of association between two variables and the direction of the relationship. In terms of the strength of relationship, the value of the correlation coefficient varies between +1 and − 1. When the value of the correlation coefficient lies around ±1, then it is said to be a perfect degree of association between the two variables. As the correlation coefficient value goes towards 0, the relationship between the two variables will be weaker. The direction of the relationship is simply the + (indicating a positive relationship between the variables) or – (indicating a negative relationship between the variables) sign of the correlation. Usually, in statistics, four types of correlations are measured: Pearson correlation, Kendall rank correlation, Spearman correlation and the point-biserial correlation. In this example, Pearson and Spearman correlations were used, and the following variables were found to be significant: agroforestry systems, age and farming experience.

As indicated in Table 9.5 and Fig. 9.27, there is a positive correlation between agroforestry systems and age variables. The significant level is at 10% for Pearson

Table 9.4 A strength, weaknesses, opportunities and threats (SWOT) analysis

Strengths	**Weaknesses**
Increased agriculture/forestry production	Management of projects is remote – many people on the ground needed and increased management costs
Availability of land	Lack of national coordination of agroforestry interventions
Diversification of income and risk reduction	Delayed benefits from agroforestry activities (long-term investment of 5–7 years) especially trees
Climate change adaptation and mitigation benefits	Lack of focussed and documented research
Monetary benefits – increased income from agroforestry adoption	Limited practical knowledge and applied research to address issues that affect agroforestry
	Lack of on-the-ground technical skills
	Skills shortage – management and administration of on-the-ground operations
	Lack of monitoring and evaluation of agroforestry efforts
Opportunities	**Threats**
Global carbon market (and other environmental service markets)	No formal government agroforestry policy/programme to support agroforestry
Increased land value – preservation of land productivity and restoration of degraded land	Climate change and climate variability
Potential linkages with conservation and climate smart agriculture	Lack of markets or incentives for ecosystem services or non-carbon benefits
Co-benefits (socio-economic) such as honey production, tourism, increased wildlife viewing	Potential risks of fire and drought
Markets for diverse goods	

Table 9.5 Correlation between potential agroforestry systems and age

		Value	Asymp. Std. Error	Approx. T	Approx. Sig.
Interval by interval	Pearson's R	0.098*	0.133	0.759	0.451
Ordinal by ordinal	Spearman correlation	0.053**	0.134	0.414	0.680
N of valid cases		65			

*10% and **5% significant levels

R and 5% for Spearman correlations. This is a clear indication that any age group can participate fully in any agroforestry system in the identified areas.

As indicated in Table 9.6 and Fig. 9.28, there is a positive correlation between agroforestry systems and farming experience variables. The significant level is at 10% and 5% for Pearson R and Spearman correlations, respectively. This is an

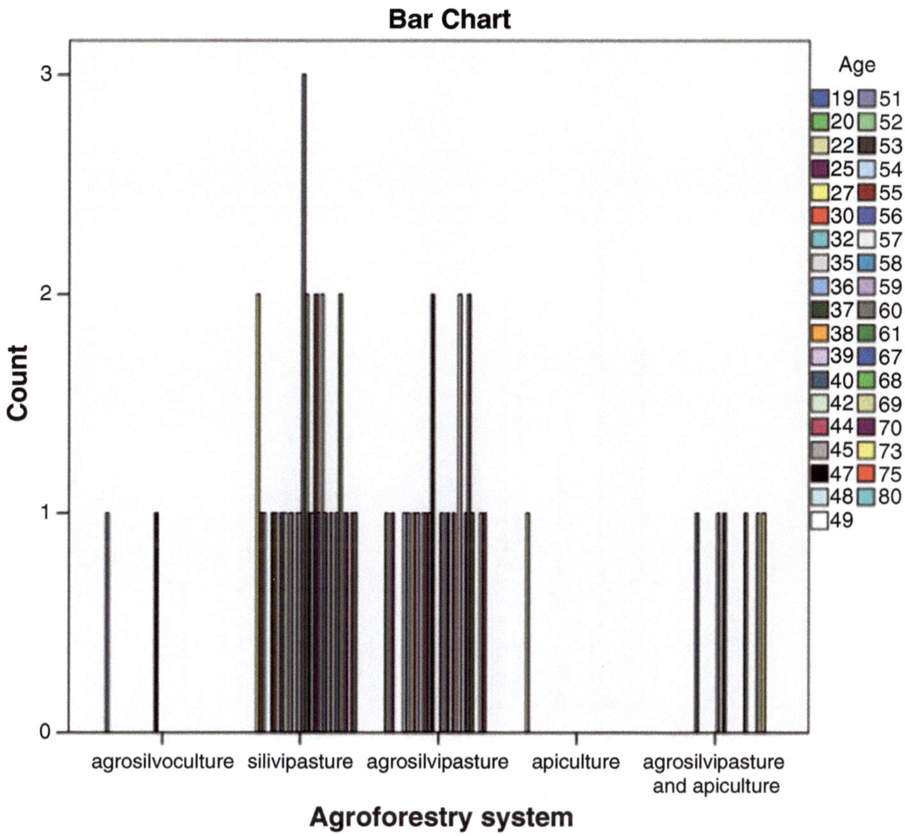

Fig. 9.27 Correlation between potential agroforestry systems and age

Table 9.6 Correlation between agroforestry systems and farming experience

		Value	Asymp. Std. Error	Approx. T	Approx. Sig.
Interval by interval	Pearson's R	0.085∗	0.121	0.658	0.513
Ordinal by ordinal	Spearman correlation	0.095∗	0.122	0.736	0.465
N of valid cases		65			

∗10% and ∗∗5% significant levels

indication that farming experience plays an important role in integrating different farming enterprises (crops and livestock) in different agroforestry systems.

The sample data variables predicted 87% for Cox and Snell (measure for binary logistics regression); 91% for McFadden (measure for multinomial and ordered logit) variation in the dependent variable was explained by the independent variables. Prediction accuracy was assessed based on the coefficient of determination

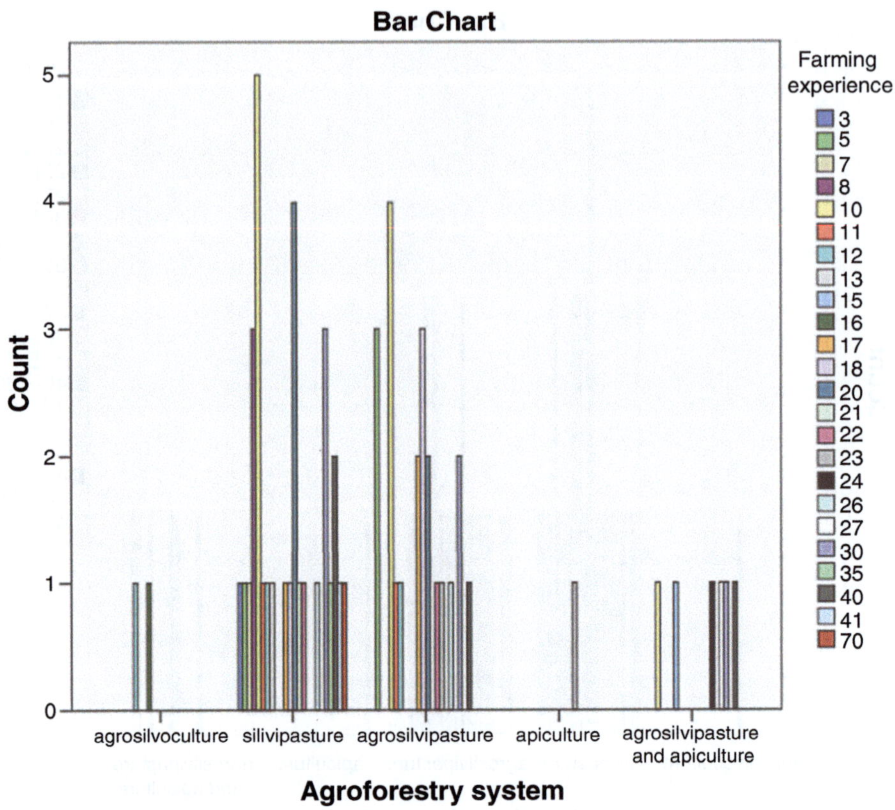

Fig. 9.28 Correlation between potential agroforestry systems and farming experience

(R^2). The coefficient of determination R- squared was used to explain the total proportion of variance in the dependent variable explained by the independent variable. The R^2 removes the influence of the independent variable not accounted for in the constructs. R-squared is always between 0% and 100%. In general, the higher the R-squared, the better the model fits the data.

The following econometric model was used to determine association of variables (Greene 1993):

$$W_i = \_ + \_ X_i + \_ i \quad (9.1)$$

$$W_i \text{ is the dependent variable value for person i} \quad (9.2)$$

$$X_i \text{ is the independent variable value for person i} \quad (9.3)$$

$$\_ \text{ and } \_ \text{ are parameter values} \quad (9.4)$$

$$\_ \text{ i is the random error term} \quad (9.5)$$

Table 9.7 Univariate analysis determining factors enhancing participation in agroforestry

Variable	Total	(%)	OR [95%CI]
Selling	65	100	1.55[0.45–3.1]1
Land size	65	100	1.33[0.67–4.555]1
Market	65	100	1.44[0.276–2.4]1
Agricultural training	65	100	1.20[0.68–3.44]1
Production inputs	65	100	1.14[0.39–2.13]1

OR odds ratio, 95% CI 95% confidence intervals, 1< no association, 1> association

$$\text{The parameter \_ is called the intercept or the value of W when } X = 0 \tag{9.6}$$

$$\text{The parameter \_ is called the slope or the change in W when X increases by one} \tag{9.7}$$

As indicated in Table 9.7, the odds of farmers' land size, selling, agricultural training, market and production inputs was more than 1. This clearly indicated a positive association with agroforestry systems. It was not surprising to realise a positive association among training, selling and market participation as a well-trained farmer is empowered to make informed decisions and to identify market opportunities where they exist. Moreover, farmers with large land sizes and production inputs could allocate their land partly for trees, livestock and bees, giving them a better position to participate in the market. These results are in line with smallholder surveys conducted by Maponya et al. (2014, 2015) in other South African districts.

9.5 Conclusion

Rainwater, temperature and soil were found to be constraints for promotion and establishment of agroforestry system in the districts of Limpopo province, viz. Waterberg, Sekhukhune and parts of Capricorn district. On the other hand, in Vhembe, Mopani and parts of Capricorn districts, factors like soil, precipitation and temperature were found to be a non-constraint for the establishment of the agroforestry system. The research indicated that potential agroforestry systems play a major role throughout human history in supporting livelihoods, assisting various communities to generate income and create job opportunities, as well as meeting food security and nutritional needs in Limpopo Province. The results further indicated an evidence of the importance of agroforestry systems especially silvipasture and agrosilvipasture for supporting food production and income generation in Limpopo Province. Some farmers in Limpopo Province highlighted that they are generating income through renting of farms for grazing, selling trees to the communities to build shelter, kraals, medicinal purposes, fuelwood, etc. Those farmers with access to water were able to grow crops and sell their produce at local communities, local municipality and international market. The farmers also indicated that they were also benefiting from nitrogen fixation, increased crop production, economic gain, soil conservation and improved soil quality and sequestration of atmospheric carbon as a result of timber-based mixed farming/agroforestry practice. The identified farmers' benefits were in line with some of the researchers' field observations. It is thus recommended that stakeholders should take note of the benefits identified by farmers in an attempt to increase agroforestry farmers' participation.

References

ARC-ISCW (2017) Agricultural Research Council – Institute for Soil, Climate and Water, Pretoria, South Africa

Backeberg GR, Sanewe AJ (2010) Towards productive water use and household food security in South Africa. Paper presented at the 6th Asian Regional Conference of ICID, Indonesia, 10–16 October 2010

Combe J (1982) Agroforestry techniques in tropical countries: potential and limitations. Agrofor Syst 1:13–27

Department of Agriculture, Fisheries and Forestry (DAFF) (2017) Agroforestry strategy framework for South Africa, March 2017, Pretoria

Google (2017) Different agroforestry systems. www.google.co.za. Accessed 25 Sept 2017

Greene W (1993) Econometric analysis, 5th edn. Prentice Hall, Upper Saddle River

Hassan R (2013) Drought management strategies in South Africa and the potential for economic policy instruments, book chapter, https://doi.org/10.1007/978-94-007-6636-5-21

LDA (Limpopo Department of Agriculture) (2010) Downloaded reports. http://www.lda.gov.za

Luedeling E, Kindt R, Huth N and Koenig N (2013) Agroforestry systems in a changing climate — challenges in projecting future performance. Curr Opin Environ Sustain 2014(6):61–67

Lundgren BO, Raintree JB (1982) Sustained Agroforestry. In: Nestel B (ed) Agricultural research for development: potential and challenges in Asia. ISNAR, The Hague

Malherbe J, Tackrah A (2003) Long term average ten daily 1km × 1km temperature, rainfall and evaporation grid surfaces modelled from weather station data with a long term recording period. ARC-Institute for Soil, Climate and Water, Pretoria

Maponya P, Mpandeli S (2013) The role of extension services in climate change adaptation in Limpopo Province, South Africa. J Agric Ext Rural Dev 5(7):137–142

Maponya PI, Modise D, Van Den Heever E, Mahlangu S, Baloyi T, Maluleke A, Carsten J, Van Der Walt M, Chauke D, Mkhari R, Sole L, Duba M, Malebana J, Mphahlele M (2014) Establishment of fruits and vegetables markets: a case of waterberg district in Limpopo province. J Agric Sci 6(6):38

Maponya P, Venter SL, Modise D, Van Den Heever E, Kekana V, Ngqandu P, Ntanjana N, Pefile A (2015) Determinants of agricultural market participation in the Sarah Baartman District, Eastern Cape of South Africa. J Hum Ecol 50(1):1–9

Maponya P, Venter SL, Du Plooy CP, Modise SD and Van Den Heever E (2016) Training challenges faced by smallholder farmers: a case of Mopani district, Limpopo province. J Hum Ecol 56(3):272–282

Mercer DE, Miller RP (1998) Socioeconomic research in agroforestry: progress, prospects, priorities. Agrofor Syst 38:177–193

Mmbengwa VM (2009) Capacity building strategies for sustainable farming SMME's in South Africa, PhD (Agricultural economics) Dissertation, University of the Free State, Bloemfontein

Nair PKR (1993) An introduction to Agroforestry. Kluwer Academic publishers, Dordrecht

Poolman E (2017) El Nino Southern Oscillation, National Disaster Management Advisory Forum (NDMAF), Centurion, South Africa

Rethman NFG, Annandale JG, Keen CS, Botha CC (2007) Water use efficiency of multi – crop agroforestry systems with particular reference. Water Research Commission, Pretoria

SAWS (2016) South African weather services. Weather media releases. www.weathersa.co.za. Accessed 10 June 2017

Agriculture, Landscape and Food Value Chain Transformation as Key Engines in Climate Change Mitigation: A Review of Some Low-Carbon Policy Options and Implementation Mechanisms

10

Louis Bockel and Laure Sophie Schiettecatte

Abstract

One of the most promising ways to mitigate climate change is through "agriculture and landscape climate solutions": the conservation, restoration and improved management of agriculture and natural land in order to increase carbon storage and/or avoid greenhouse gas (GHG) emissions in landscapes worldwide. In this perspective, the agriculture sector is facing a wide rethinking on the way to better integrate environment and climate change issues into agriculture policies.

Schools of thinking towards sustainable agriculture and agroecology are competing with new concepts such as sustainable land management, climate-smart agriculture and low-carbon agriculture. On the other hand, policymakers and donors face the issue of integration of mitigation and adaptation in agriculture policies and investment programmes. New performance indicators such as carbon footprint and carbon balance are used to select value and promote high-performing options. Similarly, food markets progressively provide products with reduced carbon footprint.

Implementing such new and marketing mechanisms slowly induce a progressive transformation of the food value chains, agriculture policies and incentives. However, high inertia does remain in the agricultural sector, slowing down the switch towards real low-carbon agriculture options. This chapter does analyse a series of shortcuts used by countries and donors to stimulate such structural change, considering appropriate appraising tools, new incentive mechanisms, new modalities of funding and acting with scaling-up perspectives. Agroforestry value chains upgrading scenarios such as cocoa or shea parkland restoration are provided as examples of upscaling paths with high-carbon-fixing performances.

L. Bockel (✉)
FAO Economist, Policy Officer in Regional Office for Africa, Accra, Ghana
e-mail: Louis.Bockel@fao.org

L. S. Schiettecatte
FAO EX-ACT expert, ESA Division, FAO HQ, Rome, Italy

© Springer Nature Singapore Pte Ltd. 2020

V. Venkatramanan et al. (eds.), *Global Climate Change and Environmental Policy*,
https://doi.org/10.1007/978-981-13-9570-3_10

The issues covered in this chapter include "environmental policy", "sustainable agriculture", "climate-smart agriculture", "climate mitigation and adaptation", "global food policy" and "low-carbon agriculture".

Keywords
Agriculture policy · Value chains · Carbon footprint · Low-carbon agriculture · Sustainable development goals · EX-ACT tool

10.1 The Food and Agricultural Sectors Are at the Centre of the Global Response to Climate Change

In July 2017, the 40th council of the Food and Agriculture Organization of the United Nations (FAO) revealed the FAO's strategy on climate change: "to provide high visibility to the collaborative work to be undertaken by FAO and its partners in the coming years around the shared aim of firmly placing the food and agricultural sectors at the centre of the global response to climate change." Feeding over 10 billion people and addressing dietary changes in this century will have to be achieved without intensification of the production whereas food systems and food security are already at threats from multiple environmental threats such as desertification, drought, soil degradation, freshwater scarcity and loss of biodiversity (FAO 2013; WRI 2018). Indeed with the expected growth of population, if the agricultural productivity is not properly managed and enhanced, it would need five times more land conversion by 2050 and would double agricultural greenhouse gas (GHG) emissions (WRI 2018).

More recently, the publication from the IPCC special report on global warming of 1.5 °C raises awareness about the urgent need for global climate action on both adaptation and mitigation. It underlines the multiple threats humanity will face if no urgent action is taken to constrain global warming within 1.5 °C. The Paris Agreement paved the road for safer climate where countries engaged themselves through their nationally determined contributions (NDCs) to reduce their own greenhouse gas (GHG) emissions. From this agreement, around 90% of the NDCs included the agriculture sector where opportunities for climate mitigation and adaptation are numerous (Smith et al. 2014; FAO 2013, 2016). Still the last IPCC's report argues that these NDCs are insufficient to ensure sustainable development and are still leading us towards a 3 °C warming pathway rather than the 1.5 °C one. Indeed, countries will have to triple their efforts to reduce emissions if there are any chances of keeping global temperature increase below 2 °C.

Maintaining the capacity of the planet's natural resource base to feed the growing world population while reducing agriculture's environmental and climate footprint is key to ensuring the welfare of current and future generations (FAO 2017).

The IPCC report highlighted numerous climate adaptation and mitigation opportunities from the food sector, that is, from production to fork. Changes in agricultural practices (including mixed crop–livestock production, improving irrigation efficiency, etc.) represent one of these adaptation opportunities. From a climate mitigation perspective, improving the efficiency of food production will have to go hand in hand with changes on the supply-side and demand-side measures.

This chapter aims at demonstrating that options to enhance agricultural productivity together with climate adaptation and mitigation are available and already applied. The agricultural sector is the second-largest GHG emitter but the potential of climate mitigation from this sector is unique and inherently coupled with other sustainable development goals (SDGs). We review here some of these examples and present the first results of a GHG appraisal of a portfolio of about 190 projects, value chains and policies on low-carbon agriculture. Results show that agriculture is a strong engine for decreasing GHG emissions. Analysed projects, value chains and policy all aimed at reducing poverty, improving livelihoods and increasing climate resilience while if climate mitigation was not necessarily a main scope, all projects contributed to tackle GHG emissions.

10.1.1 Agriculture Contribution to GHG Emissions and Potential of Climate Mitigation

Agriculture, forestry, and other land use (AFOLU) is unique among the sectors considered in this chapter because its mitigation potential derives from both an enhancement of carbon sequestration and a reduction of emissions through management of land and livestock. It is responsible for just under a quarter (approximately 10–12 giga tonne of carbon dioxide equivalent ($GtCO_2$-e) per year) of global anthropogenic GHG emissions (about 49 $GtCO_2$-e per year, 2010 data from Smith et al. 2014), mainly due to deforestation, land use changes and agricultural emissions from livestock, soil and nutrient management (Smith et al. 2014). Leveraging AFOLU mitigation potential is, therefore, paramount to contribute and meet with emission reduction targets in synergy with any other SDGs.

Agriculture has a high climate change mitigation potential both in the form of reducing emissions intensity per unit produced, as well as carbon sequestration in the biomass and the soil, and many technical options are readily available for immediate deployment. Seventy-four percent of this mitigation potential of agriculture is in developing countries. Indeed, AFOLU development projects can play an important role in climate change mitigation either by reducing emissions and/or by sequestrating carbon while being compatible with food security and other SDGs (FAO 2013, 2016; WRI 2018).

More generally, the agricultural sectors offer many approaches of low-carbon agriculture. If well managed and in a sustainable way, agriculture has high supply side mitigation potential through improvements in land use management with 89% of this potential in enhancing soil carbon sinks. This potential holds in improved grazing land management, improved cropland management, restoration of organic soils and degraded lands (Smith et al. 2007). On the demand side, whilst food waste and loss represent a third of global food production, changes in diet, lifestyle, wood consumption and reductions of losses along the food value chain have not only a mitigation potential but could also reduce environmental stress such as water and land competition (FAO 2013; Smith et al. 2014; WRI 2018).

The scientific community is increasingly focusing on estimating the mitigation potential of the agricultural sectors. From the most recent ones, Wollenberg et al. (2016) derived a feasible target for the reduction of direct (non-CO_2) emissions in agriculture in line with the 2 °C limit is 1 GtCO_2-e per year. This target could contribute up to 5% to the annual mitigation needed across sectors in 2030. Soil carbon sequestration through improved cropland and grazing land management and restoration of degraded lands could contribute to reduce GHG emissions by 1.2 GtCO_2-e per year (Blandford and Hassapoyannes 2018). The reduction of food loss and waste by 15% and changes in human diets on the basis of recommendations by the World Health Organization for health reasons could contribute to mitigation from 1.1 to 2.4 GtCO_2-e per year. Reducing the expansion of agricultural land also has a high potential to mitigate climate change and could contribute to avoid from 1.7 to 4.3 GtCO_2-e per year but at the same time it may conflict with food security (Blandford and Hassapoyannes 2018). Collectively, these options represent a mitigation potential of the AFOLU sector to about 5–9 GtCO_2-e per year (Blandford and Hassapoyannes 2018).

More recently the World Resources Institute (WRI) addressed three challenges to create a sustainable food future (WRI 2018): (1) the food gap, that is, taking measures to decrease the rate of unnecessary demand growth and adopting measures that increase supply; (2) the land gap, which is the difference between projected area of land needed to meet global food demand by 2050 and the amount of land in agricultural use in 2010; (3) the GHG mitigation gap, which is the difference between agriculture-related GHG emissions projected for 2050 and an agricultural emissions target for 2050 that is necessary to help stabilize the climate at globally agreed targets. To counter these challenges a set of 22 actions has been proposed by the WRI, many from the food sector, which benefit climate mitigation directly and indirectly by reducing GHG emissions, avoiding GHG emissions and increasing carbon sequestration. Using the GlobAgri-WRR model, the authors estimated that five courses of action could feed the world by 2050 while reducing the 2010 baseline GHG emissions from the AFOLU sector of 12 GtCO_2-e per year to 4 GtCO_2-e per year for a 2 °C target. Nevertheless as the authors stated “a sustainable food future is achievable if governments, the private sector, and civil society act quickly, creatively, and with conviction”.

10.1.2 Scenarios of Wide Rehabilitation of Landscapes and Value Chains Subjected to Heavy Degradation Process

The agricultural sector does allow significant impact on main carbon stocks, in particular biomass and soil. There is a wide range of mixed forestry–agriculture landscape systems between monocrop farming area and pure forestry area which represent different scales of forest degradation, of a mix between perennial and annual crops. Restoration of degraded land to forest or agroforestry does translate in a wide carbon fixing mechanism. It is clearly issued from the status of natural forest areas which provide the highest carbon stock while other types of land could be classified on a carbon-decreasing order such as (i) forest plantations, (ii) agroforestry, (iii) set aside land, (iv) pasture land, (v) annual cropland and (vi) degraded land.

Agriculture and landscape management activities have induced wide land use changes, starting with deforestation process towards annual crops, pastureland into annual crops and annual cropland transformed into degraded land. Some human activities did just result in degradation of existing land use such as forests (from primary forests to degraded forests). Interventions of landscape restoration through the agricultural sector may be a very powerful mechanism of carbon fixing as far as they involve wide areas, which were mismanaged for three to five decades, but could be widely rehabilitated within a 10–15 years perspective.

When natural assets have been progressively widely damaged on a wide landscape area with field practices, a value chain rehabilitation process does provide an opportunity of reshaping natural assets with high carbon fixing potential while providing other ecosystems services and benefits to local communities. For instance, in Ghana, the revival of 720,000 ha of old or virus-affected cocoa trees added to irrigation and climate-smart agriculture practices on 1,080,000 ha has a potential of mitigation of about 7.3 million tCO_2-e per year (FAO working document on a preliminary impact appraisal of cocoa value chain rehabilitation in Ghana from 2018 to 2023[1]). The restoration of 50 million hectares of parklands in the Sahel Green Wall initiative is targeting 250 million tCO_2-e fixed in 10 years (UNCCD 2015).

Dealing with such a concept of inversion of negative dynamics, that is restoration, is both an opportunity and a challenge for policymakers working on agriculture policy design. Current negative trends are usually the consequence of behaviours of a majority of stakeholders either farmers, cooperatives or all kinds of partners downstream the value chains; therefore, looking for inversion of dynamics does translate in a wide effort of mobilization of stakeholders in a transformative process.

10.1.3 Scaling Up Potential of Wide Landscape Areas Targeted

Based on FAO (2014) data, of 149 million km^2 of land in the planet, 71% are habitable land (excluding glaciers and barren land). 50% of this habitable land is used by agriculture; 37% is covered by forest. Humans use half of global habitable area for agricultural production (51 million km^2). Of the remainder, 37% is forested (39 million km^2), 11% as shrubbery and only 1% is used as urban infrastructure. Additionally more than three-quarters of our agricultural land is currently used for the rearing of livestock through a combination of grazing land and land used for animal feed production.

[1] FAO and COCOBOD study, working document, 2018

Globally, 33% of the world's farmland (about 17 million km^2) is moderately to highly degraded. This degradation concerns particularly dryland areas, affecting the quality of local people's livelihoods and the long-term health of ecosystems (FAO 2017).

The part of forest land degraded is about 50% of the forest area to be added to shrubland which is, by definition, a degraded forest. Together, they represent a global area of 31.5 million km^2 of degraded forest land. Therefore, a wide rehabilitation of degraded forest and farmland covering 30% of degraded areas could fix a critical quantity of carbon every year.

In order to estimate an order of magnitude of this potential, moderate figures of carbon fixing were derived with the help of the EX-Ante Carbon-balance Tool (EX-ACT) developed by FAO (Bernoux et al. 2010; Grewer et al. 2017). The tool is aimed at providing estimates of the mitigation impact of agriculture and forestry development projects, estimating net carbon-balance from GHG emissions and carbon (C) sequestration. We thus ran an analysis over a 20 year period of time[2] and assumed the following agro-ecological conditions, a tropical moist climate and low-activity clay soil which are rather low-carbon-fixing conditions. The following coefficients were used[3]: (i) 0.24 tonne of carbon (tC) fixed in the soil per year from an annual systems applying improved agronomic practices such as crop rotation and reduced tillage and (ii) 2.82 tonne of biomass C fixed per year for forest rehabilitation, such as switching of a tropical dry forest largely degraded (60% of the canopy is lost) to a lower degradation level (20% of the canopy is lost). Such figures applied on 30% of degraded areas could fix about 7.1 $GtCO_2$-e per year. It is positioning this type of carbon-fixing option with a global emission reduction potential of about 14%.

10.2 Climate-Smart Agriculture as a National Policy and Strategy Driver

The relationship between the food sector and climate change is hard to perceive without considering other agriculture externalities such as food security, poverty reduction and sustainable land development. Similarly hunger, food insecurity rural poverty and sustainable agriculture and natural resource management cannot ignore climate change dimension (FAO 2016).

Yet, there is still no consensus on which measures can best deliver more sustainable agriculture and how to use natural resources under the threat of climate change, but there are a wide range of options defined as climate-smart agriculture (CSA) practices.

[2] The 20 years period is the default time period for transition to reach a new equilibrium.

[3] The values used are from either the IPCC 1996 or 2006 Guidelines and are gathered from a large compilation of observations and long-term monitoring.

CSA is an approach for developing agricultural strategies to secure sustainable food security under climate change. CSA has three interrelated objectives, where the first two objectives are emphasized in low-income situations:

- Food security which is to sustainably increase crop yields and productivity and improving farmer incomes
- Improve adaptation and building farmers' resilience to climate change
- Improve mitigation (when and where possible): reducing and/or removing greenhouse gas emissions

Therefore, CSA scaling-up options usually link with wide landscape management policies, transformation of agriculture systems in agroforestry systems and sustainable value chain policies at the meso-level.

10.2.1 Sustainable Land and Soil Management Options

Appropriate soil management practices are known as field options available for farmers to adapt to the adverse effects of weather variability and climate change. Such practices are registered by the IPCC with GHG fixing coefficients. Adoption of such practices at wide scale does allow to promote both national food security and climate mitigation. Furthermore, degraded soils are at a much greater risk of erosion from the damaging impacts of climate change, driving to wide land degradation which would accentuate climate change impacts.

On the contrary, the rehabilitation of degraded soils can be achieved by simply avoiding soil compaction through zero or reduced tillage and reducing soil erosion (stone or vegetal terracing, tree lines, parcel bordering, mixed cropping), which favour enhancement of soil organic carbon and soil biodiversity, thus providing a major opportunity for climate change mitigation.

"*Integrated soil and land management practices can create optimal conditions for the sustainable production of food, fiber, fodder, bio-energy, tree crops, and livestock, and safeguard or enhance the ecosystem services agricultural production systems depend on, as well as improve livelihoods and food security of the most vulnerable population.*" (FAO 2014).

10.2.2 Promotion of Agroforestry Systems

Agroforestry is "the collective term for land-use systems and technologies in which woody perennials (e.g. trees, shrubs, palms or bamboos) and crops or grasses and/or animals are used deliberately on the same parcel of land in some form of spatial and temporal arrangement" (Choudhury and Jansen 1999).

This dynamic and ecologically based natural resource management system by inclusion of trees in farms and agricultural landscapes allows diversification and sustainability of production with economic, social and environmental benefits for both farmers and landscapes (Alao and Shuaibu 2011). Consequently, this system also shelters farmers to market fluctuations and failures that may eventually result from climate change. It is important to understand this autonomous adaptation process in order to replicate the most successful agroforestry systems in similar social, cultural and ecological circumstances.

10.2.3 Climate-Smart Landscape Approach

Improving food security and rural livelihoods within the context of climate-change adaptation and mitigation encompasses a landscape approach, which could be defined as a climate-smart landscape approach. Such an approach requires combining actions of landscape restoration, territorial development and agroforestry management through coordinated action at farm and landscape scales. This was well developed by Scherr et al. (2012).

10.2.4 CSA Promotion and Policy Innovation Platforms

The engagement of national stakeholders, including farmers, to identify pro-poor opportunities to access and engage in CSA value chains may be the best approach. For instance, the FAO Economics and Policy Innovations for Climate-Smart Agriculture (EPIC)[4] programme did test an approach with specific focus areas which included (i) bridging knowledge gap on CSA and the barriers to adoption, (ii) analysing the costs and benefits of changes in smallholder agricultural practices for costing prioritized CSA options, (iii) supporting policy framework to enhance partner countries' capacities to implement CSA and (iv) supporting capacity development targeting stakeholders and agriculture policymakers. The project did work with governments, local institutions and universities in Malawi, Vietnam and Zambia. The main output is an evidence base for identifying, developing and implementing practices, policies and investments for climate-smart agriculture.

Similarly, the Ecosystem Services for Poverty Alleviation (ESPA) is a joint initiative from the Department for International Development, Natural Environment Research Council and the Economic and Social Research Council of United Kingdom.[5] This initiative aimed at providing evidence-based lessons from 10 projects to scale up CSA. The particularity of this initiative was the strong key role of stakeholders and knowledge platforms to address opportunities and challenges for scaling up CSA. The initiative was articulated through four pillars: (1) a long-lasting partnership between government and other stakeholders including farmers, (2) a long-term investment in CSA, (3) some tailored CSA opportunities along the value chain for the poorest and the landless and (4) mainstreaming of national CSA practices where private sector and national interests merge, for instance, within the NDC, value chain and global funds.

[4] http://www.fao.org/in-action/epic/en/

[5] https://www.espa.ac.uk/

10.3 Approaches Used in Climate-Smart Agriculture Integration and Upscaling Through Policies

10.3.1 Greening the European Common Agricultural Policy CAP 2020

Within the European community, the pressures and benefits from agriculture pose an intervention dilemma between two options: extensification and intensification. Extensification benefits to semi-natural habitats include reducing local pressures on soil, water and air, but it increases the area needed for agricultural production. Intensification of production aims at the contrary but requires caution. At the global level, an average yield increase helps to avoid further deforestation, but if the yield increases are associated with enhanced pollution and disturbance of the nutrient cycle (from the use of synthetic fertilizer, for instance), the overall situation would probably deteriorate. There is, thus, a trade-off between reducing environmental pressures at the field level through extensification and maintenance of uncultivated areas at the landscape level. This has direct implications for biodiversity, and, indirectly, for the delivery of other ecosystem services such as carbon sequestration (European Environment Agency 2009).

The current common agricultural policy (CAP) reform, which ends in 2020, addresses environmental challenges by coupling agricultural subsidies to stricter cross-compliance with environmental legislation and "greening measures" such as compulsory crop diversification and maintenance of permanent grassland and ecological landscape elements. These measures aim to cover approximately 7% of the farmland ("ecological focus areas") and would be financed under the first (production-oriented) pillar. This general regime would be flanked by specific agri-environment measures under the second pillar (rural development).

The CAP reform for the period 2021–2027 is currently being discussed. It is a question of respecting "the absolute necessity (i) to secure the income of the farmers thanks to a wide range of instruments, notably insurance", while calling for the diversification of these revenues and (ii) "to upscale the answer to climate change" and to the "environmental emergency".

The contribution of the latest CAP reform was to introduce the concept of "greening", while prioritizing the protection of natural resources and biodiversity. However proposed measures demonstrated to be complex to be implemented. It is about moving away from a defensive approach to the environment. Recognizing that European agriculture provides services to society and the environment, it implies (i) to simplify environmental requirements and (ii) to acknowledge the farmers who deserve remuneration for the public goods they produce, positive externalities such as storing CO_2 in soils, to operationalize genuine payments for environmental services (PES) made by farmers, under one or other of the two "pillars" of the CAP (Gremillet et al. 2017).

Such CAP orientation does present similarities with "green agriculture" advocated by the United Nations Environment Programme (UNEP), which is an agriculture that relies on the use of on-farm resources rather than on external mineral inputs and which combine productivity increase and economic and environmental gains. Such an approach would improve the quality of soil, water and air with indirect benefits for biodiversity.

10.3.2 Agroecology Approach and Cuba Experience

The agroecology approach is "*holistic, balancing focus on people and the planet, the three dimensions of sustainable development – social, economic and environmental, while strengthening, the livelihoods of smallholder food producers, indigenous peoples, women and youth*" (FAO 2013).

Agroecology has five fundamental principles as listed by Angelo 2017: "*(1) increasing biomass recycling; (2) improving soil conditions by covering soil with mulch or cover crop; (3) reducing nutrient loss through closed system design; (4) promoting biodiversity within and between species, including landscape level biodiversity; and (5) promoting interactions and synergies among system components to encourage ecosystem function such as soil fertility and pest management, without relying on external inputs*". Agroecology is therefore a circular approach of the agricultural sector where wastes from animal production are recycled as crop fertilizers or biofuel, limiting harmful environmental externalities.

An example of such an agroecological approach has been implemented in Cuba, starting in the early 1990s. Cuba did consider the farm as a functioning ecosystem, autonomous with natural pest management through their predators and parasites, no inputs of synthetic fertilizer and use of fossil fuel and a diversity of crops and varieties to build the system's resilience.

As many other countries, Cuba is entering a period of significant economic development with increased demand for more and nutritious food, and is now facing with the choice of either adopting the industrialized agriculture approach of most of the developed world, along with its concomitant health and environmental harms, or finding ways to scale up and further advance its agroecology approach.

If Cuba can successfully expand its agroecology approach to achieve food security for its people, and perhaps enter into the international export market, the country could become a global leader in demonstrating how to achieve food security in a low-carbon economy. Additionally, it would help the country to decrease its dependence on imported food and the necessary inputs for its own agriculture, which provide only temporary food security. Such an expansion will avoid many of the health and environmental harms historically tied to industrialized agriculture (Angelo 2017; http://cait.wri.org/profile/Cuba, PANNA 2017).

10.3.3 Global Perspectives in Africa

African agriculture is well positioned for transformational change. About 200 mha of untapped and uncultivated land in the continent has potential to adopt CSA. It has been estimated that the food and beverage market in Africa will touch US$ 1 trillion in value by 2030.[6] The young population of Africa and active private sector provide impetus for the growth of agriculture. Further, 47 Sub-Saharan African countries have categorically pledged to include agriculture adaptation and mitigation actions in their Intended Nationally Determined Contributions (INDCs) (Braimoh 2018).

In order to address the climate action and development agenda, the Africa Climate Business Plan (ACBP) was launched at the Paris COP 21. The ACBP warrants about US$ 16 billion to empower and adapt the people to climate change and enhance the resilience capacity of the continent. The plan also endorses the adoption of CSA and calls for transformation in agriculture which is in line with the Malabo Declaration (Braimoh 2018).

Collaborating with governments to develop climate-smart policies is often an integral part of projects as can be seen with the East Africa Dairy Development Initiative and the drought-tolerant maize initiative for Africa (Drought Tolerant Maize for Africa) (World Bank and CIAT 2015). For some projects, the political aspects are the main target. National policy has a great influence on future investments in Africa's improved agriculture technologies. Farmers' responses to these policies will be fundamental in determining their ability to develop profitable and dynamic farming communities.

10.3.4 Low-Carbon Agriculture Policy Pathways in India

India is currently the fourth emitter of GHG, that is, 3202 million tCO_2-e in 2017, with 7% of the global GHG emissions (http://cait.wri.org/profile/India). The main emitting sectors are energy and agriculture. Through its INDC, the country engaged to reduce GDP emissions intensity by 20–25%, over 2005 levels, by 2020. It does require a scaling up of mitigation efforts, including from agriculture, while mitigation is not yet a priority in Indian agriculture. Therefore, it needs adequate policy and financial incentives for farmers and scale up investments in research and development to facilitate rapid implementation of low-carbon technologies. But the Indian agriculture sector is decentralized, and state governments are not obliged to support initiatives such as low-carbon agriculture. Additionally, there is also a lack of financial incentives for low-carbon technologies (Sonam-Wang et al. 2017). These issues make the task harder to implement a national level policy at the field level. However, the recent[7] appears as a success story of low-carbon innovation. This policy aims at maximizing indigenous production of fertilizers, making them available to farmers on time, and lessening import of urea.

Adoption and scaling up of such low-carbon technologies policies require (i) identifying and assessing technologies and management practices in agriculture that can fix carbon or reduce emissions, (ii) providing incentives to adopt such carbon-neutral practices and (iii) mobilizing farmers to maintain C sink and maximize value of C offsets at the farm levels.

[6] http://www.worldbank.org/en/news/press-release/2013/03/04/africas-food-markets-could-create-one-trillion-dollar-opportunity-2030

[7] http://fert.nic.in/new-urea-policy-%E2%80%93-2015

As financial benefits are key for farmers coming forth to adopt technologies, payment for carbon services for the agriculture sector is a prerequisite. The domestic carbon market can be quite effective if demand for carbon reduction is created. Agriculture policy must work with the financial sector to link economic incentives with technologies. It should drive carbon finance integration into public and private investment decisions and mobilization for an active participation of public sector and large domestic emitters (industry, agribusiness) in the carbon market. A systemic approach (program of investments) reviewed with rural banks and funding operators would contribute significantly towards promoting low-carbon agriculture in India (Sonam-Wang et al. 2017).

Low-carbon agriculture does generate other environmental externalities (reduced incidences of soil erosion, improved pest resistance, prevent loss of biodiversity) and productivity through sustainable use of water, land, crop rotation, cover cropping, agroforestry, use of organic fertilizer, etc. However, the switch towards low-carbon agriculture in India is slowed down by high transaction costs, enabling policy, affordable technologies and investments. Currently, the easiest low-carbon pathways to scale up in India include low tillage systems, system of rice intensification (SRI), use of high-yielding livestock breeds and measures to convert animal husbandry wastes into energy among other agricultural practices.

Agricultural policy and appropriate market incentives are still required to encourage farmers, mobilize local governments and private sectors to invest in and migrate to low-carbon agriculture without affecting food security of smallholders. It should be supplemented by a guaranteed easy access to field and agricultural inputs as well as capacity building on GHG accounting that can efficiently assess the mitigation potential in agriculture through monitoring of GHG emissions (Sonam-Wang et al. 2017).

10.4 Investment Projects as a Field Implementation Mechanism Option

In the AFOLU sector, a lot of the support provided by public sector and donors is channelled through the formulation and implementation of public field investment projects or national government investment plans. Appraising systematically the carbon balance of any step of an AFOLU project cycle provides performance indicators to compare projects in terms of carbon-fixing impact (such as in tCO_2-e per year or per hectare per year or even per household) and a carbon-fixing cost in US$ per tCO_2 fixed. A carbon-fixing performance per ha can then justify incentives for farmers in applying CSA improved practices or even payment of environment services in line with the performance of the service. A low carbon-fixing cost (per tCO2e) may attract climate mitigation funding.

Such an exercise is done using carbon-accounting tools. A multitude of those tools have been developed to screen and account for GHG emissions and carbon sequestration for the AFOLU sector, such as the EX-ACT from FAO and the Carbon Benefit Project Tool from The Global Environmental Facility, UNEP, and Colorado State University. Toudert et al. (2018) tested and compared the performances of several of those GHG tools to propose the best adapted one according to the project's sustainable land management activities, their availability, time and skills requirement, among other criteria.

The FAO's continuous GHG appraisal support on public/donor projects and policies, performed from 2010 to 2017 by the EX-ACT team, constitutes a portfolio of 77 countries of projects representative as a whole of the global profile of public investments in the AFOLU sector. The capitalization work covers 174 projects (local/regional/national), policies (national) and value chains appraised representing an aggregated investment of US$ 25 billion. The portfolio includes projects from the International Fund for Agriculture Development (IFAD), United States Agency for International Development (USAID), the World Bank, the Global Environmental Fund (GEF), Agence Française de Dévelopment (AFD), African Development Bank (AfDB), Asiatic Development Bank (ADB), World Wildlife Fund (WWF) and FAO.

Projects, value chains and policies of the present portfolio were categorized and defined according to the following classification: "(i) watershed management; (ii) sustainable forest management,; (iii) value chain commercialization; (iv) cropland & grassland rehabilitation; (v) sustainable cropland intensification; (vi) livestock intensification; (vii) climate resilience & adaptation; (viii) aquaculture & fisheries development; (ix) climate change mitigation; (x) irrigation development; (xi) agroforestry systems and (xii) wetland management".

The aggregated GHG mitigation impact of this portfolio was estimated to about 4.3 billion tCO_2-e, or about 1.29 tCO_2-e fixed per hectare per year (Table 10.1). Those investments represent a cost of US$ 5.73 per tCO_2-e fixed assuming that all project cost is accounted in the carbon-fixing cost. These data demonstrate the wide capacity of the AFOLU sector as a global GHG reduction engine for the planet with a relative low cost per tCO_2-e reduced (Table 10.1).

The analysis also demonstrated the mitigation potential of sustainable forest management projects and sustainable cropland intensification projects as compared to other categories as they generated 73% of the appraised GHG mitigation impact. In these two categories, analysed investments often relate to country- or region-wide areas with areas accounted in hundred thousand or even million hectares. Such wide landscape dimensions do guarantee scaled up impact. The typology used in the

Table 10.1 Results from the analysed portfolio with the EX-ACT tool, as covered area, projects, programmes and policies, budgets, climate mitigation potential and cost per unit of CO_2 fixed

Portfolio 2010–2017 of appraised projects	
166	Million ha
24,624	Million US$
4297	Million tCO_2-e mitigated (20 years)
215	Million tCO_2-e mitigated per year
1.29	Equivalent additional tCO_2-e mitigated per ha per year
5.73	US$ per tonne of CO_2-e mitigated

Source: FAO–ESA compilation of EX-ACT team (FAO working draft, 2018)

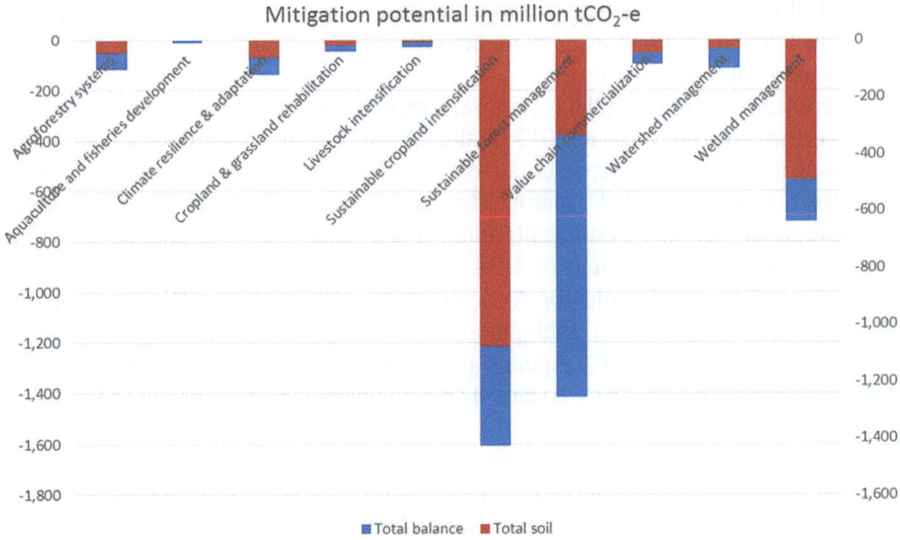

Fig. 10.1 Total GHG mitigation potential (in million tCO_2-e) according to the typology of projects. (Source: FAO–ESA compilation of EX-ACT team (FAO working draft, 2018))

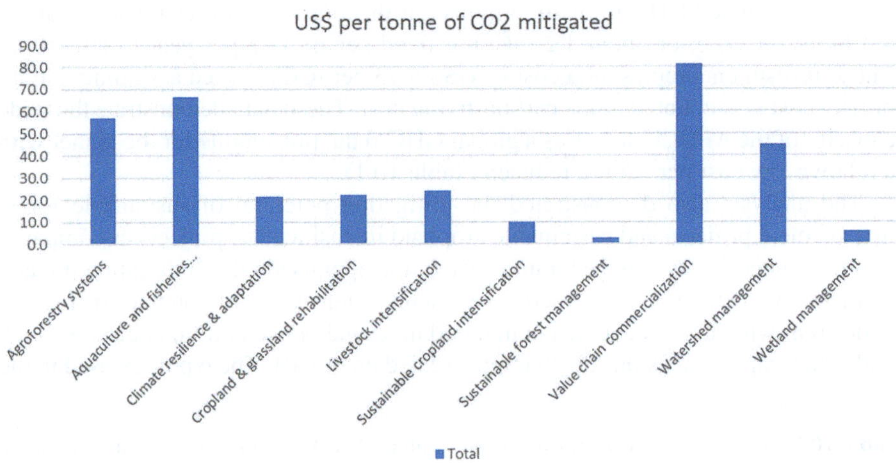

Fig. 10.2 Cost of GHG mitigation by type of investment project, programme and policy in US$ per tonne of CO_2-e reduced and avoided. (Source: FAO –ESA compilation of EX-ACT team (FAO working draft, 2018))

figure is slightly biased by the fact that some projects are inclusive of a series of diverse actions. For instance, agroforestry systems are also partly present in watershed management, value chain and sustainable cropland intensification (Figs. 10.1 and 10.2).

Translated into tCO_2-e per hectare per year, the best mitigation performances per ha were provided by projects with a focus on wetlands (potential of mitigation of about 8 tCO_2-e/ha/year), followed by agroforestry systems (7 tCO_2-e/ha/year), sustainable forest management (around 5 tCO_2-e/ha/year) and watershed management (4.6 tCO_2-e/ha/year) (data not shown)

Using total project cost divided by carbon balance, rough costs per tCO2e were worked out. The cost of GHG mitigation by type of investment project does position sustainable forest management as the most cost-efficient for AFOLU mitigation with US$ 0.7 per tonne of CO_2 fixed, followed by wetland management (US$ 3.9/ per tCO_2-e) and sustainable crop intensification (US$ 6.7 per tCO_2-e).

10.5 Value Chains as an Implementation Scaling Up Mechanism

10.5.1 Using Value Chains and Food Systems to Scale Up CSA Through Policies

Food systems contribute an estimated 19–29% of global anthropogenic GHG emissions. Most of these emissions (80–86%) are released during the production phase (Vermeulen et al. 2012). The use of fossil fuels across all stages of the food value chains, from production and processing to consumption, is critical. Therefore, a holistic approach in designing CSA strategies for sustainable food systems is paramount as each level of the food system presents many opportunities to adapt and increase resilience of food systems and enhance global food security. Such an approach requires an in-depth analysis of the food system from farm to fork and of the different practices and stakeholders and their complex interactions (Ingram 2011). These kinds of analyses can help minimize the trade-offs and harness synergies among the social, economic and environmental dimensions of the value chain sustainability. A food systems approach will also assist decision making on the potential pathways to follow to develop sustainable and low-carbon food systems.

Developing sustainable and low-carbon food systems resilient to climate change requires improved governance and specific and coordinated actions from all stakeholders in the food system. Governance, which is the vertical coordination of core value chains within food systems, can improve access to technologies, secure financing for climate-smart agriculture interventions and disseminate information about climate-smart agriculture.

Food system interventions for CSA may include advocating for changes in policies, investing in infrastructure, inputs and services, providing training on best practices and encouraging behaviour change of all food system stakeholders.

10.5.2 Carbon Footprint and Sustainable Value Chain Labelling

In the 1990s, consumers became increasingly sensitive to environmental issues and those surrounding the food they eat; the intensive mode of agricultural production attracted growing attention. Indeed, the increase in environmental awareness among consumer societies in the Europe and US pushed food suppliers and retailers to develop indicators as to the environmental impact of their production and its distribution.

To determine environmental impacts associated with agriculture, the use of a life cycle assessment (LCA) framework allows the entire production chain to be analysed for a wide variety of applications based on a quantitative method to determine environmental impacts across an entire supply chain. This method was originally developed for use in industrial operations but has later been adapted for a wider range of applications including agriculture (Caffrey and Veal 2013). Accounting for environmental impacts associated with agriculture has some distinct challenges including the wide range of agricultural activities as well as spatial (e.g., dissipated emissions and differing regional conditions) and temporal (e.g., year-to-year crop rotations and seasonal fluctuations) variations inherent to the field (Bockel and Schiettecatte 2018).

Public authorities have quickly taken over the initiative of suppliers and private retailers on the labelling of environmental impacts by providing a framework and a common methodology centred on the LCA. Research organizations, standards bodies, consultancy firms, etc., are developing similar concepts worldwide. These include the Carbon Footprint (ADEME – France), PCR ID: PA-BJ-O3 (Japan) and LCA (Ecoinvent – ISO 14040/14044).

EX-ACT VC is a tool derived from EX-ACT (EX-Ante Carbon Balance Tool) developed by FAO and retargeted for appraisal of agricultural[8] and forestry value chains. It is built with a series of Excel modules that provide co-benefit appraisals of food value chains in developing countries, on GHGs emissions, system resilience and income (Bockel et al. 2017). EX-ACT VC allows assessing agricultural value chains from cradle to shelf. It can be used for the assessment of (i) GHG impacts (carbon footprint), (ii) system resilience and (iii) socio-economic performance (value added, income, and employment generated) along all stages of food value chains. The approach is based on a specifically designed system of indicators, including also environmental indicators (water use, energy use).

A clear and detailed understanding of the current performance of the food chain can be translated into effective and efficient programmes that support or facilitate value chain development. Successful upgrading implies the upgrade is adopted by a majority of stakeholders, even if the more capable and commercially oriented smallholder farmers are often leading the transformation process (FAO 2014). Assessing carbon footprint could be of interest for three purposes: "The first one is very management oriented; the second one is highlighting possibilities of additional farmers income linked with positive externalities and the third one is addressed to market labelling" (Bockel and Schiettecatte 2018).

[8] The term agriculture also includes here livestock, fisheries and aquaculture.

10.5.3 Comparing Case Studies of Agroforestry Value Chains in Terms of Sustainable Impact

Agroforestry systems in Africa constitute the third-largest carbon sink after primary forests and long-term fallows. In Africa, Unruh et al. (1993) reported that a total of 1550 million ha are suitable for some type of agroforestry (Unruh et al. 1993). The two case studies currently in review below demonstrate this wide carbon sink potential scenarios of (i) ongoing rehabilitation of cocoa value chain in Ghana and (ii) future restoration and expansion of shea parklands in West Africa.

10.5.3.1 Example of a Ghana Cocoa Value Chain Rehabilitation Scenario (2018–2028)

Ghana is globally recognized for its cocoa cultivation and export worldwide. Cocoa is an essential contributor to Ghana's economy and as a nation at large, and the country is now the second-largest producer of cocoa beans in the world. In 2018, facing reduction of production, the Ghana Cocoa Board (COCOBOD[9]) faces the challenge of implementation of a main rehabilitation programme targeting replanting of vast areas of either virus-affected or old plantations.

The currently started renovation and rehabilitation programme is inclusive of a series of actions. For 2018, about 58,909 ha of cocoa trees on farm farmland (mix of land used as annual crops and set aside lands) should be replanted, and 320,000 ha of diseased trees and 400,000 ha of old trees should be substituted. Also, given the challenge cocoa farmers are facing in periods of drought, COCOBOD has disclosed the start of an irrigation project for cocoa farmers targeting irrigation of over 200,000 hectares of cocoa farms.

The current upper cocoa value chain (without downstream processing, retail and export) is making US$ 1.31 billion of gross product and US$ 0.9 billion of value added. The rehabilitation scenario would allow to move up to about US$ 3 billion US$ of annual gross product by 2028 while generating up to 277,000 additional employments. The yearly gross income per producer could increase from US$ 981 to US$ 2292 over the whole economical period analysis (i.e. 10 years). Such a result is making this rehabilitation scenario extremely efficient in terms of poverty reduction (Table 10.2).

Over the whole duration of the GHG analysis, that is, 20 years, without rehabilitation the scenario is emitting GHG about 1.1 million tCO_2-e per year. With the implementation of the wide rehabilitation scenario, the mitigation potential of the upgrading scenario demonstrates above 7.3 million tCO_2 fixed per year (over 146 million tCO_2-e on 20 years) (Table 10.3). Every hectare of cocoa would in average fix up to 3.9 tCO_2 per year. This co-benefit of the value chain does represent an economic value of US$ 118 per year per ha when using the social price of CO_2 (US$ 30 per tCO_2[10]).

[9] Ghana Cocoa Board.

[10] World Bank recommended price in 2016 for economic analysis of Carbon mitigation impact.

Table 10.2 Socio-economic performances of cocoa value chain in Ghana for both current situation and upgrading scenario (2018–2028)

Socio-economic performances of the value chain		**Current**	**Upgrading**	**Balance**
Production level				
	Nb of HH	0	800,000	
	Nb of employment-eq	424,800	698,950	274,150 jobs
Gross Production Value (GPV)		1,313,280	2,629,934	1,316,654,000 US$
Value Added (VA)		897,157	2,047,986	1,150,829,000 US$
Gross Income (GI)		784,891	1,833,226	1,048,336,000 US$
VA/tonne of product		1082	1233	151 US$
VA/ha		498	1102	603 US$
Gross income/HH		981	2292	1310 US$
Collectors and LBC wholesalers				
	Nb of operator eq	60	0	
	Nb of employment-eq	5466	6876	1410 jobs
Gross production value		135,709	166,569	30,860,000 US$
Value added		107,443	116,596	9,154,000 US$
Gross income		96,922	94,261	−2,661,000 US$
VA/employee		19,655	16,956	−2699 US$
Gross income/employee		17,730	13,708	−4022 US$
Downs tream processing				
	Nb of operator-eq	12	13	
	Nb of employment-eq	2001	4006	2006 Jobs
Gross Processed Production Value (GPPV)		112,734	225,757	113,023,000 US$
Value added		91,285	182,837	91,552,000 US$
Gross income		88,865	178,405	89,539,000 US$
VA/tonne of product		367	367	0 US$
Gross income/operator		7,405,437	13,723,425	6,317,988 US$
Cocobod export				
	Nb of operator eq	1	1	
	Nb of employment-eq	104	208	104 jobs
Gross production value		314,429	643,037	328,608,000 US$

(continued)

Table 10.2 (continued)

Value added		314,111	642,420	328,310,000 US$
Gross income		310,201	640,193	329,992,000 US$
VA/operator		314,110,584	642,420,138	328,309,554 US$
Gross income/operator		310,201,043	640,192,620	329,991,577 US$
Aggregated Socio-economic performances		**Current**	**Upgrading**	**Balance**
Value added		1,409,996	2,989,839	1,579,844,000 US$
Gross production value		1,876,152	3,665,297	1,789,145,000 US$
Total job generated		432,371	710,040	277,669 Jobs created

Source: Preliminary Impact Appraisal of Cocoa Value Chain Rehabilitation in Ghana (2018–2023) – Cocoa Agroforestry value chain as carbon fixing pro-poor engine, FAO COCOBOD working document (November 2018)

Table 10.3 Greenhouse gas performances and carbon footprint of cocoa value chain in Ghana for both current situation and upgrading scenario (2018–2028)

Climate mitigation dimension of the value chain	**Current**	**Upgrading**	**Balance**
GHG impact (tCO_2-e per year)	1,118,280	−6,193,355	
GHG impact (tCO_2-e per year per hectare)	0.6	−3.3	−3.9
Carbon footprint of production (tCO_2-e per tonne of product)	1.3	−4.3	−5.6
Annual tCO_2-e [emitted (+)/reduced or avoided(−)]		−7,311,635	
Annual tCO_2-e from renewable energy		0	
Equivalent project cost per tonne CO_2-e reduced or avoided (in US$ on 20 years)		6	US$
Equivalent value of mitigation impact per year (US$ 30/tCO_2-e)		219	US$ million
Equivalent value of mitigation impact per year per ha (S$ 30/tCO_2-e per year per ha)		118	US$
Carbon footprint at the different levels of the value chain	**tCO_2-e per tonne of product**		**Balance**
	Current	**Upgrading**	
Production	1.32	−4.29	−5.60
Processing	0.07	0.08	0.01
Transport	0.09	0.08	0.00
Total	1.47	−4.13	−5.60

Source: Preliminary Impact Appraisal of Cocoa Value Chain Rehabilitation in Ghana (2018–2023) – Cocoa Agroforestry value chain as carbon fixing pro-poor engine, FAO COCOBOD working document (November 2018) – Screen print from EX-ACT VC regional shea model

Such a GHG impact drives reconsidering the role of cocoa value chain as a main GHG fixing engine (146 million tCO_2 fixed on 20 years) which does fix carbon at very low public cost per tonne, that is, US$ 6 per tonne of CO_2. Considering the whole rehabilitation, a budget of around US$ 600 million will be spent by the

Government of Ghana. It will contribute together to (i) pro-poor growth of the cocoa value chain, (ii) sustaining resilience of the value chain and (iii) GHG emission reduction. The GHG being more an externality, it would be realistic to consider the part of the budget spent for GHG reduction as a maximum of 20%. In such a case, the public cost per tonne of CO_2 fixed would be of US$ 1, making cocoa investment among the best GHG reduction performance per US$ spent.

10.5.3.2 Example of a West Africa Shea Value Chain Expansion Scenario (2019–2032)

In Africa, about 16 million women are involved in shea activities in 21 countries. West Africa does account for 8 million of these women. A recent study does estimate that 4,000,000 are involved in the export value chain with 200 million US$ income generated every year in producing communities (LMC 2017). A geographic information system (GIS) spatial model (Naughton et al. 2015) does provide an estimate of 1.84 billion trees on an extensive shea tree area of 3.41 million km^2. For West Africa, the model does provide a total estimate of high stearin trees of 1.07 billion (lower tree density).

Shea parklands are managed landscapes where shea and other economic species have been favoured by local farming communities through cyclical selection and management for generations for human use. Scaling up such an approach needs to be conceived as an integrated landscape-wide approach combining a series of relevant parkland regeneration and tree improvement and management interventions in fields, fallows and bush lands. The upgrading scenario does translate into over 2.5 million ha (10% of agroforestry cropped areas), transformed over 14 years in improved agroforestry parklands (50 trees per ha) while 1.5 million ha of annuals are enriched with shea trees. Such a scenario does represent an increase of 103 million additional trees by 2032. The upgrading scenario would result in a wide increase of employment by 2032 with 396,000 additional jobs created 99% at production levels and about 3.7 million collecting women mobilized in the whole West African region. For the whole value chain, the gross production value will reach 604 million which is equivalent to an economic growth of 5.1% per year for 14 years. The added value is estimated at US$ 463 million (Tables 10.4 and 10.5). Such a growth scenario would allow reaching an annual GHG reduction impact of around 10.5 million tCO_2-e per year.

When compared to the current situation, it translates into an incremental carbon balance of 9 million tCO_2-e per year and a total carbon balance of 180 million tCO_2-e in 20 years. In this perspective, the shea value chain does provide an efficient carbon-fixing mechanism with a cost US$ 0.89 per tonne of CO_2 fixed. The economic value of such a positive externality could be around US$ 270 million per year, making the value chain a high mitigation return investment.

It is simultaneously increasing the carbon footprint (CFP) performance per tonne of shea kernel up to −8.2 tCO_2 fixed per tonne of shea kernel ready for export. This CFP is inclusive of carbon-fixed by landscape rehabilitation and improved parklands. The exceptionally high carbon footprint is fully attributed to the green carbon-fixing impact of the rehabilitation and improvement of shea parklands which are the core of an upgrading strategy. It does allow easy comparison with the current

Table 10.4 Socio-economic performances of regional shea value chain in West Africa for both current situation and upgrading scenario (2018–2032) Dec 2018

Socio-economic performance of the value chain		Current	Upgrading	Balance
Production level: collecting women and local butter processing	Nb of HH	0	3,756,943	
	Nb of employment-eq	491,714	886,639	394,925 jobs
Gross production Value (GPV)		249,238	481,441	232,203,000 US$
Value Added (VA)		177,406	389,614	212,207,000 US$
Gross Income (GI)		177,406	389,614	212,207,000 US$
VA/tonne of product		192	259	67 US$
VA/HH		48	104	56 US$
Gross income/HH		48	104	56 US$
Intermediary agents and transportation level				
	Nb of operator eq	3533	3955	
	Nb of employment-eq	3770	4392	622 jobs
Gross production value		18,330	36,779	18,449,000 US$
Value added		7456	17,304	9,848,000 US$
Gross income		5884	15,274	9,389,000 US$
VA/operator		1978	3940	1963 US$
Gross income/operator		1561	3478	1917 US$
Downstream processing actors				
	Nb of operator-eq	14	18	
	Nb of employement-eq	457	837	381 jobs
Gross processed production value (GPPV)		31,758	85,523	53,765,000 US$
Value added		15,921	56,478	40,557,000 US$
Gross income		13,513	52,054	38,542,000 US$
VA/tone of product		223	354	131 US$
Gross income/operator		965,194	2,891,908	1,926,714 US$
Aggregated socio-economic performances		**Current**	**Upgrading**	**Balance**
Value added		200,783	463,395	262,613,000 US$
Gross production value		299,326	603,743	304,417,000 US$
Total job generated		495,940	891,867	395,927 jobs created

Source: FAO working document: Regional shea value chain as key pro-poor carbon-fixing mechanism: potential and upgrading opportunities (November 2018) – preliminary analysis – Screen print from EX-ACT VC regional shea model

Table 10.5 Greenhouse gas performances and carbon footprint of regional West African shea value chain for both current situation and upgrading scenario (2018–2028) Dec 2018

Project country region budget (US$) year	Parkland expansion Please select the country of origin Please specify name 160000000 2019	Production (tonne)	Current	Upgrading
			923248	1505097
		ton/women ($t\ year^{-1}$)	0.22	0.37
		Hectares		51565178
		Households		3756943
Climate mitigation dimension of the value chain		**Current**	**Upgrading**	**Balance**
GHG impact (tCO_2-e per year)		−1,466,565	−10,497,232	
GHG impact (tCO_2-e per year per hectare)		0.0	−0.2	−0.2
Carbon footprint of production (tCO_2-e per tonne of product)		−1.7	−8.7	−7.0
Annual tCO_2-e [emitted (+)/reduced or avoided(−)]			−9,030,667	
Annual tCO_2-e from renewable energy			0	
Equivalent project cost per tonne CO_2-e reduced or avoided (in US$ on 20 years)			0.89	
Equivalent value of mitigation impact per year (US$ 30/tCO_2-e)			270,920,024	
Equivalent value of mitigation impact per year per ha (S$ 30/tCO_2-e per year per ha)			5	
Carbon footprint at the different levels of the value chain		**tCO_2-e per tonne of product**		**Balance**
		Current	**Upgrading**	
Production		−1.75	−8.73	−6.99
Processing		0.41	0.35	−0.06
Transport		0.21	0.21	0.00
Total		−1.13	−8.18	−7.04

Source: FAO working document: Regional shea value chain as key pro-poor carbon-fixing mechanism: potential and upgrading opportunities (November 2018) – preliminary analysis, screen print of FAO EX-ACT VC based shea model 2018

rehabilitation strategy of cocoa value chain in Ghana which provides a negative carbon footprint of −4.3 kg of CO_2 per kg of cocoa. Such widely negative carbon footprints demonstrate the wide climate mitigation of strategies of expansion or rehabilitation of agroforestry value chains.

10.6 Conclusion

AFOLU has a high climate change mitigation potential both in the form of reducing emissions intensity per unit produced as well as carbon sequestration in the biomass and soil, and many technical options are readily available for immediate deployment. In this perspective, the AFOLU sector is challenged as a global GHG-fixing mechanism to build appropriate implementation paths.

In line with different conceptual approaches, such as green agriculture, climate-smart agriculture, low-carbon agriculture and agroecology, implementation paths are built through policies, strategic frameworks, investment projects and sustainable value chains. Policy and financial pro-poor growth incentives along with capacity building of farmers and investments in research and development do facilitate rapid deployment of low-carbon technologies. Such an implementation exercise is done using carbon-accounting tools developed to screen and account for GHG emissions and carbon sequestration for the AFOLU sector, for most public–private actions from policies to investments.

In terms of cost efficiency of carbon fixing, the cost of GHG mitigation by type of investment project does position sustainable forest management projects as the most cost efficient for AFOLU mitigation with US$ 0.7 per tonne of CO_2 fixed, followed by wetland management (US$ 3.9 per tCO_2) and sustainable crop intensification (US$ 6.7 per tCO_2). However, through agroforestry value chain upgrading strategies, such a cost performance per tonne of CO_2 fixed is also dramatic with US$ 0.89 per tCO_2 fixed in regional shea value chains (2018–2032) and at US$ 1 for cocoa value chain rehabilitation strategies in Ghana. Such costs per tCO_2 position AFOLU low-carbon options among the cheapest alternatives.

References

Alao JS, Shuaibu R (2011) Agroforestry practices and preferential agroforestry trees among farmers in Lafia local government area, Nasarawa state, Nigeria. Water Manag Bioresour Technol 1:12–20

Angelo MJ (2017) Food security, industrialized agriculture, and a changing global climate: perspectives on the United States and Cuba. Fla J Int Law 29(1):Article 40. Available at: http://scholarship.law.ufl.edu/fjil/vol29/iss1/40

Bernoux M, Branca G, Carro A, Lipper L, Smith G, Bockel L (2010) Ex-ante greenhouse gas balance of agriculture and forestry development programs. Sci Agricola (Piracicaba, Braz) 67(1):31–40

Blandford D, Hassapoyannes K (2018) The role of agriculture in global GHG mitigation. OECD food, agriculture and fisheries papers, no. 112, OECD Publishing, Paris. https://doi.org/10.1787/da017ae2-en

Bockel L, Schiettecatte LS (2018) Chapter 18: Life cycle analysis and the carbon footprint of coffee value chains. In: Lashermes P (ed) Achieving sustainable cultivation of coffee. Burleigh Dodds Science Publishing, Cambridge, 2018 (ISBN:978 178676 152 1; www.bdspublishing.com)

Bockel L, Debrune O, Toudert A, Doro E, Lozada O, Schiettecatte L-S (2017) Practical guidance: EX-ACT value chain tool (EX-ACT VC) – co-benefit appraisal of value chain on GHG emissions, climate resilience and income. Food and Agriculture Organization of the United Nations, Roma

Braimoh A (2018) It's time to transform Africa through climate smart agriculture, September. Retrieved from World Bank: http://blogs.worldbank.org/africacan/its-time-to-transform-africa-through-climate-smart-agriculture

Caffrey KR, Veal MW (2013) Conducting an agricultural life cycle assessment: challenges and perspectives. Sci World J. https://doi.org/10.1155/2013/472431. 2013, ID 472431, 13

CAIT-WRI (2017) Climate data explorer: Cuba: country GHG emissions, ALGS. Retrieved from World Resource Institute. http://cait.wri.org/profile/Cuba [https://perma.cc/N3S2-

Choudhury K, Jansen LJM (1999) Terminology for integrated resources management and planning. Food and Agriculture Organization of the United Nations, Rome

European Environment Agency (2009) Distribution and targeting of the CAP budget from a biodiversity perspective, EEA technical report 12. Publications Office, Luxembourg

FAO (2013) Climate smart agriculture, sourcebook. Food and Agriculture Organization, Rome, 570p

FAO (2014) Developing sustainable food value chains – guiding principles. Food and Agriculture Organization, Rome, p 75

FAO (2016) The state of food and agriculture – climate change, agriculture and food security. Food and Agriculture Organization of the United Nations, Rome, 194p

FAO (2017) Future of food and agriculture – trends and challenges. FAO

Gremillet D, Gruny P, Haut C, Montauget F (2017) Politique Agricole Commune: Traverser le Cap dangereux de 2020. SENAT Republique Francaise Rapport d'Informations, 672

Grewer U, Bockel L, Schiettecatte LS, Bernoux M (2017) EX-ante carbon-balance tool (EX-ACT), quick guidance. Food and Agriculture Organization of the United Nations, Rome, 30 pp

Ingram J (2011) A food systems approach to researching food security and its interactions with global environmental change. Food Sec 3(4):417–431

LMC (2017) Socio-economic impact of shea exports. USAID – Global Shea Alliance

Naughton CC, Lovet PN, Mihelcic PR (2015) Land suitability modeling of shea (Vitellaria paradoxa) distribution across sub-Saharan Africa. Appl Geogr 58:217–227

PANNA (2017) Agroecology: resilient and productive. Retrieved from Pesticide Action Network North America. http://www.panna.org/agroecology-farming-solutions/agroecology-resilient-productive

Scherr SJ, Shames S, Friedman R (2012) From climate-smart agriculture to climate-smart landscapes. Agric Food Sec 1:12. https://doi.org/10.1186/2048-7010-1-12

Smith P, Martino D, Cai Z, Gwary D, Janzen H, Kumar P, McCarl B, Ogle S, O'Mara F, Rice C, Scholes B, Sirotenko O (2007) Agriculture. In: Metz B, Davidson OR, Bosch PR, Dave R, Meyer LA (eds) Climate change 2007: mitigation. contribution of working group III to the fourth assessment report of the intergovernmental panel on climate change. Cambridge University Press, Cambridge, UK/New York

Smith P, Bustamante M, Ahammad H, Clark H, Dong H, Elsiddig EA, Haberl H, Harper R, House J, Jafari M, Masera O, Mbow C, Ravindranath NH, Rice CW, Robledo Abad C, Romanovskaya A, Sperling F, Tubiello F (2014) Agriculture, Forestry and Other Land Use (AFOLU). In: Edenhofer O, Pichs-Madruga R, Sokona Y, Farahani E, Kadner S, Seyboth K, Adler A, Baum I, Brunner S, Eickemeier P, Kriemann B, Savolainen J, Schlömer S, von Stechow C, Zwickel T, Minx JC (eds) Climate change 2014: mitigation of climate change. Contribution of working group III to the fifth assessment report of the intergovernmental panel on climate change. Cambridge University Press, Cambridge, UK/New York

Sonam-Wang W, Woo-Kyun L, Yowhan S (2017) Low carbon development pathways in Indian agriculture. Chang Adapt Socioecol Syst 3(1). https://doi.org/10.1515/cass-2017-0002

Toudert A, Braimoh A, Bernoux M, St-Louis M, Abdelmagied M, Bockel L, Ignaciuk A, Zhao Y (2018) Smart applications of GHG accounting tools for sustainable land management – guidelines. World Bank Group, p 135.

UNCCD (2015) World leaders renew commitment to strengthen climate resilience through Africa's Great Green wall. Retrieved from https://www.unccd.int/news-events/world-leaders-renew-commitment-strengthen-climate-resilience-through-africas-great

Unruh J, Houghton R, Lefebvre PA (1993) Carbon storage in Agroforestry: an estimate for sub-Saharan Africa. Clim Res 3:39–52

Vermeulen SC, Campbell BM, Ingram JSI (2012) Climate change and food systems. Annu Rev Environ Resour 37:195–222

Wollenberg E, Richards M, Smith P, Havlik P, Obersteiner M, Tubiello F, Herold M, Gerber P, Carter S, Reisinger A, Van Vuuren D, Dickie A, Neufeldt H, Sander BO, Wassmann R, Sommer R, Amonette JE, Falcucci A, Herrero M, Opio C, Roman-Cuesta RM, Stehfest E, Westhoek H, Ortiz-Monasterio I, Sapkota T, Rufino MC, Thornton PK, Verchot L, West PC, Soussana JP,

Baedeker T, Sadler M, Vermuelen S, Campbell BM (2016) Reducing emissions from agriculture to meet the 2 °C target. Glob Chang Biol 22:3859–3864. https://doi.org/10.1111/gcb.13340
World Bank & CIAT (2015) Climate-smart agriculture in Kenya, CSA country profiles for Africa, Asia, and Latin America and the Caribbean Series. The World Bank Group, Washington, DC
WRI (2018) World resources report: creating a sustainable food future- a menu of solutions to feed nearly 10 billion people by 2050. World Resources Institute – December, p 96

Weather Based Automated Agro Advisories: An Option to Improve Sustainability in Farming Under Climate and Weather Vagaries

11

Ga. Dheebakaran, S. Panneerselvam, V. Geethalakshmi, and S. Kokilavani

Abstract

Agriculture is highly vulnerable to climate change and climate variability. It is very difficult to balance the growing interest on sophistication and climate change mitigation options. Now, we are in the stage of bolstering adaptation process to cope up our production with demand. Automated Agro Advisory Service (AAS) is a response farming tool, which helps the farmers to get timely weather based agro advisories to make necessary decision for the next few days of farm operations. The AAS requires past and forecasted weather data of 6 days from the current date for closer spatial scale, high performance computing server, faster internet service, short messaging service (SMS) and web cum mobile application. Different combinations of temperature, relative humidity, wind speed and rainfall quantity have been numbered as weather scenarios. Weather based agro advisory for multiple crops and different stages of crop growth have been developed with the help of technocrats and incorporated in the database. Farmers have to register their mobile number for advisory to their own crop, specific to crop stage. Every day, weather scenario of each block will be developed separately for past and future weather and the AAS module match the scenario, crop, stage, and advisory and send the selected advisory to the farmers' mobile as SMS. Farmers can change their crops and sowing date through web portal or mobile app. The AAS simplifies the lab to land with ICT tools and helps the farmers to get rid of weather risk and help them to increase productivity of inputs. The AAS empathises the farmers' need and acts weather smart.

Keywords

Weather based automated agro advisories · Weather smart technology · Climate change · Sustainable farming · Response farming

G. Dheebakaran (✉) · S. Panneerselvam · V. Geethalakshmi · S. Kokilavani
Agro Climate Research Centre, Tamil Nadu Agricultural University, Coimbatore, India
e-mail: gadheebakaran@tnau.ac.in

© Springer Nature Singapore Pte Ltd. 2020

V. Venkatramanan et al. (eds.), *Global Climate Change and Environmental Policy*,
https://doi.org/10.1007/978-981-13-9570-3_11

11.1 Introduction

'Jai Jawan! Jai Kisan', the slogan of honourable Mr. Lal Bahadur Shastri, the second prime minister of India epitomises the importance of farming India. United States Geological Survey (USGS) revealed that India has the world's highest net cropland area of 179.8 mha, which outranks the United States' 167.8 mha and China's 155.2 mha (India water review 2017). The Economic Survey 2017–18, released in the parliament during February 2018, indicated that about 50% of work forces in India are engaged in agriculture and allied sectors, which contributes 17–18% of the Gross Domestic Product (Sunder 2018). Considering the past 50 years, though the agriculture's contribution to Indian economy is on a declining trend, agriculture is demographically an important sector and plays a significant role in the overall socio-economic fabric of India.

Agriculture and allied sectors are both the cause and vulnerable to the climate change. Impact of climate change is inevitable and could not be controlled with immediate effect due to the growing interest of human's sophistication. Assessment Report (AR5) of IPCC (2013) had also indicated that Asian countries may experience frequent drought and floods (Fig. 11.1), where the wet area will be more wetter and dry area will be more drier. Increasing intensity of cyclonic disturbances over

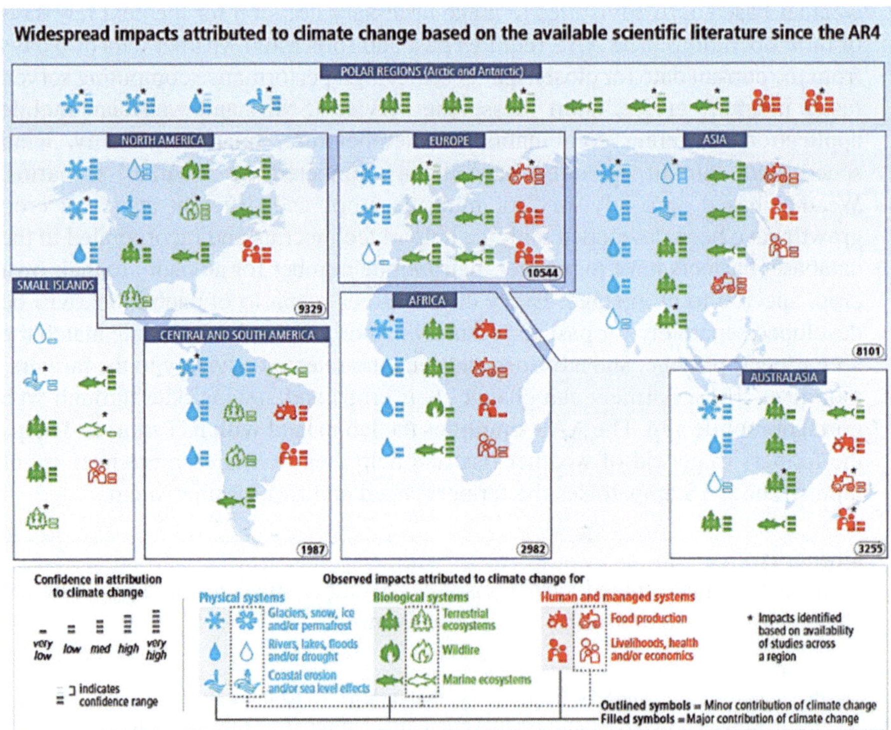

Fig. 11.1 Impact of climate change on physical and biological system of the world (IPCC 2014)

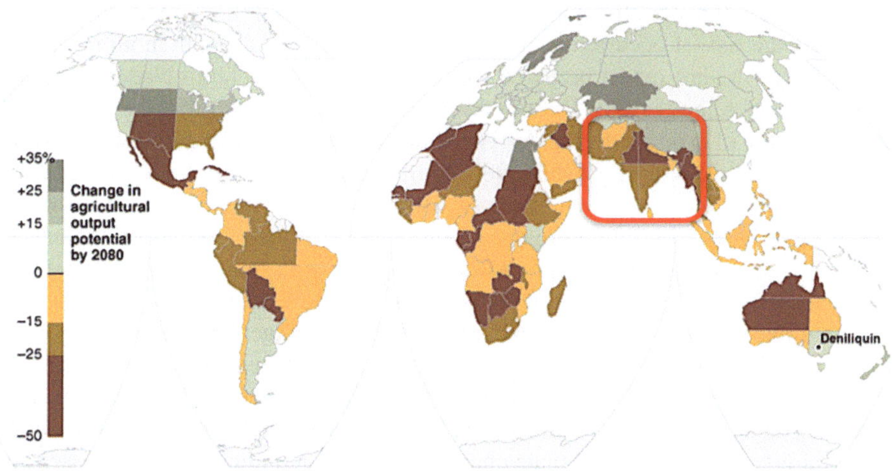

Fig. 11.2 Impact of climate change on agricultural output potential by 2080 (Cline 2007)

India and the changes in the monsoon rainfall will be ranged between 2% and 12%. In the recent past, the same is being experienced in many parts of India. Another report endorsed in IPCC (2013) with a study by Cline (2007) alarmed that Indian region would get a decline in food production of 15–25% in the southern parts and 25–50% in the northern parts of India (Fig. 11.2). The present trends of global climate change, atmospheric CO_2 and temperature levels are likely to increase in future, which will affect yields, water and nitrogen requirements of the crops in a given region.

The Food and Agriculture Organization (FAO 2007) warns that a global temperature increase of 2–4°C over pre-industrial levels could reduce crop yields by 15–35% in Africa and Asia and 25–35% across the Middle East. In India, the Indo-Gangetic Plains could become significantly heat-stressed by 2050s potentially causing losses of 50% of its wheat-growing area (Porter et al. 2014). Increases in day and night temperature were also found to negatively affect the growth, development and yield of rice and wheat crop, which are indeed the staple food crop of India (Venkatramanan and Singh 2009a, b).

Without adaptation by 2050, crops and livestock are likely to experience significant reduction in production. For instance, India's climate is projected to increase by 2–4 °C by 2050 with some marginal changes in rainfall during monsoon months and large changes in rainfall during non-monsoon months. An increase of 2 °C in temperature could decrease rice yields by about 0.75 tonnes per hectare in the high yielding areas (Chattopadhyay 2011).

In Tamil Nadu, it was observed from the long period (1950–2010) of rainfall analysis that the distribution of rainfall has become poor in all the seven agro-climatic zones. Increased rainfall intensity was supported with the reduction in rainy days and increase in quantity (Dheebakaran et al. 2016). This led to further increase in runoff, soil erosion, loss of nutrients, reduction in length of growing period, etc., which are all negative factors of food production. Climate change also increases the risk of occurrence of drought or consecutive drought in all the watershed areas. Another report prepared by the authors for the Tamil Nadu Watershed Development Agency (TAWDEVA) during 2017–18 indicated that the probability of occurrence of drought during current decade is varied once in 2 or 3 years, which was once in 5 years during 1950–1980. Shift in monsoon onset, reduction in rainy days and intermitted long dry spell is also common in most of the watersheds, resulting in frequent crop failure or less productivity. Earlier studies on the impact of climate change to biotic factors inferred that the increasing temperature and CO_2 concentration resulted in invasion of new pests and diseases, quicker generations, frequent outbreaks and temporal and spatial shifts (Ahanger et al. 2013).

In a nutshell, the spatial and temporal shift in quantity and distribution of rainfall resulted in change of length of growing season and crop potential. Further, the erosion, degradation, pollution and overexploitation of natural resources increase the vulnerability of farming to climate change. The climate induced biotic stresses add further deterioration and thus reduce the resilience of Indian farming. While major part of Indian agriculture business is dependent on resource poor marginal farmer, the coping capacity of the small and marginal farmers during climatic extremities is highly limited. The foremost concern at this stage is to forgo the impact of climate change and climatic variability, which could be possible by avoiding, escaping and adapting. Good agricultural practices, viz., suitable crop adjustment, developing proper genotypes, adopting improved management technologies against changed environment, soil and water conservation and finding alternatives, are the need of the hour.

At present, available technologies are on individual problem solving, but a holistic approach is missing. In addition, large-scale adoption of many agricultural technologies became failure due to non-consideration of weather prevailing at a specific location during that period of adoption. To lessen the impacts of climate change and variability through minimising the GHG emissions from agriculture sector, and adapting the agricultural system to changing climate, a suite of climate smart agricultural technologies are developed and of which 'weather-smart agriculture' is highly significant (Venkatramanan and Shah 2019). Response farming is a holistic approach and could be defined as farming with advisories received from the technocrats based on local weather information. Success of response farming, viz., increased productivity and reduced risk, has been already recorded in Tamil Nadu by TNAU and many other states. Response farming could be a viable option for climate change adoption strategies, as the climate change is not an abrupt one. The main reason for the success of response farming is due to both location and time specific technologies. It is the time to carry forward the success of response farming to entire farming community.

Agricultural scientists in ICAR and SAUs are working with dedication to moderate the impact of climate change on the production sector and continuously improving available technologies or developing new technologies to cope up emerging needs. These technological improvements should reach the farming community on need. If the farmer gets weather-based technical inputs at the right time, he could himself manage against the weather abnormalities and produce more for him and his country. The bottleneck in this juncture is transfer of technologies from lab to land. It is accepted that there is a failure in our extension system due to many reasons, which drew our technology's outreach. In view of the above, both Central and State Governments are running many programmes such as IMD's 'Gramin Krishi Mausam Sewa (GKMS)', Govt. of India's mKisan web portal and Tamil Nadu Government's 'Uzhavan app', etc., to enable the farmers to get timely weather-based technical inputs. Some note on these schemes is detailed below.

11.2 Present Status of Weather-Based Agro Advisories in India

11.2.1 Gramin Krishi Mausam Sewa (GKMS) and mKisan

India Meteorological Department (IMD) is implementing the GKMS programme in all the states at 127 centres. For example, 32 districts of Tamil Nadu have been grouped into 8 centres, known as Agrometeorological Field Units (AMFU). These AMFUs are located at college or research stations of State Agricultural/Animal Husbandry Universities and ICAR-KVKs. Each AMFU is led by the university scientist as technical officer. Every Tuesday and Friday, IMD issues weather forecast for next the 5 days along with past week weather data. Based on the past weather and forecasted weather information, the technical officer do the weather based agro advisory bulletin (both English and regional language) for the forthcoming week to their designated districts and disseminated the same through State Department of Agriculture, Research Stations, Krishi Vigyan Kendra and IMD Web portals (Fig. 11.3a, b). The important message is sent to farmers' mobile directly as short message service (SMS) through mKisan web portal of the Ministry of Agriculture. At present, about 11 lakh farmers are benefitted from this service, which helped the farmers not only in increasing the crop production but also reducing the losses due to inclement weather.

11.2.2 Uzhavan App

It is a bilingual mobile application, launched by the Department of Agriculture, Government of Tamil Nadu during 2018 for the benefit of farmers (Fig. 11.4). Through this 'Uzhavan app', farmers and other stakeholders receive valuable information on agricultural inputs (availability of quality seeds, synthetic fertilisers and pesticides at the Government and local private shops), weekly weather forecast and related information, availability of farm tools and equipment, latest information on farm and crop subsidies and suitable crop insurance schemes.

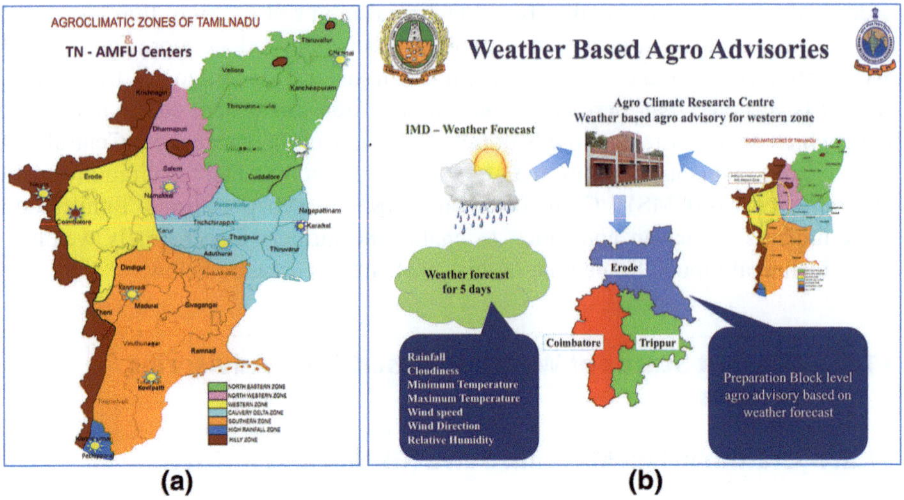

Fig. 11.3 (**a**) GKMS–AMFU centres of Tamil Nadu; (**b**) methodology of disseminating weather-based agro advisory

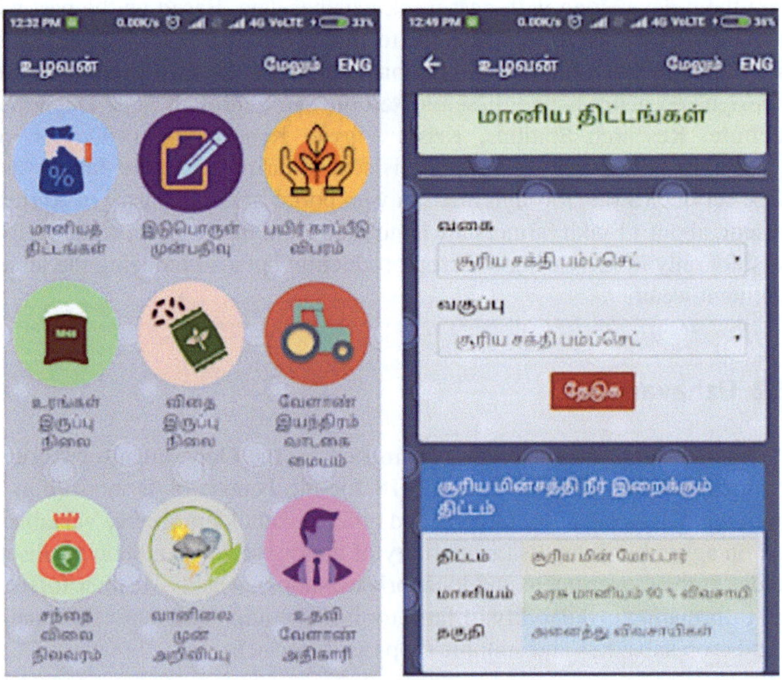

Fig. 11.4 Government of Tamil Nadu, Dept. of Agriculture's 'Uzhavan app'

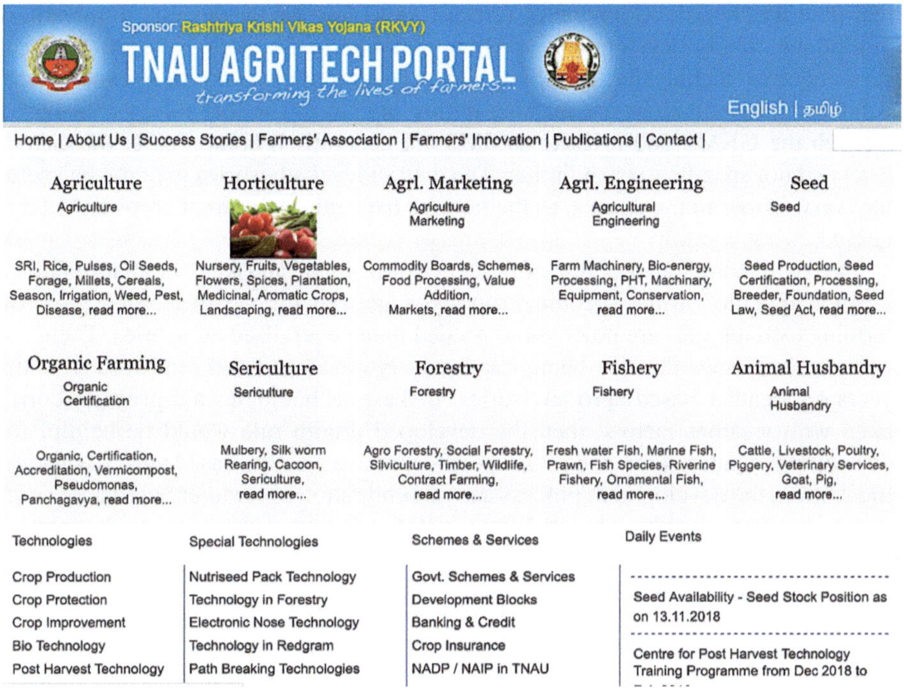

Fig. 11.5 TNAU's Agritech portal

11.2.3 Agritech Portal

Tamil Nadu Agricultural University (TNAU) has launched a web portal for farmers to get right and reliable technologies on agricultural and allied sectors (Fig. 11.5). It is also updated with recent information on agricultural policies, subsidies, weather and insurance.

Earlier, the technologies reached the end users mainly through personal contact by the extension officers during the field visits, farmer melas and training programmes. However, these types of technology transfer consume much of time and money for both the contributors and the users. The success rate is very meagre due to lack of follow-ups, leading to complete failure of agricultural extension system. Nowadays, the extension technologies have been improved along with the development in Information and Communication Technologies (ICT). Making use of the advancement in ICT, most of the technologies are being directly transferred to the farmers' mobile as SMS or WhatsApp messages. These methods are highly cost effective and quickly reach more populations on their own circulation among themselves. Still there is lot of scope in utilising the ICT tools for the transfer of weather based agricultural technologies in turn to reduce the vulnerability to climate change and climate variabilities. Though the technology transfer is easier with ICT tools, now the bottleneck is shifted to the advisories development stage. The development of weather based agro advisories requires continuous observation of real-time weather, development of forecast, involving the technocrats from various disciplines, timely development and dissemination of advisories, etc. All the process requires technically qualified human interventions and rich experiences. Advisory disseminations are missing during holidays. The technology transferred through the GKMS and mKisan as SMS are generally a broad one at the district level and not specific to each farmer. The district level advisories may not be useful for every farmer in the district, as the farmers have raised different crops and in different sowing window.

In view of complete proofing of farmers from climate and weather abnormalities, location specific, farmer specific, crop and stage specific weather based response farming technologies are much more needed than generalised advisories. There are many technologies that are being continuously developed and improved over the years on weather based agro advisories. If these technologies are properly correlated with weather factors, then the developed thumb rule would be helpful for developing location, crop and stage specific, precise weather based agro advisories. In addition, the development process and dissemination of weather based agro advisories could be automated by utilising the ICT tools, and subsequently, the farmers would get timely and precise agro advisories.

In this context, Tamil Nadu Agricultural University (TNAU), in the year 2013, has pioneered by developing 'TNAU – AAS', a Web cum Android app for the automatic development of weather based agro advisories for 108 crops and 54 weather scenarios, without human interventions. The materials and methods for the automation of weather based agro advisory development are detailed below with 'TNAU-AAS' model as an example.

11.3 Development of Automated Agro Advisory Service (AAS)

Basic requirements for developing weather based AAS are:

- Weather information:
  - Closer weather network to provide past and real time weather
  - Forecasted weather information for next one week from model run
  - Weather scenario based on past and future weather
- Preparation of weather based agro advisories
- Integration of technology with ICT and mobile tools
- Farmers' voluntary involvement

11.3.1 Real-Time and Past Weather Information

High resolution information on past and future weather is the foremost requirement of the AAS. Major weather parameters influencing agriculture are minimum temperature, maximum temperature, relative humidity during morning and evening

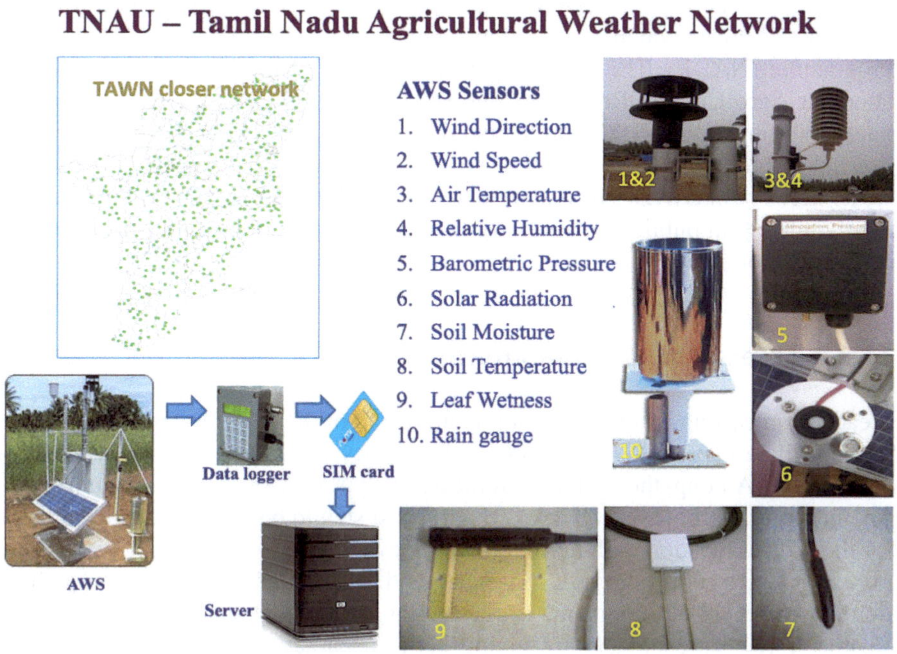

Fig. 11.6 TAWN network and weather sensors

hours and average wind speed and rainfall. If soil moisture, soil temperature and solar radiation are available, then it will add value to the agro advisory. Since farmers require precise advisory, AAS needs the above weather information of farmer's location at least 10–15 km around, which is almost equal to block/taluk level.

In general, except rainfall, all other weather parameters would not vary much within 15 km, unless there is extreme variation in the topography and local circulations. If we have rainfall information of still more higher resolution (<5 km, village level), it will give more accuracy on our advisory. These weather information should be representative of that area and reach the central server room automatically, for which installation of automatic weather station (AWS) in agricultural land is essential. The AWS sensors should be properly calibrated and maintained for accuracy as specified by the IMD. Periodical maintenance of AWS is must, and it is suggested to have annual maintenance contract (AMC) with the AWS supplier, after warranty period, that will ensure uninterrupted real time/past weather data. The weather data of past 6 days should be pooled, averaged and kept ready for further processing by every day morning 9.00 a.m.

During 2009–2014, TNAU has installed 385 AWS, each one at every block of Tamil Nadu, approximately 15 km interval (Fig. 11.6). The AWS is having sensors for 10 important weather parameters installed mainly for agricultural purpose, viz., minimum and maximum temperature, morning and evening relative humidity, solar radiation, wind speed, wind direction, leaf wetness, soil temperature and soil moisture.

This closer network of AWS is named as 'Tamil Nadu Agricultural Weather Network (TAWN)', which was developed under National Agricultural Development Programme (NADP). The weather data are regularly transferred to server maintained at Agro Climate Research Centre (ACRC), TNAU, Coimbatore, at hourly interval through GSM-SIM card technology. This type of closer network is first of its kind in India during 2009 and at present many states have installed still more closer network of AWSs. The real time weather data are being uploaded regularly at hourly interval in public domain 'tawn.tnau.ac.in' at free of cost for the betterment of farming community.

11.3.2 Forecasted Weather Information

Weather forecasting is another most important part of the AAS process, which needs lot of skill in meteorology for increased accuracy as well as ICT knowledge for automation. Among the different forecasting types, AAS needs medium range weather forecast for at least 5 days. The forecast should be calibrated and validated for the project location with minimum accuracy of 70%. Forecast with lesser accuracy won't give successful advisories and lead to 'right advisories in wrong situation'. With recent developments in forecasting methods, dynamical and ensemble weather forecasting models are giving better accuracy. The output should have higher resolution and in lined to the grid point of AWS installed for past weather information. Necessary coding for complete automation from downloading the Global Forecasting System (GFS) data, pre processing, processing, post processing and generation of outputs for our designated grid points has to be written as Linux Shell script. Forecasting process should be automated in such a way that the output should be ready by 9.00 a.m. every day.

At the Tamil Nadu Agricultural University, Weather Research Forecast (WRF) model is being used to develop medium range weather forecast (MRWF). Necessary script has been written for complete automation from downloading of GFS data to final forecast. The GFS data being downloaded are of 0.25 degree resolution, 12 hour cycle and 6 hourly interval.

The downloading link for the GFS data is https://nomads.ncep.noaa.gov/pub/data/nccf/com/gfs/prod/gfs.20190612/12/gfs.t12z.pgrb2.0p25.fxx. About 28 files with total size of 6 GB is downloaded every day from the global data centre. These data are pre processed in WRF Pre processing System (WPS) with geogrid.exe, ungrip.exe and metgrid.exe for two nested domains having grid resolution of 9 km and 3 km, respectively. The met_em outputs are then inputted in WRF model. The output from WRF model is then inputted in ARWpost, the post-processing system to get desired output grid. Then using shell script and grads, forecast of next 6 days are being developed daily by 8.00–9.00 a.m.

The forecast contains weather parameters such as minimum temperature (T_{min}), maximum temperature (T_{max}), morning and evening relative humidity (Rh1 & Rh2), morning and evening wind Speed (WS1 & WS2) and rainfall (RF) of 385 blocks of Tamil Nadu and uploaded in TAWN website by 9.00 a.m. (Fig. 11.7). The output grid has been matched to AWS of a particular block. With the help of continuous

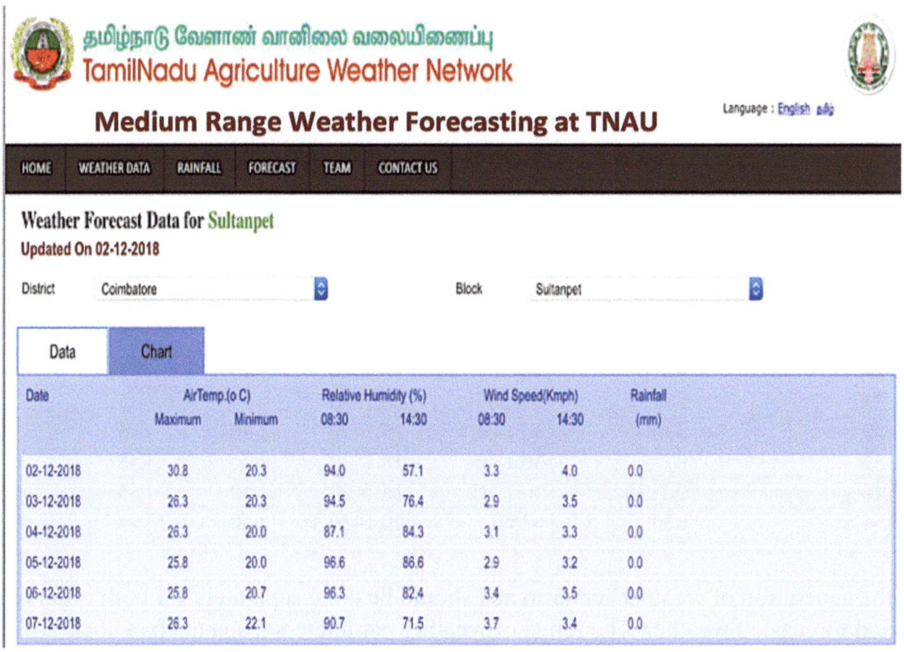
தமிழ்நாடு வேளாண் வானிலை வலையிணைப்பு
TamilNadu Agriculture Weather Network

Medium Range Weather Forecasting at TNAU

Language : English தமிழ்

HOME WEATHER DATA RAINFALL FORECAST TEAM CONTACT US

Weather Forecast Data for Sultanpet
Updated On 02-12-2018

District Coimbatore Block Sultanpet

Data Chart

Date	AirTemp.(o C)		Relative Humidity (%)		Wind Speed(Kmph)		Rainfall
	Maximum	Minimum	08:30	14:30	08:30	14:30	(mm)
02-12-2018	30.8	20.3	94.0	57.1	3.3	4.0	0.0
03-12-2018	26.3	20.3	94.5	76.4	2.9	3.5	0.0
04-12-2018	26.3	20.0	87.1	84.3	3.1	3.3	0.0
05-12-2018	25.8	20.0	96.6	86.6	2.9	3.2	0.0
06-12-2018	25.8	20.7	96.3	82.4	3.4	3.5	0.0
07-12-2018	26.3	22.1	90.7	71.5	3.7	3.4	0.0

Fig. 11.7 Medium-range weather forecast

studies, WRF model was calibrated by altering the microphysics to suit Tamil Nadu and the accuracy is now about 80%.

11.3.3 Developing Weather Scenario

The next step in the automation of AAS is preparation of weather scenario using the observed weather data of past 6 days and generate weather forecast of next 6 days. The observed and forecasted weather data have to be first checked for their error or abnormal values. Due to sensor problems or wasp combs or dust or damage by animal or human, data may have some error and abnormal values. For that, we have to give a limit or range of normal weather of that area for the month/week. If any error comes in the data, we have to moderate the data to the normal values. In forecast process, due to network issues or error in GFS, the values won't be generated in time. In such situations, necessary coding should be done to download forecast data of required specification from open sources. Here also, the forecasted values have to be checked for error and abnormal values. The verification and correction of data is unavoidable and need to be done automatically. The raw data and corrected data should be stored separately for future investigations.

After checking the data, both the observed and forecasted weather values of 6 days are to be averaged for temperature (°C), humidity (%) and wind speed (km/h), whereas cumulated for rainfall (mm/week). The averaged values have to be used for

Table 11.1 Range of weather parameters for developing weather scenarios

Level	RF (mm)	T_{max} (°C)	T_{min} (°C)	Rh (%)	WS (km/h)
I	0	<20	<15	20.1–40	<5
II	0–20	20.1–30	15.1–20	40.1–70	5.1–15
III	20–30	30.1–35	20.1–30	70.1–90	>15
IV	>30	>35	>30	>90	–

Table 11.2 Sample weather scenario and range of weather events

Scenario number	RF (mm)	T_{max} (°C)	T_{min} (°C)	Rh (%)	WS (km/h)
1	0	<20	<15	>40	<5
12	0.1–30	20–30	15.1–20	>40	<5
22	>30	20–30	20.1–30	>40	<5
23	>30	30.1–35	15.1–20	>40	<5
32	0	30.1–35	15.1–20	>40	>5
41	0.1–30	30.1–35	15.1–20	>40	>5
54	>30	>35.0	20.1–30	>40	>5

the generation of weather scenario and should be done separately for both observed and forecasted weather. The range of weather parameters (Jagannathan et al. 2014) to be used for developing weather scenarios are depicted in Table 11.1.

If needed, observed soil moisture, soil temperature and day length could be added in the scenarios without much complication for the advisory preparations. The levels and range of weather parameters could be modified or eliminated based on our requirements. Since, high-performance computing tools are used, it can be any number of weather combinations. However, too many combinations complicate the preparation of advisory for each weather window, which requires more of expertise. Hence, it is better to limit the weather scenario combinations <60 in order to reduce complications in advisory preparations. In TNAU-AAS, it is limited to 54 weather scenarios only and some examples are given in Table 11.2. Care must be taken that a weather combination should not be repeated in two different scenario numbers.

11.3.4 Preparation of Weather-Based Agro Advisories

After developing weather scenarios, technocrats from various fields such as agronomist, entomologist and pathologist are to be involved to develop weather-based agro advisories for every stage of the crop. For easy handling, based on the date of sowing, the stages of crop can be divided into land preparation (1 week before sowing), seedling, vegetative, reproductive, grain development, harvest and post-harvest (1 week after harvest). Few examples of different stages of crops are depicted in Table 11.3.

Based on the critical stages of crop, advisory for irrigation may be prepared. For example, if past and forecasted weather shows dry weather scenario, irrigation may

Table 11.3 Demarcation of crop stage based on duration (days after sowing)

Sl. no.	Crop	Seedling	Vegetative	Reproductive	Maturity	Harvest
1.	Rice – short duration	21	42	63	105	119
2.	Rice – med. duration	35	70	105	154	168
3.	Wheat	14	56	91	119	133
4.	Maize	14	42	63	105	119
5.	Red gram	14	35	63	119	133
6.	Black gram	14	35	56	70	84
7.	Groundnut	14	35	49	98	112
8.	Cotton	14	42	119	161	175

be advised. If past or future weather has rainfall, advice is given to postpone the irrigation. If past and future has rainfall, drainage may be advised. Similarly for fertiliser application, based on the crop stage, basal or top dressing is advised. If rainfall is there, advice is provided to make use of the rainfall; otherwise, advice is provided to apply fertiliser with irrigation.

Based on the crop vulnerability to pest or diseases, advisory is prepared for plant protection measures. Here, favourable weather scenario for the emergence, multiplication and outbreak of pest and diseases are to be correlated with crop stage. Care must be taken that the advisories should have limited characters, well within the limits of single SMS, i.e. 164 characters including spacing between words. As advisories are to be used by the farmers, the advisories should be prepared in regional language. In TNAU-AAS, advisories have been developed for 108 crops, 6 stages and 54 weather scenarios. Some samples of crop wise and stage wise weather based agro advisories for short duration rice are given in Table 11.4.

11.3.5 Integration of Technology with ICT and Mobile Tools

In the above text, we have learnt to prepare individual input such as observed weather, forecasted weather, weather scenario and weather-based agro advisories for the development of AAS. Next step is integration of all these inputs in a sequential way in an ICT platform. It requires a high-performance server with latest web applications, short messaging service (SMS) application and fast internet connectivity.

11.3.5.1 High Performance Server

All the process can be run in multiple servers separately, and a central unit will combine the outputs. It is better to run all the process in a single server having high capacity (i.e. having more than 4 numbers of 64 bit processors with minimum of 128 threads in total, latest Linux Operating System and internal storage of >1 TB). It should be supported with sufficient power backup during power failure. Necessary firewall and security systems have to be installed as the AAS have farmers' database. TNAU-AAS runs in a single server with 8 processor, 512 GB RAM and 1.2 TB internal storage.

Table 11.4 Weather based agro advisories for short-duration rice

Stage	Observed weather scenario	Forecasted weather scenario	Advisory
Land	17/18/26/27/44/45/53/54	17/18/26/27/44/45/53/54	Dry weather is expected. Do nursery preparation
Land	–	10/11/12/13/14/15/16/17/18/19/20/21/22/23/24/25/26/27/37/38/39/40/41/42/43/44/45/46/47/48/49/50/51/52/53/54	By utilising the anticipated rainfall, prepare nursery for sowing
Seedling	10/11/12/13/14/15/16/17/18/19/20/21/22/23/24/25/26/27/37/38/39/40/41/42/43/44/45/46/47/48/49/50/51/52/53/54	10/11/12/13/14/15/16/17/18/19/20/21/22/23/24/25/26/27/37/38/39/40/41/42/43/44/45/46/47/48/49/50/51/52/53/54	By utilising rainfall, prepare nursery for sowing. If already sown, keep water in the evening to avoid splashing effect of rain drops
Seedling	19/20/21/22/23/24/25/26/27/46/47/48/49/50/51/52/53/54	19/20/21/22/23/24/25/26/27/46/47/48/49/50/51/52/53/54	Heavy rainfall expected Provide adequate drainage to nursery
Seedling	17/18/26/27/44/45/53/54	17/18/26/27/44/45/53/54	Dry week anticipated and sucking pests may appear Spray Phosphomidan 40 SL @ 50 ml per 20 cents
Seedling	10/11/12/13/14/15/16/17/18/19/20/21/22/23/24/25/26/27/37/38/39/40/41/42/43/44/45/46/47/48/	–	By utilising past week rainfall, prepare main field for planting
Seedling	–	10/11/12/13/14/15/16/17/18/19/20/21/22/23/24/25/26/27/43/44/45/46/47/48/49/50/51/52/53/54	Heavy rainfall expected Provide adequate drainage. Postpone further transplanting for another 2 days
Seedling	19/20/21/22/23/24/25/26/27/46/47/48/49/50/51/52/53/54	19/20/21/22/23/24/25/26/27/46/47/48/49/50/51/52/53/54	Considering past week and anticipated rainfall, provide adequate drainage and postpone fertiliser application
Vegetative	17/18/26/27/44/45/53/54	17/18/26/27/44/45/53/54	Dry weather with increasing temperature favours sucking pest like thrips and mites. Spray Phosphomidon @ 500 ml/ha

Vegetative	28/29/30/31/32/33/34/35/36	28/29/30/31/32/33/34/35/36	Due to non-rainy days with heavy wind, avoid top dressing. Spray foliar nutrition non-windy during morning hours
Vegetative	1/2/3/4/5/7/8/9	1/2/3/4/5/7/8/9	Apply top dressing of urea and potash, as weather is favourable with no rain and light wind
Vegetative	17/18/26/27/44/45/53/54	17/18/26/27/44/45/53/54	Due to dry weather, swarming worm may appear. Spray Chloripyriphos @ 1250 ml/ha
Vegetative	10/11/19/20/28/29/37/38	10/11/19/20/28/29/37/38	Light drizzling with high humidity favours blast and leaf spot disease. Spray Carbendazin 50 WP @ 500 g/ha
Vegetative	46/47/48/49/50/51/52	46/47/48/49/50/51/52	Provide adequate drainage and more nitrogen fertiliser to flood affected crop
Reproductive	1/2/10/11/19/20/28/29/37/38/46/47	1/2/10/11/19/20/28/29/37/38/46/47	Low temperature is expected. Spray salicylic acid @ 120 mg/litre water
Reproductive	17/18/26/27/44/45/53/54	17/18/26/27/44/45/53/54	Due to dry weather and low humidity, stem borer and leaf hopper may appear. Spray Phosalone 35 EC 1500 ml/ha
Reproductive	–	19/20/21/22/23/24/25/26/27/46/47/48/49/50/51/52/53/54	Heavy rainfall expected Provide adequate drainage. Postpone fertiliser applications
Maturity	17/18/26/27/44/45/53/54	17/18/26/27/44/45/53/54	Due to dry weather, ear head bud may appear. Spray NSKE 5%
Maturity	1/2/10/11/19/20/28/29/37/38/46/47	1/2/10/11/19/20/28/29/37/38/46/47	Low temperature and high humidity favours blast and leaf spot. Spray Carbendazin50 WP @ 500 g/ha
Harvest	1/2/10/11/19/20/28/29/37/38/46/47	1/2/10/11/19/20/28/29/37/38/46/47	Low temperature and high humidity may cause discolouration in grains. Provide mechanical drying and store at 12% moisture

11.3.5.2 Software Platform

Usually HTML platform with PHP and MySQL backend are used in web portal applications. Any new programming tools such as Bootstrap, Jquery, JAVA spring boot can be added to improve easiness, quickness with security. The necessary license of all the applications should be obtained properly for getting future support.

11.3.5.3 Internet Connectivity and SMS Applications

Weather forecasting part of AAS requires downloading of large-sized files (6 GB) from the global network within short period of time (30 min). Continuous observed data supply from closer AWS network, SMS applications (20,000 to 30,000 SMS per day) and frequent reach by more than 1000 farmers in a day requires 1:1 leased line with >15 Mbps speed with minimum of 10 static IPs for uninterrupted two-way communications.

11.4 Design of Work in AAS Module

Step 1 Set coding to run the forecast model automatically and get weather forecast for next 6 days for required parameters by 9.00 a.m. and update the same in AAS folder. Once updated, data should not be edited or deleted.

Step 2 Set coding to past weather data from AWS database of all the AWS, calculate daily values and update in AAS folder by 9.00 a.m. Once updated, data should not be edited or deleted. Hence, check the data before uploading.

Step 3 Set programme to develop weather scenarios for both forecast and observed values separately for each locations (block/village) using weather scenario limits fed in AAS module database. Necessary options have to be given in front end for the admin to add, edit and delete weather scenarios. Provide ID for each scenario.

Step 4 Provide options for registration of farmer and technical officers.

- Active mobile number will be the user name and ID in AAS module.
- Registration form includes name, father's name, district, block, mobile number, DoB (year), education and land area.
- Farmer can generate his password and edit his details later except his mobile number.
- Need to be approved by admin/technical officer of his block before crop registration.
- After approval, a message has to be sent from AAS to his mobile number for the registration of crop and stage. Farmer can register only one crop per mobile number.
- Option to register the crop up to a month before sowing and any time after sowing.
- SMS should be limited to total duration/7 days (if 105 days = 15 numbers).

- Reminder for registration of new crop at the end of previous crop.
- Same method for technical officer's registration, but can register 10 crops of his block for easy monitoring. Registration of block officers should be approved by the admin.

Step 5 Store the weather-based agro advisories for every stage of crop with separate ID in database. For this, crop group, crop and crop stage have to be given with separate ID. Necessary option is given for the admin in front end to add, edit and delete the advisories.

Step 6 Provide adequate security for database to safeguard the information of farmer, block officers and advisories, which is amenable for misuse.

Step 7 Develop a web portal for registration of crop, profile maintenance, adding new crops, viewing advisories, past and forecasted weather, etc.

Step 8 Develop an android mobile application with maximum futures of web portal as everyone is having smartphone.

Step 9 Provide options to the admin to get reports on (i) active and inactive farmer, block officer; (ii) district/block/farmer/day-wise SMS advisory (iii) login details of every one and (iv) weather data of past and forecast for any duration with weather scenario details.

Design and work flow of AAS process is depicted in Fig. 11.8. The screenshots of TNAU-AAS web portal and Android app are depicted in Figs. 11.9 and 11.10.

Fig. 11.8 Design and work flow of AAS process

Fig. 11.9 TNAU-AAS web portal

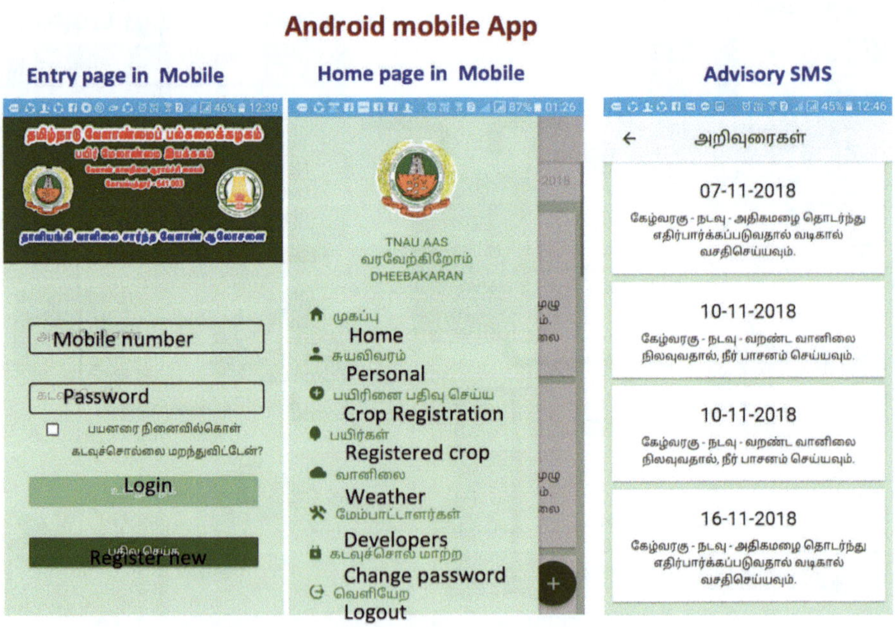

Fig. 11.10 TNAU-AAS Android apps

11.5 Stakeholder Involvement

Success or failures of any system depend on the involvement of the stakeholders. In this Automated Agro Advisory Service, there are three stakeholders, viz., farmer, technical officer and admin. Each and every one has separate tasks.

11.5.1 Farmer

As an end user of the AAS, the farmers should be actively involved and feed right information in the portal to get right advisories on time. At the same time, he should not blindly follow the advisories, and adopt the advisory based on his ground reality. Also, he should give feedback to the admin/block officers about the correctness or error in the advisories, any flaws in the weather data and any other information regarding the improvement of advisories.

11.5.2 Technical Officer

As the technical officer is the direct controller of block-level farmers, he/she should be actively involved to increase the number of beneficiaries. He should promote AAS properly to the farmers and make them to adopt the same. He should follow the adoption and monitor the success or failure of advisories. He should be the bridge between admin and farmer and have cordial relationship between them. He has to collect the feedback from the farmers and update it to the admin for the improvement of system.

11.5.3 Admin

Overall control is with the admin. The admin should consider the feedback from the end users to improvise the AAS module. He should not alter the database of disseminated advisory even if wrongly sent. He should regularly update the advisory information and upgrade the AAS module.

11.6 Usefulness of AAS

- Fully automated and lab-to-land transfer of technologies become direct and easy. Hence, reduce the work load of extension functionaries.
- Farmers can get advisories on time, and there are no holidays as it is automated.
- Expertise of technocrats is not required continuously except simple monitoring.

- Surely reduce the crop failure risk of climate-dependent farming.
- AAS is the viable option to do weather-based precision and response farming.
- Can reach the every corner of the nation without much monetary involvement.
- Farmer won't be confused with general messages as of now, and he would get only the messages pertaining to his crop and stage.

Future Thrust

- Adding new crops, perennial horticultual crops and animal husbandry.
- Incorporating dynamic weather calendar with soil moisture, and length of growing period.
- Feedback from the users for the success/failure on interactive mode.

11.7 Conclusion

Automated weather-based Agro Advisory Service (AAS) is the best option for response farming, precision farming and to exert sustainability in the farming under climate variability and climate change. It will surely help to reduce the risk of weather vagaries and increase the productivity of inputs. It will help in improving farmers' standard of living and national food security. AAS is considered to be a boon for farming community, since all agricultural operations are weather dependent. It empathises the farmers' need and acts weather smart.

Acknowledgement The authors acknowledge the National Agricultural Development Programme (GoI and GoTN) for the funding towards the 'Development of Automated Agro Advisory Service – Web cum Mobile App'. We also acknowledge the retired agronomists Dr. T.N. Balasubramanian and Dr. R. Jagannathan for their contribution in basic structure development and Dr. N. Maragatham and other TNAU scientists and research fellows who were also involved in "TNAU-AAS" development project.

References

Ahanger RA, Bhat HA, Bhat TA, Ganie SA, Lone AA, Wani IA, Ganai SA, Haq S, Khan OA, Junaid JM, Bhat TA (2013) Impact of climate change on plant diseases. Int J Mod Plant Anim Sci 1(3):105–115

Chattopadhyay N (2011) Climate change and food security in India. In: Lal R et al (eds) Climate change and food security in South Asia. Springer, Amsterdam, pp 229–250. https://doi.org/10.1007/978-90-481-9516-9_15

Cline WR (2007) Global warming and agriculture: impact estimates by country. Peterson Institute for International Economics, Washington DC, p 250. ISBN:978-08-813-2403-7

Dheebakaran Ga, Kokilavan S, Panneerselvam S, Jagannathan R (2016) Climate change on shift in rainy days and rainfall quantity over Tamil Nadu. In: Palanivelu K (ed) Proceedings of national conference on "Climate change impacts and natural resource management" Anna University, Chennai, 18–19 February 2016

FAO (2007) World hunger increasing. In: Food and Agriculture Organization (FAO). 30 October 2006. Accessed on 2011-07-07

India Water Review (2017). http://www.indiawaterreview.in/Story/Features/india-outranks-us-china-with-worlds-highest-net-croplandarea/2096/2#.XAx97idS_dd Accessed 7 Dec 2017

IPCC (2014) Climate change 2014: synthesis report. contribution of Working Groups I, II and III to the fifth assessment report of the Intergovernmental Panel on Climate Change [Core Writing Team, Pachauri RK, Meyer LA (eds)]. IPCC, Geneva, Switzerland, 151 pp

IPCC (2013) Summary for policymakers. In: Stocker TF, Qin D, Plattner G-K, Tignor M, Allen SK, Boschung J, Nauels A, Xia Y, Bex V, Midgley PM (eds) Climate Change 2013: The physical science basis. Contribution of Working Group I to the fifth assessment report of the Intergovernmental Panel on Climate Change, Cambridge/New York, Cambridge University Press

Jagannathan R, Balasubramanian TN, Geethalakshmi V, Maragatham N (2014) Basic record of weather scenario and agro advisories for TNAU's Automated Agro advisory development

Porter JR, Xie L, Challinor AJ, Cochrane K et al (2014) Food security and food production systems. In: Field CB, Barros VR, Dokken DJ, Mach KJ, Mastrandrea MD, Bilir TE, Chatterjee M, Ebi KL, Estrada YO, Genova RC, Girma B, Kissel ES, Levy AN, MacCracken S, Mastrandrea PR, White LL (eds) Climate change 2014: impacts, adaptation, and vulnerability. Part A: Global and sectoral aspects. Contribution of Working Group II to the fifth assessment report of the Intergovernmental Panel on Climate Change. Cambridge University Press, Cambridge/New York, pp 485–533. Available at: http://www.ipcc.ch/pdf/assessment-report/ar5/wg2/WGIIAR5-Chap7_FINAL.pdf

Sunder S (2018) India economic survey 2018: Farmers gain as agriculture mechanisation speeds up, but more R&D needed in www.financialexpress.com. Updated January 29, 2018 https://www.financialexpress.com/budget/india-economic-survey-2018-for-farmers-agriculture-gdp-msp/1034266/. Accessed 7 Dec 2017

Venkatramanan V, Shah S (2019) Climate smart agriculture technologies for environmental management: the intersection of sustainability, resilience, wellbeing and development. In: Shah S et al (eds) Sustainable green technologies for environmental management. Springer, Singapore, pp 29–51. https://doi.org/10.1007/978-981-13-2772-8_2

Venkatramanan V, Singh SD (2009a) Differential effects of day and night temperature on the growth of rice crop. Pusa Agric Sci 32:57–62

Venkatramanan V, Singh SD (2009b) Differential effects of day and night temperature on the growth of wheat crop. Ann Agric Res 30(1&2):49–52

Climate-Smart Agriculture: Assessment and Adaptation Strategies in Changing Climate

12

Muhammad Arif, Talha Jan, Hassan Munir, Fahd Rasul, Muhammad Riaz, Shah Fahad, Muhammad Adnan, Ishaq Ahmad Mian, and Amanullah

Abstract

Climate change is the most critical threat to food security amid increasing crop demand. This increasing demand for food has been previously tried to be met through the use of synthetic fertilizers and effective application of weed- and pest-controlling chemicals. However, these methods of increasing crop productivity rely on finite resources and are often unsustainable. They are now proven to be posing a great threat to the environment and causing a negative change in the planet's natural climate. Fortunately, the threat has been realized by scientists, and the world has started to lay the foundations for sustainable intensification of agriculture and to heighten the resilience of crops to climate change. The solutions discovered so far are numerous with many of them not yet tested. Climate change assessment is the first priority in this regard. Much of the recent researches have demonstrated a multi-scale and multidimensional nature of climate change to assess the potential effects of climate change on agriculture and the options for adaptation. These options for adaptation have been different in different regions of the world with clear differences among strategies in rich and

M. Arif (✉) · T. Jan · Amanullah
Department of Agronomy, The University of Agriculture, Peshawar, Pakistan
e-mail: marifkhan75@aup.edu.pk

H. Munir · F. Rasul
Department of Agronomy, University of Agriculture, Faisalabad, Pakistan

M. Riaz
Department of Environmental Sciences and Engineering, Government College University, Faisalabad, Pakistan

S. Fahad · M. Adnan
Department of Agricultural Sciences, The University of Swabi, Swabi, Pakistan

I. A. Mian
Department of Soil and Environmental Sciences, The University of Agriculture, Peshawar, Pakistan

© Springer Nature Singapore Pte Ltd. 2020

V. Venkatramanan et al. (eds.), *Global Climate Change and Environmental Policy*,
https://doi.org/10.1007/978-981-13-9570-3_12

poor countries. The pressure for adaptation is greatest in poor countries where the adaptive capacity is least abundant. Adaptation to climate change could be autonomous (market-driven) or planned. Both of these adaptation strategies are driven by certain measures. Some adaptation strategies are easily achieved with the help of existing technologies, some need development of new technologies while others just need policy and institutional/market reforms. Numerous researchers have tried to assess and give tools for the potential impact of climate change which are largely based on modelling techniques. Indeed, models are useful tools for assessing this potential impact and evaluating the options for adaptation, yet they do not match the level of real solutions that could be brought about by efficient adaptive human agency. The importance of agriculture as performance is useful in counterbalancing the modelling approaches towards mitigating the negative impacts of climate change. The adaptation and mitigation strategies are and should be social phenomena which need social attendance in the form of improved and sustainable agricultural practices and could help agriculture contribute less to the changing climate. This chapter will focus on numerous strategies that could be adapted to assess and cope with the negative impacts of changing climate on agriculture.

Keywords
Climate change · Climate-smart agriculture · Adaptation · Mitigation · Food security

12.1 Introduction

Climate change has emerged as a real threat to agriculture, food security and livelihood of millions of people around the globe (IPCC 2014). Major components of climate affect the actual and future projected crop yields. Some of these components determine yield potential (CO_2 concentrations, temperature and radiation), some limit these yields by the extent of their availability (nutrients and water) and others reduce it by their presence in the form of biotic (diseases, weeds and pests) and abiotic (e.g., waterlogging, salinity and ozone) stresses (Neumann et al. 2010; van Ittersum et al. 2013) Almost every climatic factor has dramatically changed in its intensity over the past few decades. Global temperature has increased by 0.74 °C over 1880–2012 and is expected to increase more by 0.2 °C per decade with a final increase of 2–4 °C by the end of the twenty-first century (IPCC 2007). CO_2 concentration has risen to 391 ppm in 2011, 40% above that of the pre-industrial era (IPCC 2014) and, presently, the CO_2 concentration is 410 ppm. Climate change has its severe effect on water availability, and it is estimated that nearly 20 million hectare (mha) area will be water-scarce by 2025 (Bouman et al. 2007). Among other abiotic stresses indirectly caused or amplified by climate change are salinity and air pollution. With changing climate, extreme weather events may increase in intensity, frequency, duration and timing. Climate change may affect agriculture directly through increased CO_2 levels or indirectly by changes in the patterns of solar radiations and the subsequent heat. Sea level rise may inundate coastal lands. Air pollution will block solar radiation thwarting crop outputs (Tang et al. 2013). Drought and floods may increase in severity. Pest and weed incidences may escalate. Precipitation in many high latitude and equatorial pacific regions may increase while midlatitude subtropical regions may become drier (IPCC 2014).

12.2 Contribution of Agriculture to Climate Change

The N_2O, CH_4 and CO_2 concentrations were 270 ppb, 722 ppb and 280 ppm, respectively in 1750 A.D. These values rose to 324 ppb, 1803 ppb and 391 ppm, respectively, by 2011 (IPCC 2014). It has been confirmed that these values are higher than at any time in the last 65,000 years (Long et al. 2015). Between 1961 and 2005, food production increased to 4.8 from 1.8 billion tons per annum while cropland increased to 1208 from 960 m ha. Average crop yield increased to 3.96 tons per ha from 1.84 tons per ha during this period. Agriculture now contributes significantly to climate change as greenhouse gas emissions due to many current farming practices. Some authors put it at 19–29% of the total greenhouse gas emissions (Vermeulen et al. 2012) with significant variability worldwide. If land-use change is also considered, agriculture is responsible for the generation of 30% of greenhouse gas production (FAO 2012). Most of this contribution is in the form of direct emissions from agriculture systems as CH_4 and N_2O emissions or indirectly in the form of energy consumption driven by agriculture (CO_2) (Vermeulen et al. 2012). By 2030, greenhouse gas emissions are predicted to rise almost 40 per cent (Smith et al. 2008).

12.3 Effect of Climate Change on Agricultural Productivity

Climate change is projected to have both negative and positive effects on agriculture systems around the world with negative impacts overweighing the positive ones (Müller 2013). Climate change may negatively impact agriculture by affecting farming structure and planning, destroying agro-climatic resources incurring agrometeorological and biological disasters and reduction in crop growth and yields (Cohn et al. 2016). Several studies have indicated that agriculture production could be adversely affected by rising temperatures (Aggarwal 2009), changes in precipitation patterns (Prasanna 2014) and variations in the intensity and frequency of extreme climatic events such as droughts and floods (Singh et al. 2017). Studies report that "the daily minimum night time temperature increased at a faster rate than daily maximum temperature in the last century" and the differential effects of increase in day and night temperatures were observed on the growth, development and yield of rice (Venkatramanan and Singh 2009a) and wheat (Venkatramanan and

Singh 2009b). This study of increases in day and night temperatures on rice and wheat is significant as the rice–wheat system drives food security in South Asia and China. The frequency and severity of floods and droughts will be increased in the coming decades, posing risks for both croppers and livestock keepers (Thornton and Gerber 2010). The risk has been further increased by climate change and weather variability. Climate change will have more impact on crop productivity in the lower altitudes (Stocker 2014).

Climate-driven diseases caused an estimated 16% reduction globally for unprotected crops (Oerke 2006). Climate change–induced yield reduction is 1–5% over the last 30 years (Porter et al. 2014). A greater impact is projected to be on cereal crops. Climate change has negatively affected maize and wheat crop yields in many regions on a global average (Knox et al. 2012) but the effect on rice and soybean is small (IPCC 2014). Wheat and maize yields have already been reduced by 5.5% and 3.8%, respectively, since 1980 (Lobell et al. 2008). Climate change is predicted to reduce crop yields by 20% in south Asia and Sub-Saharan Africa with 30% and 40% risk of crop failure for wheat and maize, respectively (Lobell et al. 2008; Thornton and Gerber 2010). The anticipated impacts of climate change on cereal crop yields in different parts of the world are estimated to be −20% for wheat, −35% for rice, −60% for maize, −50% for sorghum and −13% for barley depending on the projected year, location and future climate scenarios (Porter et al. 2014).

12.4 Effect on Food Security

World food production is expected to increase by 70 million annually (Popp et al. 2013) with the projection that the world's population will be up by two billions by 2050 (FAO 2013). The world crop area increased by only 12% over the last 50 years (1969–2009) while the population doubled, causing per capita crop land globally to fall to 0.25 ha from 0.44 ha. World demand for food has been estimated to increase by 60% by 2020 (FAO 2012). This has led to the fact that for now more than a billion people are going to bed hungry every day. Moreover, agriculture employs 1.3 billion small holders and landless workers who are likely to be severely affected by climate change. World edible crop production is under risk due to the production of biofuel crops, decrease in chemical inputs for mitigation purposes and installation of solar farms on arable lands. Increase in demand always ends up with price hike. Almost 20% of local food prices in 51 countries were affected by domestic weather variability between 2008 and 2012 (Brown and Kshirsagar 2015). The adverse effects of climate change have more severely affected the poor people in developing countries, resulting in retarded economic growth and undermining poverty alleviation (Brown and Kshirsagar 2015). In wealthier nations, the problem has a different face. In addition to increase in global population and subsequent increase in food demand, the food consumption patterns are also changing. People are getting wealthy, and they are now consuming more food and meat, hence, increasing competition for energy, land, water and other inputs in food production.

12.5 Benefits of Climate Change to Agriculture

Increase in atmospheric CO_2 concentration may boost production for various crops (Franzaring et al. 2008; Miglietta et al. 1998; Qaderi et al. 2006). Increase in CO_2 concentration might be beneficial for plant productivity with significant biomass increase for C3 plants. Other potential benefits include decreased stomatal conductance and oxidative stress, higher abiotic stress tolerance, increased root growth and high water-use efficiency (Lopes et al. 2016). For instance, Amthor (2001) compiled and compared 50 studies on the impact of increased CO_2 concentration on wheat and found that doubling of CO_2 concentration to 700 from 350 ppm enhanced wheat yield by 31%, subject to ample supply of nutrients and water. However, findings of yield enhancement in free air CO_2 enrichment are fewer compared to enclosure studies (Long et al. 2016). These findings reflect the output produced in control conditions and do not necessarily translate for productivity in agricultural ecosystems. CO_2 concentration increase will be accompanied by increased global warming. A modest warming trend (1–4 °C) may cancel out the beneficial effects of increased CO_2 levels on plant productivity (Lopes et al. 2016).

12.6 The Way Forward

Farmers across the globe, with many of them not yet realizing the climate change threat, may face difficulties adopting suitable practices in a real climate change scenario. Currently, the existing farmland too is reducing as we lose more land to salinity, waterlogging and, above all, erosion. The whole system needs a complete rethinking of food systems. Meeting world food demand needs to be met while saving the planet. It is getting increased production from existing farmland in such a way that it has little impact on the environment and which do not undermine our capacity to produce more food in future (Garnett et al. 2013). The solution lies in two concepts, namely, sustainable intensification and climate-smart agriculture. Although the two terminologies refer to two different approaches, climate-smart agriculture is the key to achieve the goal of sustainable intensification.

12.7 Climate-Smart Agriculture

Climate-smart agriculture (CSA), a relatively new concept that lies at the interface between science and policymaking, was proposed by FAO in 2010 at The Hague Conference on Agriculture, Food Security and Climate Change (FAO 2013). Two processes developed parallel soon after the Hague Conference. On one hand, the process of policymaking was initiated which resulted in the making of a "global alliance for climate smart agriculture" in 2014 and on the other hand, a scientific processes was started in the form of worldwide conferences on CSA (Saj et al.

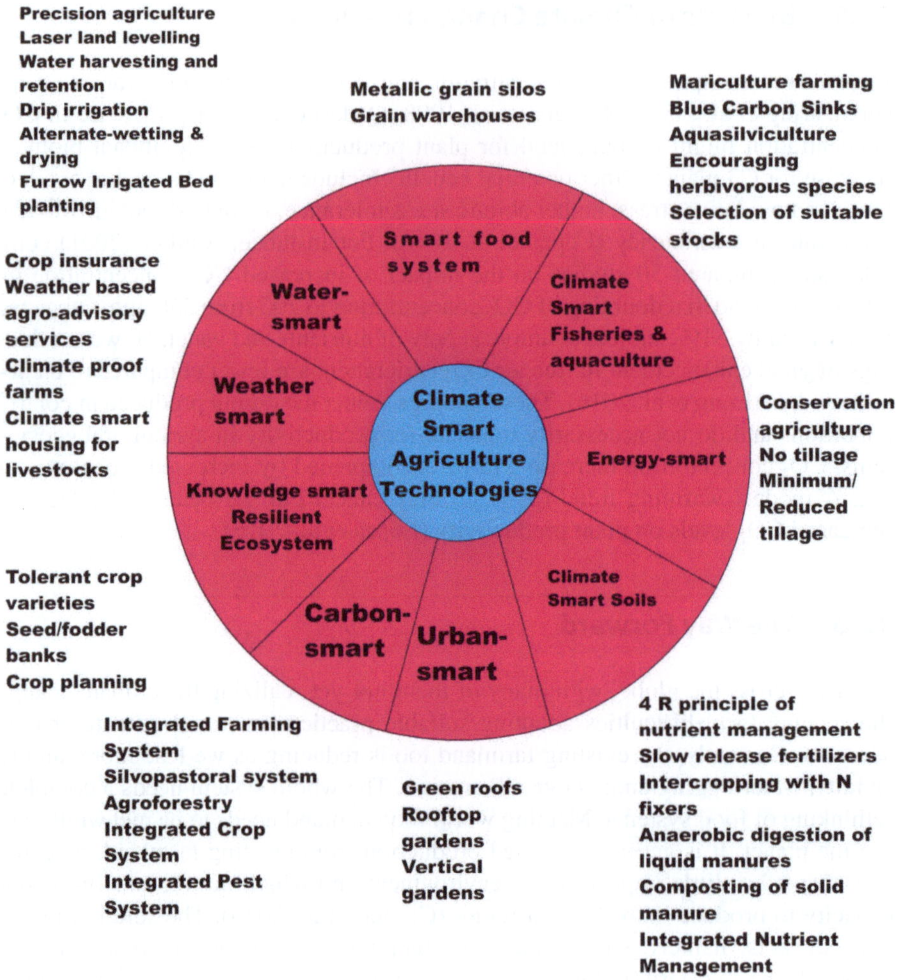

Fig. 12.1 Climate-smart agriculture technologies. (Source: Venkatramanan and Shah 2019)

2017). Climate-smart agriculture can be clearly defined by three objectives. Firstly, increasing agricultural productivity for enhanced food security; secondly, increasing the adaptive capacity of the food production system to a changing climate; and thirdly, reducing greenhouse gas emissions from current and intensification practices. To achieve the triple objectives of climate-smart agriculture, there is a dire need of innovative climate-smart agriculture technologies (Fig. 12.1), which aid in conserving agricultural resources, achieving food security and tapping the mitigation potential of the agriculture system.

12.7.1 Adaptation

Adaptation means adjustment to the expected or actual impact of climate. Adaptation includes agronomic management for soil, water, nutrients, pest- and weed-adjusting cropping system and distribution (Bonzanigo et al. 2016; Rippke et al. 2016), conservation agriculture (Powlson et al. 2014), plant breeding and biotechnology (Abberton et al. 2016), strengthening infrastructure construction and enhancing disaster prevention.

Negative impacts of climate change can be ameliorated through adaptation strategies that could range from minor changes in production practices to transformative reforms in production policies. The adaptive capacity may be built in such a way that farmers, farmers' service providers and institutions respond effectively to long-term climate variability. The adaptation strategies are diverse. The main components are efficient soil, water and plant nutrient management which could be achieved by improved irrigation systems and on-farm water management, breeding and making access to crop varieties that are resistant to drought, heat, salinity, flooding conditions and other climate-related threats, building the capacity of institutions to enhance actions, disseminate knowledge and pave the way for efficient policy reforms. Actions vary around the globe with clear differences between poor and rich countries due to differences in adaptive capacities.

12.7.2 Mitigation

In a changing climate context, two challenges, meeting global food demand and reducing the amount of greenhouse gases emitted per unit of food production, go parallel. With increasing population, going back to organic farming and still meeting the world food demand is seemingly impossible; however still, achieving lower N_2O and CH_4 emissions per unit of output is the major objective of CSA. All practices that could reduce the emissions of these gases can be termed climate change mitigation strategies. Mitigation also includes soil friendly practices to control the rate of land loss to drought, salinity, and waterlogging and soil erosion. However, keeping in mind the need for producing more, getting more out of existing farmland without damaging the environment, is the main component of climate change mitigation. Mitigation may also include a practice of not doing anything in the form of reducing land cover changes, especially wetlands and carbon-rich forests (Wollenberg et al. 2011).

12.8 Adaptation Strategies

Climate change adaptation includes "initiatives and measures to reduce the vulnerability of human and natural systems facing actual or expected impacts of climate change" (IPCC 2007). Adaptation takes into account the implementation of a range of strategies that include introduced and local biotechnology, hard technologies (equipment, machinery and tools), soft technologies (knowledge, awareness raising and capacity building) and organizational technologies (resource user organization and institution building). Climate change adaptation practices can be adopted in almost every field in agriculture with more focus on saving and harvesting water, managing soil for water and nutrient conservation, nutrient management, crop production practices and improved livestock management.

12.8.1 Water Management Practices

Water management practices for climate change mainly focus on saving water or coping drought stress and ways of harvesting water from natural sources.

12.8.1.1 Sprinkler Irrigation

Sprinkler irrigation improves water-use efficiency and contributes to food production in a changing climate characterized by heat and drought stresses. It works well under limited water supply. It also performs better in high temperatures by supplying colder water and also reduces the risk of crop freezing due to low temperatures (Hodgkinson and Smith 2018).

12.8.1.2 Drip Irrigation System

Drip irrigation system is used to ensure a constant supply of water through pipes and valves to the root of plants. It has the most efficient water supply (90%) compared to sprinkler (75%) and flood irrigation (60%). Drip irrigation works well in arid and semi-arid areas. It has not only high water-supply efficiency but also nutrients can be supplied through this system, reducing risks of leaching or volatilization (Nigussie et al. 2018).

12.8.1.3 Fog Harvesters

Fog harvesters are simple and low-cost collection systems that are an alternative source of fresh water for agricultural irrigation and domestic use in dry regions. Fog harvester works best in coastal areas with long fog periods and mountainous regions whose height range from 400 m to 1200 m (UNEP 1997).

12.8.1.4 Rainwater Harvesters

Rainwater harvesters range from small water tanks to large reservoirs and dams in areas with no surface water, where groundwater is too deep to draw or where drinking and irrigation water are too salty or acidic (Pillay and Kalu 2012).

12.8.2 Soil Management Practices

Climate change already has and is projected to affect soil negatively through increase in CO_2, changes in vegetative cover, sea level rise, changes in temperature and rainfall patterns and human activities. The climatic origin of these problems is not fully certain but still some soil management practices are found helpful (Brinkman and Sombroek 1996).

12.8.2.1 Terrace Farming

Terrace is a levelled surface used for farming in slope regions in areas where soil and climatic conditions are conducive to erosion. Benefits of terraces include control in water and wind erosion, increase in soil moisture retention and improvement in general agricultural conditions. Besides controlling erosion and retaining moisture in a changing climate, terraces also capture heat during daytime and release it at night-time, hence, protecting crops against frost (Mars 2005).

12.8.2.2 Conservation Tillage

Conservation tillage refers to soil management practices that allows establishment of crops in previous crops residues. Soil is minimally disturbed. This helps in slowing down the water movement and hence controls soil erosion. Conservation tillage is widely practiced in Latin America and has huge potential to be introduced to Europe, Asia and Africa (Derpsch 1999). Indeed, the conservation tillage that includes minimum/reduced tillage is "energy-smart technology" as it conserves not only the soil ecosystem, but also energy by minimizing soil disturbance (Venkatramanan and Shah 2019).

12.8.2.3 Integrated Nutrient Management

Although chemical fertilizers have helped attaining higher yields amid rapidly increasing population, they have environmental costs. On the other hand, organic farming has very less productivity potential and is expected to eventually fail if tried to practice globally in the context of current food demand. An integrated approach towards the combined use of chemical and organic fertilizer has great adaptation capacity. The organic part of integrated nutrient management has the potential to minimize many risks posed by climate variability such as increased pests and diseases, erosion and others. Root growth in the presence of organic matter is high and, hence, can cope with drought stress. Moreover, organic matter–containing soils have high water-holding capacity.

12.8.3 Biotechnology

Climate change is sudden and current crops and varieties at hand are mostly susceptible to it. In such a scenario, the need for crop diversification has increased manifold.

12.8.3.1 Breeding New Varieties

Breeding new varieties that are resistant to different kinds of climatic stresses such as heat, drought, salinity and pests is a promising adaptation strategy (Mottaleb et al. 2017). Resistant varieties are proposed to experience and incur minimum losses under unfavourable conditions in a changing climate. Besides high yields and resistance, varieties can also be bred for improved nutritional status, providing benefits to animals and humans alike.

12.8.3.2 Crop Diversification

It refers to the introduction of new crops to an agricultural system to increase biodiversity and gain economic return. It also helps farmers identify crop species that can thrive best in the environmental conditions of a particular locality. This minimizes the risk of total crop failure in case of sudden climate change or in times of disaster. Natural biodiversity is increased, and the ability of agroecosystem to severe stresses is strengthened.

12.8.3.3 Genetic Engineering

Conventional breeding has been proved promising in developing pest- and disease-resistant varieties along with increased yield performance. However, this process is slow and is limited to the exploitation of the existing genetic variation between crops and their nearest relatives. Biotechnology and genetic engineering are rather quick responses to this problem. The first generation of genetically modified plants was introduced in 1996 and produced genetically modified maize, soybean and cotton that had tremendous resistance to pests and tolerance to herbicides. Promising results have also been shown in transgenic canola, alfalfa, squash and papaya. Genetic modification has been limited to pest and herbicide resistance and to date no genetically modified drought- and heat-tolerant variety has been released. However, some of the related problems are being addressed, for example, transgenic rice with the introduction of HRD gene has improved its water-use efficiency and the ratio of biomass produced to the water consumed has been increased, mainly because of enhanced photosynthetic capacity and reduced respiration (Karaba et al. 2007).

12.8.4 Seed and Grain Storage

Secured seed and grain storage is among the major challenges faced by resource-poor farmers in developing countries (Wambugu et al. 2009). Seeds must be stored against biological damage by insects, rodents and microorganisms, chemical damage by acidity and physical damage due to poor post-harvest management. Recent advancement in technology has ensured safety in post-harvest operations, improved storage conditions, helped attain safe moisture levels and provided safety against insects and rodents. However, poor economic conditions in underdeveloped countries have prevented farmers of those countries to benefit from the technology. Efficient seed and grain storage ensures food security of both animals and humans in case of prolonged drought or other natural disasters. In fact, seed and grain storage conditions and reserves of food and other agricultural inputs have been used as indicators of the adaptive capacity of nations (CARE 2010).

12.8.5 Cropping System Management

Much of the productive and ecological resilience to climate change comes from managing crops and livestock diversity which help farmers maintain their agricultural productivity.

12.8.5.1 Mixed Farming

Mixed farming is composed of several practices that involve the cycling of inputs and outputs within the farms or between several farms. Forage or fodder crops are fed to livestock and in return animal excreta is used as fertilizer for the crops. Intercropping and crop rotation are also included in mixed farming wherein legumes benefit grain crops by their supply of nitrogen. Another example of mixed cropping is combined chicken–fish farming where chicken waste serves as fish feed. Mixed farming has greater adaptive capacity in a changing climate because of the diversification of crops under uncertain weather conditions. Failure of one crop can easily be compensated by another crop. Moreover, livestock is a valuable asset and in fact a walking bank that could be sold in case of crop failure due to prolonged drought or flooding.

12.8.5.2 Agroforestry

It is an integrated approach towards farming in a world of changing climate. In this system, trees and non-tree crops are sown either together at same time, in rotation or in separate plots. Materials from both crops and trees benefit one another. Agroforestry has great adaptive capacity. Trees can withstand severe climatic conditions in the form of drought and winds. Their deep-rooted nature helps them explore water and nutrients from deep grounds and bring them up. They have higher evaporation rates and hence keep the soil aerated (Martin and Sherman 1998). Trees also significantly reduce erosion. A combination of leguminous shrubs and Napier grass in contour hedgerows was reported to reduce erosion by 70% on 10% inclination slope in central Kenya (Mutegi et al. 2008).

12.8.6 Priming

Priming is the technique of pre-exposure of plants to an eliciting factor, enabling plants to cope with later stress events. It is a cost-effective strategy that improves plant tolerance to stress. In plants pre-primed with abiotic stresses, beneficial microbes or pathogens showed stronger responses against biotic and abiotic stresses as compared to non-primed plants in the same generation (Conrath et al. 2015). Various priming mechanisms induce the regulation of primary metabolism, chromatin modification, increase in levels of pattern recognition receptors and accumulation of nitrogen-activated protein kinase (Balmer et al. 2015; Conrath et al. 2015).

12.8.6.1 Priming for Heat Tolerance

Heat tolerance depends on the ability of plants to perceive the stimulus as well as biochemical and physiological adjustments (Hasanuzzaman et al. 2013). It has been found that thermopriming has successfully induced tolerance in plant species to heat that recurred in the later growth stages of plants. In wheat, when day/night temperature was increased by 8 °C as compared to control in 7 and 9 leaf stages for 2 days, enhanced heat tolerance of wheat at anthesis was observed. Primed plants showed higher photosynthetic capacity, higher activity of peroxides and glutathione reductase in mitochondria and superoxide dismutase in chloroplast. This increased activity of hormones resulted in lower damage caused to cell membrane which was indicated by lower content of malondialdehyde in the chloroplasts and mitochondria of wheat leaf. Also more starch was found in the grain of primed plants (Wang et al. 2014).

12.8.6.2 Priming for Cold Stress Tolerance

Cold stress has been found to damage plant cell membrane. Cold stress enhances the overproduction and ability of reactive oxygen species (ROS) to damage cell membrane. Cold priming enhances the activity of ROS-scavenging enzymes and hence checks their negative activity (Thomashow 1999). Exposure of plants to cold temperature before anthesis is known to reduce the effects of cold stress. This could be attributed to the accumulation of metabolites such as proline, sucrose and other osmolytes induced due to cold priming (Iba 2002). Cold priming with 10 °C (5 °C lower than the ambient temperature) for 7 days at tillering could alleviate the negative effects of cold stress at jointing stage. This can be attributed to the upregulation of genes encoding ARX, SOD and GR in chloroplast and mitochondria induced by priming. All these activities help protecting photosynthetic apparatus by protecting cell membrane (Li et al. 2014).

12.8.6.3 Priming for Drought Stress Tolerance

Many researchers have found drought priming effective against drought stress at later growth stages (Selote and Khanna-Chopra 2010; Wang et al. 2015). Selote and Khanna-Chopra (2010) exposed wheat seedlings (21 days old) to mild drought stress for 9 days and then well-watered them for 2 days. They found that the wheat plants performed excellent during a subsequent 11-day severe drought. They attributed this tolerance to the activation of enzymes and non-enzyme processes in the ascorbate glutathione cycle, maintaining plant water status through maintenance of turgor potential and redox homeostasis. Wang et al. (2015) got enhanced wheat yield due to enhanced photosynthetic activity and improved ROS-scavenging capacity in comparison with non-primed plants.

12.8.6.4 Priming for Waterlogging Stress

Waterlogging stress can be ameliorated by the formation of aerenchyma, which facilitates oxygen diffusion from air to root tips (Colmer and Voesenek 2009). Li et al. (2011) exposed wheat at 7–9 leaf stage and heading stage to waterlogged conditions and found enhanced tolerance to waterlogging at grain-filling stage. They also found higher chlorophyll content, improved photosynthetic activity and enhanced light use efficiency in waterlogged primed plants. Wang et al. (2016) conducted proteome profile on wheat plants primed for a week and found the induction of proteins related to energy production, protein storage and destination and stress defence.

12.8.7 Pest Control Adaptations

Several practices are promoted for pest control through natural predator strategies.

12.8.7.1 Reduced Tillage

Reduced tillage retains soil moisture and protects against drought (VandenBygaart 2016). It also causes a significant reduction in GHG. Conservation tillage residues also help against drought and soil erosion (Henneron et al. 2015). In the context of combating pests, reduced tillage increases ground predators (spiders and staphylinids) in maize and soybean (Crowder et al. 2010). Conservation tillage in wheat provides habitat for natural predators (Rivers et al. 2016).

12.8.7.2 Intercropping

Intercropping helps agriculture by improving yields, increasing water-use efficiency and reducing erosion (Hassanali et al. 2008). It repels insects through several ways, firstly by releasing volatiles, secondly by masking volatiles released by other crop plants and lastly by providing alternative food in the form of less important intercropped crops (Lopes et al. 2016). More than 30,000 farmers in east Africa have adapted a push–pull strategy in which, for instance, maize is intercropped with legumes with Sudan, Napiers or molasses grasses at borders. This can pull or push insects (Khan et al. 2011).

12.8.7.3 Cover Cropping

Cover cropping reduces moisture loss, erosion and reduces greenhouse gas emission from otherwise fallow land. Cover crops offer habitat to predator insects, thus increasing natural predator population. More than 39 parasitoids and 2 predator species were attracted by buckwheat in Florida (Campbell et al. 2016).

12.8.7.4 Organic Farming

Organic fertilizers such as the use of compost and manures can boost beneficial insects while suppressing others. Significantly higher predator population of beneficial insects was observed in manure-applied plots in alfalfa during maize–alfalfa rotation. The possible reason for this might be the more diverse soil biome due to lack of chemicals (Garratt et al. 2011).

12.8.7.5 Biochar

Biochar has been found to reduce greenhouse gas emissions from carbon-depleted acidic soils (Dickie et al. 2014). Fecundity and development rates of rice brown plant hopper have been reported to decrease with application of biochar (Hou et al. 2015).

12.8.8 Genetic Engineering for Adaptation

Novel genetic and epigenetic variation through transportation activation and differential methylation can be induced in plants through exposing them to stressful environments (Cavrak et al. 2014). These epigenetic changes are considered drivers of climate change mitigation and arise more frequently than genetic mutation (Becker et al. 2011). However, these changes are challenged by instability and impersistence across generations which are critical for the implementation of breeding programmes. Selection under optimal conditions prevents realization of full genetic potential of the crop while selection under stressful conditions may help plants uncover their genetic potential (Des Marais et al. 2014).

12.8.8.1 Root Architecture Manipulation

Manipulation of root architecture can make plants able to cope with multiple stresses such as water deficit and nutrient deficiency. This can be exemplified by Deeper Rooting 1 (DRO1) locus in rice. It enables rice root system to become more vertical and go deeper improving drought tolerance and enhanced nitrogen acquisition (Uga et al. 2013). Overexpression of Cytokinin catabolic enzyme CKX resulting in root specific cytokinin degradation has also reported to increase drought tolerance and accumulation of micro- and macronutrients both in tobacco and Arabidopsis. Expression of botanical RNA Chaperon CspB in maize has shown higher yield under drought conditions (Werner et al. 2010).

12.8.8.2 Stomatal Function Regulation

Regulation of stomatal functions to achieve low canopy temperature is also a target for optimal crop performance in multiple stresses including drought and heat stress as well as resistance against stomatal invading pathogens (Lawson and Blatt 2014). Higher transpiration rate can prove detrimental leading to moisture loss especially in drought conditions. In this case, stomatal closure is preferred; however stomatal closure can amplify heat stress (Blum 2015). Engineering ABA receptors to non-agonist chemical receptors can contribute to field scale manipulation of stomatal functions (Park et al. 2015).

12.8.8.3 Increasing Photosynthetic Efficiency

This process through genetic engineering might unlock a great set of unrealized crop potential and maximize positive impact of increasing CO_2 levels (Long et al. 2016). Engineering a Rubisco protein with higher carboxylase catalytic activity has been reported to increase the net photosynthetic efficiency (Lin et al. 2014).

12.8.8.4 Enhancing Fertilization

Floral fertility and sustained fruit set is of crucial importance to plant productivity under stressful conditions. Ear-specific expression of trehalose-6-phosphate phosphatase has shown to increase yield and kernel set in maize under drought conditions (Nuccio et al. 2015).

12.8.8.5 Enhancing Nutrient Status

Optimal nutritional status is of crucial importance to getting higher yields under stressful conditions. Certain genes are involved in the facilitation of uptake, transport and assimilation of nutrients. Examples are the Phosphate Efflux Transporter (PH1) and the NRT1 and NRT2 transporter family which are reported to mediate nitrogen uptake in crops (Hu et al. 2015; Schroeder et al. 2013).

12.8.8.6 Resistance to Pathogens

Building pest resistance is quite simple in comparison to abiotic stresses. Deployment of *mlo* mutants can be exemplified which has successfully shown to induce resistance against powdery mildew in various crops (Appiano et al. 2015). The first generations of commercial transgenic insect-resistant and herbicide-tolerant crops have been highly successful. They have been cultivated on more than a billion hectares while the second generation of many abiotic stress tolerant crops is struggling to make its way into the market (Klümper and Qaim 2014). The complex nature of these stresses and plant response pathways seems to pose a challenge to the spread of this generation.

12.8.8.7 Domestication of Extremophiles

Extremophiles have evolved and reproduced in severe stress environments such as drought, salinity and others. Adoption and domestication of extremophiles seem to be a feasible strategy in developing agriculture in stressful environments. They can be further improved through the use of modern genetic tools and knowledge of domestication events of other crops. Novel genes/alleles can be taken from extremophiles and used in other crops to make them tolerant too. Some extremophile species are given in this regard. *Miscanthus* spp. is a C4 plant that could be used as a biofuel and feed plant. Similarly, *Halophyte Salicornia* could be used as a vegetable and oilseed crop (Ventura and Sagi 2013), Opuntia as a fruit of high nutritive quality (Castellar et al. 2012) and *Chenopodium quinoa* as a cereal (Adolf et al. 2013).

12.8.9 Livestock Management as an Adaptation Measure

Livestock management generally takes into account disease management for susceptible livestock assets and breeding-resistant and more productive animals.

12.8.9.1 Disease Management

Climate change has led to the expansion of vector-borne diseases into cooler climates of both higher altitudes and temperate regions. Changes in rainfall pattern have also led to the expansion of vectors during the wetter parts of the year and hence resulted in larger outbreaks. Appropriate livestock management is hence required so that livestock keepers could benefit from the increasing demand in the face of changing climate. Livestock diseases can be reduced through controlled breeding, quarantining sick animals and controlling entry into the farm lots, improvement of the existing and development of new drugs such as antibiotics and vaccines, development of new diagnostic tools and vector control techniques.

Disease management constitutes two key components, that is, prevention and control. Prevention of diseases can be achieved by several measures, such as care during animal purchase, hygiene in water and food supply, vaccination schedule, observation for signs of diseases, careful disposal of dead animals and watching the movement of animals. Control measures include timely report of diseases, specimen submission and last but not the least, drug administration.

12.8.9.2 Selective Breeding

The genetic makeup of farm animals strongly influences the fitness and determines the tolerance of animals to extreme climate conditions such as heat and drought stress and, above all, diseases. Adaptation of animals includes animals' performance under poor nutritional conditions due to these stresses and diseases brought about by these stresses. Major breeding traits associated with climate adaptation include reliance on low quality feed, thermal tolerance, high disease resistance, kid survival rate, animal morphology and good body condition (Hoffmann 2008). Selective breeding enables animals to become more stress tolerant as well as produce and reproduce more. Three approaches are usually followed in selective breeding. *Outclassing* involves mating animals that are unrelated for at least four to six generations. *Line breeding* involves mating with half-brothers, half-sisters and cousins, etc., while *inbreeding* involves mating of directly related animals. Among all these methods, outclassing is the best method with outstanding results such as increased milk production, reproductive ability and kids' survivability. Selective breeding reduces the risk of losing animals to harsh weather and changing climate conditions and enables farmers to get maximum productivity.

12.9 Climate Change Mitigation Strategies

The agriculture sector plays a vital role in GHG mitigation by sinking 10% of GHG emissions. Agriculture reduces global GHG emissions by approximately 10% from reducing N_2O emissions, 42% by carbon offsets through biofuel production, 32% by absorbing CO_2 emissions and 15% by reducing methane emissions (IPCC 2007). Emission mitigation strategies are generally grouped as (1) reducing emissions from agriculture, (2) enhancing sinks for the sequestration of CO_2 and (3) avoiding emissions through prevention of land-use change and by using replacement products.

12.9.1 Crop Production

12.9.1.1 Improved Varieties

Improved varieties help in producing more biomass from lesser piece of land which could otherwise come from forests if productivity was lower or crops were disease susceptible. Genetically modified (GM) crops with Bt resistance are examples of this. It has been estimated that GM crops conserved 14,200 M kg of CO_2 in 2007 which is equivalent to removing 6 million cars from the circulation (Brookes and Barfoot 2012). However, GM technology has received stiff opposition from the consumers.

12.9.1.2 Cover Crops

Cover crops such as rye and clover offer tremendous GHG mitigation potential by sequestering carbon, leaving residues and least reliance on chemical inputs. They also provide nitrogen to the subsequent crops, hence reducing nitrogen input and resultantly N_2O emissions. In an old study, cover crops were found to sequester carbon at the rate of 0.28–2.60 Mg ha^{-1} year^{-1} (Lal 1998).

12.9.2 Nutrient Management

The efficient use of nitrogenous fertilizers offers mitigation of GHG emissions in two ways: by reducing N_2O emissions through efficient use and by reducing CO_2 emissions due to reduction in the manufacturing of chemical fertilizers.

12.9.2.1 Nitrification Inhibitors

The use of nitrification inhibitors such as S-benzylisothiouronium fluroate (SBT fluroate) and S-benzylisothiouronium butanoate (SBT butanoate) can lead to reduction in N_2O emissions by 4–5% and global warming potential of urea by 8–19% (Bhatia et al. 2010).

12.9.2.2 Slow-Release Fertilizers

Chemical modification and changing the size of fertilizer granules can lead to the slow release of fertilizer, greatly benefiting crop yield and reducing N_2O emissions. Chances of nutrient losses and N_2O emissions due to leaching and volatilization are greatly reduced due to slow-release fertilizers. Combined with zero tillage, there has been great reduction (19%) in global warming potential (Bhatia et al. 2010).

12.9.2.3 Type of Fertilizer

It has been found that ammonia-based fertilizers result in higher N_2O emissions as compared to nitrate fertilizers (Bouman et al. 2007). The relative magnitude and total emissions from different fertilizers are given as follows: more N_2O emissions were recorded from anhydrous ammonia, followed by urea and ammonium sulphate while least emissions were recorded from calcium ammonium nitrate (Tenuta and EG Beauchamp 2003).

12.9.2.4 Fertilizer Application Time

Nitrogen application requirement is lower at the beginning of plant growth, higher at vegetative growth and again reduces at maturity. Huge applications at the beginning and at maturity are a total loss and results in higher N_2O emissions (Hultgreen and Leduc 2003). N application immediately after rainfall and irrigation increases N-use efficiency and reduces N_2O emissions. In this case, N loss through leaching and volatilization is reduced.

12.9.2.5 Fertilizer Placement

The placement of N fertilizers near the zone of active root uptake results in greater N-use efficiency, hence reducing N_2O emissions (CAST 2004).

12.9.2.6 Fertilizer Application Rate

Lower application rates of N also result in lower N_2O emissions (Drury et al. 2008). Help could be sought from organic sources of nitrogen to make sure crop nutrient requirements are met.

12.9.2.7 Mycorrhiza

Mycorrhizal fungi reduce plants' reliance on nitrogenous and other fertilizers, hence reducing fertilizer input and resulting N_2O emissions. The resulting plant biomass offers increased residues with great potential of C sequestration. Glomalin is a glycoprotein produced by mycorrhizal fungi that contains several soil-improving properties, including enhanced carbon sequestration (Subramanian et al. 2009).

12.9.3 Soil Management for Mitigation

Soil acts a great sink for greenhouse gases. Minimal disturbance of soil to reduce the possible release of greenhouse gases along with massive burial of carbon in the soil is most accurately termed as the only way to save the planet.

12.9.3.1 Conservation Tillage

Conservation tillage has a significant role in reducing the release of GHG emissions from soil. It has been found that tillage stimulates microbial decomposition of soil organic matter which increases the release of CO_2 to the atmosphere. Conservation tillage conserves soil, water and crop residues. Agriculture-driven CO_2 emissions are also reduced due to less use of fossil fuels from agricultural operations. Examples of conservation tillage are strip tillage, ridge till and mulch till farming (MDA 2011).

12.9.3.2 Biochar Application

Biochar, a pyrolysed biomass of wood or other agricultural biomass, has been found to have great potential for carbon sequestration. It has several soil-improving properties such as porous structure which increases the soils' carbon sequestration capacity, enhanced water-holding capacity, nitrogen-use efficiency and microbial activity. All these properties have been found to greatly affect plant nutrient use efficiency and subsequent reduction of N_2O emissions (Lehmann 2007).

12.9.3.3 Enhancing Microflora in the Soil

Microbes can help plants maintain their health through several ways such as tolerance to abiotic (such as salts) and biotic (such as pathogens) stresses and increase plant efficiency of resources such as N uptake. Several reports have demonstrated reduction of N emissions and leaching as well as carbon emissions from the soil with use of mutualistic microorganisms and reduced reliance on agrochemicals (Dobermann and Cassman 2002; Bakker et al. 2012).

12.9.4 Biofuels

The agriculture sector has recently started to contribute to the production of solid, liquid and gaseous biofuels which substitute fossil fuels for energy delivery. Several crops such as sugarcane, corn, sorghum, soybean, oil palm, switch grass as well as crop residues, bioengineered algae, Miscanthus and Jatropha have been recently known to produce biofuels. Biofuel production spans over three generations. The first-generation biofuel crops consisting of sugarcane and maize have been highly successful. The second-generation cellulosic ethanol crops (e.g., Miscanthus) are gaining ground while the third-generation of biofuels, where micro-algae is grown on water and CO_2, are being tested to produce biofuels with research in its infancy (Harvey et al. 2014).

12.9.5 Agroforestry

Forests are important carbon sinks. Though land-use change is not encouraged, afforestation and making judicious use of variable land via occupying land through crops and trees are appreciable in the mitigation process. Trees have a greater capacity for terrestrial carbon sequestration and in the soil through root growth and incorporation of organic matter (Martin and Sherman 1998).

12.9.6 Management in Rice Production Systems

Rice cultivation is responsible for 10–16% GHG emissions. Proper nutrient management (such as use of nitrification inhibitors) to reduce N_2O and residue management to control CO_2 emissions are vital for the mitigation process. Burning rice residues is responsible for the worst CO_2 emissions worldwide and often results in the creation of smog which creates widespread health issues. Dry land zero tillage is the best strategy in this regard. Certain irrigation measures such as midseason drainage (removal of water for about 7 days towards the end of tillering), alternate wetting and drying and direct seeded rice are also found helpful in reducing methane emissions. Direct seeding of pre-germinated rice has a shorter flooding period and hence leads to lesser methane emissions. Corton et al. (2000) noticed 16–54% while Wassmann et al. (2000) observed 16–92% reduction in methane emissions with direct seeded rice. Electron acceptors such as ferrihydrite, when added to paddy soil, stimulate microbial population which slows the activity of methanogens and hence reduces methane emissions. Breeding and introducing rice cultivars with greater mitigation potential are of utmost importance in meeting the mitigation goals. Rice cultivars with small root systems, high root oxidative capacity, more productive tillers and high harvest index have been found to have greater mitigation potential.

12.9.7 Livestock Management for Mitigating GHG Emissions

The livestock sector is mainly responsible for the generation and emission of CH_4 in the environment. In 2005, it was responsible for two-third of total agricultural methane emissions and over one-third of global methane emissions. CH_4 mitigation option for livestock sector falls into the following categories:

1. Improved feeding practices of generally low-quality feed with an aim to increase its utilization as product output and less feed is converted into methane as well as using other feed additives such as oils, oilseeds and others, hence improving the pasture quality with regard to CH4 emissions. Feeding quality and practices are improved through several ways ranging from adding additives to increasing the amount of concentrates in the forages.
2. Animal breeding and management aimed to increase animal productivity per unit of output. Total emissions can be substantially reduced by reducing the number of animals and increasing the productivity per unit of output.

12.9.7.1 Straw Ammonisation

In this practice, low-value forage, such as rice straw, corn stalks and wheat straw, is ammoniated in a cement ammonization pond or water tank. Adding urea, ammonium bicarbonate and liquid ammonia completely degrades the lignin part while the nutrients are enhanced. Rumen microorganisms easily digest it (Shankar et al. 2015).

12.9.7.2 Straw Silage

It is a type of forage prepared through the fermentation of forage grass, fresh green fodder, vines and other materials by lactobacillus in an air-proof (anaerobic conditions) silage container. In this method, raw materials are converted into organic acids (mainly lactic acid). Due to airtight conditions with no microbial activities after completion of fermentation, it remains unchanged for a longer period of time (Shankar et al. 2015). These methods improve the digestibility of forage and reduce the methanogenesis process. In China, the amount of straw processed through ammonization in 2009 was about 92 million tons, accounting for 44% of the total amount of forage (MOA 2010).

12.9.7.3 Increasing the Amount of Concentrates in Feed

An animal's diet normally consists of forage and concentrates. Forage provides the fibre necessary for energy generation, sustaining microbial flora and adding milk with fat while concentrates mainly provide fats, protein, vitamins and minerals. The ratio of concentrate to forage significantly affects the growth performance, health conditions and methane emissions. A diet with more fibre forage than concentrates has a high number of methanogens, enhanced methanogenesis, and hence increased methane emissions and vice versa. So the ration of concentrates must be higher than that of fibre (Demeyer and Henderickx 1967).

12.9.8 Human Diet Adjustment

Human diet greatly affects GHG emissions. It is commonly perceived that supply affects demand. Increasing demand of animal protein results in more GHG emissions as compared to plant protein. Mutton and beef have particularly high emissions (Davis et al. 2010). Therefore, dietary change holds a large theoretical potential towards mitigating the adverse effects of GHG emissions which is obvious from several studies (Berners-Lee et al. 2012; Green et al. 2015). Consumer preferences play a vital role in the practical implementation of dietary mitigation. Dietary mitigation is limited by very few researches. Policy intervention such as consumption taxes differentiated by emission levels might help.

12.10 Conclusions

Climate change has emerged as a major threat to planet Earth as a whole and to agriculture in particular. Severe biotic and abiotic stresses have brought life on Earth to the brink of extinction. While these lines are being written, a recent speech of Sir David Attenborough is going through my mind in which he termed climate change as the greatest threat to life in thousands of years. He said "if we don't take action now, the collapse of civilization and the extinction of much of the natural world is on the horizon". Luckily this is not the first call for action. The threat has been realized and humanity has started to take action to save itself and the Nature. While much can be done by governments and industrialists, farmers also have to do their share of work. They need to do three things: produce more, adapt to the change that has already happened and reduce the share of agriculture in climate change, that is, mitigate it. All these practices together make the collective practice of climate-smart agriculture. The world is already producing more, but still, much of it goes to waste and billions still go to bed hungry. Producing more has consequences and makes agriculture to contribute significantly to greenhouse gas emissions and, hence, to climate change. On the other hand, it is agriculture that mitigates this change by absorbing much of CO_2, a greenhouse gas. But CO_2 is not alone, there are other gasses produced by agriculture such as N_2O and CH_4 which are far more powerful in their magnitude to warm the planet. It is thus of vital importance to reduce the agriculture-oriented emissions of these gasses. Adaptation to the current level of these gasses is needed to sustain the population through meeting the food demand. While collective actions are needed for adaptation, this chapter provides with ample adaptation strategies that could be adopted by individual farmers. These include a number of water management practices such as water-saving irrigation methods and water-harvesting techniques, soil management practices in highly vulnerable mountainous and plain soils, nutrient management practices, biotechnological and genetic engineering methods for developing stress-tolerant varieties, farming systems to produce more and diverse food as well as safe and efficient storage of the produced food and seeds. Breeding of livestock for stress and disease resistance is also of vital importance while considering adaptation strategies. Among mitigation strategies, which focus on reducing the emissions from agriculture sector, include breeding less-emitting crops with higher water- and nutrient-use efficiencies, nutrient management to reduce loss as emissions, efficient irrigation, making use of microorganisms to reduce reliance on agrochemicals, agroforestry, use of biofuels and efficient management of rice and livestock management systems to reduce methane emissions. At this point, I would recall a part of Sir David's speech in which he expressed his and his race's willingness to make sacrifices in their daily lives. Human dietary behaviour must change: Consuming food that require fewer emissions to produce and not letting food go to waste.

References

Abberton M et al (2016) Global agricultural intensification during climate change: a role for genomics. Plant Biotechnol J 14:1095–1098

Adolf VI, Jacobsen S-E, Shabala S (2013) Salt tolerance mechanisms in quinoa (Chenopodium quinoa Willd.). Environ Exp Bot 92:43–54

Aggarwal P (2009) Global climate change and Indian agriculture: case studies from the ICAR network project. Indian Council of Agricultural Research. New Delhi:148

Amthor JS (2001) Effects of atmospheric CO_2 concentration on wheat yield: review of results from experiments using various approaches to control CO_2 concentration. Field Crop Res 73:1–34

Appiano M et al (2015) Identification of candidate MLO powdery mildew susceptibility genes in cultivated Solanaceae and functional characterization of tobacco NtMLO1. Transgenic Res 24:847–858

Bakker MG, Manter DK, Sheflin AM, Weir TL, Vivanco JM (2012) Harnessing the rhizosphere microbiome through plant breeding and agricultural management. Plant Soil 360:1–13

Balmer A, Pastor V, Gamir J, Flors V, Mauch-Mani B (2015) The 'prime-ome': towards a holistic approach to priming. Trends Plant Sci 20:443–452

Becker C, Hagmann J, Müller J, Koenig D, Stegle O, Borgwardt K, Weigel D (2011) Spontaneous epigenetic variation in the Arabidopsis thaliana methylome. Nature 480:245

Berners-Lee M, Hoolohan C, Cammack H, Hewitt C (2012) The relative greenhouse gas impacts of realistic dietary choices. Energy Policy 43:184–190

Bhatia A, Sasmal S, Jain N, Pathak H, Kumar R, Singh A (2010) Mitigating nitrous oxide emission from soil under conventional and no-tillage in wheat using nitrification inhibitors. Agric Ecosyst Environ 136:247–253

Blum A (2015) Towards a conceptual ABA ideotype in plant breeding for water limited environments. Funct Plant Biol 42:502–513

Bonzanigo L, Bojovic D, Maziotis A, Giupponi C (2016) Agricultural policy informed by farmers' adaptation experience to climate change in Veneto, Italy. Reg Environ Chang 16:245–258

Bouman BAM, Lampayan RM, Tuong TP (2007) Water management in irrigated rice: coping with water scarcity. International Rice Research Institute, Los Banos, p 54

Brinkman R, Sombroek WG (1996) The effects of global change on soil conditions in relation to plant growth and food production. In: Bazzaz F, Sombroek WG (eds) Global climate change and agricultural production: direct and indirect effects of changing hydrological, pedological and plant physiological processes. Wiley, West Sussex, pp 49–63

Brookes G, Barfoot P (2012) GM crops: global socio-economic and environmental impacts 1996–2010 PG Economics Ltd http://www.pgeconomics.co.uk/page/33/global-impact-2012. Accessed 31 Jan 2013

Brown ME, Kshirsagar V (2015) Weather and international price shocks on food prices in the developing world. Glob Environ Chang 35:31–40

Campbell JW, Irvin A, Irvin H, Stanley-Stahr C, Ellis JD (2016) Insect visitors to flowering buckwheat, Fagopyrum esculentum (Polygonales: Polygonaceae), in North-Central Florida. Fla Entomol 99:264–268

CARE (2010) Toolkit for integrating climate change adaptation into development projects – digital toolkit – version 1.0 CARE international, with technical input by the International Institute for Sustainable Development (IISD), July 2010

CAST (2004) Council for agricultural science and technology (CAST). Climate change and greenhouse gas mitigation: challenges and opportunities for agriculture. Paustian K, Babcock B (Cochairs) report 141

Castellar M, Solano F, Obón J (2012) Betacyanin and other antioxidants production during growth of Opuntia stricta (Haw.) fruits. Plant Foods Hum Nutr 67:337–343

Cavrak VV, Lettner N, Jamge S, Kosarewicz A, Bayer LM, Mittelsten SO (2014) How a retrotransposon exploits the plant's heat stress response for its activation. PLoS Genet 10:e1004115. https://doi.org/10.1371/journal.pgen.1004115

Cohn AS, VanWey LK, Spera SA, Mustard JF (2016) Cropping frequency and area response to climate variability can exceed yield response. Nat Clim Chang 6:601

Colmer T, Voesenek L (2009) Flooding tolerance: suites of plant traits in variable environments. Funct Plant Biol 36:665–681

Conrath U, Beckers GJ, Langenbach CJ, Jaskiewicz MR (2015) Priming for enhanced defense. Annu Rev Phytopathol 53:97–119

Corton T et al (2000) Methane emission from irrigated and intensively managed rice fields in Central Luzon (Philippines). Nutr Cycl Agroecosyst 58:37–53

Crowder DW, Northfield TD, Strand MR, Snyder WE (2010) Organic agriculture promotes evenness and natural pest control. Nature 466:109

Davis J, Sonesson U, Baumgartner DU, Nemecek T (2010) Environmental impact of four meals with different protein sources: case studies in Spain and Sweden. Food Res Int 43:1874–1884

Demeyer D, Henderickx H (1967) The effect of C18 unsaturated fatty acids on methane production in vitro by mixed rumen bacteria. Biochim Biophys Acta (BBA)-Lipids Lipid Metab 137:484–497

Derpsch R (1999) Keynote: frontiers in conservation tillage and advances in conservation practice. In: Stott DE, Mohtar RH, Steinhart GC (eds) Sustaining the global farm. Selected papers from the 10th International Soil Conservation Organisation meeting held May, pp 24–29

Des Marais DL, Auchincloss LC, Sukamtoh E, McKay JK, Logan T, Richards JH, Juenger TE (2014) Variation in MPK12 affects water use efficiency in Arabidopsis and reveals a pleiotropic link between guard cell size and ABA response proceedings of the National Academy of Sciences 111:2836–2841

Dickie A, Streck C, Roe S, Zurek M, Haupt F, Dolginow A (2014) Strategies for mitigating climate change in agriculture: recommendations for philanthropy–executive summary. Climate focus and California environmental associates, prepared with the support of the climate and land use alliance

Dobermann A, Cassman K (2002) Plant nutrient management for enhanced productivity in intensive grain production systems of the United States and Asia. Plant Soil 247:153–175

Drury C, Yang X, Reynolds W, McLaughlin N (2008) Nitrous oxide and carbon dioxide emissions from monoculture and rotational cropping of corn, soybean and winter wheat. Can J Soil Sci 88:163–174

FAO (2012) Crop prospects and food situation (no. 3, October 2012). http://www.fao.org/docrep/016/al992e/al992e00

FAO (2013) Climate-smart agriculture sourcebook executive summary. Food and Agriculture Organization of the United Nations, Rome, p 570

Franzaring J, Högy P, Fangmeier A (2008) Effects of free-air CO_2 enrichment on the growth of summer oilseed rape (Brassica napus cv. Campino). Agric Ecosyst Environ 128:127–134

Garnett T et al (2013) Sustainable intensification in agriculture: premises and policies. Science 341:33–34

Garratt M, Wright D, Leather S (2011) The effects of farming system and fertilisers on pests and natural enemies: a synthesis of current research. Agric Ecosyst Environ 141:261–270

Green R et al (2015) The potential to reduce greenhouse gas emissions in the UK through healthy and realistic dietary change. Clim Chang 129:253–265

Harvey CA et al (2014) Climate-smart landscapes: opportunities and challenges for integrating adaptation and mitigation in tropical agriculture. Conserv Lett 7:77–90

Hasanuzzaman M, Nahar K, Alam M, Roychowdhury R, Fujita M (2013) Physiological, biochemical, and molecular mechanisms of heat stress tolerance in plants. Int J Mol Sci 14:9643–9684

Hassanali A, Herren H, Khan ZR, Pickett JA, Woodcock CM (2008) Integrated pest management: the push–pull approach for controlling insect pests and weeds of cereals, and its potential for other agricultural systems including animal husbandry. Philos Trans R Soc London B Biol Sci 363:611–621

Henneron L et al (2015) Fourteen years of evidence for positive effects of conservation agriculture and organic farming on soil life. Agron Sustain Dev 35:169–181

Hodgkinson JH, Smith MH (2018) Climate change and sustainability as drivers for the next mining and metals boom: the need for climate-smart mining and recycling. Resour Policy. https://doi.org/10.1016/j.resourpol.2018.05.016

Hoffmann I (2008) Livestock genetic diversity and climate change adaptation. Livestock and Global Change conference proceeding. May 2008, Tunisia

Hou X, Meng L, Li L, Pan G, Li B (2015) Biochar amendment to soils impairs developmental and reproductive performances of a major rice pest Nilaparvata lugens (Homopera: Delphacidae). J Appl Entomol 139:727–733

Hu B et al (2015) Variation in NRT1. 1B contributes to nitrate-use divergence between rice subspecies. Nat Genet 47:834

Hultgreen G, Leduc P (2003) The effect of nitrogen fertilizer placement, formulation, timing, and rate on greenhouse gas emissions and agronomic performance. Saskatchewan Department of Agriculture and Food. Final report project no.5300G, ADF#19990028. Regina, Saskatchewan, Canada

Iba K (2002) Acclimative response to temperature stress in higher plants: approaches of gene engineering for temperature tolerance. Annu Rev Plant Biol 53:225–245

IPCC (2007) Technical summary. In: Parry ML, Canziani OF, Palutikof JP et al (eds) Climate change 2007. Impacts, adaptation, and vulnerability. Contribution of Working Group II to the fourth assessment report of the Intergovernmental Panel on Climate Change. Cambridge University Press, Cambridge

IPCC (2014) Impacts, adaptation, and vulnerability. Part A: Global and sectoral aspects. In: Field CB, Barros VR, Dokken DJ et al (eds) Contribution of Working Group II to the fifth assessment report of the Intergovernmental Panel on Climate Change. Cambridge University Press, Cambridge

Karaba A et al (2007) Improvement of water use efficiency in rice by expression of HARDY, an Arabidopsis drought and salt tolerance gene. Proc Natl Acad Sci 104:15270–15275

Khan Z, Midega C, Pittchar J, Pickett J, Bruce T (2011) Push—pull technology: a conservation agriculture approach for integrated management of insect pests, weeds and soil health in Africa: UK government's foresight food and farming futures project. Int J Agric Sustain 9:162–170

Klümper W, Qaim M (2014) A meta-analysis of the impacts of genetically modified crops. PLoS One 9:e111629
Knox J, Hess T, Daccache A, Wheeler T (2012) Climate change impacts on crop productivity in Africa and South Asia. Environ Res Lett 7:034032
Lal R (1998) Soil erosion impact on agronomic productivity and environment quality. Crit Rev Plant Sci 17:319–464
Lawson T, Blatt MR (2014) Stomatal size, speed, and responsiveness impact on photosynthesis and water use efficiency. Plant Physiol 164:1556–1570
Lehmann J (2007) A handful of carbon. Nature 447:143
Li C, Jiang D, Wollenweber B, Li Y, Dai T, Cao W (2011) Waterlogging pretreatment during vegetative growth improves tolerance to waterlogging after anthesis in wheat. Plant Sci 180:672–678
Li X, Cai J, Liu F, Dai T, Cao W, Jiang D (2014) Cold priming drives the sub-cellular antioxidant systems to protect photosynthetic electron transport against subsequent low temperature stress in winter wheat. Plant Physiol Biochem 82:34–43
Lin MT, Occhialini A, Andralojc PJ, Parry MA, Hanson MR (2014) A faster Rubisco with potential to increase photosynthesis in crops. Nature 513:547
Lobell DB, Burke MB, Tebaldi C, Mastrandrea MD, Falcon WP, Naylor RL (2008) Prioritizing climate change adaptation needs for food security in 2030. Science 319:607–610
Long SP, Marshall-Colon A, Zhu X-G (2015) Meeting the global food demand of the future by engineering crop photosynthesis and yield potential. Cell 161:56–66
Long TB, Blok V, Coninx I (2016) Barriers to the adoption and diffusion of technological innovations for climate-smart agriculture in Europe: evidence from the Netherlands, France, Switzerland and Italy. J Clean Prod 112:9–21. https://doi.org/10.1016/j.jclepro.2015.06.044
Lopes T, Hatt S, Xu Q, Chen J, Liu Y, Francis F (2016) Wheat (Triticum aestivum L.)-based intercropping systems for biological pest control. Pest Manag Sci 72:2193–2202
Mars R (2005) The basics of permaculture design. Chelsea Green Publishing, Hartford
Martin FW, Sherman S (1998) Agroforestry principles. ECHO
MDA (2011) Conservation practices, Minnesota conservation funding guide, Minnesota Department of Agriculture. Available at: http://www.mda.state.mn.us/protecting/conservation/practices/constillage.aspx
Miglietta F, Magliulo V, Bindi M, Cerio L, Vaccari F, Loduca V, Peressotti A (1998) Free air CO_2 enrichment of potato (Solanum tuberosum L.): development, growth and yield. Glob Chang Biol 4:163–172
MOA, Ministry of Agriculture, Science and Education Department (2010) National survey and evaluation report of crop straw resources (in Chinese)
Mottaleb KA, Rejesus RM, Murty M, Mohanty S, Li T (2017) Benefits of the development and dissemination of climate-smart rice: ex ante impact assessment of drought-tolerant rice in South Asia. Mitig Adapt Strateg Glob Chang 22:879–901
Müller C (2013) African lessons on climate change risks for agriculture. Annu Rev Nutr 33:395–411
Mutegi JK, Mugendi DN, Verchot LV, Kung'u JB (2008) Combining Napier grass with leguminous shrubs in contour hedgerows controls soil erosion without competing with crops. Agrofor Syst 74:37–49
Neumann K, Verburg PH, Stehfest E, Müller C (2010) The yield gap of global grain production: a spatial analysis. Agric Syst 103:316–326
Nigussie Y, van der Werf E, Zhu X, Simane B, van Ierland EC (2018) Evaluation of climate change adaptation alternatives for smallholder farmers in the Upper Blue-Nile Basin. Ecol Econ 151:142–150. https://doi.org/10.1016/j.ecolecon.2018.05.006
Nuccio ML et al (2015) Expression of trehalose-6-phosphate phosphatase in maize ears improves yield in well-watered and drought conditions. Nat Biotechnol 33:862
Oerke EC (2006) Crop losses to pests. J Agric Sci 144:31–43
Park SY, Peterson FC, Mosquna A, Yao J, Volkman BF, Cutler SR (2015) Agrochemical control of plant water use using engineered abscisic acid receptors. Nature 520:545–548

Pillay V, Kalu A (2012) A sustainable and robust membrane water treatment unit for potable water production in remote rural areas. In: Accra, Ghana 4th international appropriate technology conference, 2012

Popp A, Rose S, Calvin K, Van Vuuren D, Dietrich J, Wise M, Stehfest E, Humpenöder F, Kyle P, Van Vliet J, Bauer N, Lotze-Campen H, Klein D, Kriegler E (2013) Land-use transition for bioenergy and climate stabilization: model comparison of drivers, impacts and interactions with other land use based mitigation options. Clim Chang 123:495–509. https://doi.org/10.1007/s10584-013-0926-x

Porter JR, Xie L, Challinor AJ, Cochrane K, Howden SM, Iqbal MM, Lobell DB, Travasso MI (2014) Food security and food production systems. In: Field CB, Barros VR, Dokken DJ, Mach KJ, Mastrandrea MD, Bilir TE, Chatterjee M, Ebi KL, Estrada YO, Genova RC, Girma B, Kissel ES, Levy AN, MacCracken S, Mastrandrea PR, White LL (eds) Climate change 2014: impacts, adaptation, and vulnerability. Part A: global and sectoral aspects. Contribution of Working Group II to the fifth assessment report of the Intergovernmental Panel on Climate Change. Cambridge University Press, Cambridge/New York, pp 485–533

Powlson DS, Stirling CM, Jat M, Gerard BG, Palm CA, Sanchez PA, Cassman KG (2014) Limited potential of no-till agriculture for climate change mitigation. Nat Clim Chang 4:678

Prasanna V (2014) Impact of monsoon rainfall on the total foodgrain yield over India. J Earth Syst Sci 123:1129–1145

Qaderi MM, Kurepin LV, Reid DM (2006) Growth and physiological responses of canola (Brassica napus) to three components of global climate change: temperature, carbon dioxide and drought. Physiol Plant 128:710–721

Rippke U, Ramirez-Villegas J, Jarvis A, Vermeulen S, Parker L, Mer F, Diekkrüger B, Challinor A, Howden M (2016) Timescales of transformational climate change adaptation in sub-Saharan African agriculture. Nat Clim Chang 6:605–609. https://doi.org/10.1038/nclimate2947

Rivers A, Barbercheck M, Govaerts B, Verhulst N (2016) Conservation agriculture affects arthropod community composition in a rainfed maize–wheat system in Central Mexico. Appl Soil Ecol 100:81–90

Saj S, Torquebiau E, Hainzelin E, Pages J, Maraux F (2017) The way forward: an agroecological perspective for climate-smart agriculture. Agric Ecosyst Environ 250:20–24

Schroeder J, Delhaize E, Frommer W, Guerinot M, Harrison M, Herrera-Estrella L, Horie T, Kochian L, Munns R, Nishizawa N, Tsay Y, Sanders D (2013) Using membrane transporters to improve crops for sustainable food production. Nature 497:60–66. https://doi.org/10.1038/nature11909

Selote DS, Khanna-Chopra R (2010) Antioxidant response of wheat roots to drought acclimation. Protoplasma 245:153–163

Shankar M, Prasanna Kumar M, Thimmegowda M, Shivaramu H (2015) Climate change-research, mitigation and adaptive opportunities for Indian Agriculture History 1:386–390

Singh P, Boote KJ, Kadiyala MDM, Nedumaran S, Gupta SK, Srinivas K, Bantilan MCS (2017) An assessment of yield gains under climate change due to genetic modification of pearl millet. Sci Total Environ 601-602:1226–1237. https://doi.org/10.1016/j.scitotenv.2017.06.002

Smith P, Martino D, Cai Z, Gwary D, Janzen H, Kumar P, McCarl B, Ogle S, O'Mara F, Rice C, Scholes B, Sirotenko O, Howden M, McAllister T, Pan G, Romanenkov V, Schneider U, Towprayoon S, Wattenbach M, Smith J (2008) Greenhouse gas mitigation in agriculture. Philos Trans R Soc B Biol Sci 363:789–813. https://doi.org/10.1098/rstb.2007.2184

Stocker T (2014) Climate change 2013: the physical science basis: Working Group I contribution to the fifth assessment report of the Intergovernmental Panel on Climate Change. Cambridge University Press, Cambridge

Subramanian K, Tenshia V, Jayalakshmi K, Ramach V (2009) Role of arbuscular mycorrhizal fungus (*Glomus intraradices*) (fungus aided) in zinc nutrition of maize. J Agric Biotechnol Sustain Dev 1:029–038

Tang H, Takigawa M, Liu G, Zhu J, Kobayashi K (2013) A projection of ozone-induced wheat production loss in China and India for the years 2000 and 2020 with exposure-based and flux-based approaches. Glob Chang Biol 19:2739–2752

Tenuta EG, Beauchamp M (2003) Nitrous oxide production from granular nitrogen fertilizers applied to a silt loam soil. Can J Soil Sci 83:521–532
Thomashow MF (1999) Plant cold acclimation: freezing tolerance genes and regulatory mechanisms. Annu Rev Plant Biol 50:571–599
Thornton PK, Gerber PJ (2010) Climate change and the growth of the livestock sector in developing countries. Mitig Adapt Strateg Glob Chang 15:169–184
Uga Y, Sugimoto K, Ogawa S, Rane J, Ishitani M, Hara N, Kitomi Y, Inukai Y, Ono K, Kanno N, Inoue H, Takehisa H, Motoyama R, Nagamura Y, Wu J, Matsumoto T, Takai T, Okuno K, Yano M (2013) Control of root system architecture by DEEPER ROOTING 1 increases rice yield under drought conditions. Nat Genet 45:1097–1102. https://doi.org/10.1038/ng.2725
UNEP (1997) Sourcebook of alternative technologies for freshwater augmentation in some countries in Asia. UNEP, Unit of Sustainable Development and Environment General Secretariat, Organization of American States, Washington, DC.
van Ittersum MK, Cassman KG, Grassini P, Wolf J, Tittonell P, Hochman Z (2013) Yield gap analysis with local to global relevance—a review. Field Crop Res 143:4–17
VandenBygaart A (2016) The myth that no-till can mitigate global climate change. Agric Ecosyst Environ 216:98–99. https://doi.org/10.1016/j.agee.2015.09.013
Venkatramanan V, Shah S (2019) Climate smart agriculture technologies for environmental management: the intersection of sustainability, resilience, wellbeing and development. In: Shah S et al (eds) Sustainable green technologies for environmental management. Springer, Singapore, pp 29–51. https://doi.org/10.1007/978-981-13-2772-8_2
Venkatramanan V, Singh SD (2009a) Differential effects of day and night temperature on the growth of rice crop. Pusa Agric Sci 32:57–62
Venkatramanan V, Singh SD (2009b) Differential effects of day and night temperature on the growth of wheat crop. Ann Agric Res 30(1&2):49–52
Ventura Y, Sagi M (2013) Halophyte crop cultivation: the case for Salicornia and Sarcocornia. Environ Exp Bot 92:144–153
Vermeulen S et al (2012) Climate change, agriculture and food security: a global partnership to link research and action for low-income agricultural producers and consumers. Curr Opin Environ Sustain 4:128–133. https://doi.org/10.1016/j.cosust.2011.12.004
Wambugu P, Mathenge P, Auma E, Van Rheenen H (2009) Efficacy of traditional maize (Zea mays L.) seed storage methods in Western Kenya. Afr J Food Agric Nutr Dev 9:1110–1128
Wang X, Cai J, Liu F, Dai T, Cao W, Wollenweber B, Jiang D (2014) Multiple heat priming enhances thermo-tolerance to a later high temperature stress via improving subcellular antioxidant activities in wheat seedlings. Plant Physiol Biochem 74:185–192
Wang X, Vignjevic M, Liu F, Jacobsen S, Jiang D, Wollenweber B (2015) Drought priming at vegetative growth stages improves tolerance to drought and heat stresses occurring during grain filling in spring wheat. Plant Growth Regul 75:677–687
Wang X, Huang M, Zhou Q, Cai J, Dai T, Cao W, Jiang D (2016) Physiological and proteomic mechanisms of waterlogging priming improves tolerance to waterlogging stress in wheat (Triticum aestivum L.). Environ Exp Bot 132:175–182
Wassmann R, Lantin R, Neue H, Buendia L, Corton T, Lu Y (2000) Characterization of methane emissions from rice fields in Asia. III. Mitigation options and future research needs. Nutr Cycl Agroecosyst 58:23–36
Werner T, Nehnevajova E, Köllmer I, Novák O, Strnad M, Krämer U, Schmülling T (2010) Root-specific reduction of cytokinin causes enhanced root growth, drought tolerance, and leaf mineral enrichment in Arabidopsis and tobacco. Plant Cell 22:3905–3920
Wollenberg EK, Campbell BM, Holmgren P, Seymour F, Sibanda LM, Braun JV (2011) Actions needed to halt deforestation and promote climate-smart agriculture, CCAFS policy brief 4. CCAFS, Copenhagen

Climate Change and Farmers' Adaptation: Extension and Capacity Building of Smallholder Farmers in Sub-Saharan Africa

13

Urban B. Kalimba and Richard J. Culas

Abstract

The efforts to reduce impacts of climate change have been taken by many African countries especially those which are highly exposed to the changing climatic condition and weather extremes. Many attempts have been directed in agriculture to adapt to climate change as agriculture is the main source of economy and livelihoods of the large population in these countries. Extension services, in particular, have been at the centre of the efforts taken by governments to build farmers' adaptation capacity for the impacts of climate change. This chapter reviews and analyses the current level of extension practices and the capacity building of smallholders farmers with specific reference to Tanzania and other countries such as Senegal, Malawi and Kenya. In particular, this chapter will look at how farmers can be adaptable to climate-smart agriculture (CSA) technologies. In doing so, this chapter will look at what extent climate change affects the agriculture sector of Tanzania, assess the CSA technologies' and practices' adaptation in the farming activities and examine extension approaches/methods being used to address the agricultural challenges in Tanzania and also in relation to the lessons learned from the other African countries (Senegal, Malawi and Kenya).

Keywords

Climate-smart agriculture · Smallholder farmers · Climate resilience · Adaptation · Agricultural extension · Participatory approach · Capacity building · Tanzania · Sub-Saharan Africa

U. B. Kalimba
Department of Agriculture and Cooperatives, Iringa District Council, Iringa, Tanzania

R. J. Culas (✉)
School of Agricultural and Wine Sciences, Charles Sturt University, Orange, NSW, Australia
e-mail: rculas@csu.edu.au

© Springer Nature Singapore Pte Ltd. 2020

V. Venkatramanan et al. (eds.), *Global Climate Change and Environmental Policy*,
https://doi.org/10.1007/978-981-13-9570-3_13

13.1 Introduction

Agriculture is the mainstay of the Tanzanian economy; it accounts for about half of GDP and export earnings. In 2015, the contribution of the sector to the country's national gross domestic product (GDP) was approximately 32% (CIAT and World Bank 2017). A large part of agricultural GDP comes from food crops which account for about 65% while cash crops account for about 10%. Agriculture contributes about 95% of the national food requirements (URT 2014; FAO 2017) and almost one-third of the total export revenue of the country comes from agriculture (FAO 2016). The sector employs about 13 million people, which is equivalent to 59% of the country's working population (World Bank 2016). The significant part of the agricultural work force is provided by women where the estimates show that they produce more than 70% of the country's food.

Recently, climate change has increasingly emerged as one of the most severe global problems hampering economic growth across many sectors. Those sectors include agriculture, water, fisheries, forestry and other land use, wildlife, energy, industrial processes and product use, waste management, human health and the sustainable livelihoods of both rural and urban communities (Bie et al. 2008; Magombo et al. 2012). This problem is rampant particularly for African counties as it is considered to be one of the most significant threats to sustainable development due to its impacts on food security, economic activities and physical infrastructure (IPCC 2007).

Although there is a rich base of natural resources, agriculture in Tanzania is entirely rain-fed, a situation that makes the country more vulnerable to the impacts of climate change. Climate change has resulted in a general decline in agricultural productivity, including changes in agro-diversity (URT 2017). According to URT (2005), the famine events as a result of either floods or drought have become increasingly common since the mid-1990s and are undermining food security. It has also been reported that the prevalence of crop pests and diseases have increased and are posing significant challenges to agricultural production (URT 2017). Furthermore, the studies by the Tanzania Meteorological Agency (TMA) shows that some of the previous high productivity areas such as the southern and northern highlands will continue to be impacted by decreasing rainfall, repeated drought and a significant increase in spatial and temporal variability of precipitation (URT 2017).

The 2014 annual report of Climate Change, Agriculture and Food Security suggests that the efforts to reduce impacts of climate change in agricultural production requires the development of appropriate and feasible climate-smart and climate-resilient agriculture practices to reduce hunger and improve food security and income (CCAFS 2015a). Also, Gwembane et al. (2015) suggest strategies like building sustainable food systems and improving the productivity and income of smallholder farmers as the best approaches for creating communities resilient against climate change impacts. To achieve these goals, the United Nations Food and Agriculture Organization (2013) has recently recommended the adoption of CSA.

The adoption of CSA among farmers requires a well-organized extension system supported by sound government policies and institutional arrangements to effectively address climate change impacts. In many countries, extension services have been playing a critical role in supporting CSA adoption through technology development and information dissemination, strengthening farmers' capacity, facilitation and brokering and advocacy and policy support (Sala et al. 2016). Specifically, action will be needed to focus on extension for capacity development. This chapter will consider the importance of extension-led capacity building for smallholder farmers, enabling adoption of climate-smart agriculture practices and technologies as a strategy to sustain agricultural productivity, given the expected impacts in Tanzania.

13.2 Research Problems

Climate-smart agriculture (CSA) technologies and practices are highly recommended to reduce the impacts of climate change to sustain agricultural productivity. The CSA approach aims "*to address the twin challenges of climate change and food security through sustainably increasing food security by increasing agricultural productivity and incomes; building resilience and adapting to climate change; and developing opportunities to reduce greenhouse gas emissions from agriculture*" (Williams et al. 2015; Sala et al. 2016). Achieving these objectives will require development of climate-smart agriculture technologies (Venkatramanan and Shah 2019), and transformations in the attitudes, behaviour and strategies and farming practices. Such a transformation will require efforts to improve the farmer's access to climate-resilient and smart agriculture technologies and practices (Sala et al. 2016).

For many years, agricultural extension has played a pivotal role in transferring important farming information to farmers with the purpose of enhancing their livelihoods by improving agricultural production. The emergence of climate change in recent years has increased risks in the agriculture sector which threatens its present and future performance. Rupan et al. (2018) argued that climate change is a new global challenge, giving extension agencies major challenges to their previous understanding of their client's needs. They must now revise their services in the light of growing water scarcity, increasing soil degradation and increasing climate uncertainty.

Recently, in many countries, the focus of extension provision has been altered in response to the changing nature of agriculture and needs of farmers. It has shifted away from the transfer of technologies and skills related to agricultural production from research to farmers, to more participatory approaches which emphasize the facilitation of innovations among farmers. According to Sala et al. (2016), these shifts are essential as they are aligned with the need for site-specific assessments to identify suitable agricultural technologies and practices needed for CSA. Nederlof and Pyburn (2012) have argued participatory approaches and methods have proven successful in many countries as an approach that encourage multiple stakeholders to innovate using emerging technologies.

The current state of extension services in Tanzania appears to observers to be not yet ready to take on the challenge of effectively equipping smallholder farmers to survive climate change. Here are some of the dimensions of its lack of preparedness. One immediately notices the low average ratio of extension officers to farmer households (1:630) and the limited technical capacity of local governmental authorities (public extension) to deliver agricultural information (particularly CSA), a problem that has slowed down the adoption of practices and technologies by smallholder farmers. Observers have also noted the lack of documentation of extension policies and procedures focused on increasing agricultural productivity and building climate resilience (Harris-Coble 2016; CIAT and World Bank 2017).

Additionally, "low budgetary allocation to the agriculture sector, inadequate access to reliable transportation for extensionists (particularly public extension workers), limited financial support to carry out demonstrations and field experiments, lack of working facilities and low salaries" (Daniel 2013) have contributed to low performance of this service to address climate change issues. Other challenges are mostly related to contemporary extension delivery systems (pluralistic approach) which tend to be weak and unsystematic, characterized by short-term projects and a lack of coordination between providers and advisers who lack the knowledge and skills to address the new demands (Sala et al. 2016). These challenges denote that the current extension services in Tanzania may not be sufficient to effectively deal with climate change issues.

Despite these challenges, the role of extension services will remain prominent in the efforts to address the problems of climate change in the farming environment due to its potential to act as an intermediary between research and farmers and as nodal points that bring together and facilitate multiple stakeholders to address complex problems and situations (Sala et al. 2016). However, many studies suggest the need for the improvement of extension services in Tanzania (Gwambene et al. 2015; Kangalawe et al. 2017; CIAT and World Bank 2017) to suit the requirements of climate change adaptation in the agriculture sector. This chapter will assess the current state of extension services and capacity building in Tanzania and suggest an alternative extension model to improve its performance in the face of climate change challenges.

13.3 Objectives of the Study

Generally, this study aims to assess the role and capacity of extension services to foster smallholder farmer adaptation to climate-smart agriculture practices and technologies in Tanzania and to suggest better ways of improving their performance in addressing climate change challenges for sustainable agricultural productivity. To address the existing challenges in the extension services, this chapter will specifically address the following areas:

1. To what extent climate change affects the agriculture sector of Tanzania
2. To assess the CSA technologies and practices adoption by farmers

3. To examine extension approaches/methods being used to address agricultural challenges in Tanzania
4. To suggest a better extension model to facilitate the CSA adoption among smallholder farmers

Accordingly, this chapter articulates the climate-change impacts, climate-smart technologies and a better extension model for smallholder farmers. The first part of this chapter will look at what extent climate change affects the agriculture sector of Tanzania, the second part will assess the CSA technologies and practices adoption in the farming activities, the third part will examine extension approaches/methods being used to address the agricultural challenges in Tanzania and also in relation to the lessons learned from the other African countries (Senegal, Malawi and Kenya) and the fourth part will look at the extension model that could be suggested to improve the adoption rate of CSA.

13.4 Current Agricultural Systems in Tanzania

13.4.1 Farming Systems

According to Mnenwa and Maliti (2010), farming systems in Tanzania can be classified into 10 categories based on natural resource base (including water, land, grazing areas and forest); climate (of which altitude is one crucial determinant); landscape (including slope); farm size tenure and organization; dominant pattern of farm activities and household livelihoods (including field crops, livestock, trees, aquaculture, hunting and gathering, processing and off-farm activities) and leading technologies used which determine the intensity of production and integration of crop, livestock and other activities. Another study which focused on differential ownership of assets such as land and livestock activities and incomes in Tanzanian rural areas classifies farming systems into small-scale, farmer-managed irrigation, rain-fed maize production and intensive upland fruit and vegetable production (Mnenwa and Maliti 2010). Furthermore, the study of Kisusu (2003) revealed two farming systems in central Tanzania namely dairy cattle and irrigated rice farming.

13.4.2 Impacts of Climate Change on the Agricultural Sector in Tanzania

Climate change refers to the "variation in either mean state of the climate or in its variables persisting for a long and an extended period (decades or longer). While climate variability refers to sudden and discontinuous seasonal or monthly or periodic changes in climate or its components without showing any specific trend of temporal change" (IPCC 2013; Venkatramanan and Shah 2019). Several studies have been conducted on temperature and precipitation projections in Tanzania. For instance, the study by Chang'a et al. (2017) shows that the mean temperature anomaly increased while the average rainfall anomaly decreased between 1961 and 2015.

Climate change impacts can be observed through an increasing frequency of seasons with few rain events, short irregular periods of rainfall seasons, poor rainfall distribution within the seasons and temperature changes (URT 2014; Gwambene et al. 2015; Coulibaly et al. 2015). The current climate change is associated with increased greenhouse emissions as a result of increased burning of fossil fuels. Also, large-scale changes in land use like deforestation also contribute to these emissions (URT 2017). The climate projection studies in Tanzania show that increasing greenhouse emissions have a significant impact on temperature and rainfall changes.

Various studies have revealed that the poor performance of the agriculture sector in Tanzania is impacted by overdependence on rain-fed agriculture and increasingly unpredictable climate change (Ehrhart and Twena 2006; Enfors and Gordon 2008; Müller et al. 2011). Also, the study by Rowhani et al. (2011) revealed that a seasonal increase in temperature by 2 °C as predicted by 2050 would lessen yields of rice, sorghum and maize by 7.6%, 8.8% and 13%, respectively, in Tanzania whilst a 20% increase in rainfall variability will lessen yields of rice, sorghum and maize by 7.6%, 7.2% and 4.2%, respectively, by 2050. Further, other studies by Ahmed et al. (2011), Arndt et al. (2011), and Rowhani et al. (2011) have also predicted that the future climate change may cause severe challenges for the agriculture sector in Tanzania.

Besides, the projections by the Intergovernmental Panel on Climate Change (IPCC) shows that by 2020, while about 75–250 million people in the developing countries including Sub-Saharan African countries will be vulnerable to increased water scarcity as a result of climate change, the agricultural productivity from rain-fed production systems may decline by up to 50%. It implies that many African countries will face a serious shortage of food, a situation likely to exacerbate food insecurity and malnutrition (IPCC 2007).

13.4.3 CSA and Its Prospects in Tanzania

CSA endeavours to achieve development and food security through an approach that encompasses climate responsiveness and agricultural growth and development. Like many other countries, Tanzania has initiated a government-led programme and the CSA Guideline (URT 2017) to provide favourable mechanisms to foster CSA as well as to direct public, private and international institutions towards CSA in the country. The purpose of this chapter is to examine the status of CSA in Tanzania regarding institutional arrangements and strategies towards the achievement of sustainable agriculture under climate change challenges. Furthermore, this part will seek to relate these efforts with progress in the uptake of CSA technologies and practices among smallholder farmers of Tanzania.

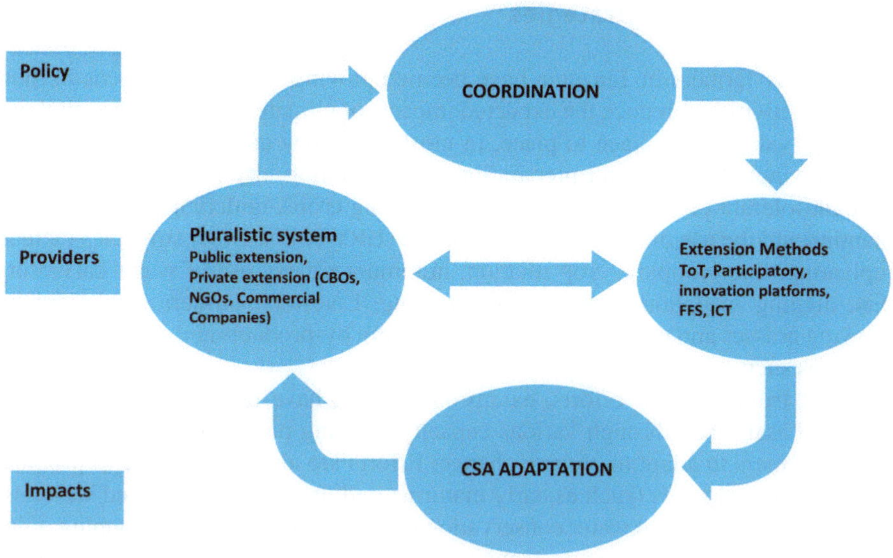

Fig. 13.1 A CSA extension model. (Source: authors' construction)

13.4.4 CSA Concept

According to FAO (2010), climate-smart agriculture aims at achieving the so-called "triple win interventions". These interventions must increase agricultural yields (food and income security), make agriculture more resilient in the face of climate extremes (adaptation) and increase the ability of the farming systems to sequester greenhouse gases (GHGs), particularly carbon dioxide (mitigation) (Fig. 13.1). In the context of Tanzania, CSA is perceived as an "agriculture approach that sustainably increases productivity and income; increases the ability to adapt and build resilience to climate change; and enhance food and nutrition security while achieving mitigation co-benefits in line with national development priorities" (URT 2017).

According to URT (2017), CSA in Tanzania has been designed to ensure sustainable agriculture through integrated water management approaches, management of land and ecosystems at landscape scale and includes techniques, such as mulching, intercropping, conservation agriculture, pasture and manure management and innovative practices, programmes and policies, such as improved crop varieties, better weather forecasting and risk insurance. Apart from that, it should emphasize agroecological approaches to soil, nutrient, water and ecosystem management with explicit attention to the importance of preserving genetic resources of crops and animals including wild relatives, which are critical in developing resilience to shocks.

13.4.5 Adaptation Strategies

Smallholder farmers in Tanzania have been employing different practices in their farming activities to reduce the expected impacts of climate change. Although those practices varied from place to place, in most cases they are similar. The common practices include early land preparation, early planting, dry planting, planting of drought-tolerant crops, planting of early maturing crops, mulching, irrigation, tree planting and the use of indigenous knowledge (URT 2017). Other strategies include replanting, intercropping, crop rotation, minimum tillage, use of water harvesting pits, digging irrigation trenches and terracing. Livestock farmers also adapt by growing grasses and perennial fodders, using farm by-products and doing additional activities such as crop farming.

Apart from farmers' efforts, extension services have been backstopping with technical assistance through various capacity-building programmes to help smallholder farmers to adapt to climate change. Efforts are also being made to promote use of improved seeds (such as early maturing and drought-tolerant seed varieties) and adoption of soil and water conservation techniques (such as intercropping, minimum tillage, water-harvesting technologies, ridging and agroforestry). Besides, training on appropriate use of industrial fertilizers and manuring are conducted, and alternative sources of income like agro-processing, mushroom cultivation and off-farm activities are encouraged to build the resilience of smallholder farmers to climate change. For livestock producers, extension agents promote the use of improved breeds and improved livestock management, artificial insemination, milk value-adding and improved fodder to help farmers to reduce the impacts of climate change.

Moreover, the research institutes have also been involved in promoting various CSA-related technologies and practices including a system of rice intensification (SRI), different water-harvesting techniques, use of weather information by knowledge sharing, face-to-face meetings, village meetings, workshops and training (URT 2017). Also, they have been conducting various climate change research initiatives and developing several types of improved seeds and breeds of crops, livestock and fish that are highly productive and drought- and disease-tolerant.

NGOs also play an essential role that contributes to promoting CSA technologies and practices through training, radios, television, mobile phones, seminars, demonstrations, FFS and integrating approaches. Most of them focus on the promotion of crop technologies and practices that include an application of organic fertilizers, use of bio-pesticides and use of improved seed varieties. In the livestock sector, they promote enhanced techniques and methods such as cross-breeding of local chicken, goats, sheep, pigs and cattle as well. In fisheries, NGOs encourage aquaculture and sustainable fishing as measures to adapt to and mitigate the impacts of climate change in the sector (URT 2017). However, URT (2017) reported that the response of the private sector in Tanzania in providing extension delivery and creating incentives for farmers to adopt new CSA methods is still low due to several factors including low participation of private industry in extension delivery, and thus small number of farmers are receiving extension services at the village level.

13.4.6 CSA Adoption Among Smallholder Farmers in Tanzania

A study conducted by Gwambene et al. (2015) in the Southern Highlands of Tanzania revealed that smallholder farmers of this area had adopted various agricultural practices to combat different environmental problems such as climate change and land degradation; they also found a low rate of adoption among many smallholder farmers. Other studies (Nyanga et al. 2011; Taneja et al. 2014) associate this low adoption with several factors including, but not limited to, a lack of appropriate implements, insufficient appropriate soil fertility management options, inadequate and sometimes inappropriate technical information and limited/poor access to credit.

The standard adopted CSA practices, particularly in crop production, include mulching, zero/minimum tillage, crop cover, contour ploughing, irrigation, crop residue incorporation, agroforestry, crop and livestock farming, terracing, mixed cropping and crop rotation (Gwambene et al. 2015). Crop rotation received a high priority as 71% of interviewed farmers had adopted it. This was followed by mixed cropping, terracing and livestock farming; approximately 40% of the respondents took each. Mulching and zero/minimum tillage received the lowest priority. Interestingly, despite the prevailing adoption level of CSA practices, the study found that most farmers were not aware of the meaning of CSA and its relevance to their farming activities. This problem may be contributed to by the lack of adequate extension services to educate smallholder farmers about CSA.

The findings of Nyasimi et al. (2017) support this argument by showing that many smallholder farmers in Northern Tanzania were willing to use CSA technologies and practices but they could not because of lack of knowledge and skills. However, the response from the smallholder farmers who were interviewed in the study of Gwambene et al. (2015) showed that most of the time, smallholder farmers tend to adopt farming practices which are perceived to be feasible and that can increase yield and food security.

Several factors cause the low adoption of CSA technologies and practices in Tanzania. Lack of appropriate technical information and the massive cost of CSA technologies are the primary constraints to the adoption (Nyasimi et al. 2017). Other limitations include a low degree of mechanization within the smallholder farming system, lack of appropriate implements, insufficient appropriate soil fertility management options and limited/poor access to credit. To a large extent, the existence of these constraints denotes the weakness of current extension services in addressing farming challenges and promoting CSA practices among smallholder farmers in Tanzania. The findings of Sala et al. (2016) support this argument by pointing out that insufficient extension services are among major limiting factors for the adoption of CSA in developing countries including Sub-Saharan Africa which is caused by inadequate and inexperienced extension workers.

Despite the existence of a well-elaborated framework for CSA in Tanzania, the approach does not seem to be performing at a satisfactory level. The low CSA adoption among smallholder farmers can justify this claim. The discussion in this chapter associates this low transformation with the existence of weak coordination and linkages between CSA actors (both public and private) and financial resource constraints that face government units and departments responsible for the promotion of CSA. It seems that the government does not prioritize the framework for CSA regarding budget allocation and also the extension agencies have not yet succeeded in linking and coordinating CSA actors for better performance.

13.5 Extension Model Currently Applied in Tanzania

Msuya et al. (2017) argue that well-targeted agricultural extension services and farmer capacity building strategies are both essential to enable rural communities to meet the challenges of the modern world. In dealing with complex farming challenges like climate change, extension services should ensure constant coordination and communication among crucial actors to facilitate the dissemination of technologies and a continuous flow of farming information in the agricultural environment. This part will assess the relevance of the current extension system applied in Tanzania. This assessment will concentrate on its capacity building role, its performance in addressing farming issues and will also analyse extension methods/approaches being used to disseminate farming information and technologies.

13.5.1 Extension Structure in Tanzania

In Tanzania, agricultural extension services are provided mainly through the Ministry of Agriculture Food Security and Cooperatives (MAFC) and the Ministry of Agricuture Livestock and Fisheries (MALF). According to the general government reforms of 1999, the responsibility for extension services in terms of policy formulation, issuance of guidelines and provision of technical assistance to local authorities is under the two directorates in the ministerial levels: the Crop Development and Livestock Development divisions, being assisted by their respective assistant directors (Rutatora and Mattee 2001). With the decentralization of extension services (as per Local Government Act No. 6 of 1999), the Ministry of Regional Administrative and Local Government has been mandated to supervise extension activities at the farm level (Mvuna 2010). Besides being a core function of the government, agricultural extension services have been and remain almost entirely financed by the government. However, both MAFC and MALF also provide room for private agencies to participate in the provision of agricultural extension services, particularly to smallholder farmers. According to Mwamakimbula (2014), this strategy has augmented the efforts of the government in transferring farming information and also to speed up the rate of adoption of various agricultural technologies and practices among farmers. Besides, it has helped to reduce the financial burden of government in funding extension services and reduce the deficit of public extension workers as well. In contrast, the study of Sala et al. (2016) found weak and unsystematic pluralistic "extension systems" in various countries, which are characterized by short-term projects, inadequate coordination between providers, insufficient financial and human resources and advisers who have insufficient knowledge and skills to deal with the new challenges.

13.5.2 Agricultural Extension in Building Smallholder Farmer Capacity to Adapt to Climate Change

Awareness creation is one of the responsibilities of extension services. This is the important first stage in introducing an intervention to address any agricultural farming challenge. Extension workers can choose to use the awareness meetings to sensitize smallholder farmers on climate change issues (Mkisi 2014). Mandleni and Anim (2011) reported that agricultural extension plays a crucial role in creating awareness among smallholder farmers of the expected effects of climate change which will help them to make informed choices on available climate change adaptation technologies and practices.

The findings of Taneja et al. (2014) and Sala et al. (2016) suggest that the CSA interventions should be location specific because to a large extent their adoption needs to be well suited to users regarding willingness, ability to practice, knowledge and their investment capacity. These findings are supported by the report of FAO (2014) which asserts that "CSA is not just a set of practices that can be universally applied, but rather an approach that involves different elements embedded in local contexts." This implies that the participatory approaches and consideration of indigenous knowledge are to be considered during the process of conducting the needs assessment as well as during the project implementation phase. The findings of Kaaria et al. (2007) and Nederlof and Pyburn (2012) reveal that extension providers in many countries have achieved positive results using participatory methods/approaches such as participatory technology development (for example, Farmer Field School).

However, it is not a must that all new developed farming technologies and practices are innovated through indigenous knowledge; they can even come from outside sources. Barakabitze et al. (2015) argued that even though farmers generally possess rich indigenous knowledge, agricultural research should continue to develop scientific knowledge and to improve technologies for agriculture. In such a situation, extension services can use demonstrations as an approach for educating farmers on newly developed techniques from research centres. The results of Okunade (2007) and Khan et al. (2009) agree that demonstrations are very useful in transferring knowledge and skills that are necessary to implement the new agricultural technology.

Sharing the experimental findings and newly developed innovations help to facilitate their adoption among farmer communities. The extension worker can use a combination of approaches throughout the process of dissemination of agricultural technology or practice. Sala et al. (2016) maintain that traditional extension approaches (such as interpersonal interaction, demonstrations, field days and printed materials) can still be used when disseminating newly developed agricultural technologies, information and practices. Mkisi (2014) also reported that field days have proven successful in publicizing the new/improved "crop varieties and livestock breeds" and "drought and disease" tolerant technologies in Malawi. Similarly, Okunade (2007) reported that field days were a practical approach in inducing attitude changes among smallholder farmers in adopting new agricultural technologies.

To minimize risks which are associated with climate change, it is imperative that accurate and well-communicated weather information be accessible to farmers as a critical component for planning and action. Agricultural extension plays a role in translating and disseminating information on the weather forecast to smallholder farmers for better planning. The report of Stigter et al. (2013) emphasizes this role of extension by suggesting the need for client-helpful weather information and early warning that could increase the preparedness of smallholder farmers, allowing them to adapt to climate change dynamically, incrementally and in a timely manner. The extension services have the capacity to do this by using different methods such as mass media and farmers' meetings, to mention two.

Climate change and variability have emerged as a new cross-cutting issue in agriculture extension service provision. The available extension staffs have either no or little understanding of this problem and thus they lack capabilities to train farmers on recommended CSA technologies and practices. CIAT and World Bank (2017) reported that the low extension service capacity to deal with climate change issues is one of the significant constraints to CSA adoption in Tanzania. According to Mkisi (2014), capacity building of extension workers is amongst the vital roles of extension services; he suggests the need for building capacity (periodic training) and creating awareness to extension staff on climate change so that they have the knowledge and skills to promote adaptation interventions.

13.5.3 Extension Approaches/Methods and Their Performance

The approaches in providing extension services to farmers have frequently been changing to meet the demands of the changing environments where it operates. To ensure sufficient provision of extension services in Tanzania, various extension approaches/methods have been used and these include, but are not limited to, the supply-driven, demand-driven, pluralistic extension systems, training and visit (T&V), non-governmental extension system (NGOs), commodity-based extension, farmer field schools (FFS), and farmer-to-farmer diffusion (Msuya et al. 2017).

The study of Nyasimi et al. (2017) showed that the government extension services are the primary source of oral information (75%) concerning climate change education to Tanzanian smallholder farmers followed by other sources which include a farmer's own experience (26%), traditional knowledge (11%), researchers (7%), neighbours (6%), and agro-service providers and seed companies. However, the study conducted by Msuya et al. (2017) found private extension (NGOs, CBOs, and private agribusiness) was perceived by extension agents to be more effective than public sector extension. According to Rutatora and Mattee (2001), this is because most of the private extension services are project oriented which tends to intensify activities and resources to the extent that shows visible and tangible results, and most of the time they supplement their interventions by integrating with other services including credit, agro-inputs and training.

The assessment of the extension model currently applied in Tanzania reveals several weaknesses that contribute to hampering the efforts towards adaptation among CSA smallholder farmers (Table 13.1). This includes limited financial resources due to low prioritization of climate change activities by the government, low and ineffective utilization of available extension methods/approaches, inadequate human resources and extension facilities. On the other side, the assessment reveals the critical roles that are being played by extension services in promoting CSA adoption among smallholder farmers. These results suggest a need to transform agricultural extension systems to meet the demand of changing farming challenges.

13.6 Experience from Other African Countries (Senegal, Malawi and Kenya)

The efforts to reduce impacts of climate change have been taken by many African countries especially those which are highly exposed to the changing climatic conditions. Many attempts have been directed at the agriculture sector because it is among the most vulnerable economic sectors to climate change while also being the sector on which the livelihoods of a large proportion of the population depend. Extension services have been at the centre of these efforts, using different approaches and methods to provide a nexus between farmers and other stakeholders of climate change. This chapter will analyse those efforts specifically in Senegal, Malawi and Kenya with a focus on climate change adaptation.

13.6.1 Senegal

The economy and livelihood of many people in Senegal are highly dependent on the agriculture sector. According to FAO (2016), the agricultural industry employs more than 70% of the workforce and represents about 17% of the country's gross domestic product. About 90% of agriculture in Senegal is rain-fed which is currently facing rainfall variability, especially in the northern region where crops are affected by uncertain climate conditions such as low rainfall and prolonged drought (Khouma et al. 2013). Like in many other countries, the frequent occurrence of these phenomena implies that the climate of Senegal has changed. In Senegal, the increased number of extreme weather events and climate shocks in recent years has pushed farmers to adopt various climate change adaptation practices and raised the need for more reliable weather information to assist them in farm management decision-making that may help minimize climatic risks to reduce frequent food shortages.

13.6.1.1 CSA Practices in Senegal

In Senegal, according to CIAT and BFS/USAID (2016), crop production typically involves the use of high-quality certified seeds and short-cycle varieties. Other practices include crop diversification, good agriculture practices (fire control, weeding), intercropping, drip irrigation, agroforestry, assisted natural regeneration, use of

Table 13.1 An analysis of extension approaches/methods being used in Tanzania

Approach/ method	Description	Strength	Weakness
Training and visit (T&V)	Training and visit (T&V) based on top-down approaches (supply-driven) was introduced and sponsored by the World Bank until the late 1990s. The primary goal of this approach was to increase food production (Swanson 2010). This approach continued to be used alongside other approaches even after the World Bank project phaseout. The study of Rutatora and Mattee (2001) explains that in Tanzania under the Agricultural and Livestock Extension Rehabilitation Project (NALERP) the former T&V (which was based on the transfer of technologies (ToT)) was modified and became more participatory and continued to be used by extension workers	One of the key strengths of this approach is the provision of access and translation of scientific innovations and information to help farmers in improving farm production. The evidence from the Agricultural and Livestock Extension Rehabilitation Project (NALERP) showed that it had achieved its goal of increasing agricultural production and increased farmer awareness of specific technical advice and increased the rate of their adoption especially to the farmers covered by the extension workers (Rutatora and Mattee 2001)	The weakness of this approach is that it does not involve farmers in identifying the challenges and adapting the research to local conditions (non-participatory). It assumes that knowledge is coming from the external source and farmers should be the recipients of the particular expertise rather than being integral to the innovation. This implies that issues of relevance, cost-effectiveness, ownership and sustainability are abandoned by this approach, a factor that predicts its failure in the future

Non-governmental extension system (NGOs)	This type of extension has increased in recent years especially after the introduction of restructuring policy (decentralization by devolution) by the government which allows private organizations to supplement public extension services	The use of participatory (demand-driven) approaches by many NGOs facilitates the emergence of innovations among farmers which help to solve challenges based on their priorities. The observations in Tanzania show that the simple organization structures of many NGOs tend to give staff members more scope to address the immediate situation on the ground and so address the local concerns. Also, NGOs can be more effective as compared to public extension services because their coverage is limited and thus the intensification of efforts can deliver better outcomes. However, the tendency of being territorial and having a higher degree of autonomy pose problems for the coordination of extension efforts (Rutatora and Mattee 2001)	Some of the weaknesses of NGOs include the reluctance to share information with public workers because they want to have all the credit for any achievement. Also they do not prefer collaboration to safeguard their independence, and apart from that they tend to concentrate on relatively small geographical coverage (Msuya et al. 2017) because they usually operate under project bases and in many cases they are biased in selecting those project areas; the issue of sustainability is critical especially when donor financing ends (MAC 2000)

(continued)

Table 13.1 (continued)

Approach/ method	Description	Strength	Weakness
Farmer field schools (FFS)	This is a group-based experiential learning approach which aims to build famer capacity to learn, understand and make informed decisions in their fields. Presently, it is a widely used extension approach in Tanzania	In a farmer field school, groups of farmers meet regularly in the field with a facilitator to observe, talk, ask questions and learn together. This participatory (interactive) kind of extension approach helps farmers to be innovative and improve their decision-making in farming activities. A group approach facilitates rapid spread of farming information among farmers rather than one-to-one paths like a contact farmer. Also, during "special" topic sessions in the FFS, farmers can learn a wide range of technical and social topics which are related to their farming activities such as water and sanitation, household livelihood security, marketing, child labour and climate change to name a few. Moreover, it helps farmers in learning by doing, reduces the cost of production, provides a systematic evaluation of different technologies, and promotes community organization (Khatam et al. 2010). According to Braun and Duveskog (2008), the networks of FFS groups in Sub-Saharan Africa have increased farmers' voice and power and access to services and markets	Despite the strengths that make it a highly recommended approach in many developing countries, it has some weaknesses which somewhat limit its adaptability. Massive expenses on the implementation of FFS appear to be a burden to the funding organizations (public/private). FFS is a time-consuming process which requires more highly committed members. Additionally, different challenges in handling a group may arise, for instance, an environment of competition between farmers during the learning process (Khatam et al. 2010)

Information and communication technology (radio, TV, mobile phones, social media) methods	Communication is crucial for addressing extension problems related to participation, integration, capacity building, decentralization and sustainability as a human dimension (World Bank 2007). In the recent years, extension delivery has increased the use of information and communications technology-based (ICT-based) methods to deliver farming messages to farmers. In Tanzania, mobile phones, radios and TV stations are now used to inform farmers about agricultural best practices, agriculture marketing, climate change issues and weather forecasts	An ICT-based approach has a more significant potential to reach many farmers at once than any other extension method. Rupan et al. (2018) argues the ICT-based method is the best extensive covering method in the environments of developing countries where the significant constraints of national extension systems are the shortage of field extension personnel and limited resources to reach large numbers of farmers spread widely across geographical areas. Furthermore, he explains the ICT-based approach is the best-suited option for awareness raising and thus it has a potential to contribute to climate mitigation, adaptation and increased food security. It is an effective method of conveying the more straightforward messages to farmers which does not necessarily require extension personnel to do this task, for example, climatic information. Also, mass media is a cost-effective method compared to other ways considering geographical coverage against costs	The ICT-based approach does not suit in disseminating complex farming knowledge which requires more practical learning processes because learning by doing is somewhat challenging with this approach. In this approach, the interactions between participants are limited as compared to other approaches such as group based. In developing countries like Tanzania, poor network coverage, especially in the rural area, could be a constraint to access information through mobile phones and, besides, massive costs of TV and its accessories could be another limitation to the farmer to use it

stone bunds for water management, application of organic fertilizer, mulching and composting and use of neem as a biological pesticide, particularly in the horticulture and arboriculture sectors. Livestock CSA practices include intensification and sedentarization of livestock and changing herd species for small ruminants. Pastoralism and, especially nomadic transhumance, is a common adaptation strategy practised in the northern parts of Senegal to cope with the climatic stresses and limited resources in the region (Diouf et al. 2014).

Apart from these practices, farmers are organizing themselves into associations to pool resources and form savings groups. However, it was observed that these CSA practices are common especially among farmers who have been exposed to government and private extension organizations which provide them with agricultural extension services, giving considerable attention to climate change mitigation (CIAT and BFS/USAID 2016). This implies that a large percentage of farmers especially those who have not received extension services are still not practising CSA.

13.6.1.2 The Extension Services and Climate Information Services in Senegal

In Senegal, the extension agents are at the centre of the entire dissemination systems to ensure the accuracy and reliability of climate information. Before the dissemination of any climate message to farmers, they make sure that it is well defined and has been interpreted to suit the local context of a particular farming system. However, Sala et al. (2016) observed that several factors limit the performance of extension services in Senegal: there is a dearth of extension workers (multidisciplinary working groups) at the local level to translate climate information to farmers, inadequate funding for operationalizing plans, capacity building of extension workers and journalists, and improving communication between actors.

13.6.1.3 Extension Initiatives to Tackle Climate Change in Senegal

The government with the support of the Research Program on Climate Change Agriculture and Food Security (CCAFS) has initiated climate information services (CIS) which provide weather forecasts helping farmers to improve their decisions about agriculture management in an environment of climate change (CCAFS 2015b; Sala et al. 2016). Another initiative for CIS which works primarily during the beginning of the rainy season was established by the national meteorological agency, Agence Nationale de l'Aviation Civile et de la Météorologie (ANACIM). It has been using some different methods such as radio broadcasting (which use local languages), direct contact with farmers, collaboration with the rural development departmental services (SDDR) and also by conducting seminars to inform farmers about the trends and projections of weather (CCAFS 2015b; Sala et al. 2016).

However, different studies in other developing countries (including Sub-Saharan Africa) have pointed out some weaknesses that could undermine the long-term sustainability of delivery and uptake of CIS. These include (i) the production of information is not locally relevant, fit for purpose and available in a timely manner (Lackstrom et al. 2014; Nidumolu et al. 2016; Singh et al. 2017); (ii) the lack of appropriate governance and institutional structures for the provision of climate information (Vaughan and Dessai, 2014); (iii) inadequate emphasis on socio-economic value in the uptake of climate information provided and subsequent decision-making; and (iv) the lack of appropriate boundary organizations to effectively communicate between information providers and users (Ouedraogo et al. 2018). This suggests a need for a determined effort to check these weaknesses to ensure effective CIS delivery particularly in developing countries like Senegal.

Furthermore, to facilitate the flow of CSA information in Senegal, a multidisciplinary working group has been established to play an essential role in disseminating climate information and advisories which significantly influences farm management decisions and other livelihood activities (Partey et al. 2018). However, Sala et al. (2016) found that there was insufficient coverage of local working groups (GTPs) across the country due to lack of funds to cover their operational costs such as meetings, transport and capacity building of GTPs, journalists and communication among actors.

13.6.1.4 Extension Methods

The studies of Lynagh et al. (2014) and D'Auria and McKune (2014) observed the existence of farmers' social networks within their communities which also play a vital role in climate information dissemination. Also, the same reports found that individuals who have active social networks were likely to get climate information which helps them to take proactive measures to minimize the risks of climate change. Settle et al. (2014) argued that creating social networks among farmers is an important strategy which helps to cluster small farmer groups (e.g. FFS groups) in which they can easily access information and also consider the adoption of new practices through farmer-to-farmer diffusion.

The report of Sala et al. (2016) on CSA and extension shows that in Senegal, apart from other extension methods, mobile phones are widely used to disseminate climate information to farmers by using short message service (SMS). Partey et al. (2018) reported that while the ICT-based extension methods have increased the availability of CIS, farmers are well informed about rainfall distribution patterns, intensity and frequency, wind storms and extreme events like droughts which enable them to plan their agricultural activities effectively and efficiently. Rapidly growing mobile phone services in Senegal have increased the access of people to climate information.

The reports of ANSD (2013) and CCAFS (2015a, b) revealed that a total of 7.4 million rural people, representing about 740,000 agricultural households, were potentially reached with climate information across all 14 administrative regions of Senegal via 82 rural community radios and SMS. Sanga et al. (2014) claimed that ICT is the best method for bridging the information gap for rural farmers concerning information related to innovative practices and technologies. For instance, the study by Veeraraghavan et al. (2009) reported that in India, mobile phone communications had been preferred as an effective means of reaching many smallholder farmers in the rural areas, transferring relevant agricultural information via SMS. However, according to Sala et al. (2016), the impact of ICT-based methods can be limited by several circumstances. They argued that despite its potential in reaching many farmers promptly, two problems arise: first, a risk of exclusion of specific population groups whose accessibility to a particular information channel is not guaranteed, and second, since ICT is a supply-based approach, it might not respond to real farmers' needs.

Another extension approach which is being used to disseminate climate messages to farmers in Senegal is field demonstrations where extension workers facilitate the emergence of innovations among farmers through participatory and interactive learning. The study of Ouedraogo et al. (2018) explains that in Senegal the demonstration approach emphasizes problem-solving and discovery-based learning which increase farmers' capability to deal with challenges of climate change, and according to the authors, this approach ensures farmers with the adoption of innovative practices which are best suited to their local farming environments. Despite the potential of this approach, the study of Franzel et al. (2018) found that in Senegal, demonstration plots are particularly useful when managed by farmers rather than extension staff; they observed that farmer-led demonstrations were better received than those presented by technical staff.

13.6.2 Malawi

In Malawi, extension has been proposed to play a centre role in dealing with climate change problems by significantly expanding farmers' awareness of the adverse impacts of the changing climatic conditions on agricultural productivity, and thus of the need for farmers to adjust agriculture management practices to adapt to climate change (Nhemachena and Hassan 2007). According to the authors, Malawian farmers who have access to climate change information have shown resilience to shocks of climate change. However, in some areas, the adoption of CSA measures by farmers seems to be low. Mkisi (2014) argued that this situation might be attributable to the lack of awareness of many smallholder farmers on the available CSA technologies and practices or lack of information overlaps with the other point about lack of awareness.

13.6.2.1 Adaptation to Climate Change

According to Mkisi (2014), there are about nine common adaptive practices used by farmers in Malawi to reduce the impacts of climate change on agricultural production. The most common practice is intercropping. It is done to reduce the threat of food shortage in case of climate variability and to maximize output as well, whereby two or more crops are grown in one field. The common combinations are maize and beans, maize and cassava, maize and pigeon peas and maize and ground nuts. The second common practice is the use of drought- and disease-resistant crops such as cassava and sorghum. The third most common practice is the use of hybrid crop varieties (mainly maize varieties) which are high yielding and early maturing. Other adaptation practices include the use of organic matter or manure, box ridging, crop diversification, crop processing, soil and water conservation measures and adoption of agroforestry practices (Magombo et al. 2012). Various studies found similar adaptation practices are being applied by smallholder farmers in other developing countries (Deressa et al. 2011; Chanika et al. 2011; Ozor and Cynthia 2011).

13.6.2.2 Limitations to Adaptation Efforts

In 2002, the Government of Malawi introduced a pluralistic extension approach to expand farmer access to extension services. However, Kakota et al. (2017) observed that this approach brought other new challenges to the agricultural communities during the implementation of programmes and projects. Lack of proper coordination and organization among extension providers led to poorly harmonized information, approaches and methods. This has further contributed to information overload on smallholder farmers and hampered their capacity to deal with farming challenges including climate change.

Additionally, lack of certified CSA training manuals by the government has resulted in contradictory information on CSA technologies and practices among extension providers and, in turn, led to the poor adoption of those practices among farmers (Kakota et al. 2017). In addition, household characteristics such as size of the households, landholding, per annum total income of household, input access and output market were also found to affect the adoption of different combinations of climate change adaptation strategies in Malawi (Magombo et al. 2012).

13.6.3 Kenya

Kenya like many other countries especially in the developing world is facing the impacts of climate change which affects its economic sectors particularly agriculture. The severity of these effects is attributed to the high dependence of Kenyan agriculture on rain-fed production systems which accounts for about 75% of the entire agricultural land (Stefanovic et al. 2017). Projections of climate change suggest that agricultural production in Kenya, especially for the critical staple crops of maize and wheat, is likely to experience declines due to increased evapotranspiration caused by increasing temperature (Bryan et al. 2012).

13.6.3.1 CSA Practices in Kenya

The results from the household survey by Deressa et al. (2011) showed that farmers in Kenya have started to use different kinds of practices in response to perceived climate change. The most common CSA practices adopted in Kenya include changing crop variety (33%), changing planting dates (20%) and changing crop type (18%). Other responses included planting trees (9%), reducing livestock numbers (7%), diversifying, improving, or supplementing livestock feeds (7%), changing fertilizer application (7%) and soil and water conservation practices (SWC) (5%).

However, the results from another household survey conducted by Bryan et al. (2012) revealed about 19% of farmer households in Kenya were not using climate change adaptive measures. Meanwhile, lack of resources, limited water access and insufficient information are mentioned to be the significant barriers to the adoption of CSA technologies and practices (Bryan et al. 2013). In contrast to these findings, a similar survey conducted in Ethiopia and South Africa found 37 and 62% of smallholder farmers, respectively, were not using any CSA measure (Bryan et al. 2013). This comparison implies that Kenya was doing better than many other African countries although the percentage of farmer households which did not adopt seemed to be high.

13.6.3.2 Extension and Adaptation to Mitigate Climate Change

The study of Silvestri et al. (2012) in the seven districts of agro-pastoral communities in Kenya has mentioned extension services as a primary determinant for implementing a successful strategy to mitigate the effects of climate change. Similarly, Stefanovic et al. (2017) cited the same factors in influencing the uptake of appropriate mitigation strategies in crop production with few exceptions such as farmers' access to non-agricultural income, the availability of farmers' groups and cooperatives and the future risk perception in case of late planting among food crop farmers.

According to Silvestri et al. (2012), poor climate change adaptation by smallholder farmers is attributed to insufficient extension services which are characterized by limited contacts between extension agents and farmer households and lack of appropriate climate change information. Furthermore, the same survey by Silvestri et al. (2012) revealed that only about 20% of farmer households in the study area had received extension visits and for about 66% of those visits, the number of the visits was limited to three times or fewer within the year. This problem may be caused by a limited number of extension workers and lack of climate change knowledge among them. However, generally, the impact of extension services was observed on changed farming practices such as selection of livestock breeds that are more resistant to new climate conditions and switching to alternative feed sources during prolonged drought (Silvestri et al. 2012).

In his study, Roncoli et al. (2010) argued that adaptation is concerned with making decisions in a situation that involves uncertainty; however, regardless of the availability of high-quality information from meteorological data, forecasts of climate or local observations, farmers will still work under a certain degree of uncertainty due to the nature of climate change. This condition adds complications in delivering extension messages. Moreover, a study into the effects of government extension services on farmers' adaptation in response to climate change reveals limited success, given climatic variability advice to farmers needs to be continuously revised in the light of changing weather conditions (Crane et al. 2011).

13.6.3.3 Extension Services Reformations to Suit CSA in Kenya

Various studies in Kenya have shown the importance of having an effective extension delivery system that will facilitate the adoption of climate change practices, a situation of strengthening farmer resilience and maintenance of agricultural productivity (Lopokoiyit et al. 2012; Silvestri et al. 2012; Stefanovic et al. 2017). Additionally, Sala et al. (2016) explain a successful CSA implementation involves effective and efficient extension providers and systems; this suggests the need for significant organization and institutional reform and capacity building at the organizational and individual levels.

With regards to these calls, the Government of Kenya has been implementing some strategies to deal with these challenges as they have been stipulated in the Kenya climate-smart agriculture strategy 2017–2026. According to GoK (2017), the first strategy is to mainstream CSA in formal training institutions at certificate, diploma, graduate and postgraduate levels to increase the capacity of extension staffs to undertake CSA. The second strategy is to enhance the ability of institutions to conduct mitigation and adaptation research. This entails availing adequate infrastructure for research; providing sufficient skilled research personnel; adopting indigenous knowledge and culture; enhancing capacity to carry out adequate measurement, reporting, and verification (MRV); and increasing operational resources for CSA research. A third strategy is to ensure the availability of data and information on CSA by establishing and maintaining a data and information management system, build capacities in data collection and information management and promote data generation and dissemination during planning, implementation, monitoring and evaluation at both national and county levels.

Recently, Kenya started to use an agro-weather tool which incorporates internationally available decision support system (DSS) devices such as crop simulation models into its system to understand and demonstrate impacts of climate change and crop management practices on specific crop yields and subsequently generate climate-smart agro-advisory. It collects, organizes and integrates all the different information required for producing a particular crop and then analyses and interprets the data and finally gives recommendations on appropriate action to maintain maximum productivity. ICT-based (mass media) communication methods/channels (such as SMS, radio messages and newsletters) have been used by extension agents to disseminate these kinds of CSA technologies in Kenya and Ethiopia for their adoption and utilization (Oladele et al. 2018).

According to Rupan et al. (2018), ICT-based communication is the best extensive covering method in the environments of developing countries where the significant constraints of national extension systems are the shortage of field extension personnel and limited resources to reach large numbers of farmers spread widely across geographical areas. In contrast, Okwu (2011) argued that ICT-based communication cannot effectively reach all groups of farmers because farmers who use mass media are those who have a reasonable level of education, belong to a relatively high-income bracket and are typically male and of a relatively high socioeconomic status, that is, not the profile of a typical smallholder farmer. Therefore, farmers' socioeconomic characteristics should be considered in planning mass media usage in agricultural information dissemination (Oladele et al. 2018).

This analysis shows that adaptation to the projected impacts of climate change is being increasingly prioritized among many African countries because of the serious threat that climate change poses for the sustainability of economic sectors, particularly, the agriculture sector. The proposed measures to mitigate climate change are similar across many developing countries with few variations depending on particular local context. There are significant relationships between socio-economic factors and the kinds of adaptation recommended while the availability and quality of extension services play a prominent role in determining the adoption of CSA technologies and practices among smallholder farmers. It is observed that lack of efforts to empower agricultural extension services to take emerging climate change issues seriously remains a critical obstacle to effective climate change action among farmers across the region.

13.7 Extension Model for Improved Action on Climate Change

13.7.1 Need for Reforming Agricultural Extension Provision in Tanzania

A successful CSA blueprint requires a well-organized extension system competent to address the urgency of climate change issues. Despite playing a role in dealing with climate change within the farming environment, the current extension model in Tanzania does not offer sufficient services which could effectively address climate change challenges. The purpose of this chapter is to suggest an extension model that may suit the current need for extension services under climate change challenges.

13.7.2 Pluralistic System

The current state of extension services, characterized by limited financial and human resources (particularly for public sector agencies), requires combined efforts of both public and private extension service providers to reduce the impact of these challenges. Also, the pluralistic systems have the potential to deal with the diversity of conditions, needs, audiences and farming systems that make up the agricultural landscape by providing an equally diverse arrangement of services and service providers. However, coordination by government departments (Fig. 13.2) will be needed to help extension actors to deliver services that utilize the relative strengths of each entity.

13.7.3 Group-Based Approaches

To build the sustainable resilience of farmers to climate change and variability requires innovations that are drawn from farmers' knowledge and experiences. Besides, to build resilience needs strong farmers' associations and networks to improve their bargaining power at the markets, to increase access to relevant information, to widen sources of finances and to have a loud voice in advocating policy and strategies that affect their activities. The group-based extension approaches have a high potential to allow farmers to self-organize to build their resilience in confronting this issue of climate change.

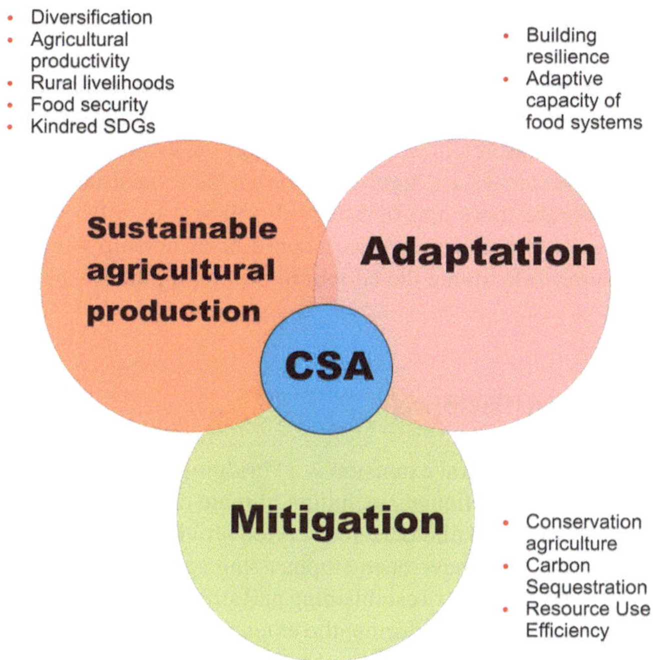

Fig. 13.2 Goals of climate-smart agriculture. (Source: Venkatramanan and Shah (2019)

13.7.4 Extension Methods

There is no single extension approach/method which is adequate to address agricultural challenges including the climate change issues effectively. This implies that to deal with the complex impacts of climate change requires more than one approach/method. There is a need for a complementary extension approach that can trap the advantages from each available approach/method to adequately address climate change. According to Sala et al. (2016), extension approaches/methods differ regarding their reach and impact potential, "the higher the reach, the smaller the impact and vice-versa".

13.7.5 Traditional Extension (ToT)

This approach suits the motive of increasing agricultural productivity by transferring farming technologies from outside of the farmers' environment. Successful CSA technologies and practices from outside sources can be effectively disseminated by using this approach and help to reduce the effects of climate change. However, implementation of this approach will require some modifications to make it more participatory to ensure the sustainability of those technologies and practices in the new environment.

13.7.6 Participatory Approaches

This approach to extension is essential due to its potential in facilitating the emergence of innovations among farmers. Addressing climate change and variability needs more specific and realistic solutions based on local conditions. Indigenous knowledge and experiences are highly recommended to facilitate the emergence of innovations which are more appropriate and practical to solve a specific climate change challenge at a local level. Also, a sense of ownership which is ensured by this approach helps to facilitate the adoption, promoting the sustainability of those innovations.

13.7.7 Innovation Platforms

One of the roles of agricultural extension is a "bridging" function. Finding solutions for most of the farming challenges including climate change requires a kind of collective effort of all rural stakeholders and service providers. More recently, in many countries, extension agents have been emphasizing agricultural innovation systems (AIS) by playing various roles in establishing and strengthening of multi-stakeholder innovation platforms. In the platforms, the extension worker is the leading innovation broker for catalysing the process and bringing other stakeholders together and facilitating interaction between them, coordinating and creating networks, ensuring the flow of information among actors and providing technical assistance.

13.8 Conclusion

The agriculture sector in Tanzania is detrimentally affected by the persisting impacts of climate change due to its high proportion of rain-fed production systems. In recent years, noticeable effects, including seasonal fluctuation of agricultural productivity, have become common in different parts of the country. Studies predict severe consequences in the future which will, in turn, affect the country's economic status due to its dependence on the agriculture sector. This suggests deliberate efforts are needed by the government in collaboration with other development partners to reduce the impacts of climate change on this sector.

At the farmer's level, one of the recommended efforts for reducing impacts of climate change is the adaptation of CSA technologies and practices in farming activities. Several field observations in Tanzania reveal that there are various CSA technologies and practices which are being used by smallholder farmers although the rate of adoption is still low due to many reasons, particularly, poor rural extension services. However, in Tanzania, CSA is still a new notion among many extension agents and smallholder farmers. This suggests efforts are needed by the extension services to build the capacity of extension agents and smallholder farmers on CSA.

The current extension services in Tanzania are faced with many challenges which affect their performance in bringing desirable outcomes in the sector. Despite the obstacles, extension remains a crucial catalyst in enhancing agricultural development due to its significant potential in dealing with farming challenges including cross-cutting issues like climate change. Among many other difficulties in extension provision, poor utilization of available approaches and methods has been affecting its performance. Many studies suggest the need for, among other things, a carefully chosen combination of approaches and methods that suit the need of a specific farming situation such as the adaptation of CSA to suit the local context.

In reforming and strengthening agricultural extension services for promoting CSA in Tanzania, there is a need to implement immediate priority actions alongside the long-term action plans. These include establishment of mechanisms at the local level for better alignment and cooperation/collaboration between public sector agricultural extension and other sectors (such as livestock, water, environment, forestry and fisheries) to strengthen the capacities of agricultural service provision and other rural stakeholders to enhance processes of innovation at organizational and individual levels and to improve the ability of service providers to identify and use a range of extension methods and approaches appropriately for sharing CSA technologies and practices with farmers.

In the long term, the government should increase investment in research, extension and training including the introduction of specialized courses in climate change, particularly CSA technologies and practices. This review also recommends short-term training of smallholder farmers on climate change and CSA technologies and practices; training and deployment of more public extension workers and short-term retraining of existing extension workers, policymakers and policy implementers in climate change and CSA thinking.

References

Ahmed SA, Diffenbaugh NS, Hertel TW, Lobell DB, Ramankutty N, Rios AR (2011) Climate volatility and poverty vulnerability in Tanzania. Glob Environ Chang 21:46–55. https://doi.org/10.1016/j.gloenvcha.2010.10.003

ANSD (2013) Rapport Definitif RGPHAE 2013. Agence Nationale de la Statistique et de la Démographie (ANSD). http://bit.ly/1NrlekU

Arndt C, Farmer W, Strzepek K, Thurlow J (2011) Climate change, agriculture, and food security in Tanzania. Helsinki, United Nations University: UNU-WIDER

Barakabitze AA, Kitindi EJ, Sanga C, Shabani A, Philipo J, Kibirige G (2015) New technologies for disseminating and communicating agriculture knowledge and information: challenges for agricultural research institutes in Tanzania. Electron J Inf Syst Dev Countries 70(2):1–22

Bie SW, Mkwambisi D, Gomani M (2008) Climate change and rural livelihoods in Malawi. Review study report of Norwegian support to FAO and SCC in Malawi, with a note on some regional implications. The Royal Norwegian Embassy, Lilongwe, Malawi

Braun A, Duveskog D (2008) The farmer field school approach-history, global assessment and success stories. Background paper for the IFAD rural poverty report

Bryan E, Ringler C, Okoba B, Koo J, Herrero M, Silvestri S (2012) Can agriculture support climate change adaptation, greenhouse gas mitigation, and rural livelihoods? Insights from Kenya. Clim Chang 118:151–116. https://doi.org/10.1007/s10584-012-0640-0

Bryan E, Ringler C, Okoba B, Roncoli C, Silvestri S, Herrero M (2013) Adapting agriculture to climate change in Kenya: household strategies and determinants. J Environ Manag 114:26–35. https://doi.org/10.1016/j.jenvman.2012.10.036

CCAFS (2015a) Annual report 2014: climate-smart agriculture – acting locally, informing globally. CGIAR research program on Climate Change, Agriculture and Food Security (CCAFS). Copenhagen, Denmark. Available online at: bit.ly/CCAFSAR2014

CCAFS (2015b) Scaling up climate advisories in Senegal and Colombia. https://cgspace.cgiar.org/bitstream/handle/10568/67903/05outcomecase.pdf?sequence=3

Chang'a LB, Kijazi AL, Luhunga PM, Ng'ongolo HK, Mtongori HI (2017) Spatial and temporal analysis of rainfall and temperature extreme indices in Tanzania. Atmos Clim Sci 7:525–539. https://doi.org/10.4236/acs.2017.74038

Chanika D, Hamazakaza P, Joubert A, Macome E, Mutonhodza C (2011) Overcoming the barriers: how to ensure future food production under climate change in Southern Africa. Oxfam International, Oxford

CIAT; BFS/USAID (2016) Climate-smart agriculture in Senegal. CSA country profiles for Africa series. International Center for Tropical Agriculture (CIAT). Bureau for Food Security, United States Agency for International Development (BFS/USAID), Washington, DC, p 20

CIAT; World Bank (2017) Climate-smart agriculture in Tanzania. CSA country profiles for Africa series. International Center for Tropical Agriculture (CIAT); World Bank: Washington, DC. p 25. https://ccafs.cgiar.org/publications/climate-smart-agriculture-tanzania#.XBx90FwzbIU. Accessed 21 Dec 2018

Coulibaly YJ, Kundhlande G, Amosi N, Tall A, Kaur H, Hansen J (2015) What climate services do farmers and pastoralists need in Tanzania? Baseline study for the GFCS Adaptation Program in Africa. CCAFS working paper no. 110. CGIAR research program on Climate Change, Agriculture and Food Security (CCAFS). Copenhagen, Denmark. Available online at: www.ccafs.cgiar.org

Crane TA, Roncoli C, Hoogenboom G (2011) Adaptation to climate change and climate variability: the importance of understanding agriculture as performance. NJAS Wagen J Life Sci 57(3–4):179–185

D'Auria Ryley T, McKune S (2014) Climate change and social networks in Senegal's peanut basin. CCAFS. http://bit.ly/1Kc3nvL

Daniel E (2013) Assessment of agricultural extension services in Tanzania. A case study of Kyela, Songea Rural and Morogoro Rural Districts. Internship report in Plant Sciences CSA. Wageningen University, Netherlands

Deressa T, Hassan R, Ringler C (2011) Perception of and adaptation to climate change by farmers in the Nile basin of Ethiopia. J Agric Sci 149(01):23–31. https://doi.org/10.1017/S0021859610000687

Diouf B, Lo HM, Dieye B, Sane O, Sarr OF (2014) Pour une agriculture intelligente face au changement climatique au Sénégal: Recueil de bonnes pratiques d'adaptation et d'atténuation. Document de travail no 85. Copenhagen, Denmark: Programme de Recherche du CGIAR sur le Changement Climatique, l'Agriculture et la Sécurité Alimentaire (CCAFS)

Ehrhart C, Twena M (2006) Climate change and poverty in Tanzania: realities and response options for CARE. Background report. CARE international poverty-climate change initiative. http://www.care.dk/multimedia/pdf/web_english/Climate%20Change%20and %20Poverty%20in%20Tanzania%20-%20Country%20Profile.pdf

Enfors E, Gordon L (2008) Dealing with drought: the challenge of using water system technologies to break dry-land poverty traps. Glob Environ Chang 18:607–616. https://doi.org/10.1016/j.gloenvcha.2008.07.006

FAO (2010) "Climate-smart" agriculture policies, practices and financing for food security, adaptation and mitigation. FAO, Rome

FAO (2013) Climate-smart agriculture: a sourcebook. Food and Agriculture Organization of the United Nations (FAO), Rome

FAO (2014) Success stories on CSA. Rome, Italy, FAO. http://www.fao.org/3/a-i3817e.pdf

FAO (2016) FAOSTAT Database, Food and Agriculture Organization of the United Nations

FAO (2017) Climate-smart agriculture guideline for the United Republic of Tanzania: a country–driven response to climate change, food and nutrition insecurity. Rome, Italy. http://www.fao.org/3/a-i7157e.pdf

Franzel S, Ndiaye A, Tata JS (2018) Senegal: in-depth assessment of extension and advisory services. In: Developing local extension capacity project. USAID, Washington, DC

GoK. (Government of the Republic of Kenya) (2017) Kenya climate-smart agriculture strategy-2017-2026. Ministry of Agriculture, Livestock, and Fisheries. Nairobi, Kenya. Retrieved from https://canafrica.com/wp-content/uploads/2017/05/MINISTRY-OF-AGRICULTURE-LIVESTOCK-AND-FISHERIES-3.pdf

Gwambene B, Saria JA, Jiwaji NT, Pauline NM, Msofe NK, Shija SMY (2015) Smallholder farmers' practices and understanding of climate change and climate-smart agriculture in the Southern Highlands of Tanzania. J Resour Dev Manag 13:37–47

Harris-Coble L (2016) Tanzania. Landscape analysis. Working document: United States Agency for International Development (USAID) and US Government Feed the Future project "Integrating Gender and Nutrition within Extension and Advisory Services"(INGENAES), p 30. https://www.agrilinks.org/sites/default/files/resource/files/ING%20Landscape%20Study%20%282016%29%20Tanzania%20-%20published%202016_05_10.pdf. Accessed 26 Dec 2018

IPCC (2007) In: Parry ML, Canziani OF, Palutikof JP, van der Linden PJ, Hanson CE (eds) Climate change 2007: impacts, adaptation and vulnerability. Contribution of Working Group II to the fourth assessment report of the Intergovernmental Panel on Climate Change. Cambridge University Press, Cambridge, 976pp

IPCC (2013) Summary for policymakers. In: Stocker TF, Qin D, Plattner G-K, Tignor M, Allen SK, Boschung J, Nauels A, Xia Y, Bex V, Midgley PM (eds) Climate Change 2013: the physical science basis. Contribution of Working Group I to the fifth assessment report of the Intergovernmental Panel on Climate Change. Cambridge University Press, Cambridge/New York

Kaaria S, Sanginga P, Njuki J, Delve R, Chitsike C, Best R (2007) Enabling rural innovation in Africa: an approach for empowering smallholder farmers to access market opportunities for improved livelihoods. In: Farmer first revisited conference, future agricultures consortium. Institute of Development Studies (IDS), Sussex

Kakota TV, Maonga BB, Synnevag G, Chonde C, Mainje M (2017) Harmonisation of extension messages on climate-smart agriculture in Malawi: do we speak with one voice, and to whom? J Agric Ext Rural Dev 9(11):255–261. https://doi.org/10.5897/JAERD2017.0905

Kangalawe RYM, Mung'ong'o CG, Mwakaje AG, Kalumanga E, Yanda PZ (2017) Climate change and variability impacts on agricultural production and livelihood systems in Western Tanzania. Clim Dev 9(3):202–216. https://doi.org/10.1080/17565529.2016.1146119

Khan A, Pervaiz U, Khan NM, Ahmad S, Nigar S (2009) The effectiveness of demonstration plots as extension method adopted by AKRSP for agricultural technology dissemination in District Chitral. Sarhad J Agric 25(2):313–320

Khatam A, Muhammad S, Chaudhry KM, Mann AA, Haq I, Amin H (2010) Strengths and weaknesses of farmers field schools approach as perceived by farmers. Sarhad J Agric 26(4):685–688

Khouma M, Jalloh A, Thomas TS, Nelson GC (2013) Senegal: In International Food Policy Research Institute (IFPRI). Discussion paper, pp 291–322

Kisusu R (2003) The impact of dairy and rice production in dairy-based and irrigated rice farming systems in Dodoma region. The Sokoine University of Agriculture

Lackstrom K, Kettle NP, Haywood B, Dow K (2014) Climate-sensitive decisions and time frames: a cross-sectoral analysis of information pathways in the Carolinas. Weather Clim Soc 6:238–252

Lopokoiyit MC, Onyango C, Kibett JK, Langat BK (2012) Human resource development in agriculture extension and advisory services in Kenya. The 8th Africa Farm Management Association (AFMA), Congress peer-reviewed papers. Moi University, Kenya

Lynagh S, Tall A, Jay A (2014) One size does not fit all: considering gender, equity, and power in climate information services. CCAFS. http://bit.ly/1Q3blaG

MAC (2000) The need for agricultural extension reform in Tanzania: review of the literature. Ministry of Agriculture and Cooperatives MAC-Extension facilitation unit, Dar es Salaam

Magombo TM, Kanthiti G, Phiri G, Kachulu M (2012) The incidence of indigenous, emerging and innovative climate change adaptation practices for smallholder farmers' livelihood security in Chikhwawa district, Southern Malawi. Working paper series no. 63. The African Technology Policy Studies Network. Nairobi, Kenya

Mandleni B, Anim FDK (2011) Climate change awareness and decision on adaptation measures by livestock farmers in South Africa. J Agric Sci 3(3):258–268. https://doi.org/10.5539/jas.v3n3p258

Mkisi RB (2014) The role of agricultural extension in smallholder farmer adaptation to climate change in Blantyre district, Malawi. Open access theses. 353. https://docs.lib.purdue.edu/open_access_theses/353

Mnenwa R, Maliti E (2010) A comparative analysis of poverty incidence in farming systems of Tanzania. Special paper 10/4 Dar-es-salaam: REPOA

Msuya CP, Annor-Frempong FK, Magheni MN, Agunga R, Igodan CO, Ndiaye A (2017) The role of agricultural extension in Africa's development: the importance of extension workers and the need for change. Int J Entomol Res 5(1):59–70. ISSN:2311-6110

Müller C, Cramer W, Hare WL, Lotze-Campen H (2011) Climate change risks for African agriculture. Proc Natl Acad Sci U S A 108:4313. https://doi.org/10.1073/pnas.1015078108

Mvuna JK (2010) Agricultural extension services delivery in Tanzania. In Kimaro WH, Mukandiwa L, Mario EZJ (eds) Towards improving agricultural extension service delivery in the SADC region. Proceedings of the workshop on information. Sharing among extension players in the SADC region, 26–28 July 2010, Dar es Salaam, Tanzania

Mwamakimbula AM (2014) Assessment of the factors impacting agricultural extension training programs in Tanzania: a descriptive study. Graduate theses and dissertations. 14227. http://lib.dr.iastate.edu/etd/14227

Nederlof ES, Pyburn R (2012) One finger cannot lift a rock: facilitating innovation platforms to trigger an institutional change in West Africa. Royal Tropical Institute, Amsterdam

Nhemachena C, Hassan R (2007) Micro-level analysis of farmers' adaptation to climate change in Southern Africa. IFPRI Discussion paper 00714 August 2007. International Food Policy Research Institute, Environment, and Production Technology Division

Nidumolu UB, Lubbers M, Kanellopoulos A, van Ittersum MK, Kadiyala DM, Sreenivas G (2016) Engaging farmers on climate risk through targeted integration of bioeconomic modeling and seasonal climate forecasts. Agric Syst 149:175–184

Nyanga PH, Johnsen FH, Aune JB, Kalinda TH (2011) Smallholder farmers' perceptions of climate change and conservation agriculture: evidence from Zambia. J Sustain Dev 4(4):73–85. http://tinyurl.com/nhokvkq

Nyasimi M, Kimeli P, Sayula G, Radeny M, Kinyangi J, Mungai C (2017) Adoption and dissemination pathways for climate-smart agriculture technologies and practices for climate-resilient livelihoods in Lushoto, Northeast Tanzania. Climate 5(63):1–22. https://doi.org/10.3390/cli5030063

Okunade E (2007) The effectiveness of extension teaching methods in acquiring knowledge, skill, and attitude by women farmers in Osun state. J Appl Sci Res 3(4):282–286

Okwu OJ (2011) Characterizing farmer users and nonusers of mass media as channels of agricultural information in Benue state, Nigeria. J Agric Food Inf 12(3–4):315–328

Oladele OI, Gitika MP, Ngari F, Shimeles A, Mamo G, Olorunfemi OD (2018) Adoption of agro-weather information sources for climate-smart agriculture among farmers in Embu and Ada'a districts of Kenya and Ethiopia. Inf Dev:1–16. https://doi.org/10.1177/0266666918779639

Ouedraogo I, Diouf NS, Ouédraogo M, Ndiaye O, Zougmoré R (2018) Closing the gap between climate information producers and users: assessment of needs and uptake in Senegal. Climate 6(13):1–16. https://doi.org/10.3390/cli6010013

Ozor N, Cynthia N (2011) The role of extension in agricultural adaptation to climate change in Enugu State, Nigeria. J Agric Ext Rural Dev 3(3):42–50

Partey ST, Zougmore RB, Ouedraogo M, Campbell BM (2018) Developing climate-smart agriculture to face climate variability in West Africa: challenges and lessons learnt. J Clean Prod 187:285–295

Roncoli C, Okoba B, Gathaara V, Ngugi J, Nganga T (2010) Adaptation to climate change for smallholder agriculture in Kenya: community-based perspectives from five districts. Report to the World Bank of the project. Adaptation of smallholder agriculture to climate change in Kenya

Rowhani P, Lobell DD, Linderman M, Ramankutty N (2011) Climate variability and crop production in Tanzania. Agric For Meteorol 151(4):449–460

Rupan R, Saravanan R, Suchiradipta B (2018) Climate-smart agriculture and advisory services: approaches and implication for future. MANAGE discussion paper 1, MANAGE- Centre for Agricultural Extension Innovations, Reforms and Agripreneurship (CAEIRA), National Institute of Agricultural Extension Management. Hyderabad, India

Rutatora DF, Mattee AZ (2001) Major agricultural extension providers in Tanzania. Afr Stud Monogr 22(4):155–173

Sala S, Rossi F, David S (2016) Supporting agricultural extension towards climate-smart agriculture: an overview of existing tools. Global Alliance for Climate Smart Agriculture (GASCA)/ FAO, Italy

Sanga C, Mussa M, Tumbo S, Mlozi MRS, Muhiche L, Haug R (2014) On the development of the mobile-based agricultural extension system in Tanzania: a technological perspective. Int J Comput ICT Res 8(1):49–67. http://www.ijcir.org/volume8-issue1/article 5.pdf

Settle W, Soumaré M, Sarr M, Garba MH, Poisot AS (2014) Reducing pesticide risks to farming communities: cotton farmer field schools in Mali. Philos Trans R Soc Lond B Biol Sci 369:20120277

Silvestri S, Bryan E, Ringler C, Herrero M, Okob B (2012) Climate change perception and adaptation of agro-pastoral communities in Kenya. Reg Environ Chang 12:791–802. https://doi.org/10.1007/s10113-012-0293-6

Singh C, Daron J, Bazaz A, Ziervogel G, Spear D, Kituyi E (2017) The utility of weather and climate information for adaptation decision-making: current uses and prospects in Africa and India. Clim Dev 10(5):389–405. https://doi.org/10.1080/17565529.2017.1318744

Stefanovic JO, Yang H, Zhou Y, Kamali B, Ogalleh SA (2017) Adaptation to climate change: a case study of two agricultural systems from Kenya. Clim Dev:1–20. https://doi.org/10.1080/17565529.2017.141124

Stigter K, Winarto YT, Ofori E, Zuma-Netshiukhwi G, Nanja D, Walker S (2013) Extension agrometeorology as the answer to stakeholder realities: response farming and the consequences of climate change. Special issue on Agrometeorology: from scientific analysis to operational application. Atmosphere 4(3):237–253

Swanson BE (2010) Strengthening agricultural extension and advisory systems: procedures for assessing, transforming, and evaluating extension systems. Agriculture and rural development discussion paper 45. World Bank, Washington, DC

Taneja G, Pal BD, Joshi PK, Aggarwal PK, Tyagi NK (2014) Farmers' preferences for climate-smart agriculture: an assessment in the Indo-Gangetic Plain. IFPRI discussion paper 01337. International Food Policy Research Institute, Washington DC, USA

URT (2005) Singida region socio-economic profile. Second eds, printed with the cooperation of National Bureau of Statistics and Singida Regional Commissioner's office, United Republic of Tanzania

URT (2014) Agriculture climate resilience plan 2014–2019. Ministry of Agriculture. Livestock and Fisheries. United Republic of Tanzania, Dar-es-salaam

URT (2017) Climate-smart agriculture guideline. Ministry of Agriculture. Livestock and Fisheries. United Republic of Tanzania, Dar-es-salaam

Vaughan C, Dessai S (2014) Climate services for society: origins, institutional arrangements, and design elements for an evaluation framework. Climate Change 5:587–603. https://doi.org/10.1002/wcc.290

Veeraraghavan R, Yasodhar N, Toyama K (2009) Warana unwired: replacing PCs with mobile phones in a rural sugarcane cooperative. Inf Technol Int Dev 5(1):81–95

Venkatramanan V, Shah S (2019) Climate smart agriculture technologies for environmental management: the intersection of sustainability, resilience, wellbeing and development. In: Shah S et al (eds) Sustainable green technologies for environmental management. Springer, Singapore, pp 29–51. https://doi.org/10.1007/978-981-13-2772-8_2
Williams TO, Mul M, Cofie O, Kinyangi J, Zougmore R, Wamukoya G, Nyasimi M, Mapfumo P, Speranza CI, Amwata D, Frid-Nielsen S, Partey S, Girvetz E, Rosenstock T, Campbell BM (2015) Climate smart agriculture in the African context. Background paper. Feeding Africa conference, 21–23 October 2015. https://hdl.handle.net/10568/68944
World Bank (2007) World congress on communication for development: lessons, challenges and the way forward. The World Bank published jointly with the Food and Agriculture Organization of the United Nations and the Communications Initiative, Washington, DC
World Bank (2016) World development indicators. http://data.worldbank.org/indicator/NV.AGR.TOTL.ZS

Climate Change and Gender Policy 14

Rashila Deshar and Madan Koirala

Abstract

As gender inequalities persist around the world, "women and men are experiencing climate change differently, affecting the ability of individuals and communities to adapt". The intersection of climate change and gender policy is more concerned with gender differences and gender mainstreaming. Though men and women are negatively influenced by the impacts of climate change, women quite often experience more burden and are indeed more vulnerable than men, due to the factors like "social status", working conditions and hardships. Indeed, the vulnerability of women owes to prevailing socio-economic conditions and social fabric. Many women in developing countries depend on the natural environment for subsistence and income. They even played a central role in agriculture and natural resource management, but due to lack of economic opportunities, women have very limited or no access in decision-making processes. Therefore, it is important that the consequences of climate change should not lead already marginalized sections of communities into further deprivation. The threats posed by climate change have failed to impress on policy-makers the importance of placing women at the heart for ensuring a sustainable future by combining development and climate change issues. Skewed participation of women in the processes involving "decision-making", "planning", "policy-making" and "implementation" results in increasing their vulnerability to climate change impacts. Gender mainstreaming is the need of the hour as it can engender viable and pragmatic solution to climate change.

Keywords

Natural resource · Decision-making · Participation · Gender policy

R. Deshar · M. Koirala (✉)
Central Department of Environmental Science, Tribhuvan University, Kathmandu, Nepal
e-mail: mkoirala@cdes.edu.np

© Springer Nature Singapore Pte Ltd. 2020

V. Venkatramanan et al. (eds.), *Global Climate Change and Environmental Policy*,
https://doi.org/10.1007/978-981-13-9570-3_14

14.1 Introduction

Gender refers to the socially constructed norms, roles and relations that a given society considers appropriate for men and women. "Gender determines what is expected, permitted and valued in a woman or a man in a determined context" (WHO 2010). Women's and men's vulnerability to the impact of extreme climate events is determined not only by biology but also by differences in their social roles and responsibilities (Easterling et al. 2000; Wisner et al. 2004). Therefore, gender impacts of climate change have been identified as an issue requiring greater attention at present. The climate change impacts are felt differently by men and women due to the reasons that are not limited to "social status", "physiological reasons" and "economic conditions". Further, the traditional household activities performed by women like "food gathering", "food production" and "food preparation and cooking" are projected to be difficult in view of the pervasive effects of climate change on food production, water resources and livelihoods. Incidentally, the climate variability like extreme weather occurrences impact the women, elderly population, children and poor people more. Some unprecedented examples are cited in the case of loss and damage caused due to the impact of climate change recently. Factors like lack of land rights and economic independence, truncated access to education and training, lesser access to decision-making process position the women in more vulnerable situation. All said, women possess knowledge on adaptation to climate change. Nevertheless, their knowledge is not yet fully utilized. In fact, their potential is untapped in climate change policy formulation. Hence gender policy tuned to climate change issue is highly significant.

14.2 Climate Change in Global Perspective

Global warming has reached unprecedented levels. It is witnessed by unprecedented rates of increase in atmospheric temperature and sea level rise (IPCC 2014). The atmospheric concentrations of greenhouse gases have increased, causing an increased level of global warming. Compared to the last thousand years, with a benchmark of period of industrialization in the recent two centuries, concentrations of greenhouse gases have increased tremendously, resulting in unprecedented level of temperature rise compared to those days, and these observed changes in the Earth's climate are at least in part due to human activities (IPCC 2018). Therefore, climate change has become a burning issue throughout the world which had affected different sectors of the environment and human population. It is now largely accepted as a real, pressing and truly global problem. The IPCC Special Report on Global Warming highlights climate impacts at the current ~1 °C global warming as well as the risks of reaching a 1.5 °C and is considered the irreversible losses if it reaches or exceeds beyond 2 °C warming (IPCC 2018).

14.3 Climate Change and Gender

Climate change manifestation is normally observed in the form of change in rainfall pattern, scarcity of drinking water, melting of the glacier, rise of sea level, destruction of the forest, tree line shift, a decrease in biodiversity and destruction of the high-altitude ecosystem (Dawson and Scott 2013). Other forms are expressed in terms of erratic rainfall, duration of drought period and changes in ecosystem characteristics. There is no question that both women and men need to face in responding to the impacts of climate change. Only the precise shape, form, scope, time scales and impacts of projected climate variability in different environmental settings vary and are unclear. Climate change impacts on biosphere ultimately may have a undesirable effect on the human population. This effect is not the same in the human population because men and women have different capacities and capabilities that will have different possibilities to cope with and adapt to the effects of climate change (Dankelman 2002). In most of the societies, women have work burden and are primary caretakers of resource management compared to men. Therefore, increased work burden with resource management push women encountering them with higher risks from the impact of climate change in the situation of poverty and lean availability of required resources. Seventy percent of the 1.3 billion people in the developing world living below the threshold of poverty are women (Denton 2002). It is important that the consequences of climate change should not lead already marginalized sections of communities into further deprivation. The threats posed by climate change have failed to impress on policy-makers the importance of placing women at the centre of their vision of sustainable development.

14.4 Climate Change Impacts in the Mountainous Regions

Species Survival Expert Group of IUCN reports that mountains cover 24% of the Earth's land surface. Mountains are home to 12% of the world's population and another 14% of the population reside in their immediate proximity. Mountains provide vital goods and services –particularly freshwater – to a significant proportion of humanity. Similarly, ecosystem services from mountains serve as key centres of biological and cultural diversity as well as important sites of traditional ecological knowledge and influence the climate at many scales. Their effective management is not only important for mountain communities, but also for a sizeable share of the global population. From the specificity perspective, mountain ecosystems are not only fragile but also inaccessible and marginal in terms of community concern (Jodha 1987). In terms of marginality, mountains are subject to both natural and anthropogenic drivers of change. These range from volcanic and seismic events and flooding to global climate change and the loss of vegetation and soils because of inappropriate agricultural and forestry practices, and extractive earth-based industries.

Communities in mountain regions face unique challenges, including a fragile ecology, natural disasters and long distances to markets, educational facilities and healthcare as well as high unemployment. Increased impact of globalization is pushing many communities in the mountain with more aging populations with out-migration of youth. Cases are being reported from the mountains of South Asian countries with emerging mental illness, drug use and alcoholism. High rates of environmental change combined with rapid economic and social changes are undermining the ability of mountain ecosystems to provide critical goods and services necessary for the well-being of mountain. Multifaceted research with both natural and social perspective is needed to understand these issues and processes and recommend effective policy measures. The ecosystem services of the mountains as carbon sinks and water towers further attach immeasurable value to these landscapes. Mountain communities suffer from high vulnerability factors such as poverty, isolation and hunger, while the fragile ecosystem struggles with present climate change and environmental degradation (Ahuja 2016).

14.4.1 Climate Change in the Hindu Kush Himalaya Region

The "Hindu Kush-Himalayan (HKH)" region extends over 4 million km^2 (about 2.9%) of the global land area and covers about 18% of the mountain area and a length of 3500 km from Afghanistan to Myanmar. The "Hindu Kush-Himalayan (HKH)" region includes countries, viz. "Afghanistan, Bangladesh, Bhutan, China, India, Nepal, Myanmar and Pakistan". Climate change had been observed in the Hindu Kush Himalayan region due to the "high altitudinal variation" and fragile ecosystem (Anup et al. 2013; Anup and Ghimire 2015). Climate change indicators in this region are increased frequency of natural hazards, change in precipitation pattern and shifting of tree line (Eugenio-Martin and Campos-Soria 2010; Anup and Ghimire 2015). Impacts of climate change are exacerbated by geographical location (rugged topography), socio-political-economic condition, lack of skilled manpower, illiteracy, nature dependent livelihood and poverty (Gurung and Bhandari 2009; Anup et al. 2013). A study done in the Tibetan Plateau across 1500 to 5000 masl gradient has shown increasing temperature rise against the higher elevation with the highest record of 3.5 °C at 3000 masl (Liu and Chen 2000).

14.4.2 Gender and Mountain Issues

Gender-based role is markedly different in the mountainous regions (Khadka et al. 2014). In fact, women are engaged in addition to household work, agricultural activity as well. Notably, women are the primary caretakers of water and forest resource management. Due to traditional and religious beliefs, women from mountain regions undertake most of the labour associated with agriculture and livestock production while men handle most of the decision-making (Dong et al. 2003; Deshar and Koirala 2016). Studies done in HKH region share findings of immense women's contribution towards the conservation and management of forests, ecologically sensitive areas, water springs and biodiversity resources (Gurung et al. 2011; Karki and Gurung 2012; Khadka and Verma 2012; Khadka et al. 2014; Talukdar 2012). However, their participation in decision-making and benefit sharing is either nonexistent or poor throughout the region (Parajuli et al. 2010; Bhasin 2011). Most women have unequal access to productive resources such as land, enterprise, education, skills, information and decision-making (Gurung et al. 2011).

Gender Contribution in Carbon Management and Their Decision-Making Power: A Case Study From a Transhumant Pastoral Society of the Himalayan Region of Nepal

The study in Gatlang village of Rasuwa District, Nepal, where the transhumance High Himalayan grasslands livelihood community existed (Fig. 14.1). In total, 23 different agro-pastoral related works were found to be carried out in the community, where 57% of work related to agro-pastoral were mainly done by the female and 39% by both male and female and 4% by only male (Table 14.1).

The study on the role of male and female in carbon management differ noticeably. There are many agro-pastoral activities that contribute to carbon release. The carbon release activities such as cooking, burning, boiling, milk warming, litter collection and firewood collection are mainly carried out by the female. Similarly, another source of carbon input in the carbon cycle was seen from the sale of livestock products such as milk, meat, leather and woollen cloths to areas inside and outside the community, which are carried out by the female. On the other hand, the dung collected from the grazing land by female helps in capturing methane gas by converting into manure which contributes to greenhouse gas reduction (Fig. 14.2).

Fig. 14.1 Transhumane pastoral movement of Sheep Gatlang VDC, Rasuwa, Nepal. (Photograph by Rashila Deshar 2016)

Table 14.1 Gender-based agropastoral activities of the Gatlang VDC

Work related to agro-pastoral activities	Genderwise labour
Cooking lunch and dinner, cooking food of livestock, fetching water, washing, firewood collection from forest, grass collection from forest, leaf litter collection from forest, dung collection, scouring wool and spinning wool, weaving, seeding and labour exchange	Female
Chopping wood, milking, transporting of milk to cheese factory, grazing livestocks, raising and collecting livestocks, sheep shearing, fertilizing, cultivation and harvesting	Male and female
Ploughing	Male

Fig. 14.2 A pile of cow dung collected by women near range land of Gatlang village, Rasuwa. (Photograph by Rashila Deshar 2016)

Further, the study shows that women's autonomy in household decision-making in purchasing daily household goods is free, but for a major household decision, wives need to take consent from husbands. In terms of ownership to fixed assets and property such as land and building, women lack access to those. Furthermore, in terms of property right, women do not have rights to land and property ownership. Therefore, women might have no influence on the major decision-making process. In many other mountain areas, the role of women and men differ substantially (Shang et al. 2016; Khadka and Verma 2012; Radel and Coppock 2013) and their contribution to carbon balance can differ greatly (Cecelski 2000; IFAD 2004; Stevens 2008). Further, the physical actions of women are integrally involved in carbon balance through their roles in livestock husbandry and fuelwood management much more than men. However, their role related to agropastoral activities have played a major role in carbon management, but their importance is not yet realized. The daily activities of women are the key to the carbon cycle in high Himalayan regions. Therefore, women's role in carbon management and decision-making should be given greater prominence.

14.5 Climate Change Impacts on Women

Climate change and variability including weather extremes affect differently and immoderately the vulnerable population like the women. Climate change is also reported to affect the human health. Climate change-induced human health issue will be more prevalent among women due to the factors including but not limited to household work pressure, fertility-related issues and lack of medical and family planning facilities (WHO 2010). Climate change through their effects on the natural ecosystem can exacerbate the negative impacts on women. For instance, the increased frequency of drought or flood in agricultural system can increase the food insecurity and malnutrition, reduce the purchasing power, increase the prevalence of vector-borne diseases, etc. (Nwoke and Ibe 2014; IPCC 2014). The vulnerability of women to climate change impacts is also due to the physiological reasons and womanhood (Mutunga and Hardee 2010; WHO 2010). As regards the food insecurity, the women at the household experience negative fallout of crop failure, decreased agricultural production and increased food prices (ADB 2013). Further, they may forego food or a meal so that every other member of the family gets the food to eat (Tirado et al. 2010). Table 14.2 shows the summary of different studies related to climate change impact on women from different part of the world.

Few case studies observed in the Nepal Himalaya are worth mentioning in the context of climate change impacts on women. Pakarwas village of Ramechhap District, Nepal, has experienced drying of natural water springs, and it takes about a whole or half a day for the women to fetch the water. So women do not prefer getting married in this village. Consequently, in the recent years, the socio-economic and cultural composition of the village was affected. Similarly, the population of few villages of Manang District has started moving out or migrating to downstream villages due to drying of natural water resources, due to climate change.

14.6 Gender Consideration Through Policy Perspectives to Respond to the Climate Change Impacts

Skewed participation of women in the processes involving “decision-making”, “planning”, “policy-making” and “implementation” results in increasing their vulnerability to climate change impacts. Earlier in the climate change discourses, the focus has been on the mitigation of climate change and adaptation to climate change. The social dimension of climate change took back seat during the negotiations and discourses. Nevertheless, the climate change impacts more the vulnerable sections that include the elderly population, the women and the children. It is almost necessary to factor in their voices during the gender policy-making. Gender mainstreaming is the need of the hour as it can engender viable and pragmatic solution to climate change (Masika 2002).

Community Forestry Management practiced in Nepal through Community Forestry User Groups has positive policy implications to respond to the impacts of climate change. Women managed CFUGS across the mid-hills and mountains are

Table 14.2 Summary of climate change impact on women from different countries

Climate signal	Country	Author and published year	Findings
Droughts and floods	Bangladesh and Uganda	Quisumbing et al. (2011)	Higher involvement of women in agriculture production risked women's assets when drought and floods hit the country
Drought	Tanzania	Nelson and Stathers (2009)	Change of cropping practices due to drought towards more drought tolerant crops increased weeding work for women in some instances
Drought	Zimbabwe	Hoddinott and Kinsey (2000) and Serna (2011)	Women were adversely affected by the 1994–1995 drought in Zimbabwe in terms of body mass compared to men due to lack of adequate food
			Reduction of meal intake by individuals especially amongst women, thus increasing their risk of health problems as well as that of children and lactating mothers
Heat wave	Europe and the United States	WHO (2009)	Women more likely to die than men
			Older women likely to die
			Older men more at risk to die, due to social isolation
Flood and landslide due to El Nino	Peru	Reyes (2010)	Women and children were more susceptible to diseases such as malaria, cholera and dengue due to less food consumption
Flood	India	Mitchell, Tanner and Lussier (2007)	Higher psychosocial problem in women due to lack of social and capital security
Drought	Vietnam and Ethiopia	Shaw et al. (2008) and Asheber et al. (2010)	More water collection burden and increased more physical labour
Less rainfall	Senegal	Dankelman et al. (2008)	Women spent more time checking water levels in boreholes with poor quality of water
Drought	Nepal	Leduc (2008)	Decreased agricultural production increased deforestation rate. Women have to spend more time in search of fuelwood
Unpredictable rainfall and climate change	Tanzania, South Africa and Nigeria	Nelson and Stathers (2009), Babugura (2010)	Male migration with increased workload and family responsibilities to women

(continued)

Table 14.2 (continued)

Climate signal	Country	Author and published year	Findings
Flood/hurricanes	India	Ahmad and Fajber (2009)	Women more vulnerable to disasters due to "gendered" nature of early warning information; women lack the ability to swim compared to men because traditional taboos prevent them from learning swimming
Asian tsunami (not climate related, but viewed as proxy)	Andaman and Nicobar	UNISDR (2008) and Gokhale (2008)	Women died due to putting their own life at risk during disaster to protect children and the elderly

Fig. 14.3 Women managed community forest user group in mid hill Nepal. (Photo CFUG)

contributing to the livelihood enhancements (Fig. 14.3). Further, their role is being acknowledged in lowering the impact of climate change as Government of Nepal is in the process to submit a proposal to Global Carbon Fund against the carbon sequestered by CFUGs.

As regards the future direction in the global policy instrument, the "Conference of Parties" to the "United Nations Framework Convention on Climate Change" has categorically acknowledged the significance of factoring in both men and women in the discourses, negotiations and the UNFCCC processes. Gender mainstreaming has been one of the important aspects mentioned in the Paris Agreement as well. The Subsidiary Body of Impact (SBI) of UNFCCC, recalling decision 18/CP.20, paragraph 15, invited Parties and observer organizations to provide information on progress made in meeting the goals of achieving gender balance and gender responsive climate policy (FCCC/SBI/2016/L.16, UNFCCC COP 18 Webpage). This submission focuses on enhancing gender balance under the UNFCCC. It builds upon the prior submission by the Mary Robinson Foundation-Climate Justice (the Foundation), dated 29 August 2016, on upscaling the "Lima Work Programme". It should also be viewed in the context of the wider programme of work required to advance comprehensive, gender-responsive climate policy-making and the mainstreaming of gender equality under the UNFCCC.

14.7 Conclusion

As the women and men experience impacts of climate change quite differently, the policy on climate policy should factor in the voices of gender, recognizing the gender differences arising due to the socio-economic conditions and social fabric, increased burden on gender on account of climate change, dependency of women on natural resources, recognizing their contribution to agricultural development and sustainable ecosystem management and significance of utilizing the leadership capabilities of women in implementing the sustainable practices in the neighbourhood and at the local level. In effect, the intersection of climate change and gender policy is more concerned with gender differences and gender mainstreaming. Gender mainstreaming is the need of the hour as it can engender viable and pragmatic solution to climate change.

References

ADB (2013) Gender equality and food security-women's empowerment as a tool against hunger

Ahmed S, Fajber R (2009) Engendering adaptation to climate variability in Gujarat, India. In: Terry G (ed) Climate change and gender justice. Practical Action Publishing, Oxford

Ahuja N (2016) Taking climate justice to the Himalayan heights: a proposed adaptive mountain governance framework to address climate change impacts in the Indian Himalayan Region

Anup KC, Ghimire A (2015) High-altitude plants in era of climate change: a case of Nepal Himalayas. In: Ozturk M, Hakeem KR, Faridah-Hanum I, Efe R (eds) Climate change impacts on high-altitude ecosystems, vol 1. Springer, Cham, pp 177–189

Anup KCA, Bhandari G, Joshi GR, Aryal S (2013) Climate change mitigation potential from carbon sequestration of community forest in mid hill region of Nepal. Int J Environ Prot 3(7):33–40

Asheber SA (2010) Mitigating drought: policy impact evaluation: a case of Tigray region, Ethiopia. University of Twente Faculty of Geo-Information and Earth Observation ITC. Available at: http://www.itc.nl/library/papers_2010/msc/gsim/asheber.pdf.

Bhasin V (2011) Status of women in transhumant societies. J Sociol Soc Anthropol 2(1):1–22

Cecelski E (2000) The role of women in sustainable energy development. National Renewable Energy Laboratory, Golden. https://doi.org/10.2172/758755

Dankelman I (2002) Climate change: learning from gender analysis and women's experiences of organizing for sustainable development. Gend Dev 20(2):12

Dankelman I, Alam K, Ahmed WB, Gueye YD, Fatema N, Mensah-Kutin R (2008) Gender, climate change, and human security: lessons from Bangladesh, Ghana, and Senegal. WEDO with ABANTU for Development in Ghana, ActionAid Bangladesh, and ENDA Senegal

Dawson J, Scott D (2013) Managing for climate change in the alpine ski sector. Tour Manag 35:244–254. https://doi.org/10.1016/j.tourman.2012.07.009

Denton F (2002) Climate change vulnerability, impacts, and adaptation: why does gender matter? Gend Dev 10(2):10–20. https://doi.org/10.1080/13552070215903

Deshar R, Koirala M (2016) Gender-wise contribution in carbon management by high himalayan ethnic group of Gatlang VDC, Rasuwa, Nepal. Nepal J Environ Sci 4:39–46

Dong SK, Long RJ, Kang MY (2003) Milking and milk processing: traditional technologies in the yak farming system of the Qinghai-Tibetan Plateau, China. Int J Dairy Technol 56(2):86–93

Easterling DR, Evans JL, Groisman PY (2000) Observed variability and trends in extreme climate events: a brief review. Bull Am Meteorol Soc 81:417–425

Eugenio-Martin JL, Campos-Soria JA (2010) Climate in the region of origin and destination choice in outbound tourism demand. Tour Manag 31:744–753. https://doi.org/10.1016/j.tourman.2009.07.015

Gokhale V (2008) Role of women in disaster management: an analytical study with reference to Indian society. The 14th World Conference on Earthquake Engineering, October 12–17, 2008, Beijing, China

Gurung GB, Bhandari D (2009) Integrated approach to climate change adaptation. J For Livelihood 8(1):90–99

Gurung J, Giri K, Setyowati AB, Lebow E (2011) Getting REDD+ right for women: an analysis of the barriers and opportunities for women's participation in the REDD+ sector in Asia. United States Agency International Development, pp 1–113

Hoddinott J, Kinsey B (2000) Adult Health in Time of Drought||, Food Consumption and Nutrition Division. Discussion Paper 79. International Food Policy Research Institute, Washington, DC

IFAD (2004) Enhancing the role of indigenous women in sustainable development. Third session of the permanent forum on Indigenous Issues. International Fund for Agricultural Development. Accessible online at: http://www.ifad.org/english/indigenous/pub/documents/indigenouswomenReport.pdf. Accessed 2 Sept 2016

IPCC (2014) Summary for the policymaker. In: Field CB, Barros VR, Dokken DJ, Mach KJ, Mastrandrea MD, Bilir TE, Chatterjee M, Ebi KL, Estrada YO, Genova RC, Girma B, Kissel ES, Levy AN, MacCracken S, Mastrandrea PR, White LL (eds) Climate change 2014: Impacts, adaptation, and vulnerability. Part A: Global and sectoral aspects. Contribution of Working Group II to the fifth assessment report of the Intergovernmental Panel on Climate Change. Cambridge University Press, Cambridge/New York, pp 1–32

IPCC (2018) Summary for policymakers. In: Masson-Delmotte V, Zhai P, Pörtner HO, Roberts D, Skea J, Shukla PR, Pirani A, Moufouma-Okia W, Péan C, Pidcock R, Connors S, Matthews JBR, Chen Y, Zhou X, Gomis MI, Lonnoy E, Maycock T, Tignor M, Waterfield T (eds) Global warming of 1.5°C. An IPCC special report on the impacts of global warming of 1.5°C above pre-industrial levels and related global greenhouse gas emission pathways, in the context of strengthening the global response to the threat of climate change, sustainable development, and efforts to eradicate poverty. World Meteorological Organization, Geneva, 32 pp

Jodha NS (1987) A case study of degradation of common property resources in India. In: Blaikie P, Brookfield H et al (eds) Land degradation and society. Methuen, London

Karki S, Gurung M (2012) Women's leadership in community forestry in the middle hills of Nepal. Gender and biodiversity management in the Greater Himalayas: towards equitable mountain development. ICIMOD, Kathmandu, pp 25–27

Khadka M, Verma R (2012) Gender and biodiversity management in the Greater Himalayas: towards equitable mountain development. International Centre for Integrated Mountain Development (ICIMOD)

Khadka M, Karki S, Karky BS, Kotru R, Darjee KB (2014) Gender equality challenges to the REDD+ initiative in Nepal. Mt Res Dev 34(3):197–207

Leduc B (2008) Case study: gender and climate change in the Hindu Kush Himalayas of Nepal. WEDO Gender and Climate Change Workshop in Dakar, Senegal, June 2008. Available at: http://www.wedo.org/wpcontent/uploads/nepalcasestudy.pdf

Liu X, Chen B (2000) Climatic warming in the Tibetan Plateau during recent decades. Int J Climatol 20:1729–1742

Masika R (2002) Gender, development, and climate change, Oxfam

Mitchell T, Tanner T, Lussier K (2007) We know what we need: South Asian women speak out on climate change adaptation. ActionAid International and Institute of Development Studies, University of Sussex, Brighton. http://www.eldis.org/go/display&type=Document&id=35433

Mutunga C, Hardee K (2010) Population and reproductive health in national adaptation programmers of action (NAPAs) for climate change in Africa. Afr J Reprod Health 14(4):127–139

Nelson V, Stathers T (2009) Resilience, power, culture, and climate: a case study from semi-arid Tanzania, and new research directions. Gend Dev 17(1):81–94

Nwoke EA, Ibe SNO (2014) Climate change impact on the health of African women and adaptation strategies. Int J Glob Warm Clim Change 1(1)

Parajuli R, Pokharel RK, Lamichhane D (2010) Social discrimination in community forestry: socio-economic and gender perspectives. Banko Janakari 20(2):26–33

Radel C, Coppock DL (2013) The world's gender gap in agriculture and natural resources: evidence and explanations. Rangelands 35(6):7–14

Reyes RR (2010) Gendering responses to El Niño in rural Peru. Gend Dev 10(2):60–69

Serna JM (2011) Drought assessment Northern Eastern Kenya (Wajir East, South and Mandera). Save the Children. Available at: http://www.disasterriskreduction.net/fileadmin/user_upload/drought/docs/SC%20Drought%2 0Assessment%20Report%20-%20Northern%20Eastern%20 Kenya-%2011- 24%20April%202011.pdf

Shang Z, White A, Degen AA, Long R (2016) Role of Tibetan women in carbon balance in the Alpine Grasslands of the Tibetan Plateau: a review. Nomad People 20(1):108–122

Shaw R, et al (2008) Drought-management considerations for climate-change adaptation: focus on the Mekong region. International Environment and Disaster Management (IEDM) and Oxfam, Vietnam. Available at: http://www.oxfam.org.uk/resources/policy/climate_change/downloads/vietnam_climate_change_report.pdf

Stevens C (2008) Gender and sustainable development: maximising the economic, social and environmental role of women. Organization for Economic Co-operation and Development. Accessible online at: http://www.oecd.org/social/40881538.pdf. Accessed 2 Sept 2016

Talukdar T (2012) Women as key manager of biodiversity in shifting cultivation in Bangladesh. In: Khadka M, Verma R (eds) Gender and biodiversity management in the Greater Himalayas: towards equitable mountain development. ICIMOD [International Centre for Integrated Mountain Development], Kathmandu, pp 21–24

Tirado MC, Clarke R, Jaykus LA, McQuatters-Gollop A, Frank JM (2010) Climate change and food safety: a review. Food Res Int 43:1745–1765

UNFCCC, COP 18 Webpage

UNISDR (2008) Climate change and disaster risk reduction. Briefing Note 1. Available at: http://www.unisdr.org/files/4146_ClimateChangeDRR.pdf

WHO (World Health Organization) (2009) Gender, climate change and health: draft discussion paper

WHO (World Health Organization) (2010) Gender, climate change, and health. https://www.who.int/globalchange/GenderClimateChangeHealth

Wisner B, Blaikie P, Cannon T, Davis I (2004) At risk: natural hazards. People's vulnerability and disasters. Routledge, New York

Climate Change and Agriculture: A Review of Crop Models

15

S. Mulla, Sudhir Kumar Singh, K. K. Singh, and Bushra Praveen

Abstract

The clear evidence of climate change impact demands proactive role by scientists, agronomist and meteorologist for upscaling agricultural production, precision forecast and food safety, especially in the tropical region. The crop simulation model suggests probable growth, development and crop yield for soil-plant-atmosphere dynamics assessment. Decision Support System for Agro-Technology Transfer Model (DSSAT) is an application-based model that gives the best-suited recommendations to achieve sustainability in the agriculture by means of simulation of users' minimum experimental data that includes weather data pertaining to site, crop growth period, and data concerning the soil, crop management practices, etc.

Identification of the weather and climate-sensitive problems due to extreme weather events on agriculture in any region can be achieved by crop model. Validation and calibration of crop simulation model is necessary with the help of field experimental data which will contain sensitive analysis, impact of epochal (temperature time period), various temperature ranges, different levels of radiation and CO_2, different dates of sowing and various nitrogen and water treatments. Extreme climate change impacts on phenological stages, the growth of a plant, dry matter partitioning to different plant organs for all seasons also need to

S. Mulla
India Meteorological Department, Pune, Maharashtra, India

S. K. Singh (✉)
K. Banerjee Centre of Atmospheric and Ocean Studies, University of Allahabad, Prayagraj, Uttar Pradesh, India

K. K. Singh
India Meteorological Department, New Delhi, India

B. Praveen
IIT Indore, Indore, Madhya Pradesh, India

© Springer Nature Singapore Pte Ltd. 2020

V. Venkatramanan et al. (eds.), *Global Climate Change and Environmental Policy*,
https://doi.org/10.1007/978-981-13-9570-3_15

be studied. Validation, linking and analysis of climate change data for different Representative Concentration Pathway (RCP) with bias-corrected climate change data and crop model data using Decision Support System for Agro-Technology Transfer Model (DSSAT) and probability distribution model will be required to investigate climate change impacts on crops. Based on these results, the formulation of:

(a) A multi-pronged plan of using local coping machinery, wider adoption of the existing technologies and/or concerted research and development efforts for evolving new technologies needed for adaptation and mitigation in rainfed and irrigated areas.
(b) More precise weather-based agromet advisories for soil, crop yield, crop condition on a spatial and temporal basis to minimize losses and increase the economy of farmers and country. Optimized inputs like land preparation, selection of crop and cultivars, date of sowing, date of harvesting, irrigation scheduling, pesticide and fertilizer application, crop growth, extreme weather events, adaptation and development of flexible and dynamic Farm Management Information System (FMIS) strategies and other value-added services, etc. can be provided for farming community.

The research can also be extended by doing a detailed analysis of estimation of soil moisture, evapotranspiration, insolation, vegetation index, growing degree days, standard precipitation index, and land surface temperature. The statistical study of estimated values of above said parameters using probability distribution model, root means square error and bias value of simulated data will be helpful for the development of a hydro-meteorological model, agriculture applications, irrigation planning over arid/semi-arid zones and forecasting systems.

Keywords
Decision Support System for Agro-Technology Transfer Model (DSSAT) · Crop simulation model · Climate change · Adaptation · Mitigation · Sustainability · Agromet advisories · Farm management

15.1 Introduction

The successful application of crop development models in agricultural meteorology and climate change studies are important in the present scenario. Agriculture is susceptible to climate inconsistency and predictable climate change. The combination of the crop simulation model and worldwide climate models predict an increase of global warming leading to the exposure of major crops to temperature stress and a decrease of yields of an uncertain magnitude for every region. Unlike in the fields of physics and engineering, universal models does not exist within the agricultural sector. Therefore, crop yields expressed as a polynomial or exponential mathematical function of defined variables, with regression coefficient analysis, are site-specific. The information obtained can only be applied to a site where climate, soil parameters and crop management practices are similar in nature to those used during developing of original functions. Several models are built to simulate crop to achieve an enhanced harvestable yield. Application-based crop simulation model which incorporates intensive data on biological and crop phenology provide more accurate results. Most of the farmers in developing countries particularly in India experience socio-economic backwardness, fragmented land holdings, rainfed farming and limited adaptive capacity impel the farmers to yearn for sustainable agricultural practices. The issues of agricultural sustainability, food security, adaptation methods and strategies to increase food productivity, farmer's income, conservation of water/environment, policy decisions, researcher, economics, environmental studies, etc. can be articulated with climate modelling and crop simulation modelling. This chapter provides a comprehensive overview of the crop simulation model, climate change assessment and mitigation and adaptation strategies and policies.

Agriculture sources are accountable for 18% of worldwide greenhouse gases (GHGs) emissions and are the potential threat to the stratospheric ozone layer. An estimation is made that emission ranges from 32.84 Gg (1980–1981) to 93.82 Gg (2000–2001) per year, and it may occur in the Indian continent which is likely due to rise in the expansion of cropping area, use of nitrogen (N)-fertilizers and animal population (Bhatia et al. 2004). The regional patterns of climate change are altered by radiative forcing associated with anthropogenic emissions and related negative effects of climate conditions. Variations at all levels include modification of the length of growing season, water availability and the creation of disturbance by extreme weather regimes (heat/cold waves, flood, drought, fires, pest outbreaks and diseases). World's livestock and paddy cultivation are the major sources of emission of methane gas, and about 35% of gas liberation is mainly contributed by India (Bhatia et al. 2004). Deforestation is also a key source of Greenhouse Gases (GHGs). During the past 100 years, more than 60% of greenhouse gases are added to the atmosphere (WRI 2001) amongst which Indian agriculture emanates about 0.23% (Bhatia et al. 2004).

The changes in rainfall pattern (south-western/north-eastern monsoon) will result in the reduction in yields over rainfed areas because of increased crop water demand. The subdued effects of climate change on the quality of fruits, vegetables, tea, coffee, aromatic, cereal crops and medicinal plants have been observed. The more frequent extreme weather events such as floods, droughts, cyclones and heat/cold waves will have unfavourable effect on agricultural productivity. The occurrence of cold waves and frost events may decrease in the future due to global warming and it would lead to a decreased probability of yield loss associated with frost damage. Climate change will check the nutritional and food security of agriculturally important animals like livestock because an increase in temperature would increase the lignifications of plant tissues, lower the digestibility, increase the water scarcity, etc. The decrease or scanty rainfall, the rise of mean temperatures, sea level, severe frequent occurrences of drought, cyclones, floods, heat/cold waves are also a great threat to agricultural biodiversity. The comprehensive rise of CO_2 concentration is remunerative because it increases photosynthesis in most of the crops, especially for C3 plants like wheat and rice. Amongst the majority of cereals crops, wheat crop was observed to show marked reduction in yield due to decrease of grain filling duration, higher respiration rate of the plant and/or lowering of rainfall/irrigation supplies. Amount of severe soil erosion or frequent change in it may happen due to alteration in the rainfall pattern, volume and frequency as well as wind intensity. The aggravation of heat stress in dairy animals will negatively affect their reproductive performance. Alteration in the rainfall pattern may influence and enhance vectors during wetter years causing large epidemics of diseases. In India, organic matter content of the soil is low and it will be further degraded. As regards the higher C:N ratio in crop plants, the decomposition rate and nutrient supply rate were found to be reduced markedly. High soil temperatures will enhance the nitrogen mineralization and its accessibility may lower due to nitrogen losses through volatilization and de-nitrification process. Conventional agricultural practices in the coastal region will be hampered badly by an increase in the sea level caused due to inundation of salty water in the coastal lands. The rise in the temperature of sea, ocean and river may likely affect the breeding, migration and harvests of fishes. Also, the cyclonic activities would have an impact on their capture, production and marketing.

15.2 Impact of Climate Change on Agriculture

Worldwide majority of the population mainly depends on rice as a daily food intake. There is evidence of negative impacts of climate variability on crop growth and yield. Mean temperature above threshold value results in the reduction of rice yield. In case of rice crop, the variability in minimum temperature is found to be more significant than the variation in the maximum temperature. "For every 1 °C increase in the growing season's minimum temperature above 32 °C lowers the grain yield by 10%" (Pathak et al. 2003). Aggarwal (2008) showed that rice yield in Punjab (India) decreases by 5.4%, 7.4% and 25.1% for every increase in temperature by 1 °C, 2 °C and 3 °C, respectively, by keeping all other climatic variables constant. Further, growth and development of rice and wheat crop are found to be negatively affected by increases in minimum and maximum temperature (Venkatramanan and Singh 2009a, b). During maturation and ripening phase of wheat crop, high temperature has proven to be disadvantageous for wheat crop production because of terminal heat stress. Intermittently more than normal temperature regimes, scanty rainfall or poor irrigation condition restricts the grain maturity phase and resulted in shrinking grain size. Degradation of wheat yield (q/ha) around 10–15% is caused by terminal heat stress. Each growth stage of plant receives a certain amount of heat energy necessary for growing called as "Heat Unit" and depends upon the base temperature. Early sown crop like wheat, barley, peas, oats, etc. have a base temperature of 40 °F, for corn 50 °F and for cotton 60 °F, and at these temperatures crop growth is appreciable. Crop base temperature is subtracted from actual daily mean temperature. Summation of this daily heat unit is known as "day-degrees/degree-days/heat-unit/thermal unit".

When the standard mean temperature is more than the base temperature, the heat unit gets accumulated. As regards the sugarcane crop, the base temperature of 70 °F is found to be suitable for crop establishment. Within the range of 80–90 °F growth becomes optimum at which absorption of nutrients and growth are at the best level. Below 70 °F, growth is reduced, phosphorous and nitrogen intake is decreased, and above 100 °F, the growth is detrimental. Komuscu et al. (1998) suggested a rise of 4–43% soil water deficit by experiencing the warming of 2 °C and 8–91% for a warming of 4 °C by applying the Thornthwaite water balance model plus boosting of evapotranspiration considering hydrological simulation model. It is also projected that the climate change will increase the frequency of weather hazards which will further decrease the mean yield of vulnerable crop plants.

15.3 Role of Crop Model in Climate Change Scenario

In tropics and sub-tropics, the useful effects of carbon dioxide will be compensated by means of the boosting in temperature, resulting in crop yield loss and increased demand for irrigation. An appropriate understanding of climate change impact will help aid to scientists, agronomist, meteorologists to advise farmers in a suitable way and provide crop managing decisions, viz. choice of crops, probable dates of sowing and irrigation scheduling. Farmers' suitably modify and adopt appropriate agricultural technologies that can in effect lessen the negative impacts of climate change. The crop simulation model has potential to be used in climate change studies to understand the potential impacts of changing climate on food system. DSSAT CROPGRO is a primary package which modifies weather simulation generators. Ignoring few limitations of GCMs, it would be in the better interest of world's farming community that the DSSAT modellers consider precise plus acceptable weather generator parameters, i.e. outcomes of GCMs for further application in the simulation models. This will facilitate in finding out the solution under the climate change scenario for creating an excellent quality crop production, specifically in underdeveloped and developing countries.

It is necessary to understand the relationship between crop growth and yield behaviour under climate change. Modelling approach could contribute towards a more efficient research model in form of a tool as crop planning. Crop modelling is increasing day-by-day and is applicable in research, teaching, farm and resource management, policy analysis, crop yield forecasting, etc. Simulation model permits the integration of knowledge across crops and in disciplines for a specific crop, e.g. productivity analysis, alteration of soil fertility status over time. A particular and precise knowledge can be achieved on genetic traits of economic yield which includes integrated genetic improvement programs already available in the model. Presently, a crop simulation model and its results became the boon in agricultural research and for farming applications. The two famous models frequently used in agro-meteorological studies are the De Wit School of models and the IBSNAT and DSSAT models. A brief discussion about simulation models and their utility for yield analysis are explained below.

15.3.1 The CANEGRO Model

CANEGRO is a leading sugarcane crop simulation model and has been used extensively in agronomic research and management. The model has been under development since the late 1980s at the South African Sugarcane Research Institute (SASRI).

15.3.2 The Erosion-Productivity Impact Calculator (EPIC) Model

Beginning in 1981, a mathematical model called the erosion-productivity impact calculator (EPIC) model was developed to determine the relationship between soil erosion and soil productivity throughout the USA. Soil attrition risk caused by cropping practices and tillage can be evaluated by the EPIC model. The model provides five evapotranspiration equations, namely, Penman-Monteith, Penman, Priestley-Taylor, Hargreaves and Baier-Robertson. Optional choice is reserved with EPIC modeller to opt for any suitable and appropriate simulation exercise. Using the same single data file, the model is capable enough to simulate and observe the growth of many crops. The model is applicable in the tropical areas and it is possible to do the multi-and-intercropping rather than mono-cropping. The model run provides the output of yields for early and late crops, chief and small crops, both heliophytic (light loving) and sciophytic (shade loving) crops, creeping and climbing crops. EPIC also supplies information of simultaneous modelling for changes in the crop environment like moisture and nutrients as these are the limitations on the productivity of tropical agricultural crops. The EPIC crop model is used for estimation of agricultural yield per unit area as well as total crop area of land planted and assessment of strategies for management of climate variability and climate change on harvest yields, the vulnerability of agricultural production and adaption option.

15.3.3 The NTKENAF

NTKENAF (Version 1.1) simulates the growth of kenaf under rainfed conditions in tropical Australia. In daily time steps, the model simulated the phenology, leaf area development, biomass accumulation and partitioning, soil water balance and dry matter yields of kenaf plants based on climatic and management inputs. The model assumes adequate nutrition and no effect of pests and diseases. The model uses daily maximum and minimum temperature, solar radiation and rainfall. The duration from sowing to flowering is predicted using temperature and photoperiod. Leaf growth is described as a function of node production (as determined by temperature), leaf area per node and leaf area senescence. Potential daily biomass is predicted from the leaf area index, the light extinction coefficient and radiation use efficiency and partitioned to the economic stem yield. Soil evaporation is predicted using a two-stage evaporation model, and plant transpiration is predicted from the daily biomass accumulation, a transpiration efficiency coefficient and predicted daily vapour pressure deficit.

15.4 DSSAT Model

Crop simulation model mainly DSSAT is the most suitable and capable of superior functioning compared to other models like an analytical, statistical, empirical and combination of two. Historic simulation models used to simulate only two parameters such as photosynthesis and carbon balance. Moreover, the statistical models were used to correlate the approach and were giving yield estimates for a large area but only restricted to use final yield data for correlation with mean weather parameters on a regional basis. To overcome these shortcomings of old models, the DSSAT model was invented which facilitates maximum input details for all crops. Thus, the total model approach has been updated in the DSSAT crop simulation model.

15.4.1 Features of DSSAT Simulation Model

Agriculture is a backbone of major parts of the world and livelihood of millions of people. In the direction to meet up these necessities, IBSNAT (International Benchmark Sites Network for Agrotechnology Transfer) has started to use the model from 1982 onwards under the agreement received from the "U.S. Agency for International Development" to "the University of Hawaii" at Manoa, USA. IBSNAT is the project on analysis and simulation system to give various choices to user end regarding alteration which ultimately focus on the production as "decision support system" competent of simulating the future threats and on the sequence of optional choices, through a multi-institute and multidisciplinary approach. To run the DSSAT model needs a number of data in order to simulate outcomes, testing and implementation of output to resolve the global agricultural issues for particular sites. DSSAT has maximum utility with respect to other simulation models.

15.4.2 DSSAT as a Tool

DSSAT used as a driver to propose for crop management, examine ecological similarities, the natural constraint of crops, material features of soil plus giving them the management alternative for better land use preparation and matchless sustainability issues.

15.4.3 DSSAT as a Business Tool

It develops effectiveness and improvements in input marketing. While, the conventional testing is time taking and expensive, use of DSSAT reduces the cost and time involved in evaluation of newly developed cultivars. In the projected climate change situation, fresh outcomes for prediction of accurate cultivar coefficient and dynamic of crop models will identify narrow crop gaps in biophysical aspects. Such effect will surely increase agricultural productivity.

The two-stage approach in DSSAT model is adopted to evaluate the crop simulation model as sensitivity analysis and validation. The sensitivity evaluations facilitate to find out as whether the crop model under consideration possibly will be used for examination of an a priori *hypothesis*. The evaluation shows the estimated output harvest specified by the model is acceptable although by doing modifications in the environmental parameters, viz. appropriate use of fertilizers by which enhancement of yield production may be expected.

15.4.4 Calibration and Validation in DSSAT

Calibration, or parameter estimation, is a difficult but critical aspect of a crop modelling project. Predictions from a model are heavily dependent on calibration of models. Adjustment of input parameters is required in order to match the simulated results with observational results. Validation is a precise test which compares the original interpretation with the model output results obtained with similar environments existing at the time of original data were prepared. For further precise as well as reasonable outcome, no gaps amongst observations and forecast are typically illustrated within the model calibration. Validation method includes qualitative as well as a quantitative comparison of virtual production and experiential production. As regards the water-soil-crop model, it is very crucial to authenticate the extractable wetness and leaf region constituents because the biomass accreditations are mainly dependent on these factors. Crop data reveals the soil heterogeneity along with the changes in environmental parameters during the growing phase of the crop.

15.4.5 Potential Heat Unit Calculation of DSSAT

A meticulous quantity of potential heat units (PHU) is compulsorily needed for the different growing stage and in the ripening stages of any crop variety under consideration. The choice of PHU is therefore depended upon the user at the time of generating the new operations schedule file (OPS). In case of unknown or anonymous crop type, the modeller can reduce the gap amongst the virtual and experimental yield observation thereby regulating the amount of PHU in the operation schedule file (OPS) (Easterling et al. 2000). Simultaneously modeller can add phenological data of the unknown cultivar type.

15.4.6 Sensitivity Analysis in DSSAT

The assumptions of work are concerned about whether the model possibly will repeat the sensitivity of the original information systems or not. During the sensitivity experiment, archival of weather data is used. However, for climate change scenario, those weather data/variables will no more exist. In such circumstances, mock data can be accessed via GCM and these data are within the tolerance and in the satisfactory range. Since there exist a few uncertainties related to the prediction of climate change, researchers make use of projected climate scenarios to give an approximation of how the climate is likely to affect a particular targeted system. In case of farming production, scenarios are being derived from GCMs and random sensitive tests, for example, +2 and + 4 °C temperature changes and +/−10% precipitation change are suggested for the approximation of probably expected changes in yield and other important relevant variables of agronomical type.

15.4.6.1 Sensitivity Analysis for Seasonal Rainfall Data

For sensitivity analysis of seasonal rainfall data, one has to obtain climate and daily weather data records from the station. For example, in the simulation model, if the particular crops are sown on 1 June for each year and then the same crop is harvested on 30 August. The normal practice of farmer is to wait for the initial intense rainstorm to sow the crops. Additional delay may create the risk of harvest failure as a consequential effect of sudden termination of the rainy season later on in the current season of the year. Furthermore, during the progressive rainy season, there is less reception of solar radiation at the earth's surface causes the lessening effect of vegetative and reproductive growth which ultimately give a considerable decrease in yield. Therefore, farmers can choose an option to simulate crop for early-maturing varieties which will be popular on local levels. For example, a few types of maize crop grown-up within 120 days from sowing and other varieties mature within 90 days.

DSSAT sensitivity analysis is not restricted to rainfall data only but also includes temperature, radiation and carbon dioxide concentration using anticipated incremental scenarios for climate change impact assessments. In view of this, it is suggested that "incremental changes in temperature and precipitation is supposed to be combined with the baseline climate data to create incremental scenarios".

15.4.7 Application of DSSAT Model

DSSAT applications range from real-time decision support for crop management to assessing the potential impact of climate change on global food security. Crop models are also invaluable as heuristic devices that help to identify research problems where our current knowledge has limits and further research is needed. The ability of crop models to simulate how different weather years or soil conditions affect crop performance make models especially useful in research involving climatic uncertainty or geospatial variation. Recent advances in field phenomics and crop genomics are opening opportunities for crop models to support research in fundamental plant science.

15.4.7.1 CERES-Maize Model

CERES-Maize model is modified to improve the simulation of the site-specific crop development and yield. It is a prognostic and deterministic model intended towards simulation of maize development. Soil, water, temperature and soil nitrogen dynamics are at a field level for the single growing season. CERES-Maize model in DSSAT needs the set of six numbers of cultivar specifications and its relevant parameter used for calibration purpose. Amongst six cultivars, four types, namely, P1, P2, P5 and PHINT, manage the timing of phonological phase, while the remaining two cultivar types, namely, G2 and G3, are the characteristics to characterize the prospective yield under most favourable environment. The GLUE model is being run and simultaneously performed the sensitivity analyses with the same crop coefficients. Also, the tested cultivar's coefficients were calculated through the calibration process available in the model.

Upon the introduction of hybrid genotypes, maize has an important position amongst the cereal crops. Under the future climate change assessment, it is essential to identify the detail crop growth behaviour of the hybrid crop model for a variety of environment. The DSSAT model is calibrated and validated by using the actual field data; therefore, the crop growth model is vital in describing the study of changeability performance. The model performance is assessed through phenology, biomass at harvest phase, leaf area index and grain yield performance. The simulated outcomes were in good agreement by means of the experimental value and also these are within the statistical significance limit. Biomass, to some level, was the above forecasted value in the model simulation output but within the significant limit. Simulated and practical phenologies, as well as yields, were also in close concurrence with the experimental values. Decision Support System for Agro-Technology Transfer Model (DSSAT) version 4.6 have been successfully applied for simulation of growth and yield of hybrid maize developed under different biotic and abiotic stresses, including evaluation of the climate change impact.

15.4.7.2 CANEGRO Model

Singels et al. (2008) have shown that there is a satisfactory forecasted value of Brazilian sugarcane in southern Brazil (Marin et al. 2011; Nassif et al. 2012). The model calculated the daily increase of entire biomass by utilizing the radiation use efficiency approach (Singels and Bezuidenhout 2002), CO_2 concentration and CO_2 fertilization effect algorithm. In addition to the photosynthesis, DSSAT/CANEGRO also simulates the effect of CO_2 on stomata resistance and transpiration (Long et al. 2004) method of Allen et al. (1985).

15.4.7.3 SALTMED Model

Models are convenient tools in agricultural aspects of an irrigation scheduling, estimation of crop water requirement, prediction of yields and soil salinization for distinct irrigation systems, soil categories, crops and trees. The current version, SALTMED 2015, contains extra sub-models, crop growth as per the heat units/degree days, crop rotations, nitrogen dynamics, soil temperature, dry matter and yield, subsurface and deficit irrigation with the partial root drying (PRD), drainage flow to tile or open drain systems, presence of shallow groundwater and evapotranspiration (ET) using Penman-Monteith equation. It has numerous options for acquiring the canopy conductance. The current version permits up to 20 fields or treatments to run simultaneously.

15.5 Uncertainties in the Assessment of Climate Change

The key areas of uncertainty that are possibly to impact on worldwide scales are agricultural productivity, meteorological, hydrological and plant physiological as per the illustration received from the model performance. In global-scale assessments, it is found that there is inadequate capability to confine the uncertainty in projected climate extreme weather hazard and that of in pests and diseases. Agricultural reliance on the regional rainfall, glaciers melting and pollution has added extra complexity in the climate uncertainties. Meandering impact of an increase in sea level, storm surges, diseases to plants and human have also not yet been quantified. In the current scenario, the collective effect of climate change on a global scale agricultural efficiency cannot be consistently quantified.

Under climate variability scenario, scientists, meteorologists and researchers have a great challenge to know the requirement of crop growth, yield production, identification and potential use of agricultural land in order to maintain the sustainability of agriculture. These weather scenarios are highly uncertain about their occurrences and impacts. The literature showed that these high winters (October–December) or post-monsoon temperature detriments the rabi crops production especially wheat crop due to terminal stress. Also a remarkable increase in temperature during winter season, intense and more foggy days and extreme rainfall (cloud burst, snowfall, hailstorm etc.) are projected impacts caused by climate change and climate variability during the second half of the twenty-first century. Indian states and city, viz. Kerala, Uttarakhand and Mumbai city, recently showed extreme rainfall occurrences due to cloud burst and vigorous SouthWest-monsoon systems which caused a lot of living and non-living losses. As such, there is no considerable change or enhancement in the temperatures during SouthWest-monsoon season (June–September). But, the erratic and uncertain behaviour of monsoon rainfall will adversely affect the agriculture, livelihood and economy.

15.6 Mitigation and Adaptation of Agriculture

With the help of national governments, mitigation and adaptation are the main tasks as considered by international bodies (Ali et al. 2017). Equally important mitigation and adaptation methods are needed in the present scenario to diminish the impacts of climate change. Mitigation refers to the prevention or reduction of emission of heat-trapping GHGs in the atmosphere. Other meaning of mitigation is to apply advance technology or renewable energy sources for improvement in efficient agricultural management system.

Adaptation is not at all a substitute for mitigation because we know that quick climate change is still sustained and stabilization of GHG concentrations will go beyond an optimistic evaluation capacity to regulate. Adaptation is equally important and in numerous cases, it provides a significant decrease in adverse impacts. A measure of such adaptation offers co-benefits to progress goals by making them principally more attractive. The way to lower "Urban Heat Island Effect" is to enhance greenery by tree plantation. Urban heat island effect also increases the threat from severe heat waves to biodiversity (Kikon et al. 2016; Nhemachena and Hassan 2007).

Modification according to present & future climate change is known as "adaptation". Adaptation includes increase capacity to face adverse climate change impact and lowering the susceptibility of agriculture. However, there are scientific and economic constraints which impede the timely needed actions of adaptation. The main aim is to diminish susceptibility which has a destructive impact specifically on agriculture caused by climate change. Mitigation and adaptation to climate change depend on availability and access to quality seeds, resource-efficient crop cultivation practices, conservation and management of agricultural biodiversity and pro-active extension system delivering innovative technologies from lab to land and empowering the farmers through awareness and training.

15.7 Conclusion

Climatic changes and climatic variability are the greatest challenge for soil, food and water system. Uses of crop simulation models and climate change scenarios with higher spatial and temporal resolution are significant to dissect the impacts of climate change on water availability, crop yield, crop water productivity and soil water balance. Crop yield will be restricted by crop selection, sowing zones, soil deficiency, changing climate and water accessibility throughout the crop growth period. DSSAT is a deterministic, dynamic, specified and optimized crop model. DSSAT identifies optimise crop management plan to stimulate biomass accretion, forewarns pest & diseases, compares actual and simulated grains yield, provides detailed crop characteristics etc.

References

Aggarwal PK (2008) Global climate change and Indian agriculture: impacts, adaptation and mitigation. Indian J Agric Sci 78(11):911–919

Ali SA, Tendone L, Mastro GD (2017) Climate variability impact on wheat production in Europe: adaptation and mitigation strategies. In: Quantification of climate variability, adaptation and mitigation for agricultural sustainability. Springer, Cham, pp 251–321. https://doi.org/10.1007/978-3-319-32059-5_12

Allen LH, Jones P, Jones JW (1985) Rising atmospheric CO_2 and evapotranspiration. Proceedings of the national conference on advances in evapotranspiration, Chicago, IL, 16–17 December, ASAE, St Joseph, MI, USA, pp 13–27

Bhatia A, Pathak H, Aggarwal PK (2004) Inventory of methane and nitrous oxide emissions from agricultural soils of India and their global warming potential. Curr Sci 87(3):317–324

Easterling DR, Meehl GA, Parmesan C, Changonon SA, Karl TR, Mearns LO (2000) Climate extremes: observations, modeling, and impacts. Sci 289(5487):2068–2074. https://doi.org/10.1126/science.289.5487.2068

Kikon N, Singh P, Singh SK, Vyas A (2016) Assessment of urban heat islands (UHI) of Noida City, India using multi-temporal satellite data. Sustain Cities Soc 22:19–28
Komuscu A, Erkan A, Oz S (1998) Possible impacts of climate change on soil moisture availability in the Southeast Anatolia development project region (GAP): an analysis from an agricultural drought perspective. Clim Chang 40:519–545. https://doi.org/10.1023/a:1005349408201
Long SP, Ainsworth EA, Rogers A, Ort DR (2004) Rising atmospheric carbon dioxide: plants FACE the future. Annu Rev Plant Biol 55:591–628
Marin FR, Jones JW, Royce F, Suguitani C, Donzelli JL, Pallone WJF, Nassif DSP (2011) Parameterization and evaluation of predictions of DSSAT/CANEGRO for Brazilian sugarcane. Agron J 103(2):304–314
Nassif DSP, Marin FR, Filho WJP, Resende RS, Pellegrino GQ (2012) Parameterization and evaluation of the DSSAT/Canegro model for Brazilian sugarcane varieties. Pesq agropec bras, Brasília 47(3):311–318
Nhemachena C, Hassan R (2007) Micro level analysis of farmers adaptation to climate change in southern Africa. Internationl Food Policy Research Institute (IFPRI), Washington, DC
Pathak H, Ladha J, Aggarwal P, Peng S, Das S, Singh Y, Singh B, Kamra S, Mishra B, Sastri A, Aggarwal H, Das D, Gupta R (2003) Trends of climatic potential and on-farm yields of rice and wheat in the Indo-Gangetic Plains. Field Crop Res 80:223–234
Singels A, Bezuidenhout CN (2002) A new method of simulating dry matter partitioning in the Canegro sugarcane model. Field Crop Res 78(2):151–164
Singels A, Jones M, van den Berg M (2008) DSSAT v4.5 Canegro sugarcane plant module: scientific documentation. SASRI, Mount Edgecombe, South Africa, 34pp
Venkatramanan V, Singh SD (2009a) Differential effects of day and night temperature on the growth of rice crop. Pusa Agric Sci 32:57–62
Venkatramanan V, Singh SD (2009b) Differential effects of day and night temperature on the growth of wheat crop. Ann Agric Res 30(1&2):49–52
WRI (2001) https://www.wri.org/publication/world-resources-2000-2001. Accessed date 21 Sep 2018

MIX
Papier aus verantwortungsvollen Quellen
Paper from responsible sources
FSC® C105338

If you have any concerns about our products,
you can contact us on
ProductSafety@springernature.com

In case Publisher is established outside the EU,
the EU authorized representative is:
Springer Nature Customer Service Center GmbH
Europaplatz 3, 69115 Heidelberg, Germany

Printed by Libri Plureos GmbH
in Hamburg, Germany